THE SCIENCE AND TECHNOLOGY OF AEROSOL PACKAGING

THE SCIENCE AND TECHNOLOGY OF AEROSOL PACKAGING

EDITED BY

JOHN J. SCIARRA, B.S. Pharm., M.S., Ph.D.

*St. John's University, College of Pharmacy
and Allied Health Professions, Jamaica, New York*

LEONARD STOLLER, B.B.A.

Givaudan Corporation, Clifton, New Jersey

A WILEY-INTERSCIENCE PUBLICATION

JOHN WILEY & SONS, New York • London • Sydney • Toronto

Library of Congress Cataloging in Publication Data:

Sciarra, John J   1927–
    The science and technology of aerosol packaging.

    "A Wiley-Interscience publication."
    Includes bibliographies.
    1. Pressure packaging.   2. Aerosols.   I. Stoller,
Leonard, 1923–   joint author.   II. Title.

    TS198.A3S33   621.7′57   73-14688
ISBN 0-471-76693-3

Printed in the United States of America

10 9 8 7 6 5 4 3 2 1

This book is dedicated to our wives and children
for their patience and endurance.

To my wife Barbara
and our sons, Christopher, John,
Gregory, and Brian

JOHN J. SCIARRA

To my wife Renée

LEONARD STOLLER

ADVISORY EDITORS

VICTOR DIGIACOMO, B.S.

Givaudan Corporation, Clifton, New Jersey

WALTER V. JONES

Givaudan Corporation, Clifton, New Jersey

CONTRIBUTORS

THOMAS CIFELLI, *A.B., LL.B., J.D., Attorney-at-Law, Maplewood, New Jersey*

SAUNDERS T. CRAIGE, *B.S., M.S., Aerosol Sales Manager, Peerless Tube Company, Bloomfield, New Jersey*

DONALD A. DAVIS, *B.A., Editor, Drug & Cosmetic Industry, New York, New York*

VICTOR DIGIACOMO, *B.S., Vice President of Fragrance Laboratories, Givaudan Corporation, Clifton, New Jersey*

LLOYD T. FLANNER, *B.S., Director of Research & Development, Aerosol Systems, Inc., Macedonia, Ohio.*

WILLIAM A. GREGG, *B.S., Director of Research and Development, Precision Valve Corporation, Yonkers, New York*

WILLIAM E. HEDGES, *B.S., LL.B., Attorney-at-Law, Philadelphia, Pennsylvania*

THEODORE R. HERST, *Regional Manager, Foster-Forbes Glass Company, Division National Can Company, Forest Hills, New York*

ANTHONY IANNACONE, *B.S., M.S., Consultant, Hawthorne, New Jersey*

HARRY H. INCHO, *B.A., Senior Research Entomologist, Agricultural Chemical Division, FMC Corporation, Middleport, New York*

MONTFORT A. JOHNSEN, *B.S., M.S., Vice President, Research and Development, Peterson/Puritan, Inc., Danville, Illinois*

WALTER V. JONES, *Director of Technical Services, Givaudan Corporation, Clifton, New Jersey*

MICHAEL J. KAKOS, *B.S., M.S. Sales Manager, Amcel Company, Inc., Pan Amcel Affiliate of Celanese Corporation, Newark, New Jersey*

GUS W. LEEP, *B.ChE., M.ChE., Vice President, Research and Development, Illinois Bronze Powder & Paint Company, Lake Zurich, Illinois*

LLOYD A. MCDONALD, *B.S., Research Chemist, Agricultural Chemical Division, FMC Corporation, Middleport, New York*

HARRY OBARSKI, *B.S., Field Technical Service Coordinator for Aerosols, Continental Can Company, Inc., Chicago, Illinois*

ROBERT L. RAYMOND, *B.S. Pharm., Senior Manager, Toiletries Product Development, The Gillette Company, Toiletries Division, The Gillette Corporation, Boston, Massachusetts*

PAUL A. SANDERS, *B.S., Ph.D., Research Associate, "Freon" Products Laboratory, E. I. du Pont de Nemours & Company, Inc., Wilmington, Delaware*

JOHN J. SCIARRA, *B.S. Pharm., M.S., Ph.D., Chairman, Department of Allied Health and Industrial Sciences and Professor of Pharmaceutical Chemistry, St. John's University, College of Pharmacy and Allied Health Professions, Jamaica, New York*

LEONARD STOLLER, *B.B.A., Director, Public Relations and Advertising, Givaudan Corporation, Clifton, New Jersey*

CHARLES O. WARD, *B.S. Pharm., M.S., Ph.D., Associate Professor of Pharmacology, St. John's University, College of Pharmacy and Allied Health Professions, Jamaica, New York*

PREFACE

The science and technology of aerosol packaging have changed tremendously since the first pressurized product was produced, and many more improvements, in both container and contents, are certain to be made in the years to come. In a comparatively short span of time, however, the aerosol has achieved a very strong position, and its progress has been much faster than that of any previous packaging medium. Statistics on the use of aerosol products are evidence of the popularity of pressurized cosmetics, toiletries, and household specialties. Moreover, many items are on store shelves today that would not be there were it not for the availability of aerosol packaging.

During the past two decades, the literature of aerosols has been voluminous and several books on the subject have already been published. In planning for this new compendium, we as editors have attempted to make a book as complete as possible so that it can serve not only as a reference for the industry but as a text for educational purposes as well. In this way we will contribute to the continued development of the aerosol industry. Therefore we have collected the contributions of knowledgeable and authoritative representatives in the industry into what we believe to be the most comprehensive work thus far produced on aerosol packaging.

Among the contributors to this volume are men who have long been associated with aerosol manufacturing and filling, as well as with the fragrance, cosmetic, toiletry, pharmaceutical, household specialties, food, and other industries in which aerosol packaging is used. They have contributed to the progress of the aerosol and, even as their chapters were being printed, were making further improvements in their individual areas of interest.

To these contributors, we express our heartfelt thanks for the work they have done. Appreciation is also due to our typists for the preparation of the manuscripts and to those who assisted in the library research. We especially would like to express our thanks to Jean Cerruti and Regina Albright for their aid throughout the editing of this textbook.

We believe that the aerosol package will continue to have a promising future, and we hope that this volume will serve to promote its continuing success.

JOHN J. SCIARRA

LEONARD STOLLER

Jamaica, New York
Clifton, New Jersey
June 1973

CONTENTS

THE SCIENCE AND TECHNOLOGY OF AEROSOL PACKAGING

THE HISTORY OF PRESSURIZED PACKAGING

The true meaning of "aerosol" has very little relationship to the wide usage of the word today. Webster's definition, "a suspension of fine solid or liquid particles in air or gas," was further explained by Why-tlaw-Gray and Patterson (1), who defined the word as referring to a colloidal system consisting of very finely subdivided liquid or solid particles dispersed in and surrounded by gas. The definition was further amplified by Sinclair (2), who specified that the particles should be smaller than $50\,\mu$ in size and usually less than $10\,\mu$.

During the early years of the World War II, when insecticides were dispensed in pressurized containers, the word "aerosol" received a new meaning. As a result, a subcommittee on definitions of aerosol products of the Chemical Specialties Manufacturers Association adopted a new definition, which took into consideration the varied applications of this packaging medium (3):

A self-contained sprayable product in which the propellant force is supplied by a liquefied gas. This definition includes space, residual, surface coating, foam and various other types of products but does not include gas-pressurized products such as whipping cream. The term aerosol as used here is not confined to the scientific definition.

Herzka and Pickthall (4) suggested use of the term "pressurized pack," which they defined as a "self-contained pack which contains the product and the propellant necessary for the expulsion of the former." They added, "This definition includes packs which utilize compressed gases as propellants and, by discarding the term 'aerosol,' acknowledgment is made of the fact that this word has a definite scientific meaning."

Although this suggestion had merit, it was not accepted, and the word "aerosol" has become associated, among the consuming public, with any product that is dispensed with the aid of a propellant.

EARLY HISTORY OF AEROSOLS

In his discussions of the history of aerosols, Shepherd (5) cited the patent of Helbing and Pertsch (6) as the oldest reference to a product's being propelled by energy derived from the pressure of one of its own ingredients. According to the patent, which covered the "preparation and application of coating solutions," the inventors used methyl and ethyl chlorides, making a solution with either or both, with or without alcohol or ether. They further stated:

This solution is placed, with a considerable excess of methyl or ethyl chlorid (sic), if required, in a receiver of glass or metal having a suitable orifice which can be hermetically sealed at will with a close-fitting cap. When it is required to make use of the solution, the cap is removed and the vessel inclined as may be necessary. The heat of the hand holding the vessel, or even of the surrounding atmosphere, if of the ordinary temperature of a room, immediately causes the ethyl or methyl chlorid inside the receiver to begin to evaporate. This evaporation causes internal pressure and the solution is thereby ejected through the orifice in a fine jet or spray. By properly directing the orifice of the receiver the solution can be distributed over the part required.

In 1901, Gebauer was issued the first of his patents (Figure 1.1) (7), in which he proposed to overcome the objections to previous methods of dispensing liquids and "to eject the liquid in the form of a spray, which will extend over a surface and will cover the same in such a thin film that evaporation must take place almost immediately and practically at the surface of the skin." A little more than a year later, Gebauer obtained a second patent (Figure 1.2) (8), in which certain changes in his original container were made. The improvements made the unit less expensive and permitted ejection of the contents as either a spray or a stream.

What may have been the first patent (Figure 1.3) pertaining specifically to a perfume aerosol was issued to Russell W. Moore in 1903 (9) and was entitled "Perfumery Atomizer." In an apparatus similar to those in use today, Moore utilized carbon dioxide as the propellant.

After the Moore patent there seemed to be a period of inactivity in this area. The next patent, issued in 1921 to Mobley (10), covered a method of dispensing liquid antiseptics, using carbon dioxide as the propellant. Then followed the two important patents of Rotheim, issued in 1931 (11) and 1933 (12). The first, entitled "Method and

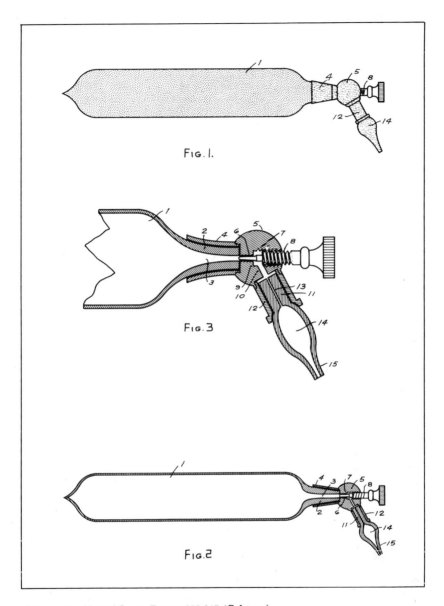

Figure 1.1.   United States Patent 668,815 (Gebauer).

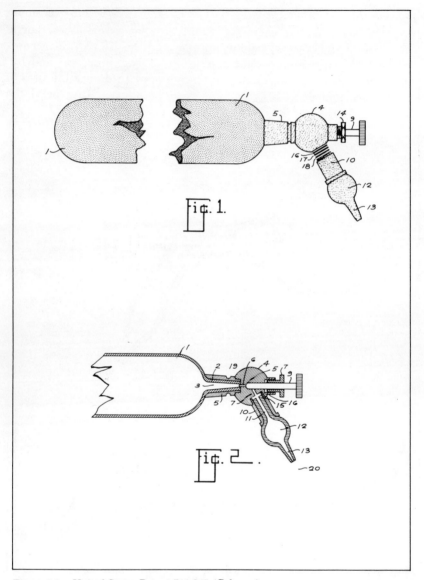

Figure 1.2.   United States Patent 711,045 (Gebauer).

Means for the Atomizing or Distribution of Liquid or Semiliquid Materials," utilized dimethyl ether as the propellant and specifically mentioned cosmetic products. The inventor wrote, "When the desired method is to be used for example for eau de Cologne this material obtains the novel property of giving a spray which is considerably cooled in relation to the atmospheric temperature as a consequence of the expansion of the added condensed gas." He added that "cosmetic products such as, for example, liquid or solid brilliantines, pomatums, vaselines, creams, toilette liquids, and the like are in accordance with the present method handled in a more practical and hygienic manner than at present." He also foresaw the possible uses of other gases and other applications when he wrote, "In general the

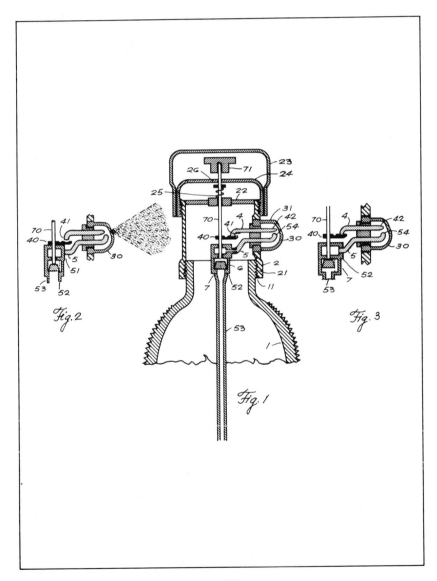

Figure 1.3.   United States Patent 746,866 (Moore).

method according to the invention can be used in connection with any substances (in any state of aggregation) which are able to be brought into sprayable condition by means of dimethyl ether or other suitable gas."

In his second patent, Rotheim made several improvements and specifically mentioned several condensable gases as possible propellants: "dimethylether (CH_3OCH_3), methyl chloride (CH_3Cl), isobutane and other hydrocarbons, for example, low boiling petroleum distillates such as rhigolene and cymogene (with boiling points about $0°C$.), methyl nitrite (CH_3ONO) with a boiling point of $12°C$., and vinylchloride (ethylene chloride) with a boiling point of $15°C$. to $18°C$." He also mentioned new uses for this method of dispensing and indicated newer valve designs upon which many later developments may have been based. He also noted, for perhaps the first time, the cooling effect when colognes are sprayed: "When the method is used for example for eau de Cologne the new result is attained that the spray becomes considerably cooler than the atmosphere on account of the expansion of the added condensed gas."

The first use of the fluorinated hydrocarbons as propellants was described in a patent covering a "process for preventing fire by nontoxic substances," issued in 1933 to Midgley, Henne, and McNary (13). They wrote that their "flame arresters . . . may be used in pressure devices in which a low boiling compound is employed which creates sufficient pressure to expel itself from the apparatus." They referred to dichlorodifluoromethane as the propellant, making use of materials that had been developed as refrigerant gases to replace the toxic ammonia and sulfur dioxide.

The work on fire-extinguishing compounds continued, and in 1935 a patent relating to the "use for fire extinguisher purposes of halo-fluro derivatives of hydrocarbons" was issued to Bichowsky (14). Specifically, the inventor used fluorinated hydrocarbons as propellants to expel fire-fighting materials which could not provide their own propusion.

It was not until the years of World War II, however, that major progress in the field of aerosols was made. As a result of the research of L. D. Goodhue and W. V. Sullivan, working with the U.S. Department of Agriculture on methods of insect control in the various battle areas throughout the world, a method of dispensing insecticides using the same liquefied refrigerant gases previously employed in the pressurized fire extinguishers was developed. Goodhue and Sullivan were issued a patent (15) that became the one contribution to aerosol technology recognized as the basis for almost all later developments. Although their high-pressure unit was somewhat cumbersome, millions were used during World War II and served to introduce this new method of packaging to the American public. It was also as a result of the work of Goodhue and Sullivan that the search for newer containers and valves that were less expensive and easier to make began.

POST-WORLD WAR II PROGRESS

There followed a period of development resulting in newer aerosol containers to meet the needs of a growing market. In 1950, a patent that covered "emulsion containing a liquefied propellant gas under pressure and method of spraying same" was issued to Boe (16). The inventor cited several uses, including a pine oil deodorant, an insecticide, and a mothproofing compound. In 1953, Abplanalp (17) was issued a patent covering a new valve mechanism, and Spitzer, Reich, and Fine (18) obtained a patent for the preparation of shaving lathers and other pressurized foam products which led to the successful marketing of such items.

Among the other developments, two are worth mentioning. In 1954, Croce (19) received a patent covering "a perfume atomizer in which the perfume will be ejected from the container in response to gas pressure" and "in which the liquid to be dispensed and the dispensing gas are contained in separate reservoirs." A year later, a British patent (20) was issued to the Metal Box Company, Ltd., for a device that would "provide a means of dispensing flowable products in which the propellant and product are maintained separate one from the other within the container thus leading to considerable economies inasmuch as lesser quantities of propellants are required to achieve the desired results."

Many other changes have been made in the aerosol field to achieve the designs we know today. Different metals have been used, and containers have also been made of glass and plastic. Metered valves were designed to limit the amount of product dispensed at one time, and numerous lining materials are utilized to prevent corrosion and other incompatibilities. Many of these developments will be cited in other chapters in this book. Here it suffices to say that the aerosol evolved in a comparatively short span of time from the somewhat cumbersome insecticide bomb of the war years to the light, attractively designed, and quite functional container of today.

REFERENCES

1. R. W. Whytlaw-Gray, and H. S. Patterson, *Smoke*, Edward Arnold and Company, London, 1932.
2. D. Sinclair, *Handbook on Aerosols*, Washington, D. C., 1950, p. 64.
3. Glossary of Terms Used in the Aerosol Industry, Chemical Specialties Manufacturers Association; New York, Aug. 25, 1955.
4. A. Herzka and J. Pickthall, *Pressurized Packaging (Aerosols)*, Butterworths Scientific Publications, London, 1958, p. 3.
5. H. R. Shepherd, in E. Sagarin (Ed.), *Cosmetics: Science and Technology*, Interscience Publishers, New York, 1957, p. 788.
6. H. Helbing and G. Pertsch, Patent 628,463 (July 11, 1899).
7. C. L. Gebauer, Patent 668,815 (Feb. 26, 1901).
8. C. L. Gebauer, Patent 711,045 (Oct. 14, 1902).
9. R. W. Moore, Patent 746,866 (Dec. 15, 1903).
10. L. K. Mobley, Patent 1,378,481 (May 17, 1921).
11. E. Rotheim, Patent 1,800,156 (Apr. 7, 1931).
12. E. Rotheim, Patent 1,892,750 (Jan. 3, 1933).
13. T. Midgley, Jr., A. L. Henne, and R. R. McNary, U. S. Patent 1,926,396 (Sept. 12, 1933).
14. F. R. Bichowsky, U. S. Patent 2,021,981 (Nov. 26, 1935).
15. L. D. Goodhue, and W. N. Sullivan, U. S. Patent 2,321,023 (June 8, 1943).
16. C. F. Boe, U. S. Patent 2,524,590 (Oct. 3, 1950).
17. R. H. Abplanalp, U. S. Patent 2,631,814 (Mar. 17, 1953).
18. J. G. Spitzer, I. Reich, and N. Fine, U. S. Patent 2,655,480 (Oct. 13, 1953).
19. S. M. Croce, U. S. Patent 2,689,150 (Sept. 14, 1954).
20. Metal Box Company, Brit. Patent 740,635 (Nov. 16, 1955).

AEROSOLS: A DYNAMIC
GROWTH INDUSTRY

In attempting to describe accurately the first quarter century in the history of the aerosol package, it is difficult to avoid such adjectives as "dynamic," "fast-growing," "impressive" and even such superlatives as "tremendous." Ever since the early 1950s, when the low-pressure aerosol began to find ready acceptance in such areas as household, personal, hardware, and industrial products, the growth of this industry has been a constant source of astonishment to packaging experts.

From the beginning, the rapid popularity of aerosols seems to have caught many segments of the packaging industry completely unprepared. For example, the major suppliers of packaging machinery were technologically unready for burgeoning demands for such then-unusual devices as gassers, crimpers, gasser-shakers, hot-water leak test baths, cappers, and valve buttoners. New suppliers with names like Mojonnier (later, Kartridg-Pak), J. G. Machine Works, Alpha, Colton, Nalbach, and General Kinetics had to jump into the gap, replacing the jerry-built equipment designed in many cases by the fillers and marketers themselves. Makers of fitments and closures were unable to adapt their devices to provide for the unusual internal-pressure and leakage potential of the self-pressurized can, and a whole new group of package component manufacturers—the valve suppliers—came into existence.

And so it went in every category; from individual components like dip tubes and valve gaskets to cans and cover caps, from machinery to services like contract manufacturing and consulting, there was a considerable gap between supply and demand. In some cases, lack of availability of components or filling capacity or technical advice kept the industry from expanding at an even greater rate, though at the same time it may have acted constructively as a deterrent to impractical product concepts or marketing ideas. At any rate, it was not until the mid-1960s that many of the large custom fillers were able to cut back from three to two shifts, and that the suppliers of valves and caps caught up on back orders to a point where they could begin expanding into new geographical locations or into such technological innovations as codispensing, "spray-anyway" valves and valve-cap combinations with real style.

The logistical shortages of the 1950s gave way to more sophisticated technological problems in the 1960s. Any account of the industry's recent history based on its production figures must be correlated with technical improvements, evidence of consumer acceptance or rejection, and documented instances of misjudgment or miscalculation.

Like their military predecessors, the first aerosols produced after World War II were high-pressure insecticides in heavy steel containers fitted with a machined brass or steel valve. Though highly effective in performance, their obvious lack of style gave them little potential of ever becoming important items in supermarkets or other self-service outlets. In fact, most of the three to four million made in 1946 were sold in hardware stores or army-navy outlets, probably to ex-G.I.s with good memories of how well they performed on some South Pacific atoll.

The year 1947 marked a significant milestone in aerosol history, for it witnessed the introduction of the low-pressure aerosol. The first cans were delivered by Crown Cork & Seal Company for filling insecticides; the valves continued to be relatively expensive, awkward variations of the steel and/or brass monstrosities conceived by Bridgeport Brass and Westinghouse under wartime conditions. It was not until 1948 that product proliferation began—first with room deodorants and colognes, then paints, whipped toppings, shave creams, and hair sprays, and subsequently a myriad of household and personal products. By the end of the next decade, 1958, there were more than 100 different types of products on the market, and the annual volume in the United States and Canada had mushroomed to more than 485 million units.

Apart from this highly significant product proliferation, generated by the first stirrings of developmental research in the laboratories of du Pont, Crown, Allied Chemical, and Bridgeport Brass, the first few years of the low-pressure can were not happy ones for the new marketers and fillers. Valves continued to be expensive and not very reliable, and the second supplier of tin-plate cans, Continental Can, chose as one logical solution the selling of a can-valve "package," with the rubber and steel valve staked into the convex cap of the beer can variation then being sold as the standard pressure can.

It was not until 1949 and 1950 that standard pressure valves became commercially available, first from the new Precision Valve Corporation, but also from Oil Equipment, Aerosol Research, Engine Parts Manufacturing Company, and a few others. The availability of a general-purpose valve, with a standard 1-in.-opening tin-plate can to fit it into, really opened the way for a fantastic statistical climb in aerosol production figures over the next two decades. The versatility provided by a good, dependable aerosol can, in combination with a valve that could be varied as to actuator design and function, proved to be the major factor in opening up hundreds of product applications.

THE DIZZYING CLIMB OF THE AEROSOL INDUSTRY

By 1954, the production of insecticides (including mothproofers) had risen to 47 million, almost a tie with shaving lather. Room deodorants also began to pick up, reaching 11.3 million. On the other hand, the future product champion, hair sprays, probably represented just a small fraction of the 43 million units recorded for a catch-all category called "other personal products." Paints and other surface coatings also began to gain momentum, with 13.1 million. All aerosol production in 1954 amounted to 188 million.

Through the mid-1950s—from 1955 through 1958—the aerosol industry spurted forward at an incredible rate. The 188 million units in 1954 grew to 215.1 million in 1956, and within two more years had soared to over half a billion (to 513 million). During this period hair sprays gained a strangle hold on the individual product championship, and this eminence went relatively unchallenged until 1971, when personal deodorants and antiperspirants put up a strong challenge and finally surged ahead. During this same 4-year period (1954–1958), the growth rates of both insecticides and shaving lathers slowed somewhat, but a big gain for room deodorants (61.8 million by 1958) and paints (50 million) buoyed up the industry's total production figures.

The mid-1950s to the mid-1960s were the golden age for custom fillers. They sprang up all over the country, many starting in garages and old loft buildings with a weird assortment of home-made fillers and crimpers and a good deal of then-cheap hand labor. Some bigger companies also got into custom filling as a result of miscalculation; they were persuaded to buy filling lines for their own products but soon found that excess production, newly acquired know-how in a rather tricky packaging operation, and knowledge about the availability of components paid off better in servicing other national marketers than in sugging it out in limited regional markets. A few of the old-line fillers of bottles and jars successfully made the transition to the more complicated aerosol can or bottle. And some major marketers were able to turn excess productive capacity into profits, even filling for their competitors in the marketplace.

This proliferation of filling facilities became a significant factor in the continued growth of the aerosol industry. Marketers were able to find experienced fillers in most major population centers, and utilized them to get new products on the market quickly or to accommodate unexpected sales surges in old ones. More importantly, the laboratories of the filling companies began to augment those of the propellant and can companies with new product ideas, dispensing systems, and even container and valve innovations. From 1955 through 1965, the custom filler became a major source of "seed ideas," particularly in the areas of personal and household products.

However, in 1958 fillers, marketers, and component suppliers all participated in a technological blunder that was to dog the industry for the next decade. Everyone became highly excited over nitrogen-propelled toothpaste, most individuals forgetting that any compressed-gas-dispensed product is especially sensitive to misuse, resulting in quick loss of the necessary dispensing pressure. Add to this the fact that toothpaste is a product subject to unsupervised use by children, and one is left with enough factors to create packaging disaster.

Toothpaste proved to be such a disaster. After the first frenzied rush to meet obvious consumer interest—and to fill shelves in supermarkets and drugstores all over the country—an ominous stillness descended on the scene. There were many first sales but few repeats, and returns were heavy. Much the same thing happened with nitrogen-powered chocolate syrups, also designed for use by children, and introduced right after the toothpaste.

In many industries, a double disaster of the dimensions of the toothpaste and chocolate syrup fiascos would have caused a major recession. Proof of the vitality and generally good resilience of the aerosol industry is to be found in the fact that its growth rate continued unabated. Despite these misadventures (on top of a mild national recession in 1958), the production gain was an impressive 19.3% over 1957. And during the next 7 years the industry racked up annual growth percentages of 11.8, 16.8, 17.3, 16.5, 11, 13.5, and 31.9.

But something else changed in the 1960s, and the new trend seems likely to continue right through the next decade. The strong growth of and emphasis on household products during the first half of the 1960s gave way to a growing superiority by personal products. In fact, by 1969 personal products accounted for an even 50% of all U.S. aerosol production, with household products falling off to 16% and other products to 14%. This shift in emphasis is difficult to assess, without becoming inexorably entangled in such factors as the rise in numbers of working women, the development of better home-care electrical appliances, and the invention of finishes and fabrics that require less manual upkeep. Other competitive factors influencing the shift were (a) the continued popularity of hair sprays, at least until 1971; (b) an unexpectedly strong surge in the use of personal deodorants and antiperspirants, which made its presence felt in the last 3 years of the 1960s and into 1971; and (c) a leveling off in regard to room deodorants and waxes, for which a continued growth of starches could not compensate.

In connection with household products, it is of interest to examine what happened in the important period of 1959 through 1969. From 1959 through 1963, room deodorants grew from 63 million to 119 million and then suddenly plummeted to 109 million in 1964. Though there was a rebound to 154 million the next year, that was the beginning of a plateau; the category reached only 155 million in 1967 and 160 million in 1969. What happened to cause the sudden slump, the rebound, and then a slow plateau effect? One logical theory is that a mass conversion to water-based (and slightly less effective) formulations in the early 1960s was quickly detected by the consumer and caused a decline in repeat sales. The surge may have been attributable to the fortuitous introduction of disinfectants (typified by Lehn & Fink's Lysol spray), which from the beginning of the Chemical Specialties Manufacturers Association (CSMA) Aerosol Product Survey have been included in the room deodorant category.

Another plateau situation seemed to characterize the sales of waxes and polishes, which from a small base of 11 million in 1959 leaped to 41 million in 1961 and to 73 million in 1966. A sudden surge to 91 million units in 1967 may have been attributable to the addition of a lemon-fragrance product to the lines of most major marketers. Extensive television advertising to stress this new feature created a heavy demand for the lemon version and probably also stimulated sales of the conventional product. At this point in time, that is, in the first half of the 1970s, waxes and polishes seem firmly fixed on a comfortable plateau of 100 million units.

The strong-to-moderate plateau effect discernible for room deodorants and waxes (and, to a lesser extent, starches) also is detectable in other major categories of aerosols—paints and similar coatings, insecticides, and automotive products. During the 1960s, coatings recorded an unspectacular growth from 88 million to 140 million. Apart from the normal saturation experienced in all product types, what factors can account for a slight slowing down thereafter? Obviously, one was erosion of the novelty aspects of the aerosol package in the mind of the home handyman, probably coupled with growing awareness that decorating by means of the aerosol can is far more expensive than any other method of applying a coating. Another factor could have been an influx of cheap, less effective products, which contributed to a general lowering of product quality for all aerosol paints in the mind of the consumer. At any rate, into the mid-1970s, barring some exciting new dispensing or decorating idea, paints seem secure in the 130–160 million range.

Insecticides, the pioneer aerosol of all, have displayed more of a plateau effect than most other products. The 79 million units recorded in 1959 rose unimpressively to 87 million 1 year later; this in turn fell off to 73 million in 1964, recovered to 79 million in 1966, rose sharply to 95 million the next year, and reached about 108 million in 1971. But problems have dogged this category, despite the development of nonstaining pyrethrums and more pleasant solvent systems. There was no "bad bug year" until 1969 (followed by another in 1971). And by that time the effect of Rachel Carson's *Silent Spring* and subsequent outcries about pesticide safety had succeeded in creating a generally apprehensive public attitude that seemed to be reflected in a reluctance to promiscuously spray even the "safe" insecticides in kitchens, dining rooms, or front porches. The prospect of any big surge in production or sales seems unlikely, unless, of course, there is a series of summers marked by extraordinarily large insect populations.

Mainly because of disappointing performances by deicers and car waxes, automotive products also have not moved forward much in recent years. As with room deodorants (and, to a lesser extent, paints) part of the reason for the modest 11-year gain (through 1971) of from 18 to 58 million units probably was an influx of cheap, ineffective, private-brand deicers that failed to function in severe icing situations. Aerosol car waxes suffered along with conventional paste waxes, for such products the demand simply did not materialize because the major car manufacturers have adopted new finishes that do not require frequent waxing.

PERSONAL PRODUCTS

In terms of recent statistical records, new product development, continued marketer interest, and effectiveness of advertising, personal products really are where the action is! Of this category's five major subgroupings—personal deodorants, hair sprays, colognes and perfumes, shave lathers, and pharmaceuticals—the first continues very strong, the second, third, and fourth have hit plateaus, and only the last has begun showing a loss.

Hair Sprays

Despite radical changes in women's hair styles and a health scare during which hair spray inhalation was blamed for a form of cancer, hair sprays have maintained a remarkable record. Production peaked in 1968 (at 488 million units), slipping slightly to probably about 440 million in 1971 as the "natural look" became popular among both women and teenagers. As remote as the connection may seem, hair spray sales seem inexorably linked to trends in the fashion industry. The natural look in clothes, the rejection of the dictates of the couturiers of New York and Paris, the versatility in hemlines, and the popularity of pants—all influence hair styles and, hence, hair spray usage. At this writing it seems unlikely that hair spray marketers will ever again enjoy a mass return by women to the high, bouffant look of the mid-1960s, a style that required repeated and heavy applications of hair spray. More likely the first user will be the young adult who seeks to adopt some sort of manageable style when she grows out of the teen-age long, natural look. From such modest use grows the heavier buying needed to maintain the more tailored, more affluent look of the woman in her thirties or forties.

In assessing the total hair spray market, something needs to be said about hair sprays for men, which from 1969-1970 probably have preserved the total category from a severe slump. At the same time that the natural look was curtailing sales of this product to women, the long hair favored by men was responsible for a whole new audience for this type of product. The key to the success of male hair spray has been emphasis on its "dry" quality and its prevention of the "wet look" characteristic of older forms of hair tonic.

Technical factors also contributed to the steady increase in popularity of hair sprays. New hair spray resins from National Starch & Chemical Corporation and Barr-Stalfort Company, improved formulation technology from General Aniline & Film (on the copolymers of polyvinyl pyrollidone), and design of more glamourous valve-cap combinations improved both product performance and consumer package appeal. Although mother-daughter combinations (selling a large reservoir and home-use can in conjunction with a purse-size unit refillable from the larger container) failed, production of hair sprays continued to hold up well until the early 1970s.

Personal Deodorants

Personal deodorants in spray form were tried first in the early 1950s, but container corrosion and valve clogging made early efforts costly disasters. Even in the early 1960s, Gillette's initial effort with Right Guard ran into consumer resistance or, more accurately, disinterest. One reason may have been the slightly unpleasant chilling effect of fluorocarbon propellants on the human armpit. Another may have been the male orientation of the advertising, in a mass market where most of the buyers are women. But with marvelous perseverance and a huge advertising outlay, Gillette was able to overcome this resistance by stressing (a) the convenience of the spray can, (b) the fact that fingers are not required to spread the often messy product, and (c) the refreshing effect of the cooling spray. Gillette was very nearly alone on the market for the first 1–3 years; and only after the product clearly became a success with the male market did the competition come in with advertising and promotional campaigns. And from this base Gillette was able to push firmly into promoting both family and female use for its product.

No production figures are available for personal deodorants until 1965, when the success of Right Guard and its new competitors compelled the CSMA Aerosol Product Survey Committee to remove deodorants from the catch-all category called "other personal products." That first year deodorants registered an astonishing 166 million units. This spiraled rapidly to 187 million the next year, 114 million in 1967, 319 million in 1968, 375 million in 1969, and more than 400 million in 1971.

The sudden success of Right Guard in the mid-1960s precipitated all kinds of technical activity among marketers and component and propellant suppliers. The propellant companies were able to help by coming up with fluorocarbon blends that proved to be considerably less chilling to the human armpit. The valve companies developed actuators that did not clog with slightly large amounts of solids. The can companies improved the inner coating systems to permit slightly more corrosive ingredients. Moreover, by 1968 new aluminum-salt-based antiperspirants appeared on the market and began to seriously challenge the hexachlorophene-based deodorants. Products with such names as Extra-Dry Ban and Arrid Extra-Dry rather quickly forced Gillette to bring out an antiperspirant version of Right Guard and a new women's antiperspirant called Soft 'N Dry. By 1969 there seemed little doubt that antiperspirants might well take over the major share of the deodorant market.

That same year was marked by another phenomenon that served to recruit a score of new marketers to the deodorant business. Feminine personal deodorants, designed for application to the vaginal area and tastefully advertised in print and on television, suddenly took hold during 1969. Basically, the main active ingredients have been hexachlorophene and a pleasant perfuming agent, with a propellant blend selected to provide the minimum chilling effect. This product subcategory grew to between $40 and $50 million in 1971 (with one or two market prognosticators suggesting that it may reach 5 or 6 times that figure within 5 years).

Suddenly, however, there arose a furor over feminine personal deodorants among physicians (some of whom reported sensitization and irritation in the areas on which the products were used), consumerists (who attacked these deodorants as unneeded products being foisted on an unwary public by subtly misleading advertising), women's liberationists (who cited these products as an example of the exploitation of women through denigration of their sexual status), and regulatory officials (who expressed concern about the effects of frequent spraying of hexachlorophene in areas where mucous membranes would be inclined to absorb the chemical). This concentrated attack from four different directions caused a major setback for feminine personal deodorants in late 1971 and probably removed some of the appeal from the entire personal deodorant/antiperspirant category.

Coupled with the assault on feminine hygiene sprays was a regulatroy crisis involving hexachlorophene for all types of personal products. Based primarily on Food and Drug Administration charges that the bactericide is absorbed into the body (and, in test animals, produces some temporary degeneration of the central nervous system), the crisis was resolved after a fashion when the FDA issued a preliminary edict setting 0.75% hexachlorophene as the limit for products that can be sold without prescription.* Even products containing less than 0.75%

---

* *Editor's Note:* On September 17, 1971, the *Federal Register* published specific regulations pertaining to the use of hexachlorophene. In accordance with the orders issued by the FDA, the use of hexachlorophene in over-the-counter, nonprescription products is limited to 0.1% as a preservative. Products containing higher concentrations of hexachlorophene for special purposes are to be made available on a prescription basis only and are subject to an approved new drug application.

Since this order was issued after the completion of the various chapters in this book, the reader should keep the new rules in mind in referring to chapters containing information on the use of hexachlorophene.

hexachlorophene must bear label warnings: "Caution: Contains hexachlorophene." Issurance of this edict changed the emphasis in both product formulation and promotion. Henceforth many products were formulated to emphasize antiperspirant qualities, and promotion began stressing "contains no hexachlorophene."

Shave Creams

Shave creams in pressurized form have not been an outstanding success, with production only doubling (75 to 151 million) in the period from 1959 through 1968. Although part of the problem has been heavy competition for the shaving market from electric razors, an offsetting factor has been the introduction of stainless steel, chrome-plate, and finally platinum-plated blades for conventional double-edge mechanical razors. Radical technical improvements (beyond the addition of lime perfuming or lanolin) in aerosol shave creams have been slow in coming. Early in the 1960s an effort by Schick and later by Carter-Wallace to popularize a top-of-the-can heat exchanger (which utilized hot tap water to heat the shaving foam) failed when male users found that they were unable to obtain a consistently hot product from the heat exchanger. The next serious effort to produce a commercially acceptable, inexpensive hot product came in 1968–1969 via a system called codispensing. Codispensing means the simultaneous delivery of two components from an aerosol can; in this case peroxide from a small inner packet is permitted to interact in the actuator with a reductant in the shave cream, the resulting chemical reaction producing a hot foam. The first commercial product utilizing this principle was Gillette's The Hot One, introduced early in 1969; a considerably more expensive version was tried out by Gillette several months earlier.

Minor technical problems still plague the codispensing system. The temperature of the delivered foam seems to vary, probably because of unequal blending of peroxide and foam. To gain a lead on the competition for this product, Gillette adopted the unusual procedure of designing and manufacturing its own valves. In some of the other early commercially available valves, excessively flimsy peroxide bags and lack of precision in the metering orifices of the valve seemed to instill special caution in marketers who had been contemplating early introduction of their versions of a hot-foam product. Tabbed as far more promising an application of codispensing than hot shave cream is a hair dye, called Magic Moment, produced in 1969 by Gillette's Toni Company Division.

For shave creams, an additional problem is recognition of hot-shave products. Since the contours of can and top are exactly the same as those of the conventional competition, how is the prospective customer to recognize the difference? Gillette and its two competitors nearest to having a market-ready product have been trying to emphasize the difference with bright red colors, special introductory tags, and descriptive brand names. However, what seems to have been lack of movement for the Gillette product early in the 1970s could indicate that this approach is only an indifferent success, and lack of advertising support may herald abandonment of the product within a few years.

Another late innovation in shave creams was the first toiletry entry of S. C. Johnson & Sons, for many years the champion marketer of household and garden aerosols. In 1969, Johnson began a heavy regional marketing program for a product called Edge. This is packaged in a Sepro can, in which the ingredients are enclosed in a plastic collapsible bag to keep them separated from the propellant. In concept, then, the product is similar in performance to the old brushless shave cream once packaged in collapsible tubes. Despite this, consumer response reportedly has been favorable.

Pharmaceuticals

At the beginning of the 1960s, most observers estimated that the 11-million-unit total recorded for pharmaceuticals could well be multiplied by 10 during the next decade. By 1968, however, this category had grown to only about 61 million units, thus making it (along with foods) one of the major disappointments of the decade.

The principal explanation for the slow development of pharmaceutical aerosols has been the Kefauver amendments to the Food, Drug, and Cosmetic Act. This has mandated a profound tightening of FDA requirements regarding product efficacy, interaction of package components and ingredients, accuracy of dosage, and allied factors. As a result, what seemed to be serious efforts to perfect the aerosol as a means of administering systemic drugs and of delivering sophisticated topical, surgical, and other medication destined for the various body cavities have been greatly curtailed. Another complication, first noted in British medical journals in 1969, was the possibility that asthma treatment sprays may contribute to an increase in the death rate of asthmatic patients. Since asthma medication inhalers have been the bellwether of the pharmaceutical product category, a profound reduction in the use of these products could affect future production figures rather drastically. However, early 1971 findings suggested that the

British problem was not being duplicated in the United States and Canada.

Though optimists can still be found in out-of-the-way places, there have been subtle indications that the major pharmaceutical manufacturers are far too preoccupied with proving the efficacy and/or safety of conventionally packaged products to have the time or money to expend on the far more complicated aerosol. This is indicated by a dwindling of serious inquiries directed at valve and container suppliers, less starry-eyed optimism by consultants and package designers, and seemingly a decline in business for the enterprising custom fillers who went to the expense of installing special drug-filling facilities.

AEROSOL FOODS

Nearly every mistake that has been made with all other product types has been made to a greater degree in the food category. Still to be seriously challenged is a comment by this author in the February 1970 issue of *Drug & Cosmetic Industry*:

It would be safe to say that more developmental money was wasted, more promotional money lost, more production scrapped, and more reputations squandered on this than on any other aerosol product category. A combination of technological shortcomings (sometimes attributable to sloppy or too accelerated product testing), too costly propellants, failure to appreciate and act upon consumer complaints, and lack of regard for competing packaging developments, all were contributing factors.

Even du Pont's Freon Products Division, whose Freon C-318 was touted as the solution to the food propellant problem, has attempted to buttress the disappointing production figures for foods in a subtle way. Instead of continuing to record the dwindling aerosol food figures on its annual market reports on the same basis as before, in 1965 du Pont began including in this category refillable containers of whipped cream sold mainly to restaurants and institutions. But this ploy too failed, for even with this way of figuring the totals continued to decline: 1961, 55 million; 1961, 63 million; 1963, 67 million; 1964, 75 million; 1965, 130 million; 1966, 113 million; 1967, 116 million; 1968, 116 million. Without the refillable units, for 1971 the CSMA estimates that food production was 90 million. Even this figure is subject to question, however, since it includes what could be an optimistic "adjustment" of 60 million for companies that failed to respond to the association's annual Aerosol Product Survey that year.

Though aerosol statisticians seem to be in agreement that by far the best statistics available in this field come from the CSMA Aerosol

Product Survey (which is presented formally each May at the Association's midyear meeting), even the untrained layman can discern some flaws in the way the figures are presented. For instance, in the 1971 survey, released in May 1971, the following adjustments were made:

Household products	150 million
Personal products	111 million
Food products	55 million
Automotive products	3 million
Insect sprays	68 million
Industrial products	11 million
Miscellaneous	3 million

Simple addition quickly shows that the adjustment added to the 1971 total was a staggering 669 million, more than 38% of the 255 billion recorded for all production that year. Though these adjustments are made with care by comparing reported production figures with the numbers of valves, cans, and overcaps made in the same year, the distortion possible with "fudging" of such proportions must be disturbing indeed to a sober statistician or economist.

Even more bothersome than the size of the adjustment is the way it has been permitted to affect the method of organizing the figures. Buried in the fine print of the final survey report is language such as this to justify the way the adjustments are made: "The commanding positions in specific products attained by some of the nonreporting companies would cause any adjustment by product to be too revealing." The far-from-subtle implication is that the entire survey (or at least many of the important categories) has been altered to protect or otherwise accommodate special-status companies who do not choose to report. This policy is carried to the lengths of not making a valid adjustment for room deodorants; instead, such adjustment is transferred to the less specific "household products" category to protect two major marketers who do not report.

A PROVOCATIVE THEORY

Many aerosol industry "watchers" were intrigued by a paper delivered by Harvey D. Hirsch of Fry Consultants, Inc., at the CSMA meeting in Washington on December 10, 1968. He suggested that the aerosol market is approaching saturation since "industry growth is increasing at a decreasing rate." He then related this to the widely accepted product life cycle theory, which postulates that products and product categories inevitably move through a series of four stages: introduction, growth, maturity, and decline. The "declining growth rate." he argued, is a clear signal that aerosols are "moving out of their growth stage into maturity," the obvious implication being that decline will come in the foreseeable future.

To reinforce this idea Hirsch cited further evidence to be found in other trends typically associated with the maturity stage:

Price competition is becoming more common on all levels of industry. Marketers, fillers, and component and machinery suppliers are increasingly aware of price competition and attempt to maintain margins by passing their costs on to others. For example, marketer-fillers still maintain some link with custom fillers to ensure an outside research and development capability. Marketers partially rely on component manufacturers for market research and other marketing assistance. Another method of protecting margins is to produce internally that which was previously acquired externally. This too is occurring in the aerosol industry. Increasingly marketers are satisfying most of their own production needs.

Secondly, overcapacity is beginning to plague fillers and component suppliers. Supply, which lagged demand during the growth stage, is now beginning to outpace it. One response to overcapacity is forward integration, a response some component manufacturers have made. However, this response leads to greater competition, thereby intensifying industry problems and speeding movement to the maturity stage. Thirdly, soaring sales volume, which covered a variety of ills in the growth stage, is leveling, and as a result, a variety of problems is being uncovered. For example, on the filler level undercapitalization and weak management are only two of the problems revealed by a slowing growth rate. On the marketer level the need for stringent quality control has become evident. This marks an important shift by marketers from the production-oriented philosophy of the growth stage to the marketing-oriented philosophy of the maturity stage....

The sharpness of Hirsch's perception about the management and overcapacity problems of fillers and component suppliers is undeniable, but in one aspect at least—equating a product life cycle with the life cycle of an industry—he can fairly be faulted for mixing apples and oranges. But in regard to product life cycle alone, industry production figures suggest that his provocative theory has validity. For instance, it seems logical to conclude that personal deodorants/antiperspirants and probably household cleaners and mouth fresheners are still well within the growth phase and have ample growing room in the 1970s.

At the other end of the curve, experiencing the unhappy aspects of decline, are foods and pharmaceuticals. And scattered at various stages of the maturity phase are hair sprays, colognes and perfumes, coatings, automotive products, and waxes and polishes, with shave lathers, insecticides, and room deodorants approaching the end of maturity just before entering the gentle but painful decline.

Of course, an obvious stimulus to continued strong aerosol growth would be a constant reseeding with new products, especially those with the potential of achieving heavy use, such as a starch, a hair spray, or a personal deodorant. But the best aerosol minds among marketers, fillers, component and propellant suppliers, independent consultants, and research agencies have not been able to come up with much that is really new since the mid-1960s. Even the two products that manifested the most glowing promise at the close of the 1960s—feminine personal deodorants and mouth fresheners—are not radically new ideas. The first is really a minor variation of the widely used personal deodorant formulation, admittedly with more sophisticated perfuming and better packaging design. The second is approximately the same mouth freshener formulation, again with more convenient packaging, as that introduced by the Graham Chemical Company to the dental trade in the mid-1950s.

Clearly, the mid-1970s will provide a major test of the vitality of the aerosol industry, for by then some of the leading-volume products are likely to go into slow but worrisome decline. Consumerism, with its insistence on value and safety, already has affected the sales of feminine hygiene sprays, paints, foods, and perhaps personal deodorants. The sophisticated public, ever in search of something new, tends to redirect its fickle loyalties whenever a novelty is introduced. Therefore involved technical questions have a validity in assessing the industry's future. For example, can codispensing, Venturi systems and other more sophisticated approaches to dispensing open up important new product categories? Will recently improved technology in warm sprays, powder dispensing, and metering provide a new impetus for pharmaceuticals? Can the growing public awareness of ecology, product safety, family planning, or "natural" products be translated into a more vigorous market for "safe" pyrethrum-based insecticides, aerosol contraceptives, or products based on natural oils and resins? Is the industry mature enough to sustain regulatory pressures and still go on to achieve new production records?

INTRODUCTION TO AEROSOL SCIENCE AND TECHNOLOGY

DEFINITIONS

Aerosol products first appeared after World War II, and since that time the term "aerosol" has taken on several different meanings. According to the classical definition advanced by the physical chemist, aerosols are "those particles or droplets, ranging in size from about 0.15 to 5 microns, which are suspended or dispersed in a gaseous media such as air" (1). As such, smoke, dust, and water droplets in the form of dew, as well as salt in the air at a seashore resort, are all considered to be aerosol particles. The term "aerosol generator" refers to the equipment necessary to mechanically produce particles or droplets in this size range. The particles that are dispersed can be either solids, such as dust, or liquids, such as moisture and dew, suspended in air.

Aerosol Technology

Technology may be defined as a systematic knowledge of the industrial arts and includes the terminology used in the arts and sciences. Technology has also been defined as an applied science. Aerosol technology may be defined as a systematic study of the scientific principles involved in the development, preparation, manufacture, and testing of products that depend on the power of a liquefied or compressed gas to expel the contents from the container. This study of necessity will include the research and development of the product and study of all components, as well as the manufacture, quality control, testing, and shipping of these products. The area of interest can also be extended to include the physical and chemical properties and the physiological,

25

pharmacological, and toxicological properties of both the finished aerosol system and the propellants.

This definition is quite different from that of the classical term "aerosol" but includes the types of aerosols that are under discussion. Other definitions have also been advanced for the products referred to as aerosols (2, 3).

Goodhue and Sullivan (4) were confronted with the problem of producing an insecticide that could be dispersed in air rather quickly and that would produce minute particles. In turn, the particles would remain suspended for a relatively long period of time. Dicholorodifluoromethane was utilized as the propellant to supply the push for the insecticidal agent to be dispensed from the container, through a valve, and into the atmosphere. When the liquid came into contact with the air, the propellant quickly vaporized, causing the insecticide to be dispersed.

The particles produced by the ealy systems approached the size indicated in the classical definition. As newer products were developed, however, larger and larger particles were produced, giving rise to products dispensed in forms ranging from fine to wet sprays or mists to foams and semisolids. Although the latter products did not produce particles, there were classified as aerosols since a pressurized system was used (5).

Propellants

The propellant supplies the push needed to expel the contents from the container in the desired form. A propellant has been defined as a "liquefied gas with a vapor pressure greater than atmospheric pressure at a temperature of 40°C (105°F)" (6). This definition does not include trichloromonofluoromethane (Propellant 11) or the compressed gases such as nitrogen, carbon dioxide, or nitrous oxide. Although Propellant 11 is never used alone as a propellant, it is a common component of a propellant blend. The compressed gases are used as propellants only in the compressed state, and they supply the power to expel the contents from the container. The use of these compounds as propellants is covered in Chapter 6.

RELATIONSHIP OF AEROSOL COMPONENTS TO EACH OTHER

An aerosol package is really no different from any other type of preparation. Aerosols are unique, however, in that the various components are related to one another, resulting in a package that is functional. For example, the formulation on a nonaerosol emulsion preparation, whether it be a pharmaceutical, cosmetic, or paint, does not involve to any great extent the container that will be used to package the product. Nor does the closure affect the formulation to any great extent. If extremely volatile materials are used, however, the cap has to be such that it will prevent the escape of the material. The preparation of a pharmaceutical liquid to be packaged in pint or gallon bottles is not necessarily related to the container or closure except for the fact that the lining of a plastic cap may react with the product, or that the glass must be of a specific type. Formulation of a nonaerosol liquid preparation involves only the product and not, to any great extent, the container or closure.

Aerosol formulation, on the other hand, involves a consideration of all of the components. Although it is true that certain individuals are responsible only for packaging components for aerosols, others only for the manufacturing and packaging aspects, a third group only for the labeling, and so on, the formulating chemist is not only concerned with the formulation but is also deeply invloved with the propellant, container, and valve. Not only must he be familiar with the availability of containers and their usage, but also he must have a knowledge of pressure limitations as well as construction features. The same is true in regard to the propellant, valve, and other components. In aerosol science and technology, one must be familiar with the entire package and the way in which each component functions. This is true to a greater degree with aerosols than with any other type of preparation.

All aerosols consist of certain basic components, which include the propellant, container, valve (including an actuator and dip tube), and product concentrate. There are many variations of these components, however, and only when all of the components are selected and assembled properly, does an aerosol product having the proper characteristics result.

Propellant

The propellant is said to be the "heart" of the aerosol and, as previously indicated, serves to supply the power or push to expel the product and to deliver it in the proper form.

A liquefied gas propellant can be likened to any solvent that is used in a nonaerosol preparation. The propellant is unique, however, in that it exerts a vapor pressure at room temperature. A propellant which is useful in aerosols has a relatively low boiling point and a relatively high vapor pressure as compared to solvents such as water, alcohol, or acetone. For example, alcohol can be poured from one container to another at room temperature and atmospheric pressure. Propellants, on the other hand, require special handling techniques, since they are in the vapor state at room temperature and atmospheric pressure.

Whereas a nonaerosol product consists only of the product and not the means for application (paint product and brush, perfume and atomizer, medicinal agent and nebulizer), aerosols are self-contained products and no special applicators other than those attached to the product are required. The propellant, together with the valve, must deliver the product to the site of action in a form in which it can be utilized. The propellant actually performs the work normally performed by the individual using a nonaerosol product. The whipping of cream with a mechanical mixer or egg beater, the use of nebulizer and atomizer, and the application of paint by means of a brush have been replaced by use of an aerosol product.

An ointment preparation packaged in a tube required further work on the part of the user in that the tube must be squeezed to emit the product and then it must be rubbed into the skin. One of the advantages of an aerosol preparation is the fact that use of the product does not require further effort; the propellant does the work. If the product (e.g., asthma preparations) is for inhalation therapy, the medication not only is delivered to the site of action, which is the lungs, but also is delivered in a given particle size range. Particles delivered in other than the right size would be useless.

Container

All aerosol containers are made to withstand a certain amount of pressure. Although containers can be made of many materials (7), the commonly used container is made of steel plate coated with tin. This is similar to a beer can and may or may not have an organic coating on the inside. In addition to tin-plated steel, aluminum and stainless steel are also used, although the latter is restricted to perfumes and a few medicinals in fairly small containers. Glass is also used to fabricate aerosol containers. These containers, which generally have thicker walls than conventional glass containers, are either uncoated glass or plastic-coated glass. Plastic has also been used but presently is of limited value.

Valve, Actuator, and Dip Tube

An aerosol is made up of different components, and it is impossible to indicate their relative importance. The valve is useless without the propellant; the propellant is useless without the valve; and both the valve and the propellant are useless without the container. A cough syrup packaged in a conventional bottle fitted with a screw cap can still be used in the event that the cap breaks. However, if a valve on an aerosol container fails to operate, the product becomes useless because it cannot be dispensed. Similarly, if a valve does not spray properly and becomes clogged, there is no way of getting the material out of the container. An aerosol valve is multifunctional. It serves as a physical closure for the container and must be capable of being opened and closed without loss of content. It also serves to deliver the product in the desired form. Emulsions are delivered as a foam or spray, depending on the formulation; hair preparations are delivered as fine to wet sprays; and oral pharmaceuticals are delivered in a very fine state of subdivision. For example, a hair spray has to be delivered as a fine or wet spray so that the particles will be deposited and retained on the hair. If the spray is too fine, the particles will become air-borne. If the spray is too wet, it will not be deposited properly nor will it dry properly, so that particle size becomes an important consideration. This property is related to the valve as well as to other factors (8).

Product Concentrate

Various preparations have been formulated as aerosols, and it is the product concentrate that contains the active ingredient. In many cases, this component is similar to its nonaerosol counterpart. Through proper choice of propellant and valve system, the product concentrate can be delivered from the aerosol container in its proper form. Succeeding chapters in this book will discuss the specific product concentrates used in the formulation of various types of aerosol products.

OPERATION OF THE AEROSOL PACKAGE

The mode of operation of an aerosol package is dependent on the type of propellant used. Aerosols are different from other products in that they are packaged under pressure, that is, there is energy in the container to push the contents out. This can be accomplished in many different ways. If all that is desired is to push the material out of the container, energy can be supplied in the form of a gas or some mechanical means such as a spring. If the product characteristics are to change upon dispensing, additional energy in the form of a liquefied gas and/or mechanical breakup systems are required.

Liquefied Gas

Liquefied gases are commonly used to supply the energy required for the proper functioning of the aerosol container. Liquefied gases, as the name indicates, are materials that at room temperature and atmospheric pressure exist in the gaseous or vapor state and are capable of being liquefied at relatively low pressures or temperatures. Although most gases can be liquefied, only those that can be liquefied at a relatively low pressure or at a temperature close to room temperature are useful as aerosol propellants. For example, nitrogen, nitrous oxide, and carbon dioxide can be liquefied. However, in order to liquefy some of these gases, pressures of over 700 psig are required. There is a group of hydrocarbons and fluorinated hydrocarbons that can be liquefied at relatively low pressures. These materials, at a fairly low pressure (15–80 psig), are present in the liquid state. They function as aerosol propellants in that, because of their relatively low vapor pressure, they can be maintained in the liquefied state in a fairly inexpensive, thin-walled metal container that is capable of withstanding the pressures normally used for aerosol products.

Some of the commonly used liquefied gas propellants have boiling points below 0°C. If the temperature of the liquefield gas is brought below the boiling point, the propellant will remain in the liquid state. When this liquid, held at the low temperature, is placed in an open container at room temperature and is exposed to atmospheric pressure, immediately some of the molecules will leave the liquid state and be converted to the gaseous state, since the surrounding temperature is higher than the liquid temperature. In so doing, the vaporized molecules escape and this process continues with increasing velocity as the temperature of the liquid reaches its boiling point. This temperature is maintained until all of the liquid has vaporized.

If the container is sealed so that the vaporized molecules cannot escape, the space between the top of the liquid level and the top of the container is occupied by the vaporized propellant. Eventually, a point is reached where the space is completely filled with vapor. As more and more vaporized molecules squeeze into this space, there is an increase in pressure until an equilibrium is attained. At this point, the pressure (now referred to as the vapor pressure) remains constant as long as there is no loss of either liquid or vapor phase and the temperature remains constant. When part of the vapor phase is allowed to escape, however, the equilibrium is upset and there is a drop in pressure due to loss of some of the vaporized molecules. This temporary drop in pressure allows some of the liquid phase to be converted to the vapor phase, and eventually the equilibrium is re-established at the same pressure.

If a hypodermic syringe and needle is inserted through the top of the container and the needle is extended down to the liquid phase, it can be seen that the plunger of the syringe will rise because of a greater pressure inside the container than atmospheric pressure. However, the vapor is exerting a pressure upon the liquid. In exerting this pressure, some of the liquid will be pushed into the syringe unless additional force is placed upon the plunger. As the liquid phase escapes, the volume occupied by the vapor phase will increase, resulting in a drop in pressure. However, the pressure will again be re-established, as previously indicated.

Replacement of the plunger and barrel of the hypodermic syringe by a conventional aerosol valve and the needle by a dip tube results in an aerosol package fitted with the conventional components. When the container described above is inverted, the opposite phase is released. By placing an additional small opening in the valve in the vicinity of the head space, both liquid and vapor phases can be discharged simultaneously (the vapor tap valve will be discussed later). As long as liquid phase is present, the pressure will remain constant. It is for this reason that a liquefied gas propellant is referred to as a constant-pressure propellant. This is an essential difference between liquefied gas and compressed gas propellants.

Depending on whether the product concentrate is dissolved, suspended, or emulsified with the propellant, the final product will be discharged as a spray or foam. The same material, formulated differently, can be discharged in one case as a spray and in another as a stream or foam. The final form is dependent on the product concentrate and the valve. Both fluorinated hydrocarbons and hydrocarbons function in a similar manner, but there is a difference in the properties of the two groups. Such characteristics as density, solubility, pressure, heat exchange, and flammability may limit the use of some of them.

Since every aerosol consists of a product concentrate and a propellant, some consideration must be given to the effect of the solvent on the aerosol system. When a mixture of equal parts of alcohol and propellant is made, a solution will result. When this mixture is discharged through an aerosol valve, using a liquefied gas, it will emerge as fairly large droplets, depending on the type of actuator used. Both propellant and alcohol are in the liquid state the moment they are emitted from the nozzle of the valve. As soon as the mixture hits atmospheric pressure, however, the liquid propellant tends to revert to the vapor state. In doing so, it will expand. In expanding, it will also disperse the alcohol into minute particles. The size of the particles will depend on the quantity of the propellant, the nature of the valve, and also the type of propellant used.

If, in place of alcohol or a soluble material, an insoluble powder such as talc or starch is used, the same steps as outlined above take place. The only difference is that the insoluble powder remains behind as a solid rather than a liquid particle.

Similar principles are involved when an emulsion-type product is formulated. It can be so formulated that either the emulsion is coated with the propellant or the propellant is coated with the emulsion. This can be likened to an oil-in-water (o/w) or a water-in-oil (w/o) emulsion, water being the continuous phase in the former and the dispersed phase in the latter. In the case where the propellant is the continuous phase, the propellant surrounds the particles of emulsion, and as the product is discharged, the propellant (in the external phase) vaporizes, leaving behind an undisturbed droplet of emulsion. In reality, the emulsion is dispersed into tiny droplets that resemble a spray. Some of the water-based room deodorants and insecticides exemplify this system, which will be covered in greater detail later. When the emulsion is on the outside and the propelant is on the inside, a foam results. As the product is discharged, the propellant starts to vaporize, but this time, in vaporizing, it has to travel through the emulsion. In going through the emulsion (the situation is similar to an egg beater and whipping cream), the propellant whips the emulsion into a foam, as exemplified by shaving foam. Depending on the structure, a quick-breaking to stable foam can be achieved.

The internal pressure of a liquefied gas propellant, a solvent, or any other material is an indication of the inner forces (attractive forces) that are present within a molecule. Calculation of the internal pressure of a propellant enables one to determine the degree of physical and chemical similarity of materials. This will be a subject for further discussion, since it is one of the factors that accounts for deviation from ideal behavior in regard to the vapor pressure of an aerosol system.

Compressed Gases

An aerosol utilizing a compressed gas as the propellant operates essentially as a pressure package in that the pressure of the gas forces the product from the container in its original state. Under limited conditions, a spray or foam-type product may be achieved. A compressed gas propellant is essentially a gas which has been placed in a container under pressure, for example, nitrogen at 90 psig. Because of the difference between atmospheric pressure and 90 psig, as soon as the valve is opened, the gas forces the contents out the valve. The gas can be nitrogen, nitrous oxide, or carbon dioxide. Only the product is discharged; the gas remains behind, occupying the head space. The gas will then expand because of the increase in volume. As it expands, the pressure will drop from the originally level of, say, 90 psig to 80, 70, or 60 psig. The drop in pressure is related to the amount of material discharged and may be calculated from the gas laws. Since there is no propellant in the liquid phase (unless some solubility is present), the pressure decreases as the contents are utilized.

Other Mechanical Systems

Certain systems have been developed which depend upon mechanical means for supplying the pressure. These will be covered in the next chapter.

APPLICATION OF THE PRINCIPLES OF AEROSOL SCIENCE AND TECHNOLOGY TO PRODUCT FORMULATION AND DEVELOPMENT

Types of Products

Aerosol products have been classified under seven broad categories, which include insect sprays, coatings and finishes, and household, personal, animal, industrial, and food products. Personal and hygienic deodorants and antiperspirants, hair sprays and other hair products, paints and related products, household cleaners, starches and laundry products, room deodorants and disinfectants, and shave lathers lead the list of aerosol products in the given order of popularity. Other aerosol products that are available but have not been listed above include waxes and polishes, shoe dyes and polishes, fuels, shampoos, pharmaceuticals (both oral and topical), colognes and perfumes, suntan preparations, depilatories, breath fresheners, and windshield deicers.

Advantages of Products Packaged as Aerosols

The chief advantage found for most aerosol products is their convenience; they are easy to use and can be applied directly from the container with no further treatment. Other advantages include the fact that the aerosol package is a completely sealed unit so that the product is never exposed to the deteriorating effect of air and there is no danger of contamination with foreign materials. Certain products (e.g., hair sprays and foams) are available only as aerosols.

Aerosol packaging has often been referred to as "convenience" packaging, since in most cases a more convenient form of a commonly used product results from its development as an aerosol. One need only examine shave lathers, whipped creams, deodorants and antiperspirants, perfumes and colognes, paints, insecticides, and room deodorants. It is no accident that these are representative of some of the more popular categories of aerosol products.

PERSONNEL FOR THE AEROSOL INDUSTRY

The development of aerosol products has become more complicated and involved. Ever since Goodhue and Sullivan assembled the first aerosol on a Sunday afternoon in 1942, investigators throughout the world have developed a large number of aerosol products. However, the "easy" products were developed "yesterday," the "difficult" ones are being developed "today," and the "impossible" aerosols will become available "tomorrow." With increasing demands for containers that will withstand corrosive products and higher pressures, valves that will dispense a controlled amount of product in a variety of different ways, propellants that will be miscible with water and other innocuous solvents, and new aerosol systems for a host of different materials and applications, the aerosol chemist has found himself immersed in a far-ranging technology quite distinct from the one that existed 15 to 20 years ago.

The aerosol chemist of today must not only be familiar with the principles involved in the formulation, development, manufacture, and testing of aerosol products, but must also be knowledgeable in metallurgy, plastic and glass technology, and engineering, to say nothing of legal and commercial considerations. With the increasing imposition of governmental controls and regulations on aerosol products, the aerosol chemist must of necessity become involved in practically all aspects of an aerosol development program.

The aerosol industry has grown rapidly in the past 25 years, and as this growth continues, the need for properly qualified and trained individuals becomes ever more apparent. In many instances, individuals trained in other fields have been called upon to assume related positions in the aerosol industry. Given "on-the-job" training, in a relatively short period of time they have become well versed in aerosols. Although this is perhaps not the most efficient means of obtaining recruits for the aerosol industry, it has been, in most instances, the only way.

Education is the key to success and can well determine the present and future development and growth of all industries. Without education, scientific and technological advances cannot take place to any great extent. The present-day economy, with its ever-expanding industries, requires individuals with special knowledge and skills. As industry expands and grows, however, education and training sometimes fail to keep pace with rapid developments. Certainly the aerosol industry is no exception. Whereas many of the older industries could recruit personnel from various colleges and universities, the aerosol industry was limited in that it had to take individuals trained in the basic sciences and then give them the specialized training required of aerosol chemists. It should be pointed out that the cosmetic industry continues to face the same problem, since there is no one program specifically designed for cosmetic chemists.

When one considers the manufacturing, quality-control, shipping, and marketing aspects of aerosol products, the need for many individuals with diverse training and background becomes apparent. Individuals with backgrounds in business administration, engineering, pharmacy, chemistry, biology, physics, and mathematics have been found to be extremely useful in the aerosol industry. Recently a need for pharmacologists and toxicologists has become apparent because of the abuse of propellants by a certain segment of the population and the various adverse reports made in regard to certain aerosol products.

REFERENCES

1. R. W. Whytlaw and H. S. Patterson, *Smoke*, Edward Arnold and Company, London, 1932.

2. Glossary of Terms Used in the Aerosol Industry, *Aerosol Guide*, Chemical Specialties Manufacturers Association, New York, 1971, p. 7.

3. J. J. Sciarra, Chapter 8 in J. B. Sprowls (Ed.), *Prescription Pharmacy*, 2nd ed., J. B. Lippincott Company, Philadelphia, 1970, pp. 281-282.

4. L. D. Goodhue, and W. N. Sullivan, U. S. Patent 2,321,023 (1943).

5. *Package for Profit, E. I. du Pont de Nemours and Company, Wilmington, Del.,* *1968, pp. 6–8.*
6. Glossary of Terms Used in the Aerosol Industry, *Aerosol Guide,* Chemical Specialties Manufacturers Association, New York, 1971, p. 8.
7. J. J. Sciarra, Chapter 90, in J. E. Hoover (Ed.), *Remington's Pharmaceutical Sciences,* 14th ed., Mack Publishing Company, Easton, Pa., 1970, pp. 1743–1744.
8. P. A. Sanders, *Principles of Aerosol Technology,* Van Nostrand Reinhold Company, New York, 1970, pp. 46–52.

TYPES OF AEROSOL SYSTEMS

An aerosol system consists of a product concentrate, propellant, container, and suitable valve. The concentrate can be of the solution, dispersion, emulsion, or semisolid type, while the propellant can be either a liquefied or a compressed gas. Depending on the nature of the propellant and product concentrate, as well as the combination of these two components with the appropriate valve and actuator, the product can be dispensed as a spray, foam, or semisolid.

CLASSIFICATION OF AEROSOL SYSTEMS

According to Sanders (1), all aerosol systems can be classified as either homogeneous or heterogeneous. When the concentrate and propellant and/or solvents are mutually miscible, a homogeneous system is obtained, whereas immiscible components or components containing insoluble substances yield heterogeneous systems. Hair sprays, room deodorants, and perfumes and colognes are examples of the former type; shave foams, foot powders, and some antiperspirants, of the latter.

In addition to the above broad classification, other systems, based either on the type of spray produced or on a physical description of the final aerosol product (2), have been used to classify aerosols. A combination of several systems is used in this textbook.

Definitions

There are various ways in which the aerosol components can be brought together to form a finished aerosol product. The product concentrate, which may be a solid or a liquid, can be dissolved or mixed with the propellant so that a true solution is formed. A cosolvent may be added in cases where the propellant and concentrate are immiscible, although this is not always possible. The concentrate can also contain insoluble solids which can be dispersed throughout the propellant. It may also be possible to disperse the propellant throughout the powder. If the propellant and concentrate phases are immiscible, it may also be possible to bring them together by means of an emulsifying agent or surfactant, resulting in the formation of an emulsion. This emulsion can be dispensed as a foam or as a spray, depending on the nature of each phase and the design of the valve and actuator. Additional systems that serve to simply push the contents out of the container or else to separate the propellant from the concentrate are also available.

Aerosol Systems

Aerosol systems will be discussed in greater detail under the following headings:

Liquefied Gas Aerosols.
Compressed Gas Aerosols.
Separation of Propellant from Concentrate Systems.
Other Systems (CoDispensing System and Mechanical System).

LIQUEFIED GAS AEROSOLS

A liquefied gas aerosol, as the name indicates, utilizes a liquefied gas as the propellant. Of the liquefied gases used, there are generally two basic types, fluorinated hydrocarbons and hydrocarbons. From the viewpoint of the aerosol system itself one need not be concerned whether the propellant is of the fluorinated hydrocarbon or hydrocarbon type; both groups of propellants function in a similar manner.

In a liquefied gas aerosol system, the propellant is present in the liquid state. As a result, there is a buildup of pressure, referred to as the vapor pressure, in the container; this pressure is constant throughout the life of the aerosol product or remains appreciably constant as long as some of the liquid phase is present in the container. When all of the liquid is used, the pressure will drop. In the liquefied gas aerosol system the pressure is constantly regenerated; that is to say, as the level of the liquid falls, the propellant in the vapor state will occupy a greater volume, causing the pressure to drop temporarily. However, as the pressure drops, some of the liquid propellant will vaporize to re-establish the original vapor pressure. Hence liquefied gas aerosols are said to have a "reservoir" of potential pressure.

Solution Systems

This type of aerosol system consists of two distinct phases, a liquid and a vapor phase. The system may be defined as a "solution of active ingredients in pure propellant or a mixture of propellant and solvent." The solvent is used to dissolve the active ingredients and/or to retard the evaporation of the propellant. Solution aerosols are relatively easy to formulate, provided that the ingredients are soluble in the propellant. However, the liquefied gas propellants are nonpolar in nature and in most cases are poor solvents for some of the commonly used aerosol ingredients. Through use of a solvent that is miscible with the propellant, one can achieve varying degrees of solubility. Ethyl alcohol has found use for this purpose, along with other solvents, such as acetone, glycols, dipropylene glycol, ethyl acetate, mineral spirits, acetone, and methylene chloride, the choice depending on the type of product to be formulated. Figure 4.1 illustrates this system.

Solutions have been used to formulate almost all types of aerosols, including hair sprays, colognes, perfumes, insecticides, and room deodorants. These products are generally formulated to consist of from 50 to 95% propellant and from 5 to 50% of active ingredients and cosolvent. The greater the amount of propellant present, the greater will be the degree of dispersion and the finer the spray. As the concentration of propellant decreases, the wetness of the spray increases. The pressure of such a solution varies between 15 and 70 psig.

When the valve of a solution aerosol is depressed, a mixture of active ingredients, solvents, and propellants is emitted into the atmosphere. As the liquid propellant hits the warm surrounding air, it tends to vaporize and, in so doing, breaks up the active ingredients and solvents into fine particles. Depending on their size, the particles remain suspended in air for relatively long periods of time. The particles of spray can vary from as small as 5 to 10μ or less to as large as 50 to 100μ. Solution aerosols can also be called "surface" or "space" sprays. A surface spray contains relatively large particles, since the particles are intended to coat a surface and not to become suspended in air. Examples of surface sprays include residual insecticides, coatings, and hair sprays. Space sprays are formulated to contain relatively smaller particles that will remain suspended in air for longer periods of time. In

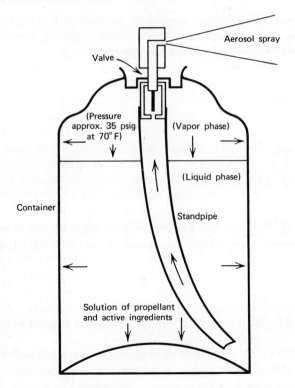

Figure 4.1. Solution-type aerosol system.

order to remain suspended, the particles must be in a relatively fine state of subdivision. Room deodorants and space insecticides are examples of space sprays.

Space sprays can be differentiated from surface sprays on the basis of the amount of propellant present and the valve system. A typical space spray may have as much as 85 to 90% propellant, whereas a coating spray may contain only 50% propellant or less. A space spray generally has a higher percentage of propellant, as well as a higher percentage of the propellant with the higher vapor pressure. In a surface-type spray a greater percentage of a propellant with the lower vapor pressure is used, and the valve, particularly the actuator, is capable of breaking the particles.

As the valve of a solution-type aerosol is depressed, the product is forced up the dip tube, through the valve and actuator, and into the atmosphere. The propellant is trapped within the liquid concentrate, which has been partially dispersed in passing through the valve and actuator. The propellant starts to vaporize and, in so doing, reduces the size of the liquid droplets. The size of these droplets will depend on the nature of the propellant, the amount of propellant, the nature of the product concentrate and the valve design.

Suspension Systems

For substances that are insoluble in the propellant or the mixture of propellant and solvent, or in cases where a cosolvent is not desirable, the active ingredients can be suspended in the propellant vehicle. When the valve is depressed, the suspension is emitted, followed by rapid vaporization of the propellant, leaving behind the finely dispersed active ingredients. This system has been successfully used to dispense antiasthmatic aerosols, body talcs, feminine hygiene sprays, and spot removers. However, the formulation of this type of aerosol is not without difficulty. Problems involving caking, agglomeration, particle-size growth, and clogging of the valve arise. Valves are available that can be used to dispense powders. Some of the more important factors which must be considered in formulating this type of system include:

Moisture content of ingredients.
Particle size of solid.
Solubility of active ingredients.
Density of both propellant and active ingredients.
Use of surfactant or dispensing agent.

Both active ingredients and propellant must be anhydrous when used with this type of system. It is generally agreed that the total moisture content of the system should be below 300 ppm (3). The propellants can be dried by passing them through desiccants, and the other ingredients can be dried by the usual method.

The initial particle size of the insoluble ingredients should be in the micron range, generally from 1 to 5 μ and not more than 10 μ, depending on the amount of powder to be dispensed. A jet pulverizer or ball mill can be used to reduce the particle size.

The derivative of the active ingredient having minimum solubility in the propellant and solvents should be selected. It is the slight solubility of the active ingredients in the propellants and solvents that contributes to particle size growth.

By adjusting the density of the propellant and/or the insoluble material so that they are approximately equal, the rate of sedimentation can be substantially reduced. This can be accomplished by using a mixture of different propellants of varying densities as well as by the addition of an inert powder to the active ingredients.

Final consideration should be given to the use of a surfactant or dispering agent. Isopropyl myristate (which is not a surfactant) has been employed primarily for its lubricating properties, and mineral oil has been used in a similar manner. In addition, many surfactants have been used for this purpose (4).

Emulsion Systems

Water and hydrocarbon or fluorinated hydrocarbon propellants are not miscible. In order to formulate a suitable aerosol using these materials, various techniques can be used. An emulsion aerosol consists of active ingredients, an aqueous or nonaqueous vehicle, a surfactant, and a propellant. Depending on the choice of ingredients, the product can be emitted as a stable or quick-breaking foam or as a spray. Approximately 7 to 10% fluorocarbon propellant or 3 to 4% hydrocarbon propellant is used in conjunction with 90 to 97% emulsion concentrate when a foam is desired.

THREE-PHASE (TWO-LAYER) SYSTEM

This system consists of a propellant and an aqueous solution which contains the active ingredients. One way to prepare such a system is to mix the two components together. One material will layer out on top of the other, depending on the density of the propellant and the nonmiscible liquid. If the propellant is of the fluorinated hydrocarbon type, it will sink to the bottom of the aqueous layer, since the specific gravity of most of the fluorinated hydrocarbons are greater than 1. If a hydrocarbon such as butane or propane with a specific gravity of less than 1 is used, however, the propellant will float on top.

Figure 4.2 illustrates this system, which has been used for many years for a mothproofing product. In operation, the liquid propellant, whether it is on the top or the bottom, will vaporize and fill the vapor space. Since the vapor is not soluble in the water layer, it will go right through the aqueous layer and form a pressure within the container in the head space. The nature of the propellant will determine the pressure. These have been referred to as three-phase systems because three distinct phases—a vapor phase, an aqueous phase, and a propellant phase—can be discerned. When the valve is actuated, the pressure will push down on the aqueous solution and allow it to come to the valve. As long as the dip tube does not touch the propellant layer (when the propellant is on the bottom), only aqueous solution will be dispensed. By using a mechanical breakup actuator, the product can be dispensed as a wet, coarse spray. The system is limited to the formation of this type of spray.

FOAM SYSTEM

The propellant used in an emulsion is an important part of this system and determines the type of foam produced. It should be indicated that the propellant is generally considered to be part of the immiscible phase and as such can be in the internal or external phase. When the propellant is included in the internal phase, a typical stable or quick-breaking foam is emitted. When the propellant is in the external phase,

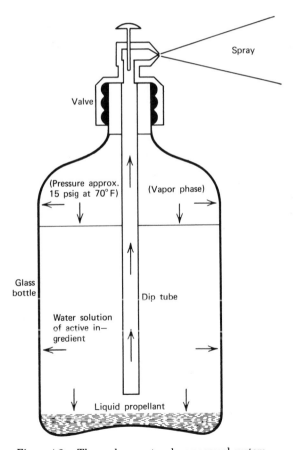

Figure 4.2.   Three-phase or two-layer aerosol system.

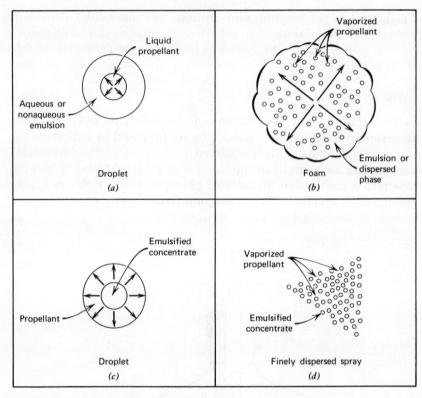

Figure 4.3. Typical emulsion with propellant in the internal or external phase. (a) Propellant in internal phase; (b) formation of aerosol foam; (c) propellant in the external phase; (d) formation of a wet spray.

the product is dispensed as a spray. Figure 4.3 illustrates these two types of emulsions.

As can be seen from Figure 4.3a, when the propellant is in the internal phase, the propellant vapor must pass through the emulsion formulation in order to escape into the atmosphere. In traveling through the emulsion, the trapped vaporized propellant forms a matrix for a foam to develop, as seen in Figure 4.3b. Depending on the nature of the formulation and the propellant, the foam can be stable or quick-breaking.

Figure 4.3c illustrates a typical emulsion system in which the propellant is in the external or continuous phase. As the liquefied propellant vaporizes, it escapes directly into the atmosphere, leaving behind droplets of the formulation which are emitted as a wet spray (Figure 4.3d). This system is typical of many water-based aerosols.

Stabilized Foams

A typical emulsion system leading to the dispensing of a stabilized foam is shown in Figure 4.4. In an emulsion system where the propellant is in the internal phase (generally part of the oil phase of the emulsion), water makes up the external or dispersed phase, although nonaqueous solvents can also be used. The propellant is generally used to the extent of about 7 to 10% of the total weight. When a hydrocarbon propellant is used (such as isobutane/propane blends), as little as 3 to 4% propellant is sufficient to produce a suitable foam. This propellant is emulsified with the aqueous emulsion; however, some of the propellant will vaporize and be present in the head space to produce the necessary vapor pressure. The pressure will be approximately 40 psig, depending on the propellant that is used. When the valve is depressed, the pressure will force the emulsion up the dip tube and out the value. Depending on the formulation, either a stable foam, such as would be expected in shave cream, or a quick-breaking foam will occur. A quicking-breaking foam will be dispensed as a

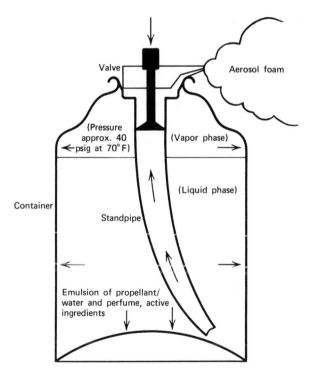

Figure 4.4.   Emulsion system (stable foam).

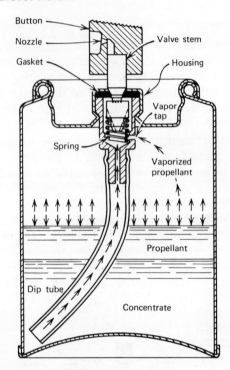

Figure 4.5. Water-based system. [From *Aerosol Technicomment*, **6,** 1 (April 1963), p. 3.]

foam but will collapse in a relatively short period of time. This type is not desirable in a shave cream, but may be advantageous in preparations where, for example, one may wish to have the material dispense as a foam so that it can be applied to a limited area and then collapse. For example, one or two after-shave lotions have been formulated in this manner.

Quick-breaking Foams

These foams consist of ethyl alcohol, water, and a surfactant that is soluble in either alcohol or water but not in both. Other miscible solvents can be used in place of alcohol and water. The surfactant can be nonionic, anionic, or cationic. The product is dispensed as a foam but quickly collapses upon coming into contact with the skin. This is particularly advantageous in pharmaceutical aerosols for topical application, since the foam will hit the affected area and then collapse, so that there is no further injury by mechanical dispersion of a material. Steroid, burn, and other topical preparations can be applied in this manner. Nail polish remover has also been formulated as a quick-breaking foam.

One advantage of a foam system over a spray system is the fact that the area with which the product can come into contact is controlled. For example, preparations containing irritating ingredients may be dispensed in this manner. The incidence of air-borne particles can be substantially reduced, thereby lowering the toxicity of spray products that cause irritation on release.

Spray Emulsions

The base for this type of product is a water-in-oil emulsion. A fairly large amount of propellant (about 25 to 30%) is miscible with the outer oil phase so that the propellant remains in the external phase of the final emulsion. When this system is dispensed, the propellant vaporizes, leaving behind droplets of water-in-oil emulsion with no foaming, as shown in Figure 4.3c and 4.3d. Figure 4.5 illustrates this type of aerosol system (5). Since the propellant and concentrate phase tend to separate on standing, products formulated using this system must be shaken before use. A hydrocarbon propellant or a mixed hydrocarbon/fluorocarbon propellant is preferred for this system, since the specific gravity of the propellant will be less than 1 and the propellant will float on the aqueous layer. In addition, such systems use a vapor tap valve, which tends to produce finely dispersed particles. These systems have been used to formulate insecticides and room deodorant sprays.

COMPRESSED GAS AEROSOLS

Insoluble Gas

As far as compressed gas systems are concerned, they were originally developed simply to provide the means of expelling a product from its container. The product to be dispensed was severely limited in that its consistency had to be free flowing. When the valve of the container was depressed, the compressed gas would push the concentrate out the valve, as shown in Figure 4.6. If the product is a dental cream, ointment, or any other semisolid, it will be dispensed as such. If a liquid concentrate, such as an aqueous solution of active ingredients or a nonaqueous solution, is used with a compressed gas propellant, the product will be dispensed as a spray when a mechanical breakup actuator is used. Nitrogen is insoluble in most materials and is generally used as the propellant.

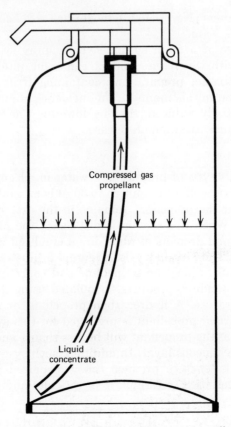

Figure 4.6.   Compressed gas aerosol using insoluble gas as propellant.

Although these products are not intended to be shaken before use, shaking will cause some of the insoluble gas to be dispersed, not dissolved, in the liquid concentrate. This will tend to disperse the product to a greater extent, but will result in an excessive loss of propellant. If enough propellant is lost, the product will become inoperative.

Soluble Gas

A soluble gas such as nitrous oxide or carbon dioxide can also be used as a propellant in a compressed gas system. These gases have an advantage in that they are soluble in the concentrate and will supply a reservoir of propellant. When the product is dispensed, a mixture of the liquid concentrate with the gas dissolved in it is expelled. The dissolved gas tends to expand and to escape into the atmosphere, resulting in dispersion of the liquid into fine particles. Since the pressure decreases as the product is dispersed (the volume occupied by gas increases), some of the gas will come out of solution and partially restore the original pressure. The application of the gas laws to these systems is discussed in the next chapter.

A modification of this system can produce "solid-stream dispensing." However, in this particular case, the container is used in the inverted position. Such a preparation consists of a product concentrate together with a soluble compressed gas. Generally, soluble gases are preferred to nonsoluble ones in order to hold a greater amount of gas in the container. When the valve is depressed, the material is emitted as a stream. By special design of the valve system, one can control the velocity of the stream so that a concentrate added to water or milk contained in a glass is self-mixing, resulting in a homogeneous mixture. This system was developed to be used with pressurized beverage products (6). However, the concentrate can consist of any type of liquid useful for foods, cosmetics, or pharmaceuticals.

Foam Dispensing

In order to produce a satisfactory foam, a soluble compressed gas (nitrous oxide or carbon dioxide, or a mixture of both) is used together with an emulsion concentrate. A mixture of the two gases is desirable because of their collective solubility characteristics and physical and chemical properties. In addition, carbon dioxide tends to have an acid reaction, which may or may not present compatibility problems when this gas is used alone.

It should also be pointed out that with all of these preparations the valve plays an important role in determining the manner in which the product is dispersed. For example, an emulsion preparation that supposedly should be dispensed as a foam will produce a fine stream when used with a spray-type valve. This system has been used primarily to dispense whipped cream and toppings, as well as several veterinary foams.

SEPARATION OF PROPELLANT FROM CONCENTRATE SYSTEMS

Although the greater number of aerosol products in use today utilize the typical aerosol systems, there are many additional products for which these systems are impracticable. The viscosity of the product, incompatibility of the product concentrate and propellant, and desired dispensing characteristics of the finished product represent a few of the reasons why the typical aerosol systems may be unsuitable. For such cases, several additional systems characterized by provision for a physical separation of propellant and product are available.

Many designs have been submitted for these systems, but only two basic ones have been accepted to date: the barrier packs and the siphon systems. Barrier packs utilize a plastic bag or piston to separate the product from the propellant, while siphon systems utilize a propellant cartridge and an outer, nonpressurized container which holds the product. Both of these systems are available commercially. Although both systems separate the propellant from the product, they differ in several respects. In barrier packs only the product is delivered, and there is never any contact between product and propellant (unless the propellant or product permeates through the plastic bag or piston). Since the propellant is located within the product container in siphon systems, the outer container no longer need be made from the usual materials used in the manufacture of conventional aerosol containers. With the absence of pressure, the container can be made of plastic and other similar materials, in addition to metal and glass. Also, there is a greater freedom in the design of the container since it does not have to withstand any pressure.

The following packages have been developed utilizing these principles:

Barrier Packs
1. MiraFlo® (American Can Company).
2. Sepro® (Continental Can Company).
3. Powr-Flo® (American Can Company).

Atomizer Systems
1. Preval® (Precision Valve Corporation).
2. Innovair® (J. R. Geigy S. A.).

Barrier Packs

MIRAFLO

This system, consisting of a "free" piston fitted into a two-piece aluminum container, can be used for viscous, semisolid products (7). It has been used successfully to package cheese spreads, cake-decorating icings, and pharmaceutical ointments. Other products utilizing this system have included caulking compounds, grease and lubricants, and other household and industrial items. Many other uses with different products have been suggested from time to time. The top of the piston (generally polyethylene) is contoured to fit the top of the valve so that, when the piston is pushed to the top, the product will be expelled, and, since the piston fits snugly against the top of the valve, theoretically, one would have dispensed most, if not all, of the material. In use, the product is filled through the 1-in. opening of the metal container and occupies the space above the piston. The valve is then sealed into place. After the gas is added through a small hole in the bottom of the can, the opening is sealed with a rubber plug and pressurized with nitrogen to about 90 psig or with about 5 to 10 g of Propellant 12.

When the valve is tilted (for containers used in the upside-down position) or depressed (for containers used in the upright position), the pressure pushes against the piston and forces the product out of the container in its original consistency. For products of semisolid consistency this system offers an advantage over the conventional compressed gas system in that, as the piston moves, it scrapes the walls of the container and gradually clears the product from them.

The liquefied gases are advantageous in that a constant pressure is maintained. However, these gases cannot be used with food products, since there may be some seepage of vapor through the piston or between the walls of the container and the piston. Valves with relatively large openings, such as foam valves, have been used. Since the products that use this system are semisolid and viscous, they are dispensed as a lazy stream rather than as a foam or spray.

This system is limited to viscous materials, since limpid liquids, such as water or alcohol, will pass between the walls of the container and the piston. Therefore, the MiraFlo system cannot be used for products where a spray is desired or necessary. Figure 4.7 shows a typical piston can system. For all products that are semisolid in nature this type of container is suitable.

SEPRO CONTAINER (8)

The Sepro container consists of a collapsible plastic bag fitted into a standard three-piece, tin-plated can (202 × 214, 202 × 406, or 202 × 509) as shown in Figure 4.8. The product is placed within the bag, and the propellant is added through the bottom of the container, which is fitted with a one-way valve. Since the product is placed in a plastic bag, there is no contact between the product and the container walls unless any product escapes by permeation through the plastic bag. There is no limitation on the viscosity of the product, as all products

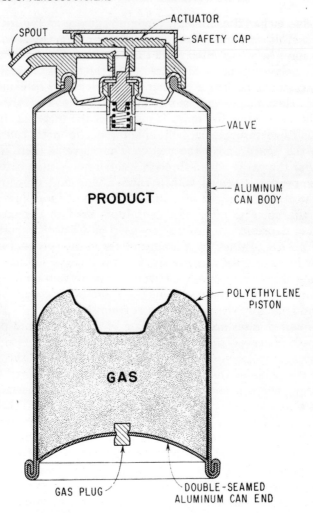

Figure 4.7.   Piston-type system. [From *Modern Packaging,* **34,** 12 (August 1961).]

from water to heavy viscous materials can be dispensed. A limpid liquid can be dispensed either as a stream or a fine spray, depending on the type of valve used. A viscous material must be dispensed as a stream. Since there is no admixture of propellant and product, most foams cannot be dispensed by this system. A recent development by Monson (9) makes possible the dispensing of postfoaming shaving gels.

The compatibility of the product with the plastic bag must be considered. Several different materials are available and have been used successfully. The Sepro system has been used for caulking compounds, postfoaming gels, and depilatories.

About 8 to 10 g of fluorinated hydrocarbon or 3 to 4 g of hydrocarbon propellant is used in this system. Nitrogen can also be used. An interesting example of the application of the Sepro container is illustrated by the postfoaming gels, which, as mentioned above, have been used for shaving products. The propellant is placed both inside and outside of the bag. Hexane, pentane, or Propellant 114 is placed in the bag and

Figure 4.8.   Sepro system. (Continental Can Co.).

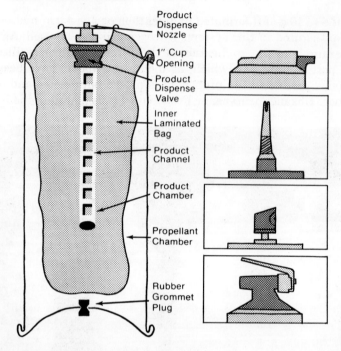

Figure 4.9.   Powr-Flo system (American Can Co.).

mixed with the gel. Propellant 12 or hydrocarbon propellant is placed outside and pushes the gel out of the container. Once the gel is released, the propellant trapped within it escapes, causing the gel to foam.

Powr-Flo Container

This system consists of a plastic bag in a can (10). The product is placed in the bag, which is made of a blend of low-density polyethylene films. A container with a 1-in. opening and a nominal capacity of 8.3 fl oz is used. The propellant is added through a special valve located in the bottom of the container. Liquefied gas is added through this valve, using a hypodermic needle-type gasser. Since pressures of about 30 psig are sufficient to dispense most products, the liquefied gas propellants are ideal for this application. From 1 to 4 g of propellant is used for each container. Figure 4.9 illustrates this system.

A variety of products can be dispensed by the Powr-Flo system. Although the major advantages of this type of system are obtained with semisolid preparations, it can also be used to dispense limpid liquids.

However, it should be kept in mind that, with all of the barrier pack systems, in order to obtain a spray a mechanical breakup actuator must be used, as no propellant is dispensed along with the product. In the case of limpid liquids, a wet, residual-type spray, at best, can be achieved. Viscous food products, pharmaceutical creams, and similar products can be used with this system.

Since the inner bag is made of low-density polyethylene, the valve must be fitted with a special dip tube that has a series of holes. As a result, the inner bag cannot be sealed, preventing the further discharge of product, by the action of the propellant.

Atomizer Systems

These systems are based on the Venturi effect, as can be seen from Figure 4.10. A stream of air or vapor is passed over a small opening, creating an area of decreased pressure or partial vacuum. The difference in pressure is sufficient to cause a liquid to rise through a tube and be carried with the air stream. The liquid can then be dispersed into relatively small particles. These systems are used primarily to dispense liquids, but they are suitable also for powders and possibly foam products. The propellant cartridge that supplies the vapor for the air stream is generally made of aluminum or glass. The cartridge is attached to a valve and placed in the outer container, which holds the product. Product and propellant vapor are mixed in the actuator, and the product is dispensed as a fine mist.

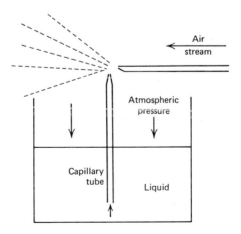

Figure 4.10.   Venturi effect for spraying liquids.

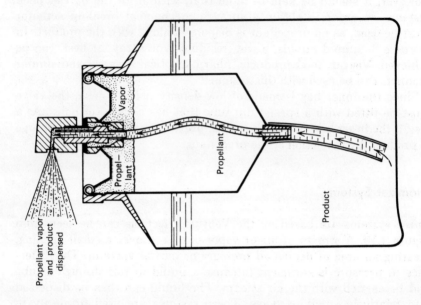

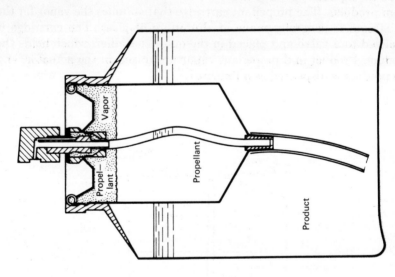

Figure 4.11. Preval system (Precision Valve Corp.).

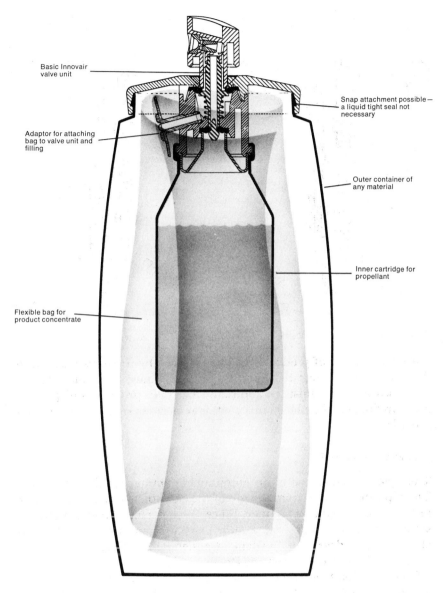

Basic Innovair
valve unit

Snap attachment possible —
a liquid tight seal not
necessary

Adaptor for attaching
bag to valve unit and
filling

Outer container of
any material

Inner cartridge for
propellant

Flexible bag for
product concentrate

Figure 4.12.   Innovair system (Geigy); note inner noncorrosive bag unit.

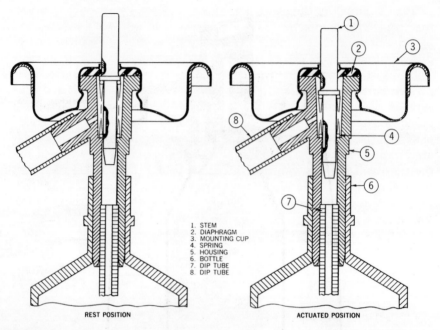

1. STEM
2. DIAPHRAGM
3. MOUNTING CUP
4. SPRING
5. HOUSING
6. BOTTLE
7. DIP TUBE
8. DIP TUBE

REST POSITION   ACTUATED POSITION

Figure 4.13.   VCA co-dispensing valve.

A large number of products can be dispensed by these systems. Since
the propellant is separated from the product, aqueous products can be
dispensed as a fine mist. In addition, since only product is dispensed,
more product can be placed in a container as compared to conventional
aerosols.

PREVAL SYSTEM (11)

This system makes use of an aluminum container for the propellant.
The dip tube extends from the top of the container through the bottom
and into the product container. Since the outer container is nonpressurized, it can be made from a number of different materials. There is
also a greater degree of freedom in designing the shape of the outer
container. Figure 4.11 shows the Preval system.

The propellant cartridge consists of Propellant 12. The dip tube
allows for the flow of the product when the valve is opened. However,
the valve is so designed that, when the specially designed actuator is
depressed, the propellant is allowed to vaporize and passes over the dip
tube, creating a low pressure and allowing the product to be drawn up
the tube. In the actuator, the propellant and product are mixed and
atomized by controlling the size of the various orifices. In use, the

product is added to the container, and the valve cartridge (containing the propellant) is inserted. The unit can be held in place permanently by crimping, or, if it is desirable to reuse the unit to dispense many different materials, it can be screwed into place.

This system can be used to dispense many types of products. The present cartridge contains about 4 oz by weight of propellant and can serve to dispense the contents of a 16 to 24 oz container. The amount dispensed (ratio of product to propellant) is dependent to a great extent on the size of the orifice. This, in turn, is related to the degree of atomization. Generally speaking, the finer the spray, the less will be the ratio and the less material will be dispensed by a given weight of propellant. This is due to the fact that a greater amount of vaporized propellant must be mixed with the product in order to dispense it in a fine state of subdivision. When coarse or wet sprays can be tolerated, a smaller amount of propellant is required.

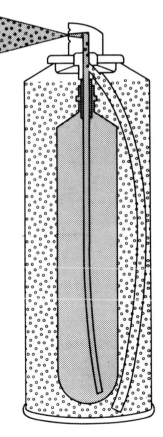

Figure 4.14.   VCA codispensing valve (VCA Corp.).

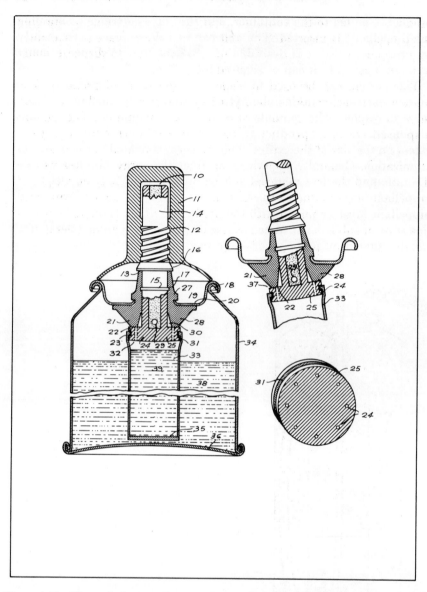

Figure 4.15. Thermal shaving foam utilizing flexible, polyethylene bag. (From U.S. Patent 3,326,416.)

There are many advantages to this type of system. The propellant and product are completely separated and do not come into contact with one another except at the time of dispensing. Aqueous products can be easily dispensed. Since the propellant is contained in the aluminum tube, only this part of the system is under pressure. The outer container, which holds the product, does not have to be a pressure container. This permits greater versatility in the design of this component.

INNOVAIR SYSTEM

Figure 4.12 illustrates the Innovair unit, which consists of an inner aluminum cartridge filled with propellant, generally Propellant 12.

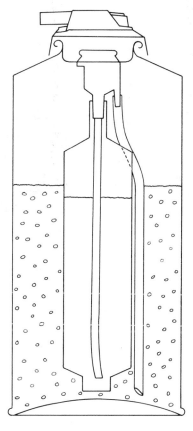

Figure 4.16.   OEL dual-dispensing valve.

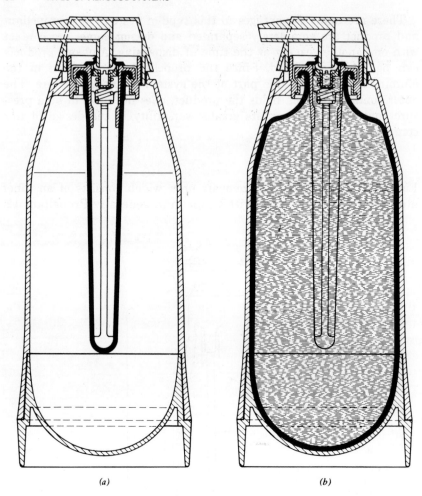

(a)   (b)

Figure 4.17.   Selvac unit. (a) Unfilled; (b) filled (Plant industries, Inc.).

This cartridge is attached to a special valve system that allows for the simultaneous dispensing of a vaporized propellant and the product. The product is contained in a flexible bag, which is attached to the valve unit. When the unit is sprayed, concentrate is drawn from the bag and into the valve and is mixed with vaporized propellant (12). This mixture is then sprayed into the atmosphere.

This system is advantageous in that the product is contained in the bag and does not come into contact with the inner aluminum container, eliminating the possibility that corrosion will take place. Product leakage is also eliminated, since the product concentrate does not come into contact with the air inlet. This system can be used for the same applications as the Preval unit.

OTHER SYSTEMS

Many other systems for dispensing aerosols have been developed; two of them will be described.

CoDispensing System

Several years ago, the codispensing system for the dual dispensing of liquids became available. It was first used for "hot-shave foam," of which several types are available. Recently, hair dyes and colors, as well as antiperspirants, have been packaged using this system. Two incompatible ingredients can be separated by packaging one of them (hydrogen peroxide in the case of thermal foams) in a bag attached to the valve. The other is mixed in with the concentrate (potassium sulfite or a similar reducing agent). In use, the two materials are dispensed separately and are mixed in the valve. The desired reaction, which, in the case of thermal foams, generates heat, then occurs (13). Other uses for the system involve the separation of corrosive ingredients or aqueous solutions of ingredients that are immiscible with the propellant phase.

Although several different systems are available (Figures 4.13 to 4.16), they all serve to separate the incompatible ingredients and to bring them together at the time of use. Both sprays and foams can be dispensed from these systems.

Mechanical System

This system, termed Selvac (Plant Industries, Inc.), consists of an elastomeric prestressed bladder made of natural rubber. The product is added into this bladder much as a balloon is filled with air. As the valve is released, the pressure of the expanded bladder dispenses the product (14). Figure 4.17 illustrates this system.

The bladder is a two-part unit. The inner part is a synthetic compound of permeation-resistant material. The outer part, which provides the energy to dispense the product, can be made of plastic or any other material since it is not required to be pressure resistant. This system can be used for many viscous or semisolid products, such as ointments, creams, or lotions.

REFERENCES

1. P. A. Sanders, *Principles of Aerosol Technology,* Van Nostrand Reinhold Company, New York, 1970, pp. 4-5.
2. J. J. Sciarra, Aerosols, in J. E. Hoover, (Ed.)*Remington's Pharmaceutical Sciences*, 14th ed., Mack Publishing Company, Easton, Pa., 1970, p. 1729.
3. J. J. Sciarra, Pharmaceutical Aerosols, in L. Lachman, H. A. Lieberman, and J. L. Kanig (Eds.), *The Theory and Practice of Industrial Pharmacy*, Lea and Febiger, Philadelphia, 1970, pp. 616-617.
4. A. Herzka, Powder Aerosols, *J. Soc. Cosmet. Chem.*, 21 (1970), 553-562.
5. F. Presant and C. Carrion, Jr., Water-Based Aerosols and Their Cosmetic Applications, *Aerosol Technicomment*, 6, 1 (April 1963), 1-6.
6. I. Fox and S. Palley, Packaging and Dispensing Beverage Concentrates, U.S. Patent 2,977,231 (March 1961).
7. R. S. Schultz, Free Piston Container, *Soap Chem. Spec.*, 38, 3 (March 1962), 127.
8. L. F. Irland, and J. W. Kinnary, The Sepro Can, *Drug Cosmet. Ind.*, 101, 2 (August 1967), 42.
9. J. A. Monson, Package Containing a Post-Foaming Gel, U.S. Patent 3,541,581 (November 1970).
10. J. P. Rem, A Review of the Sterigard System, *Aerosol Age*, 16, 5 (May 1971), 36.
11. Introducing the Preval, *Aerosol Age*, 11, 1 (January 1966), 29.
12. Second Generation Geigy Innovair System, *Aerosol Age*, 15, 1 (January 1970), 82.
13. H. Boden, Hot Shave Technology, *Aerosol Age*, 13, 3 (March 1968), 19.
14. Selvac Is Back, *Aerosol Age*, 17, 6 (June 1972), 27.

PHYSICOCHEMICAL ASPECTS OF AEROSOL SCIENCE AND TECHNOLOGY

Aerosol products are identical to other types of products, except that they are packaged under pressure. Regardless of the source of the pressure, aerosol products depend on the power of the gas causing this pressure for their proper operation. In the absence of the gas, the product ceases to function.

Physicochemical Properties of Aerosols

Either a compressed or a liquefied gas is used in all types of aerosol products. In addition to the properties of the gas, other physicochemical properties play an important role in the formulation, development, and other aspects of aerosols. Several of these physicochemical properties are solubility, surface tension, density, vapor pressure, and viscosity. Those pertaining to the gaseous state and to solubility considerations are considered in this chapter. Several of the other properties will be covered in Chapter 6.

Aerosol systems can consist of solutions, dispersions, emulsions, or semisolids. Depending on the type of system, valve, and propellant, the product can be dispensed as a spray, foam, or solid stream. These systems were covered in greater detail in Chapter 4.

Physical Aspects of the Operation of Aerosols

FORMATION OF SPRAY

Aerosol sprays are dispensed as liquid droplets by the rapid expansion and vaporization of a liquefied gas. When the propellant surrounds the liquid particles in solution, a fairly wet spray is produced. When the propellant is miscible with the liquid within the droplet, a fine spray results. In the former case, the droplets are emitted from the aerosol valve as a spray. However, the liquid propellant, which is in the outer phase of the droplet, quickly vaporizes, leaving behind a fairly large droplet that consists of less volatile solvents and active ingredients. These active ingredients can be either liquids or solids. On the other hand, in the case of the droplet with the propellant surrounded by less volatile solvents and active ingredients, the liquid propellant vaporizes, and since the vaporized propellant must pass through the liquid phase to escape, it disperses the liquid into smaller particles and an extremely fine spray is obtained. This breakup can also be aided by use of a mechanical breakup actuator attached to the valve. In fact, such actuators are essential for the breakup of aerosols propelled by compressed gases.

FORMATION OF FOAMS

A foam consists of a coarse dispersion of gas in a liquid. The volume of gas is generally larger than that of the liquid. Two types of foams and a spray may be obtained, depending on the nature of the emulsion, propellant, and valve.

Stable Foams

Stable foams are formed from emulsion systems in which the propellant is in the internal phase (o/w). As the emulsion is released, the liquid propellant vaporizes within the surfactant/water mixture or surfactant/nonaqueous mixture, forming a tiny bubble. As the propellant continues to vaporize, the bubbles expand, forming a thin film. This film keeps the many bubbles separated and forms the basis for the foam structure (1).

Unstable or Quick-Breaking Foams

These foams vary in stability, depending on the amount of water, solvent, and surfactant present. The surfactant must be soluble in either the water or the solvent, but not in both. In addition, the solubility of the surfactant is such that the propellant is necessary to achieve complete solubility in the mixture.

As this type of emulsion is dispensed, the propellant vaporizes and causes a decrease in the solubility of the surfactant. This produces a bubble containing vaporized propellant surrounded by a film of solid surfactant. Depending on the strength of this film, the foam can be stable or collapse rather quickly. Generally, these emulsions contain water, alcohol, and a self-emulsifying nonionic wax; Polawax® (Croda) is generally used for this purpose. The effect of liquid crystalline phases on the stability of emulsions was described by Friberg et al. (2, 3).

Emulsions Emitted as Sprays

A spray, rather than a foam, can be obtained from an emulsion system when the propellant is located in the outer phase (w/o). In this case, as the emulsion is expelled, the propellant simply vaporizes, leaving behind the aqueous or nonaqueous droplet.

THE GASEOUS STATE

Since every aerosol system is pressurized and contains a gas, much of the behavior of these systems can be predicted or described from a knowledge of the gaseous state.

A gas may be defined as "consisting of molecules traveling in a straight line at random and at high speeds within the containing space of a container and colliding frequently with other molecules on the walls of the container" (4). The gaseous state may be characterized as fulfilling the following conditions:

- Volumes change considerably with changes in temperature and pressure.
- Gases do not possess any definite shape or volume of their own; they take the shape and volume of the outer container.
- Gases are generally completely miscible with one another; they have relatively low densities and viscosities.
- The space occupied by the gas molecules is very small; if air in a $20' \times 10' \times 10'$ room were liquefied, the liquid volume would be approximately 2.4 ft^3.

Gases can be classified as (a) ideal and (b) nonideal or real. Ideal gases behave according to the gas laws or else can be made to conform to these laws under restricted conditions.

In aerosol or pressurized products, the gas occupies the head space of the container. It is within this head space that the gas exerts its pressure, causing the product to be emitted from the container after the valve has been opened. In this regard, compressed and liquefied gases show essentially the same activity, and this activity can be predicted from a consideration of the kinetic theory of gases and the gas laws.

Kinetic Theory of Gases

Bernoulli first proposed this theory in 1738, and although it has been modified and extended since then, its essential assumptions remain unchanged. Five of these are as follows:

1. Gases are composed of minute discrete particles (molecules), all having the same mass and size. This mass and size may differ, however, depending on the gas. These molecules are perfectly elastic so that no speed is lost when one molecule collides with another.

2. These particles travel in a chaotic continuous motion and follow Newton's laws of motion. As the particles hit the sides of the container or collide with one another, a force is exerted, and the force per unit area is referred to as *pressure*.

3. Gas has intrinsic energy that manifests itself as heat.

4. As the gas is compressed, the number of molecules per unit volume is increased. This increase in molecules per unit volume brings about an increase in the bombardment of particles with each other and with the walls of the container, resulting in an increase in pressure.

5. If the temperature of the gas system is raised, the velocity of the molecules becomes greater, thereby increasing the force of these collisions with a resultant increase in pressure.

On the basis of these considerations, various gas laws have been developed.

Gas Laws

BOYLE'S LAW

At constant temperature, the volume of a gas varies inversely as the pressure; mathematically expressed, at $\Delta T = 0$,

$$V \approx \frac{1}{P} \qquad \text{or} \qquad V = k\frac{1}{P}$$

Since $PV = k$, the equation can be modified so that

$$P_1V_1 = k = P_2V_2$$

where P_1 and V_1 represent the pressure and volume, respectively, under one set of conditions, and P_2 and V_2 represent these same units under a different set of conditions. Figure 5.1 illustrates this relationship.

THE CHARLES OR GAY-LUSSAC LAW

In 1802, Gay-Lussac noted that all gases increase in volume with an increase in temperature at constant pressure. He noted that this increase was $1/273.17$ of the volume for each degree centigrade. On the basis of this observation, we can show this relationship as follows: at $\Delta p = 0$,

$$V = kT \quad \text{or} \quad \frac{V_1}{V_2} = \frac{T_1}{T_2}$$

where T_1 and T_2 are the absolute temperatures at both conditions,[*] and V_1 and V_2 are the respective volumes. Figure 5.2 shows the relationship between volume and temperature.

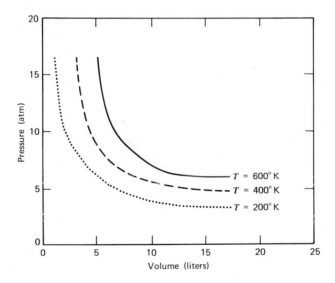

Figure 5.1.   Isothermal plot of pressure versus volume.

[*] Absolute temperature $= T = {}^\circ\text{C} + 273.17$; ${}^\circ\text{F} = 9/5\,{}^\circ\text{C} + 32$, and ${}^\circ\text{C} = {}^\circ\text{F} - 32 \times 5/9$.

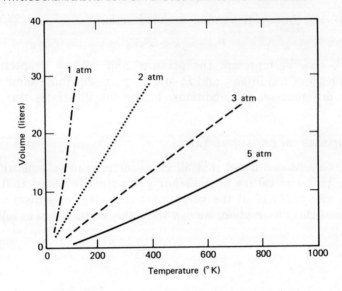

Figure 5.2.   Charles' law plot of volume versus temperature.

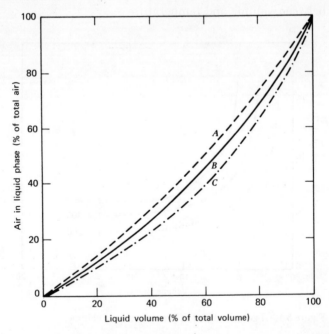

Figure 5.3.   Distribution of air between liquid and vapor for various amounts of liquid at 70°F. Curve A: Freon 12 propellant; B: Freon 12/Freon 11 (50:50); C: typical insecticide formulation using Freon 12/Freon 11 (50:50) as the propellant. (From Freon® Aerosol Report FA-15, E. I. du Pont de Nemours and Co.)

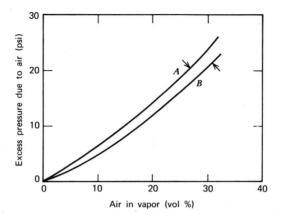

Figure 5.4. Excess pressure due to air related to concentration of air in the vapor phase at 70% loading and 70°F. Curve A: Freon 12/Freon 11 (50:50); B: typical insecticide formulation. (From Freon® Aerosol Report FA-15, E. I. du Pont de Nemours and Co.0

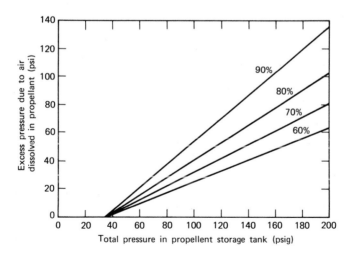

Figure 5.5. The effect of air dissolved in the propellant on the pressure in aerosol containers. The total pressure includes the vapor pressure of the propellant and applied air pressure. The four curves represent different liquid loadings in the container. The temperature is 70°F. (From Freon® Technical Bulletin FA-15, E. I. du Pont de Nemours and Co.)

DALTON'S LAW

This law states that the pressure exerted by a mixture of gases is equal to the sum of the pressures that each gas would exert separately. This can be shown as

$$P_T = p_1 + p_2 + p_3 + \text{etc.}$$

where p_1, p_2, p_3, etc., represent the individual pressures of all of the gases present.

Dalton's law is especially applicable to aerosols in that it accounts for the increased pressure observed in a container of an aerosol product that contains some entrapped air. The air remaining in the container after the valve is sealed into place is compressed as the propellant is added, resulting in the increased pressure. In the case of products with pressures already near the maximum allowable levels, the excess pressure may cause the containers to be in violation of the shipping regulations and may place added stress on the containers.

Figures 5.3 to 5.5 show the excess pressure due to air which may be present. The calculations necessary to determine the pressure have been published (5).

Combined Gas Laws

In order to utilize the gas laws, it is possible to combine Boyle's and the Charles/Gay-Lussac laws into a single equation. This can be expressed as the ideal gas equation or as the van der waals or a similar equation.

IDEAL GAS EQUATION

This equation can be expressed as

$$PV = nRT$$

where P = pressure (atm), V = volume (liters), T = absolute temperature (degrees K), R = universal gas constant,* and n = number of moles present.

A given quantity of gas, P_1, V_1, T_1, compressed to pressure P_2 at constant temperature, will occupy a new volume, V_x. According to Boyle's law,

$$\frac{V_x}{V_1} = \frac{P_1}{P_2} \quad \text{or} \quad V_x = \frac{P_1 V_1}{P_2} \qquad [1]$$

---

* This constant can be expressed as follows: 0.08205 liter atm/(deg)/(mole), 8.314×10^7 ergs/(deg)/(mole), 8.314j/(deg)/(mole), 1.987 cal/(deg)/(mole).

If this gas is then heated to temperature T_2 at constant pressure, V_x will expand to V_2. This new relationship can be expressed as

$$\frac{V_2}{V_x} = \frac{T_2}{T_1} \quad \text{or} \quad V_2 = \frac{T_2 V_x}{T_1} \tag{2}$$

Substituting for V_x in [2] the value from [1], we obtain

$$V_2 = \frac{T_2}{T_1} \left(\frac{P_1}{P_2} V_1 \right)$$

and rearranging terms gives

$$\frac{P_1 V_1}{T_1} = \frac{P_2 V_2}{T_2}$$

Since $P_2 V_2 / T_2$ is constant under any set of conditions, the value can be represented by R, which is the universal gas constant. Dropping the subscripts, we obtain

$$\frac{PV}{T} = R \quad \text{or} \quad PV = RT \tag{3}$$

Equation [3] holds for 1 mole of gas.
Generally: $PV = nRT$

THE VAN DER WAALS EQUATION

Real gases differ from ideal gases in that real gases are not composed of infinitely small and perfectly elastic particles. Van der Waals' equation tends to correct this difference by the introduction of two constants, a and b; a corrects for the attractive force between the molecules, and b for the volume occupied by the gas molecules. These factors are related to the basic ideal gas equation as follows:

$$\left(P + \frac{a}{v_2} \right) (V - b) = nRT$$

As can be seen, when V is very large, a/v_2 is very small and both a and b can be neglected. At low pressures a is predominate; at small volumes, b is predominate. There is an intermediate point where a and b balance each other. Table 5.1 illustrates these values for some commonly used aerosol gases (6).

Other equations that allow for additional corrections are also available (7).

TABLE 5.1.   VAN DER WAALS' CONSTANTS FOR GASES

Gas	a (liters)^2 atm/(moles)^2	b (liters/mole)
Carbon dioxide	3.592	0.04267
Nitrogen	1.390	0.03913
Nitrous oxide	3.782	0.04415

Application of the Gas Laws to Aerosol Systems

The application of the gas laws may best be illustrated by considering the following examples.

SEMISOLID AEROSOLS USING INSOLUBLE COMPRESSED GAS

Examples of this system include aerosols in which the product and the propellant are separated. The propellant is insoluble in the product. Dental creams, ointments, and products packaged in compartmented aerosol containers are examples. In these cases, the propellant serves simply to expel the contents from the container.

Six fluid ounces of a viscous ointment are placed into an 8-fl-oz aerosol container at 25°C. The valve is crimped into place, and the container is pressurized with nitrogen to 90 psig. Determine the weight of nitrogen in the container. After 3 fl oz of product has been dispensed, what is the pressure? Finally, what is the pressure remaining in the container when all of the product has been dispensed?

Weight of nitrogen

$$PV = nRT, \qquad n = \frac{PV}{RT}$$

$$\text{Pressure (atm)}^* = \frac{90 + 14.7}{14.7} = \frac{104.7}{14.7} \text{ atm}$$

$$\text{Volume} = (29.57 \text{ ml/fl oz}) \times 2 = 59.14 \text{ ml}$$

$$\text{Temperature} = 273.2 + 25 = 298.2°\text{K}$$

$$n = \frac{[(104.7/14.7) \text{ atm}] (59.14 \text{ ml})}{[82.05 \text{ ml atm/(deg)(mole)}] (298.2 \text{ deg})} = 0.017 \text{ mole}$$

$$\text{Weight of N}_2 = (0.017)(14)(2) = \underline{0.476 \text{ g N}_2}$$

* The unit psig can be converted to psia by adding atmospheric pressure (14.7) to psig. Pressure in terms of psia can then be converted to atmospheres by dividing the pressure by 14.7 psi/atm.

Pressure when 3 fl oz has been dispensed

The gas now occupies a volume of 5 fl oz so that

$$\frac{V_1}{V_2} = \frac{P_2}{P_1}$$

Since the volumes are changing in the ratio of 2 to 5, the pressure will change in the ratio of 5 to 2:

$$\frac{2}{5} = \frac{P_2}{104.7}$$

$$P_2 = \underline{41.9}\,\text{psia}$$

$$41.9 - 14.7 = \underline{27.2}\,\text{psig}$$

Pressure when all of the product has been dispensed

The volume occupied by the gas at this time is 2 fl oz plus 6 fl oz, which is 8 fl oz. As above,

$$\frac{V_1}{V_2} = \frac{P_2}{P_1}$$

$$\frac{2}{8} = \frac{P_2}{104.7}$$

$$P_2 = \underline{26.2}\,\text{psia or 11.5 psig}$$

EFFECT OF TEMPERATURE ON PRESSURE (FOR COMPRESSED GAS AEROSOLS)

1. What is the pressure in an aerosol container at 130°F* if it has a pressure of 100 psig at 70°F?† In the calculations, neglect the expansion of the product.

$$\frac{P_2}{P_1} = \frac{T_2}{T_1}$$

$$\frac{P_2}{100 + 14.7} = \frac{327.5}{294.2}$$

$$P_2 = \underline{127.7}\,\text{psia or 113 psig at 130°F}$$

2. An aerosol container is completely emptied of product and propellant at 70°F. What pressure would develop in the container if the

---

* $T_2 = 130°F = 54.44°C = 327.5°K$.

† $T_1 = 70°F = 21.11°C = 294.2°K$.

container was dropped into an incinerator where the temperature was 800°F?‡

Under these conditions, the pressure within the empty container would be equal to atmospheric pressure or 14.7 psi. The pressure at 800°F can be calculated as follows:

$$\frac{P_2}{P_1} = \frac{T_2}{T_1}$$

$$\frac{P_2}{14.7} = \frac{699.8}{294.2}$$

$$P_2 = 34.97 \text{ psia or } 20.3 \text{ psig at } 800°F$$

From these values one can see that, if the aerosol container is completely empty, an increase in temperature does not have a great effect on the pressure. However, when one considers the effect of a rise in temperature on the pressure of a liquefied gas, the increase is more dramatic. This will be discussed in a later portion of this chapter under the heading "The Vapor State."

FORMULATIONS WITH SOLUBLE COMPRESSED GASES

Soluble compressed gas propellants also follow the gas laws except that one must consider the solubility of the gas in the liquid product. According to Henry's law, "the amount of gas dissolved by a liquid is directly proportional to the pressure of the gas." The pressure in an aerosol container is due essentially to the gas located in the head space of the container. Dissolved gas contributes very little, if any, to the total pressure of the system. In a soluble compressed gas system, as the contents are used, the volume occupied by the gas increases, causing a drop in pressure. However, according to Henry's law, a drop in pressure is accompanied by a decrease in gas solubility, so that some of the dissolved gas is released from solution and goes to the head space, resulting in a partial restoration of the original pressure. This is considered in the next section.

THE LIQUID STATE

Liquefied gas propellants can be treated as volatile liquids. Mixtures of liquefied gas propellants can be considered as solutions. The vapor

‡ $T_2 = 800°F = 426.7°C = 699.8°K$.

pressure of the solution is dependent on the escaping tendency of the molecules in going from the liquid to the vapor state. Two bodies are said to be in thermal equilibrium when they are at the same temperature. At this point the change in free energy* of the system is equal to zero, or

$$\Delta G = 0$$

when two bodies are at different temperatures, heat will flow from the hotter body to the cooler one, or the cooler body will take heat from the hotter one, until an equilibrium is established. The heat in the hotter body is said to have a greater escaping tendency than the heat in the cooler body. The quantitative measure of the escaping tendency of a system is known as free energy or G. When ΔG of a system is negative. the process can be expected to take place.

For example, consider a system consisting of water and ice at $0°C$. At this temperature, the system is in equilibrium since both ice and water have the same escaping tendency. At temperatures above $0°C$, however, ice has a greater escaping tendency than water and therefore the ice will melt:

$$\Delta G = G_{\text{liq}} - G_{\text{ice}}$$

Below $0°C$, ice will not melt since ΔG of the system will be positive, as the escaping tendency for the ice is less than that for water.

The same is true of liquids where the escaping tendency is dependent on the volatility or vapor pressure of the system.

Raoult's Law

Raoult's law is concerned with the vapor pressure of ideal solutions and is dependent on the vapor pressure of the individual components. According to Raoult's law, the vapor pressure of the solution is equal to the sum of the mole fractions of each component present times the vapor pressure of the pure component at the desired temperature. Mathematically, this can be expressed as

$$p\hat{A} = \frac{n_A}{n_A + n_B}\, p_A{}^0 = N_A p_A{}^0$$

where p_A = partial vapor pressure of component A in solution, $p_A{}^0$ = vapor pressure of pure component A, N_A = mole fraction of component A, and

$$p\hat{B} = \frac{n_B}{n_B + n_A}\, p\hat{B}^0 = N_B p_B{}^0$$

where B represents corresponding values for component B.

* Formerly, the free energy of a system was indicated by F.

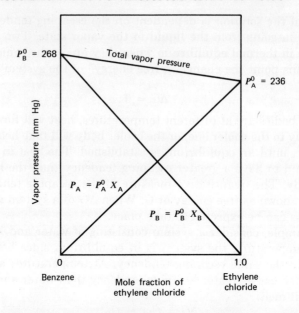

'Figure 5.6.   Vapor pressure versus composition for ideal two-component system (10).

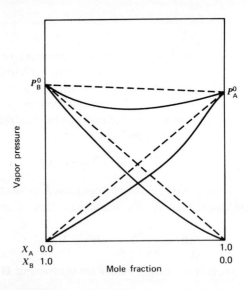

Figure 5.7.   Binary system illustrating negative deviation from Raoult's law (10).

The total vapor pressure of the system is expressed as

$$P = p_A + p_B$$

When the mole fraction of one component is large, the mole fraction of the other component is relatively small and therefore has little, if any, effect on the nature of the solution, so that the system can be treated as ideal. Figure 5.6 illustrates a typical system showing ideal behavior.

Depending on the nature of the components, both negative and positive deviation can be noted from Raoult's law. Figure 5.7 shows a system having negative deviation. In such systems, the attraction of different molecules for one another is greater than the attraction of molecules of the same kind. Positive deviation takes place when the interaction between different molecules is less than that between molecules of the same substance, resulting in an increase in escaping tendency and, consequently, an increase in vapor pressure, as compared to the result expected from Raoult's law. This is seen in Figure 5.8.

According to Flanner (8), propellants and the commonly used solvents always exhibit positive deviation. Several of these systems are seen in Figures 5.9 and 5.10. Flanner attributes positive deviation to

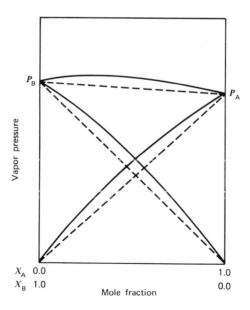

Figure 5.8.   Binary system illustrating positive deviation from Raoult's law (10).

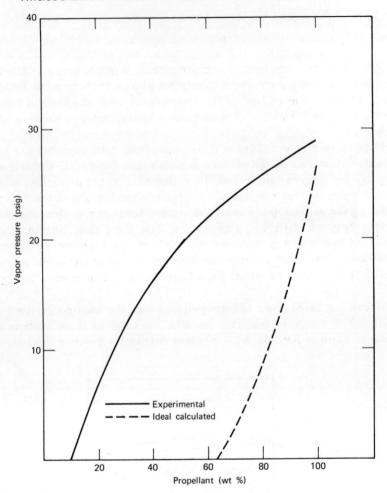

Figure 5.9. Vapor pressure versus composition at 70°F for mixtures of Genetron® .12/144 (20:80) and ethyl alcohol. (From L. Flanner, *Vapor Pressure of Solvents and Propellant Mxtures,* Allied Chemical Corp.)

the following factors:

Internal pressure of the two components in the propellant/solvent mixture.
Polarity.
Length of the hydrocarbon chain or analogous grouping.
Association in the liquid phase of either component.

Flanner also indicated that a measure of the nonideality of the system can be obtained from

$$\gamma = \frac{PY_1}{p_1{}^0 X_1}$$

where PY_1 is the partial pressure of either component in the vapor phase, and $P_1\ ^0X_1$ is the expected partial pressure of the same compo-

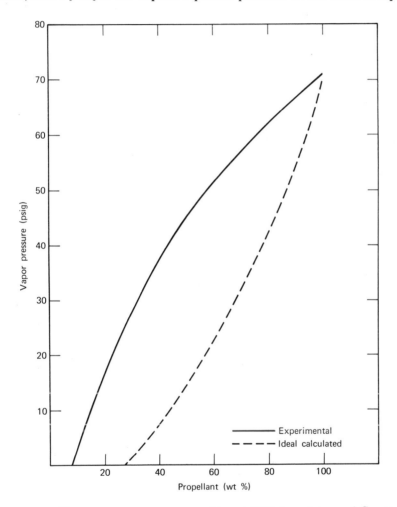

Figure 5.10. Vapor pressure versus composition at 70°F for mixtures of Genetron® 12 and isopropyl alcohol. (From L. Flanner, *Vapor Pressure of Solvents and Propellant Mixtures,* Allied Chemical Corp.)

TABLE 5.2.   SOLUBILITY OF GASES IN WATER—
HENRY'S LAW CONSTANT

Gas	$K \times 10^6$
Nitrogen	5.34 (25°C)
Carbon dioxide	1.25 (25°C)
Nitrous oxide	0.21 (30°C)

nent according to Raoult's law. This ratio has been termed the "activity coefficient." Calculations involving Raoult's law as applied to the determination of the vapor pressures of propellant blends are covered in Chapter 6.

Solubility Considerations

SOLUBLE COMPRESSED GASES

When a dissolved substance evaporates into a closed space, the concentration of its molecules in the vapor phase increases, and the rate at which these molecules return to the solution also increases until a dynamic equilibrium is attained. This system can also pertain to solutions of gases in liquids.

According to Henry's law,

$$p_{solute} = KN_{solute}$$

that is, the partial pressure of the solute is proportional to its mole fraction in solution. For ideal solutions, the value of K is equal to p^0 in Raoult's law. For nonideal solutions, the value of K is determined by measuring the vapor pressure of the solute in dilute solutions at known mole fractions:

$$K = \frac{p_{solute}}{N_{solute}} = \frac{p_A}{N_A} = \frac{\text{pressure of gas (mm)}}{\text{mole fraction of gas in solution}}$$

Table 5.2 illustrates these values for some commonly used gases.

Bunsen Coefficient or Solubility

This coefficient can be used to calculate the solubility of a gas in a solvent such as water. By definition, Bunsen solubility is the "number of liters of gas reduced to 0°C and 760 mm pressure which will dissolve in 1 liter of solvent under a pressure of 760 mm at a specified temperature." The use of this value can be seen in the following example.

Express the solubility of carbon dioxide in water at 1 atm pressure and 25°C in terms of moles per liter, assuming that a liter of solution contains practically 1000 g of water.

$$K = \frac{p_{gas}}{n_{gas}}\left(n_{gas} + \frac{\text{grams of solvent}}{\text{mol wt of solvent}}\right)$$

From Table 5.2, k for carbon dioxide is 1.25×10^6:

$$K = 1.25 \times 10^6 = \frac{760}{n_{CO_2}}\left(n_{CO_2} + \frac{1000}{18.02}\right)$$

Since n_{CO_2} may be considered negligible in comparison with the number of moles of water,

$$1.25 \times 10^6 = \frac{760}{n_{CO_2}}\left(\frac{1000}{18.02}\right)$$

$$n_{CO_2} = \frac{(760)(55.49)}{1.25 \times 10^6} = 3.38 \times 10^{-2}\,\text{mole/liter}$$

This solubility can be expressed as the Bunsen coefficient as follows:

$$\text{Bunsen coefficient} = (3.38 \times 10^{-2}\,\text{mole/liter}) \times$$

$$(22.41\,\text{liters/mole}) = 0.757$$

Henry's law is effective within 1 to 3% for pressures of about 1 atm. According to Dalton's law, the solubility of the individual gases in a mixture of gases is directly proportional to their partial pressures, the solubility of each gas being nearly independent of the presence of the other gases. When the solubility of the gas is an important consideration, greater total solubility can be achieved by using a mixture of gases.

Ostwald Solubility

According to Hsu and Campbell (9), the Ostwald coefficient can be used to determine the total pressure of a compressed gas system. The Ostwald solubility coefficient (λ) is defined as the ratio of the volume of gas dissolved in a liquid to the volume of liquid at a specified temperature and pressure:

$$\lambda = \frac{\text{volume of gas dissolved}}{\text{volume of liquid}}$$

TABLE 5.3.   OSTWALD SOLUBILITY COEFFICIENT AT 70°F

	GAS	
Solvent	Nitrous Oxide	Carbon Dioxide
Water	0.6	0.85
Acetone	5.92	6.95
Ethanol	2.96	2.84
Methanol	3.20	3.51
n-Pentane	4.13	1.94
Glycerin	1.2	0.03
1,1,1-Trichloroethane	4.96	3.74

This coefficient is advantageous to use in calculations since it does not vary with pressure but does vary with temperature. Table 5.3 lists Ostwald values for several solvents used with nitrous oxide and carbon dioxide.

In cases where the Ostwald coefficient is not available, it can be easily determined (8) using the following equation:

$$\lambda = \frac{1}{x}\left(\frac{WRT}{V_cMP} + x - 1\right)$$

where λ = Ostwald solubility coefficient, x = volume of liquid per volume of container, W = weight of propellant gas (g), R = gas constant [ml atm/(deg)/(mole)], T = absolute temperature, V_c = volume of container (ml), M = molecular weight of propellant, and P = total pressure (atm) (absolute). This equation has been derived from the ideal gas equation (8).

Consider the following example.

An aqueous aerosol product is packaged using carbon dioxide as the propellant. A 12-fl-oz container (overflow capacity) is filled with 10 fl oz of aqueous product, air is evacuated, and the container is pressurized with carbon dioxide to a pressure of 100 psig at 70°F. (a) Neglecting the solubility of the gas in the product, calculate the weight of carbon dioxide required to pressurize the product. (b) Repeat (a), taking into account the solubility of carbon dioxide in the aqueous product (assume water as the total solvent).

(a) We can utilize the ideal gas equation

$$PV = nRT$$

and solve for n:

$$n = \frac{PV}{RT}$$

$$P = 100 \text{ psig} + 14.7 = 114.7 \text{ psia}$$

$$= \frac{114.7 \text{ psia}}{14.7 \text{ psi/atm}} = 7.80 \text{ atm}$$

$$V = (2 \text{ fl oz}) (29.57 \text{ ml/fl oz}) = 59.14 \text{ m}$$

$$R = 82.06 \text{ ml atm/(mole)(deg)}$$

$$T = 70°\text{F} = 21.1°\text{C}$$

$$°\text{K} = 21.1 + 273.1 = 294.2°\text{K}$$

$$n = \frac{(7.80)(59.14)}{(82.06)(294.2)}$$

$$= 0.0191$$

$$g = (n \times \text{mol wt}) = 0.0191 \times 44$$

$$= \underline{0.84 \text{ g}}$$

Under these conditions, 0.84 g of carbon dioxide would be used, assuming there is no solubility of gas in the liquid.

(b) We use the concept of Ostwald solubility and solve the equation for W:

$$W = \frac{V_c M P}{RT} (x\lambda + 1 - x)$$

$$V_c = (12)(29.57) = 354.8 \text{ ml}$$

$$M = 44 \text{ g/mole}, \qquad P = 7.8 \text{ atm}$$

$$R = 82.06 \text{ ml atm/(mole)(deg)}$$

$$\lambda = 0.85 \text{ (from Table 5.2)}$$

$$T = 294.2°\text{K}$$

$$x = \frac{10 \text{ fl oz}}{12 \text{ fl oz}} = 0.83$$

$$W = \frac{(354.8)(44)(7.8)}{(82.06)(294.2)} [(0.83)(0.85) + 1 - 0.83]$$

$$= (5.04)(0.87)$$

$$= \underline{4.41 \text{ g}}$$

As one can note from the calculations, a considerably greater amount of soluble gas than of insoluble gas can be incorporated into an aerosol product. The greater the Ostwald solubility coefficient, the more pronounced will be the solubility of the gas in the solvent.

One of the advantages to the use of a soluble compressed gas is the fact that the pressure drop as the contents of the container are utilized is not as dramatic as that expected with insoluble gases. As shown previously, the pressure drop of an insoluble gas can be calculated utilizing the ideal gas equation (see the example on p. 74) Hsu and Campbell (9) derived equations that can be used to calculate the pressure drop encountered with soluble compressed gases. They developed the following equation:

$$P_2 = \frac{P_1}{1 + [\Delta x/(x\lambda - \Delta x\lambda + 1 - x)]}$$

where P_1 and P_2 are the pressures (atm) in the container before and after dispensing, respectively, Δx is the amount of product dispensed as a fraction of the container volume, and λ is the Ostwald solubility coefficient.

The example on p. 75 can be utilized to illustrate the type of calculation involved.

Six fluid ounces of ethyl alcohol are placed into an 8-fl-oz container at 70°F. The valve is crimped into place, air is evacuated, and the container is pressurized with nitrous oxide to 90 psig. What is the pressure in the container after 3 fl oz of alcohol has been dispensed?

$$P_2 = \frac{P_1}{1 + [\Delta x/(x\lambda - \Delta x\lambda + 1 - x)]}$$

$$P_1 = \frac{90 + 14.7}{14.7} = 7.12 \text{ atm}$$

$$\Delta x = \tfrac{3}{8} = 0.375, \qquad X = \tfrac{6}{8} = 0.750$$

$$\lambda_{\text{eth alc}} = 2.96$$

$$P_2 = \frac{7.12}{1 + \{0.375/[(0.750)(2.96) - (0.375)(2.96) + (1 - 0.75)]\}}$$

$$\frac{7.12}{1 + \{0.375/[(2.22) - (1.11) + (0.25)]\}} = \frac{7.12}{1.276} = 5.58 \text{ atm}$$

$$= (5.58)(14.7) = 82.03 \text{ psia or } \underline{67.3 \text{ psig}}$$

Under these same conditions, had nitrogen been used as the propellant, the remaining pressure would have been 27.2 psig.

LIQUEFIED GASES

The behavior of liquefied gases as solvents for other substances can be noted from a consideration of solutions of liquids in liquids, where the liquefied gas is considered to be the solvent and the other liquid substance is regarded as the solute. On the basis of the chemical nature of either component, the resultant solution can be considered to be ideal or nonideal. It has previously been noted that mixtures which follow Raoult's law over the entire range of composition are said to be ideal solutions. In the case of real or nonideal solutions, either positive or negative deviation can occur. Generally, negative deviation takes place between polar compounds and results in an increase in solubility, whereas with positive deviation, which takes place between nonpolar and semipolar materials, a decrease in solubility is noted. With negative deviation, the change is attributed to the effect of hydrogen bonding, solvation, and other attractive forces that may be present. The attractive forces that may occur are referred to as "internal pressure."

Internal Pressure

For ideal solutions, the internal pressure may be calculated from

$$P_i = \frac{\Delta H_v - RT}{V}$$

where ΔH_v is the heat of vaporization, and V is the molar volume at temperature T.

For water, ΔH_v is 10,500 cal at 25°C, and the molar volume is 18.01 cm^3. The internal pressure can be calculated from

$$P_i = \frac{10,500 - (1.987)(298.2)}{18.01} = 550 \text{ cal/cm}^3$$

Since 1.987 cal/(deg)/(mole) = 82.05 ml atm/(deg)/(mole),

$$P_i = 22,700 \text{ atm}$$

For Propellant 12, where $\Delta H_v = 39.47$ cal/g, molecular weight = 120.93, and density = 1.311 g/cm^3, the internal pressure can be obtained as follows:

$$\text{Molar volume} = \frac{\text{mol wt}}{\text{density}} = \frac{120.93}{1.311} = 92.24 \text{ cm}^3$$

$$\text{Molar heat of vaporation} = (39.47 \text{ cal/g})(120.93 \text{ g/mole})$$

$$= 4773 \text{ cal/mole}$$

$$P_i = \frac{4773 - (1.987)(298.2)}{92.24} = 45.32 \text{ cal/cm}^3$$

$$= 1870 \text{ atm}$$

Where there is a great difference between the internal pressures of two substances, the molecules of one cannot easily mix or mingle with the other, resulting in a decrease or partial solubility. From the above example of water and propellant, one would not expect these two substances to be miscible, since the internal pressure of water is high whereas the propellant shows a fairly low internal pressure. These nonpolar substances with low internal pressures are generally "squeezed out" by the polar substances with higher internal pressures (10).

Solubility Parameter

This value assigned to solvents is a measure of the amount of work that must be done to overcome the intermolecular forces of attraction, or to remove a molecule from the solute phase and deposit it in the solvent.

Solubility parameters are derived from a consideration of free energy, as shown by Sanders (11) and Reed (12):

$$\Delta G = \Delta H - T \, \Delta S$$

where ΔG is the change in free energy of the system, ΔH is the heat of mixing, T is the temperature, and ΔS is entropy. Entropy is a function of the state of a system and is a measure of the disorder or randomness of the system. It is also a measure of the freedom of movement of molecules; for example, gas has high entropy and solids have low entropy. Ice, in changing to water, will show an increase in entropy.

Sanders (11) has also shown that the square of the difference between the solubility parameters of two substances is proportional to ΔH and that, when the solubility parameters approach each other, ΔH becomes equal to zero and

$$\Delta G = 0 - T \, \Delta S$$

so that ΔG is negative and solubility will take place. The solubility parameters for propellants and other solvents are given in Chapter 6.

THE VAPOR STATE

Liquefication of Gases

In order to liquefy, a gas must be below its critical temperature and the pressure must be great enough to cause liquefication (the lower the temperature, the lower is the required pressure). In liquefying, the incoming gas is cooled by the outgoing gas, which in turn has been

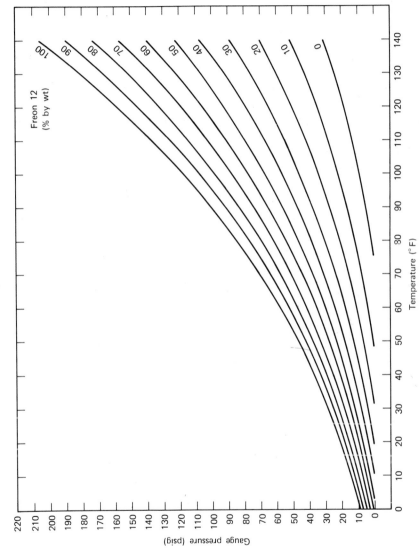

Figure 5.11. Freon 12/Freon 11 vapor pressure versus temperature (E. I. du Pont de Nemours and Co.).

cooled by expansion. Cooling of the gas is brought about by the external work done by the gas, as well as the expansion of the gas against the force of attraction between the molecules.

Vapor Pressure

DEFINITION

Vapor pressure is defined as the pressure existing at the point where there is an equilibrium between molecules in the vapor and liquid state. Vapor pressure is independent of quantity but is dependent on temperature and atmospheric pressure. The temperature of the substance at which the vapor pressure is equal to atmospheric pressure (14.7 psi or 760 mm Hg) is known as the boiling point.

A plot of vapor pressure versus temperature, as shown in Figure 5.11, yields a smooth curve. However, a plot of the logarithmic function versus $1/T$ should produce a straight line, as shown in Figure 5.12.

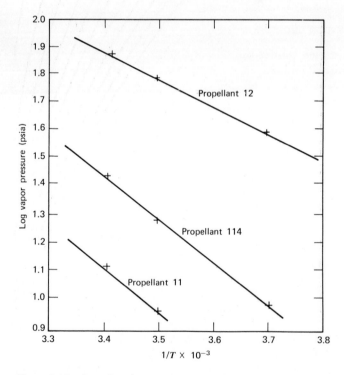

Figure 5.12.   Log plot of vapor pressure versus $1/T$.

From Figure 5.12 the equation of a straight line can be obtained:

$$\log p = A + \frac{B}{T}$$

where P = vapor pressure, A = y intercept, B = slope of line. After evaluation of the constants A and B, the vapor pressure at any temperature can be determined.

If a known amount of volatile liquid is vaporized, one can determine the vapor pressure, assuming ideal behavior, from the ideal gas equation:

$$p = \frac{gRT}{MV}$$

where P = vapor pressure, g = grams of volatile liquid vaporized, R = gas constant, T = absolute temperature of system, M = molecular weight, and V = volume occupied by grams of volatile liquid vaporized.

CLAPEYRON EQUATION

This equation relates the rate of change in vapor pressure to changes in temperature and can be derived from the free-energy equation:

$$\frac{dp}{dT} = \frac{\Delta H_{vap}}{T(V_v - V_l)}$$

where dp/dT is the rate of change of vapor pressure with temperature, ΔH_{vap} is the heat of vaporization (cal/g), T is the absolute temperature, and V_v and V_l are the volumes occupied by 1 mole of the vapor and the liquid, respectively.

In order to understand the use of this equation, consider the following example.

What is the rate of change in the vapor pressure of water at 100°C? The ΔH_{vap} is 539.7 cal/g, the molar volume is 18.02, and the molar volume of steam is 30.199 liters at 100°C and 1 atm pressure.

$$\frac{dp}{dT} = \frac{(539.7 \text{ cal/g}) (18.01 \text{ g/mole}) \left[\dfrac{0.08205 \text{ liter atm/(deg)/(mole)}}{1.987 \text{ cal/(deg)/(mole)}} \right]}{(373 \text{ deg}) (30.199 \text{ liters/mole} - 0.01802 \text{ liters/mole})}$$

$$= 0.0356 \text{ atm/deg} \quad \text{or} \quad 0.52 \text{ psi/deg}$$

For Propellant 12, the rate of change of vapor pressure with temperature would be calculated as follows:

$\Delta H_{vap} = 39.47$ cal/g at $-29.79°C$

$$V_v = \frac{\text{mol wt}}{\text{density}_{vap}} = \frac{120.93}{6.33} = 19.10 \text{ liters/mole at boiling point}$$

$$V_l = \frac{\text{mol wt}}{\text{density}_{liq}} = \frac{120.93}{1.311} = 92.24 \text{ cm}^3/\text{mole} = 0.09224 \text{ liter/mole}$$

$$\frac{dp}{dT} = \frac{(39.47)(120.93)(0.0413)}{(273 - 29.79)(19.10 - 0.09224)} = 0.0427 \text{ atm/deg}$$

$$= 0.63 \text{ psi/deg}$$

CLAUSIUS-CLAPEYRON EQUATION

If one assumes that the vapor of a liquefied gas follows the ideal gas laws,

$$\frac{dp}{dT} = \frac{p\,\Delta H_{vap}}{RT^2}$$

and

$$\frac{d\ln p}{dT} = \frac{\Delta H_{vap}}{RT^2}$$

Since R is constant, and if we assume that ΔH_{vap} remains constant over a given temperature range, then

$$\ln p = -\frac{\Delta H_{vap}}{RT} + C \quad \text{or} \quad \log p = -\frac{\Delta H_{vap}}{2.303RT} + C$$

A plot of $\log p$ against $1/T$ should yield a straight line, and the slope of the line would be equal to $-\Delta H_{vap}/2.303R$ or

$$\Delta H_{vap} \text{ (cal)} = -\text{(slope)}(2.303)(R)$$

From this plot one can then determine ΔH_{vap} for the system.

Integrating this equation between limits of p_1 and p_2 and T_1 and T_2, and rearranging terms, one obtains

$$\log \frac{p_2}{p_1} = \frac{\Delta H_{vap}(T_2 - T_1)}{2.303RT_2T_1}$$

A more exact equation can be obtained by the addition of another term, such as

$$\log p = -\frac{\Delta H_{vap}}{2.303RT} + \frac{\Delta C_p}{R}\log T + C$$

where ΔC_P is the change in heat capacity of the propellant at constant pressure.

Equations of State

An equation relating vapor pressure and temperature can be developed for each of the commonly used propellants.

Propellant 11 (13)

$$\log p = 39.0481 - \frac{4147.11}{T} - 11.7406 \log T + 0.0035694T$$

Propellant 12 (14)

$$\log p = 39.8838 - \frac{3436.6322}{T} - 12.4715 \log T + 0.004730T$$

Propellant 114 (15)

$$\log p = 57.7688 - \frac{4763.34}{T} - 18.6011 \log T + 0.0062959T$$

These equations can be used to calculate the vapor pressure of the propellant at any temperature.

Applications

In a previous portion of this chapter we illustrated the effect of an increase in temperature on the pressure of a compressed gas aerosol containing nitrogen (see p. 75). In this example, a compressed gas aerosol having a pressure of 100 psig at 70° F would develop an internal pressure of 113 psig at 130° F and a pressure of 258 psig* at 800° F.

* This pressure was calculated as in the example given:

$$\frac{P_2}{P_1} = \frac{T_2}{T_1}$$

$$T_2 = 800°F = 426.7°C = 699.8°K, \qquad T_1 = 70°F = 21.11°C = 264.2°K$$

$$\frac{P_2}{114.7} = \frac{699.8}{294.2}$$

$$P_2 = 272.8 \text{ psia or } 258.1 \text{ psig at } 800°F$$

In the case of a liquefied gas, the change in pressure with a change in temperature can be calculated from the appropriate equation of state or from the Clausius-Clapeyron equation. The following example illustrates the use of the equation of state.

Calculate the vapor pressure of Propellant 12 at 800° F if its vapor pressure is 84.9 psig at 70° F.

$$\log p = 39.8838 - \frac{3436.6322}{T} - 12.4715 \log T + 0.004730T$$

$$T = 699.8°K$$

$$\log p = 39.8838 - \frac{3436.6322}{699.8} - (12.4715)(2.8450) + (0.004730)(699.8)$$

$$= 39.8838 - 4.9109 - 35.4752 + 3.3101$$

$$\log p = 2.8078$$

$$p = \underline{642.4} \text{ psia at } 800°F$$

REFERENCES

1. P. A. Sanders, *Principles of Aerosol Technology,* Van Nostrand Reinhold Company, New York, 1970, pp. 210–216.

2. S. Friberg, L. Rydhag, and G. Jederstrom, Liquid Crystalline Phases in Aerosol Formulations. I: Phase Equilibria in Propellant Compositions, *J. Pharm Sci.,* **60,** (1971), 1883–1885.

3. G. Jederstrom, L. Rydhag, and S. Friberg, Liquid Crystalline Phases in Aerosol Formulations. II: Influence of Liquid Crystalline Phases on Foam Stability, *J. Pharm. Sci.* (in Press).

4. S. H. Maron, and C. F. Prutton, *Principles of Physical Chemistry,* 3rd ed., The Macmillan Company, New York, 1958, p. 7.

5. *The Effect of Air on Pressure in Aerosol Containers,* Freon® Aerosol Report FA-15, E. I. du Pont de Nemours and Company, Wilmington, Del., 1960.

6. *Handbook of Chemistry and Physics,* 49th ed., Chemical Rubber Company, Cleveland, Ohio, 1968–1969, p. D-107.

7. F. Daniels and R. A. Alberty, *Physical Chemistry,* John Wiley and Sons, New York, 1961, pp. 26–29.

8. L., Flanner, *Vapor Pressures of Solvents and Propellant Mixtures,* Allied Chemical Corporation, Morristown, N.J.

9. H. Hsu and D. Campbell, Calculations for Formulations with Soluble Gas Propellants, *Aerosol Age,* **9,** 12 (December 1964), 34.

10. A. N. Martin, J. Swarbrick, and A. Cammaratu, *Physical Pharmacy,* 2nd ed., Lea and Febiger, Philadelphia, 1969, pp. 295–299.

11. P. A. Sanders, *Principles of Aerosol Technology,* Van Nostrand Reinhold Company, New York, 1970, pp. 94–96.

12. A. B. Reed, Jr., Solubility Parameters and Their Application to Aerosols, *Chem. Spec. Mfr. Assoc. Proc., 46th Ann. Meet.,* New York, December 1959.

13. *Thermodynamic Properties of Freon® 11,* No. T-11-B, E. I. du Pont de Nemours and Company, Wilmington, Del., 1963.

14. *Thermodynamic Properties of Freon® 12,* No. T-12S, E. I. du Pont Nemours and Company, Wilmington, Del., 1956.

15. *Thermodynamic Properties of Freon® 114,* No. T-114 B, E. I. du Pont de Nemours and Company, Wilmington, Del., 1944.

PROPELLANTS AND SOLVENTS

In the early days of the aerosol industry, a propellant was defined as a liquefied gas with a vapor pressure greater than atmospheric pressure (14.7 psi) at 105°F. This definition did not include compressed gases because they were used to only a limited extent in the aerosol industry at that time. Now, however, little distinction is drawn between liquefied and compressed gases as far as the word "propellant" is concerned. Both phrases, "compressed gas propellant" and "liquefied gas propellant," are commonly used in the aerosol industry today. As a matter of fact, even the distinction between solvents and propellants has become hazy, and methylene chloride is sometimes considered to be a part of the propellant system.

LIQUEFIED GAS PROPELLANTS

There are three general classes of liquefied gas propellants—fluorinated hydrocarbons (fluorocarbons), hydrocarbons, and miscellaneous propellants. The fluorocarbons are by far the most widely used, followed by the hydrocarbons. Compounds falling in the last group (vinyl chloride and dimethyl ether) are not used individually as propellants but are employed only as components of propellant blends.

The specific properties of the individual propellants and propellant blends are discussed in detail later in the chapter. At this point, however, some of the characteristics that are desired in a propellant are considered in order to present a broad picture of a liquefied gas propellant.

BOILING POINT AND VAPOR PRESSURE

The propellant (or propellant blend) must have a boiling point below room temperature if it is to vaporize when discharged from the aerosol container. The boiling point should not be so low, however, that the compound cannot readily be liquefied by pressure or cooling.

The pressure in the aerosol produced by the propellant should not exceed certain allowable limits.

COST

Comparatively low-cost propellants are necessary if aerosol products are to be priced for the average consumer. A propellant could have every other property required, but if the manufacturing cost were so high that aerosol products incorporating this propellant had to be priced beyond the reach of the general public, the propellant would have only limited use.

FLAMMABILITY

Although it is not necessary for a propellant to be nonflammable, this property is very desirable. This is particularly true from the point of view of the propellant manufacturer and aerosol loader, who must handle and store large quantities of propellants. In addition, some aerosol regulations are specifically concerned with flammability. Although a number of aerosols are formulated with flammable propellants, usually the quantity of propellant in the product is sufficiently low, or enough other nonflammable propellants or solvents are present, so that the flammability of the product is minimized.

ODOR

Propellants with low levels of odor are a necessity for most aerosol products. For cosmetic products, such as perfumes and colognes, the propellant must have essentially no odor at all. An undesirable odor can be masked in many cases, but to do this is time consuming and often expensive.

PURITY

Propellants for aerosol products must consistently be of high purity. Both the chemist who develops the products and the marketer must be able to depend on the quality of the components in the aerosol, including the propellants. The presence of significant amounts of impurities in a propellant could result in variable product quality, off-odors, container corrosion, and inoperable valves. Ultimately these defects could lead to withdrawal of the product from the market.

SOLVENT PROPERTIES

Since the propellants are present as liquids in aerosol products, they are part of the solvent system. The solvent system must have adequate solubility for the active ingredients in the product. Many propellants have poor solvent characteristics, and this must be considered in formulation work. A water-soluble liquefied gas would have a multitude of uses, but such a propellant is not known at present.

STABILITY

A high degree of stability, both chemical and physical, is required. A propellant has to be unreactive under the conditions encountered in aerosol products. A reactive propellant not only would lose its own effectiveness but also could affect the active ingredients.

TOXICITY

Low toxicity is one property that a propellant must possess. Many aerosol products, such as hair sprays and room deodorants, are inadvertently inhaled during use, and others are applied directly to the skin. Some pharmaceutical products are designed for oral inhalation therapy. The necessity for a propellant with a low order of toxicity is obvious.

Fluorocarbon Propellants

HISTORY

One of the necessary starting materials for the preparation of fluorocarbons is anhydrous hydrofluoric acid (1). The preparation of a fluorocarbon by the addition of HF to amylene was first reported by Young in 1881 (2). During the next few years other organic fluorine compounds were prepared from halogenated compounds by replacement of halogen with fluorine. Metallic fluorides were used as catalysts. In most cases, however, the reactions were slow and incomplete. One of the most important discoveries was made around 1890 by Swartz, who found that the addition of a small amount of pentavalent antimony to SbF_3 produced a catalyst which gave complete and rapid exchange of fluorine with other halogens. The Swartz reaction, as it is known today, is the basis for most of the commercial processes for preparing the fluorocarbon propellants.

The development that was destined to create the aerosol industry actually occurred in 1928 in the refrigeration field (3). Up to that time, the commercial refrigerants consisted of compounds such as sulfur dioxide, methyl chloride, and ammonia. These materials were effective refrigerants but were very toxic. The need for a refrigerant of low toxicity was apparent, and the events leading up to the preparation of the first fluorocarbon refrigerant have been described by Thomas Midgley (4) as follows:

I was in the laboratory one morning and called Kettering [Charles F. Kettering, General Motors, Vice President for Research] in Detroit about something of minor importance. After we had finished this discussion, he said, "Midge, I was talking with Lester Keilholtz [Chief Engineer for Frigidaire] last night, and we came to the conclusion that the refrigeration industry needs a new refrigerant if they ever expect to get anywhere."

Within 3 days, Midgely and his coworkers, Henne and McNary, had synthesized dichlorodifluoromethane. By October 1929, a small experimental plant was producing sufficient quantities for detailed investigations. In 1930 Midgely presented the results of his work at a meeting of the American Chemical Society (5). During the presentation of his paper Midgely inhaled sufficient quantities of the new refrigerant to extinguish a burning candle when he exhaled, thus dramatically demonstrating the safe properties of the new compound. From 1931 to 1934 other refrigerants were produced commercially.

Until World War II the major use of fluorocarbons was in the refrigeration and air conditioning industries. In 1942, however, the work of Goodhue and Sullivan (6), in developing an aerosol insecticide for the armed forces, opened up an entirely new industry. The basic General Motors patents on the fluorocarbons expired in 1951, and between 1952 and 1965 other companies started production of fluorocarbons.

PROPELLANT TERMINOLOGY

Fluorocarbon propellants fall into two groups, depending on their flammability. The companies that manufacture the fluorocarbons designate their nonflammable propellants by using the company trademarks. Although originally the term "propellant" or "fluorocarbon" was reserved for use with flammable propellants, the former is generally recognized today as referring to all types of liquefied gases used as aerosol propellants.

The propellants are manufactured by companies all over the world. The trademarks of a number of these companies are listed in Table 6.1.

TABLE 6.1. WORLDWIDE FLUOROCARBON PRODUCERS

Country	Company	Trade Name
United States	E. I. du Pont de Nemours & Co.	Freon
	Allied Chemical & Dye Corp.	Genetron
	Kaiser Chemicals	Kaiser
	Pennwalt Chemical Co.	Isotron
	Racon, Inc.	Racon
	Union Carbide Corp.	Ucon
Argentina	Ducilo S.A.I.C.	Freon
	Fluoder S.A.	Algeon
	I.R.A., S.A.	Frateon
Australia	Australian Fluorine Chemicals Pty., Ltd.	Isceon
	Pacific Chemical Industries Pty., Ltd.	Forane
Brazil	Du Pont do Brasil S.A.—Industries Quimicas	Freon
	Hoechst do Brasil	Frigen
	Quimica e Farmaceutiça S.A.	
Canada	Du Pont of Canada, Ltd.	Freon
	Allied Chemical (Canada), Ltd.	Genetron
Czechoslovakia	Slovek Pro Chemickov A Hutni Vyobu,	Ledon
	Ustinad Cabem	
England	Imperial Chemical Industries, Ltd.	Arcton
	Imperial Smelting Corp., Ltd.	Isceon
France	Products Chimiques PECHINEY-SAINT-GOBAIN	Flugene
	Societe d'Electro-Chimie d'Electro-Metallurgie et des Acieres Electrique d'UGINE	Forane
East Germany	V.E.B. Alcid Fluorwerk Dohna	Frigedohn
	DIA Chemie-Anlagen Nunkritz	Forane
West Germany	Chemische Fabrik von Heyden AG	Heydogen
	Farbwerke Hoeschst AG	Frigen
	Kali-Chemie AG	Kaltron
India	Everest Refrigerants, Ltd.	Everkalt
	Navin Fluorine Industries	Mafron
Italy	Montecatini-Edison	Algofrene
		Edifren
Japan	Mitsui Fluorochemicals Co., Ltd.	Freon
	Daikin Kogyl Co., Ltd.	Daiflon
	Asahi Glass Co., Ltd.	Asahiflon
Mexico	Halocarburos S.A.	Freon
	Quimobasicos S.A.	Genetron
Netherlands	Du Pont de Nemours (Nederland) N.V.	Freon
	Uniechemie N.V.	Fresane
	Noury van Der Lande N.V.	FCC
	(Zinc-Organon)	
Russia	Not Known	Eskimon

TABLE 6.1.   CONTINUED

Country	Company	Trade Name
South Africa	African Explosives & Chemical Industries, Ltd.	Arcton
Spain	Ugine Quimica de Halogenos, S.A.	Forane
	Kali-Chemie Iberia S.A.	Kaltron
	Electroquimica de Flix S.A.	Frigen
Switzerland	Du Pont de Nemours International S.A.	Freon

NUMBERING SYSTEM

The trademark for a propellant indicates the manufacturer. In addition, a number that identifies the propellant follows the trademark. Three of the most commonly used propellants have numbers 12, 114, and 11. The numbering system was developed so that the chemical structure of the compound could be determined from the number (7). This system is now used by all the manufacturers of fluorocarbons for both refrigerants and aerosol propellants. Although the numbering system also applies to hydrocarbons and chlorinated hydrocarbons, it is generally used only with the fluorinated hydrocarbons.

The structural formulas for the propellants can be determined from their numbers by applying the following rules:

Rule 1: The first digit on the right is the number of fluorine (F) atoms in the propellant.

Rule 2: The second digit from the right is one more than the number of hydrogen (H) atoms.

Rule 3: The third digit from the right is one less than the number of carbon (C) atoms. This digit is omitted from the number when it is zero.

Rule 4: The number of chlorine (Cl) atoms in the compound is obtained by subtracting the sum of the fluorine and hydrogen atoms from the total number of atoms connected to the carbon atoms. The total number of attached atoms is four for one carbon atom and six for two carbon atoms.

Rule 5: When isomers exist, each has the same number but the most symmetrical is indicated by the number alone. As the isomers become more unsymmetrical, the letters a, b, c, etc., are added. Symmetry is determined by dividing the molecule in two and adding the atomic weights of the groups attached to each half. The closer the total weights are to each other, the more symmetrical is the molecule.

Rule 6: For cyclic derivatives, the letter C is used before the identifying numbers.

Rule 7: For unsaturated compounds, the preceding rules apply and the number 1 is added as the fourth digit from the right to indicate an unsaturated double bond.

The formula for Propellant 12 is determined from the preceding rules as follows:

a. The first digit on the right in "Propellant 12" is 2. Therefore the compound contains two fluorine atoms in the molecule (Rule 1).

b. The second digit from the right is 1. Therefore there are no hydrogen atoms in the compound, since 1 is one more than zero (Rule 2).

c. The third digit from the right has been omitted from the number and therefore is zero. This indicates that the compound contains one carbon atom (Rule 3).

d. Since the compound contains one carbon atom, there are four other atoms attached to the carbon atom. The number of chlorine atoms is found by subtracting the sum of the fluorine atoms and the hydrogen atoms from four. There are two fluorine atoms and no hydrogen atoms. Therefore there are two chlorine atoms in the compound (Rule 4).

Since Propellant 12 contains one carbon atom, two fluorine atoms, and two chlorine atoms, the formula is CCl_2F_2.

By use of the above rules, the following formulas are found:

Propellant 114: $CClF_2CClF_2$

Propellant C-318: $\begin{array}{c} CF_2 - CF_2 \\ | \qquad | \\ CF_2 - CF_2 \end{array}$

Propellant 1140 (vinyl chloride): $CH_2 = CHCl$.

PROPERTIES

Formulas, Boiling Points, and Vapor Pressures

The various fluorocarbon propellants, together with their formulas, molecular weights, boiling points, and vapor pressures, are listed in Table 6.2. Fluorocarbon 113, although not considered a propellant because of its high boiling point, is included to complete the series.

Propellants are available with vapor pressures ranging from subatmospheric to 121.4 psig at 70°F. The propellants listed in Table 6.2 are all miscible with one another, and therefore a propellant blend with any desired vapor pressure within the available range may be obtained by using the appropriate mixture. This is a considerable advantage,

TABLE 6.2.   PROPERTIES OF THE FLUOROCARBON PROPELLANTS

Propellant Number	Formula	Molecular Weight	Boiling Point (°F)	Vapor Pressure (psig) 70°F	Vapor Pressure (psig) 130°F
11	CCl_3F	137.4	74.8	(13.4 psia)	24.3
12	CCl_2F_2	120.9	−21.6	70.2	181.0
114	$CClF_2CClF_2$	170.9	38.4	12.9	58.8
152a	CH_3CHF_2	66.1	−11.0	63.0	176.3
142b	CH_3CClF_2	100.5	14.4	29.1	97.3
115	$CClF_2CF_3$	154.5	−37.7	103.0	254.1
113	CCl_2FCClF_2	187.4	117.6	...	...
21	$CHCl_2F$	102.9	48.1	8.4	50.5
22	$CHClF_2$	86.5	−41.4	121.4	296.8
C-318	C_4F_8 (cyclic)	200.0	21.1	25.4	92.0

since the formulator is not limited to the specific vapor pressures provided by the individual propellants. The most commonly used propellants are Propellant 11, Propellant 12, and Propellant 114. Vapor pressure curves over a wide temperature range for combinations of Propellants 12 and 11 and Propellants 12 and 114 have been published and are illustrated in Figures 6.1 and 6.2 (8).

Calculation of Vapor Pressure

Vapor pressures have been determined for many propellants and propellant blends (8–19). Sometimes, however, it is desirable to know the vapor pressure of a new blend for which experimental values are not available. This is particularly true for three-component blends. In these cases, it is helpful to be able to calculate the vapor pressure. If the vapor pressures of the individual propellants are known, this can be done using Raoult's law. This law, applied to propellant blends, states that the vapor pressure of a propellant mixture is equal to the sum of the partial pressures of the individual propellants. The partial pressures of the individual propellants in the mixture are obtained by multiplying the mole fraction of the propellant in the liquid phase by the vapor pressure (absolute) of the pure propellant.

The procedure for calculating vapor pressures is shown below for a Propellant 12/11 (30:70) blend.

a. Assume that the total weight of propellant blend is 100 g. Therefore there are 30 g of Propellant 12 and 70 g of Propellant 11.

b. The number of moles of each propellant in the blend is obtained by dividing the weight of the propellant by its molecular weight.

$$\text{Moles Propellant 12} = \frac{30}{120.9} = 0.248$$

$$\text{Moles Propellant 11} = \frac{70}{137.4} = 0.509$$

$$\text{Total moles} = \qquad 0.757$$

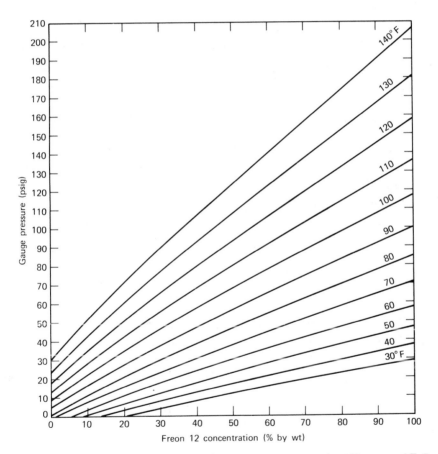

Figure 6.1. Freon 12/Freon 11 vapor pressure versus concentration. (Courtesy of E. I. du Pont de Nemours and Co,)

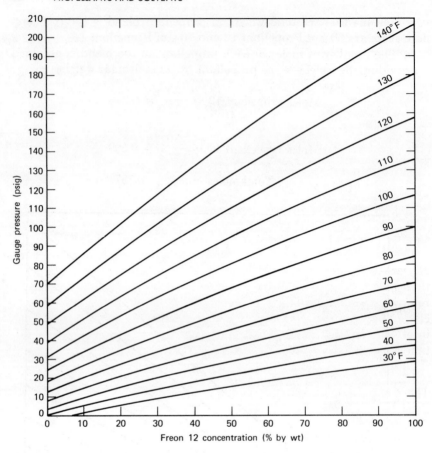

Figure 6.2.   Freon 12/Freon 114 vapor pressure versus concentration. (Courtesy of E. I. du Pont de Nemours and Co.)

c. The mole fraction of each propellant is obtained by dividing the number of moles of the propellant by the total number of moles.

$$\text{Mole fraction Propellant 12} = \frac{0.248}{0.757} = 0.33$$

$$\text{Mole fraction Propellant 11} = \frac{0.509}{0.757} = 0.67$$

d. The partial pressure of each propellant is obtained by multiplying the absolute vapor pressure of the propellant by its mole fraction. The vapor pressures of Propellants 12 and 11 at 70°F are 84.9 and 13.4 psia, respectively (8).

Partial pressure of Propellant 12 = 0.33 $\times$ 84.9 psia
= 28.0 psia

Partial pressure of Propellant 11 = 0.67 $\times$ 13.4 psia
= 9.0 psia

e. The total vapor pressure of the blend is obtained by adding the partial pressures of the individual propellants.

Total pressure of blend = 28.0 psia + 9.0 psia
= 37.0 psia or 22.3 psig

The vapor pressures of propellant blends containing three or more components can be calculated in the same way.

The calculated values of the fluorocarbon propellant blends generally are fairly close to the experimental values. In some cases, however, there is a considerable difference between the two. This difference is referred to as a deviation from Raoult's law. It occurs because of association between the molecules of one or more of the components. This, in effect, changes the molecular weight of the component. The deviation is particularly noticeable with propellant alcohol mixtures.

Vapor-Phase versus Liquid-Phase Composition

The composition of a vapor phase over a single propellant, such as Propellant 12, is the same as that of the liquid phase, that is, 100% Propellant 12. However, in a blend of propellants with different boiling points and different vapor pressures, the composition of the vapor phase will be different from that of the liquid phase. Since the propellant with the lowest boiling point vaporizes more readily than the higher-boiling propellants, the vapor phase will have a higher proportion of the lower-boiling propellant than is present in the liquid phase.

This is illustrated in Figures 6.3 and 6.4, which show the composition of the vapor phase in equilibrium with the liquid phase for Propellant 12/11 and Propellant 12/114 mixtures. Thus, when the composition of the liquid phase is 50% Propellant 12 and 50% Propellant 11, the composition of the corresponding vapor phase is 87% Propellant 12 and 13% Propellant 11. It should be emphasized that the data in Figures 6.3 and 6.4 apply only to a closed system where the vapor phase is in equilibrium with the liquid phase. When the vapor phase is removed continuously by evaporation or distillation, the compositions of both the liquid and vapor phases change. Initially, the vapor phase has

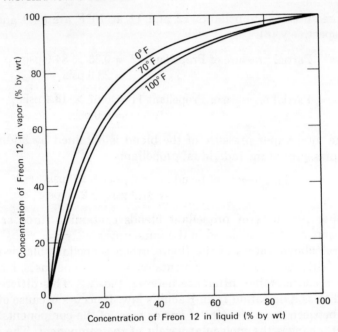

Figure 6.3.   Composition of the vapor above Freon 12/Freon 11 solutions. (Courtesy of E. I. du Pont de Nemours and Co.)

the highest concentration of the lowest-boiling component in the liquid phase. As evaporation proceeds, however, the liquid phase ultimately becomes richest in the highest-boiling propellant. Finally, both the liquid phase and the vapor phase consist mainly of the higher-boiling component.

Density, Flammability, and Surface Tension (8–20)

Density. The density of a propellant is important because it determines the amount that can be loaded into a cylinder, tank car, or other container. The propellant density also affects the density of an aerosol product, and this in turn determines the maximum weight of product that can be packaged in an aerosol container. The Department of Transportation regulations require most aerosol products to be heated to 130°F without evidence of distortion or leakage. The latter can occur if too much product is packaged in a container at room temperature. The decrease in density and consequent expansion in volume as the aerosol is heated may cause the container to become liquid-full at

130°F. This can be avoided if the density of the product is known at both 70°F and 130°F.

The liquid densities of the fluorocarbon propellants at 70°F and 130°F are given in Table 6.3; Propellant 152a is the only fluorocarbon listed with a density less than that of water. The variation of density with temperature for Propellant 12/11 and Propellant 12/114 is illustrated in Figures 6.5 and 6.6 (8). These data, in combination with the density of the aerosol concentrate, are sufficient to calculate the density of a particular aerosol product. The density of the concentrate can be obtained simply by determining the weight of a known volume of the concentrate.

Flammability. A flammable propellant is one that forms explosive mixtures with air. However, a flammable gas will form explosive mixtures only at certain concentrations in air. The lowest concentration of the propellant at which an explosive mixture is formed is referred to as the lower flammability limit of the propellant, and the highest concentration as the upper flammability limit. Flammability limits are

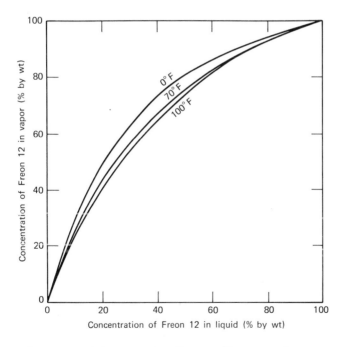

Figure 6.4. Compsotion of the vapor above Freon 12/Freon 114 solutions. (Courtesy of E. I. du Pont de Nemours and Co.)

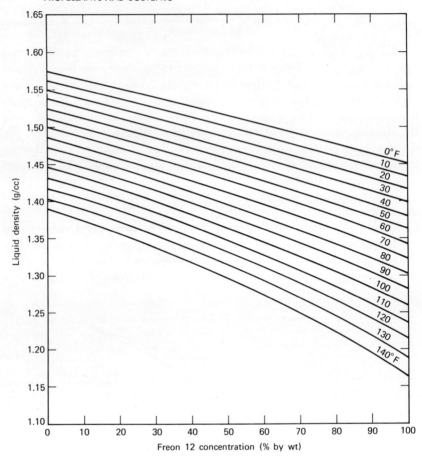

Figure 6.5.   Freon 12/Freon 11 liquid density versus concentration. (Courtesy of E. I. du Pont de Nemours and Co.)

reported in volume per cent. Flammable propellants form explosive mixtures at all concentrations between the upper and the lower flammability limits.

As shown in Table 6.3, several values for the flammability limits of Propellants 142b and 152a have been reported in the literature. There are several explanations for the different results. For example, they could be due to differences in the size of equipment used to obtain the flammability limits. Haenni et al. (21) reported that mixtures of 84% Propellant 12 and 16% ethylene oxide were nonflammable when tested in a conventional eudiometer tube. However, when the mixtures were evaluated in a 13.1-ft³ autoclave, the concentration of ethylene oxide had to be reduced to 12% to obtain a nonflammable mixture.

Another factor that can affect the results is the type of ignition source employed. Coward and Jones (22) noted that, if the ignition source were not hot enough, some flammable mixtures could not be ignited. They pointed out that the ignition source should be sufficiently strong so that the flammability of the gas mixture is tested, and not the capacity of the ignition source to initiate flame.

For most purposes in the aerosol industry, it is usually sufficient to know whether the propellant will form an explosive mixture in air at

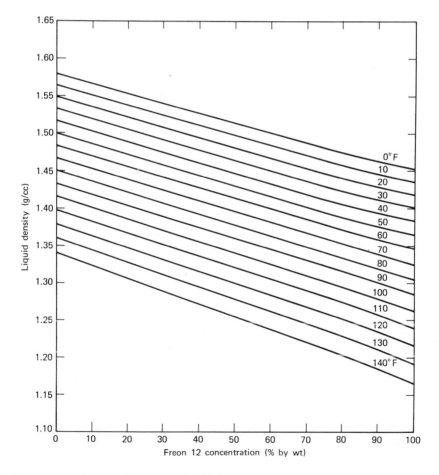

Figure 6.6.   Freon 12/Freon 114 liquid density versus concentration. (Courtesy of E. I. du Pont de Nemours and Co.)

TABLE 6.3.   PHYSICAL PROPERTIES OF THE FLUOROCARBON PROPELLANTS

Propellant Number	Liquid Density (g/cc)		Flammability Limits in Air (vol %)		Estimated Surface Tension at 80°F (dynes/cm)
	70°F	130°F	Lower Limit	Upper Limit	
11	1.49	1.40	Nonflammable		18
12	1.33	1.19	Nonflammable		9
114	1.47	1.36	Nonflammable		12
152a	0.91	0.81	3.7–5.1	17.1–18.0	...
142b	1.12	1.03	6.0–9.1	4.8–15.0	...
115	1.31	1.11	Nonflammable		5
113	1.57	1.49	Nonflammable		19
21	1.38	1.29	Nonflammable		18
22	1.21	1.06	Nonflammable		8
C-318	1.51	1.37	Nonflammable		7

any concentration. A knowledge of the actual flammability limits is generally not necessary.

Surface Tension. The low surface tension values for the propellants in Table 6.3 reflect the fact that the attractive forces between the propellant molecules are relatively weak. A low surface tension indicates that a propellant can be emulsified with comparative ease but does not indicate the stability of the resultant emulsion.

Odor

No quantitative odor data are available for the fluorocarbons. However, the odor levels of Propellants 12 and 114 are so low that they are used for most aerosol perfumes and colognes. Propellant 142b has been reported to enhance the fragrance of certain perfumes (23) but is not used to any extent in such products. Propellant 11 has a slight but definite odor which has limited its use in perfumes. However, Propellant 11 is satisfactory for products such as hair sprays, where the perfume may be an important part of the product but is not the major consideration.

Purity

The fluorocarbon propellants, as originally used, were required to have a degree of purity seldom found in organic compounds produced on a large scale. Since the aerosol propellants are the same compounds that are used as refrigerants, these propellants of necessity are as pure as the refrigerants of the same structure.

The organic purities of the propellants generally exceed 99.9%. The specifications guaranteed by most manufacturers for the major propellants are listed in Table 6.4.

Solvent Properties

The fluorocarbon propellants are present in aerosol formulations as liquids and therefore are part of the solvent system. The solubility properties of the propellants determine their compatibility with other solvents. They also affect the capacity of the solvent system to dissolve the active ingredients. Many unnecessary experiments can be avoided if the solubility properties of the propellants are known.

Several physical properties of liquids are useful for predicting the miscibility of one liquid with another or for estimating the solvent power of a liquid relative to that of other liquids. Two of these, the kauri-butanol values and solubility parameters, are particularly helpful in the propellant field. Many of these values have been determined for the propellants.

Kauri-Butanol Values. A kauri-butanol value is used to estimate the relative solvent power of a liquid. The kauri-butanol value of a solvent is the number of milliters required to produce a specified degree of turbidity when the solvent is added to 20 g of a standard solution of kauri resin in *n*-butyl alcohol at 25°C. Toluene is the standard, and the concentration of kauri resin in butyl alcohol has been adjusted so

TABLE 6.4. PROPELLANT SPECIFICATIONS

	Propellant		
Property	11	12	114
Maximum water content (ppm by wt)	10	10	10
Maximum nonabsorbable gas (vol % in the vapor)	. . .	1.5	1.5
Boiling point at 1 atm pressure (°F)	74.8	−21.6	38.4
Maximum boiling range (°F)	0.5	0.5	0.5
Maximum high-boiling impurities (vol %)	0.01	0.01	0.01
Chloride ion content	None	None	None

TABLE 6.5. SWELLING OF ELASTOMERS BY VARIOUS COMPOUNDS (18)

			Increase in Length at Room Temperature (%)		
Compound	Solubility Parameter	Kauri-Butanol Value	Neoprene GN	Buna N	Natural Rubber
Methylene chloride	9.5	136	37	52	34
Chloroform	9.1	208	43	54	45
Carbon tetrachloride	8.6	113	36	11	44
Propellant 21	8.0	102	28	48	34
Propellant 11	7.5	60	17	6	23
Propellant 113	7.2	31	3	0.5	17
Propellant 152a	7.0	11	2	1	2
Propellant 142b	6.8	20	3	3	5
Propellant 22	6.5	25	2	26	6
Propellant 114	6.2	12	0	0	2
Propellant 12	6.1	18	0	2	6
Propellant C-318	5.6	10	0	0	0

Solubility parameter of neoprene GN = 9.2
Solubility parameter of buna N = 9.4
Solubility parameter of natural rubber = 8.3

that 105 cc of toluene will produce the required turbidity. Kauri butanol values can be used only for relatively nonpolar solvents, such as aliphatic hydrocarbons, aromatics, and halogenated compounds. Oxygenated compounds, such as alcohols, would have infinite kauri-butanol values since they could be added indefinitely to the kauri-butanol solution without causing turbidity.

In order to determine the kauri-butanol value of a propellant, 20 g of standard kauri-butanol solution is placed in a suitable pressure bottle, such as a 4-oz. aerosol bottle. The bottle is capped with a standard valve without a dip tube and placed over a sheet of No. 10 print (newspaper print is satisfactory). The solution is then titrated with propellant, using a pressure burette. The end point is reached when the sharp outlines of the newspaper print, as viewed through the liquid in the bottle, become obscured or blurred because of precipitation of the kauri resin from solution.

The kauri-butanol values for the fluorocarbon propellants are listed in Table 6.5. The values for carbon tetrachloride, chloroform, and methylene chloride, all excellent solvents, have been included for comparison. The data show that Propellant 21, with a kauri-butanol value of 102, has the best solvent properties of the fluorocarbons. This propellant is of interest in the pharmaceutical industry for this reason. Propellant 11, with a kauri-butanol value of 60, is next, followed by Fluorocarbon 113, with a value of 31. The rest of the fluorocarbons have lower kauri-butanol values and are poor solvents. The kauri-butanol values of a wide variety of propellants are available in the literature (24, 25).

Solubility Parameters. The concept of solubility parameters was originated by Hildebrand and Scott (26). The practical aspects of solubility parameters in solving problems in the coating industry have been presented by Burrell (27, 28). In a sense, solubility parameters are physical constants for compounds. The solubility parameter is an indication of the strength of the intermolecular forces in a compound, and this is related to its solubility properties. Solubility parameters can be calculated or determined experimentally. Values for a large number of compounds, including organic solvents, elastomers, resins, and propellants, are available in the literature (27–29).

Solubility parameters can be used to predict whether solvents will be miscible or whether an amorphous solid will dissolve in a liquid. If two liquids have approximately the same solubility parameters, the chances are that they will be miscible. Therefore, by comparing the solubility parameter of one liquid with that of any other given liquid, it is possible to predict whether the two liquids will be miscible. If a resin has about the same solubility parameter as a liquid, the resin usually will dissolve in the liquid. If a propellant or solvent has about the same solubility parameter as an elastomer, the propellant or solvent will affect the elastomer. Solubility parameters of the propellants are listed in Table 6.5.

Effect of Fluorocarbons on Elastomers. The swelling effect of the fluorocarbon propellants on elastomers is directly related to the solubility parameters of the fluorocarbons and the elastomers. This is illustrated by the data in Table 6.5, which shows the relationship between the solubility parameters of a series of solvents and their effects on neoprene, buna N, and natural rubber. Liquids with solubility parameters of 7.2 or lower have comparatively little effect on the elastomers.

Although the kauri-butanol value is not a fundamental property of a liquid, there appears to be a relationship between the solubility parameters and kauri-butanol values for the paraffin hydrocarbons (27). A similar relationship exists for the fluorocarbon propellants, as shown in Table 6.5. Consequently, a relationship between the kauri-butanol values of solvents and their effects on elastomers will also exist. As shown in Table 6.5, compounds with kauri-butanol values of 31 or lower have little effect on elastomers.

Effect of Fluorocarbons on Plastics. Plastics vary widely, depending on the structure of the polymer, the molecular weight, and the additives that are present. As a result, the effects of fluorocarbons on plastics also vary considerably, even with plastics having the same molecular structure. The compatibility of any given plastic with the fluorocarbons should be determined for each specific application. In addition, other factors, such as temperature, can change the resistance of the plastic to the fluorocarbon.

Some generalizations can be made, however. Most plastic materials are not appreciably affected by fluorocarbon compounds. An exception is polystyrene, which is dissolved by Propellants 11, 21, and 22. The effect of the fluorocarbons on plastics decreases as the amount of fluorine in the molecule increases. Propellant 12, for example, has less effect than Propellant 11.

A brief summary of the effects of fluorocarbon compounds on various plastics is given below.

Cellulose Acetate or Nitrate
Suitable for use with fluorocarbons.

Delrin Acetal Resin
Generally suitable for use with fluorocarbons.

Epoxy Resins
Suitable for use with fluorocarbons unless highly plasticized.

Lucite Acrylic Resin (methacrylate polymers)
Usually suitable for use with Propellants 12 and 114 for short periods but may crack, craze, or become cloudy on longer exposures. These polymers should not be used with Propellant 22 or 11. Their use with Fluorocarbon 113 is questionable.

Nylon
Generally suitable for use with fluorocarbon compounds but may become brittle at high temperatures in the presence of air, water, or alcohol.

Orlon Acrylic Fiber
Generally unaffected by fluorocarbons.

Phenolic Resins
Normally not affected by fluorocarbons. However, phenolic resins vary widely in composition, and samples should be tested.

Polyethylene
Suitable for many applications but should be tested. Results differ, depending on the particular sample tested.

Polystyrene
Generally not suited for use with fluorocarbon compounds.
Polyvinyl Alcohol
Not affected by the fluorocarbons, with the exception of Propellant 21. Very sensitive to water.
Teflon TFE Fluorocarbon Resin
Did not swell when submerged in fluorocarbon liquids. Some diffusion was observed with Propellants 12 and 22.
Vinyl Compounds
The effects of the fluorocarbons vary considerably, depending on the vinyl type and plasticizer present. Samples should be tested before use.

The specific effects of various fluorocarbons on a number of plastics are shown in Table 6.6, which lists the maximum linear swell of the plastic after immersion in the fluorocarbon at room temperature (30).

Brown (31) also investigated the effects of propellants on the plastic materials in valve compounds and calculated a "swelling index" for

TABLE 6.6. EFFECT OF FLUOROCARBON COMPOUNDS ON PLASTICS AT ROOM TEMPERATURE[a]

Plastic	Linear Swell Submerged in Liquid Phase (%)					
	P-11	P-12	P-21	P-22	P-113	P-114
Cellulose acetate	0	0	D	...	0	...
Cellulose nitrate	1	0	D	...	0	...
Chlorotrifluoroethylene polymer	0–3	2	...	1	0	...
Delrin® acetal resin	0	1	...	3	0	0
Lucite acrylic resin	0	0	D	D	0	1
Mylar® polyester film	...	...	...	...	...	...
Nylon (Zytel® 101)	0	0	0	1	0	0
Phenolformaldehyde resin	0	0	0	...	0	...
Polyethylene	6	1	5	2	2	1
Polyethylene, linear	...	...	...	1	2	...
Polypropylene	...	...	...	...	...	...
Polystyrene	D	0	D	...	0	...
Polyvinyl alcohol	0–3	−1	13	...	0	...
Polyvinyl chloride	0	0	15	...	0	...
Polyvinylidene chloride	0–3	0	1	4	0	...
Teflon® TFE fluorocarbon resin	0	0	0	1	0	...

[a] D = disintegrated; ... = not shown.

TABLE 6.7. SWELLING INDICES FOR
FLUOROCARBONS

Fluorocarbon	Swelling Index
Propellant 11	513
Propellant 12	94
Propellant 114	32
Propellant 21	1901
Fluorocarbon 113	71
Methylene chloride	2215

each fluorocarbon. This value was obtained by summing up the percentage changes for each plastic in various propellants. The percentage change for samples that disintegrated was taken as 100. The swelling indices for the fluorocarbons are given in Table 6.7, with methylene chloride included for comparative purposes.

Some typical diffusion rates of the fluorocarbons, oxygen, and nitrogen through various plastic films are given in Table 6.8 (30). The diffusion rates vary for a given plastic, depending on the molecular weight and additives that are present.

Stability

Chemical Stability. Generally the fluorocarbon propellants have a high degree of stability, particularly in nonaqueous systems. However, the stabilities of the propellants can vary, depending on the particular propellant involved and the type of system. This makes it desirable to know the conditions that can lead to instability, so that the propellant most suited for the product can be selected with a minimum of effort.

The hydrolytic stability of the fluorocarbons has been of considerable interest to both the refrigeration and aerosol industries. In the refrigeration industry, this parameter is important because of the possibility of corrosion of the metal components in the refrigerating system and the effect of acidic reaction products on the insulation. The hydrolytic stability of the propellants is important in the aerosol industry because an unstable propellant could change the odor of perfumes, destroy the active ingredients, and cause container corrosion and possible leakage.

TABLE 6.8. PERMEABILITY OF PLASTIC FILMS[a]

Plastic Film	Temp. (°F)	Permeability [(ml at STP) (mil)/(day) (100 in.²) (atm)]						
		N_2	O_2	P-11	P-12	P-21	P-22	P-114
Mylar® polyester	77	1.0	10.3	...	<0.004	...	...	...
	130	3.8	37	...	<0.004	...	...	...
Mylar® polyester, aluminized	130	0.29	1.3	...	<0.007	...	0.24	...
Nylon	130	5	...	...	0.04	...	...	...
Polyethylene	78	130	460	...	1300	...	...	...
Polypropylene	77[b]	...	...	...	...	...	...	...
	130[c]	...	...	...	...	...	...	...
Polyvinyl alcohol	130	0.65	...	...	<0.07	...	...	...
Polyvinyl chloride Genotherm U.G.	...	...	...	0.15	<0.03	15.5	0.13	0.03
Polyvinylidene chloride Saran	77	0.3	0.9	...	<0.004	...	...	...
	130	2.3	5.4	...	1.9	...	...	...
Teflon® 100-X FEP	130	>47	...	0.04	...	...	...	...
Urethane, polyester	76	0.9	...	...	...	...	...	...
	130	14	...	1	0.1	...	...	...

[a] ... = not tested.
[b] No diffusion with Freon® 116 or Freon® C-318.
[c] Quite permeable to Freon® 116 and Freon® C-318.

Parmelee and Downing (32) found that factors such as pH, temperature, and the type of metals present could have profound effects on the rate of hydrolysis.

Experimentally determined hydrolysis rates of the fluorocarbon propellants are listed in Table 6.9. The hydrolysis rates are expressed in units of grams of propellant hydrolyzed per liter of water per year. A description of the experimental procedures used will clarify these units. Basically, a liter of distilled water at 86°F in a glass flask was saturated with the vapor of the propellant. In order to be certain that any propellant that disappeared as a result of hydrolysis was replaced, a condition of saturation was maintained by keeping the flask under a slight positive propellant pressure at all times. At intervals, portions of the solutions were withdrawn and analyzed for chloride ion. From these data, the number of grams of propellant that hydrolyzed per liter of water per year was calculated. The experiments were carried out for a minimum of a year.

The data illustrate that replacing a chlorine atom by a fluorine atom increases the stability of the molecule. This is shown by a comparison of the hydrolysis rates of Propellants 12 and 11. The presence of hydrogen in the molecule, as in Propellants 21 and 22, decreases the stability in alkaline systems.

Although some progress has been made in establishing the mechanism of hydrolysis of Propellant 11, not all of the reactions involved have been determined. Church and Mayer (34) aged combinations of Propellant 11, various metals, and water at temperatures up to 150°C. They postulated three possible reactions (M = metal):

$$2CCl_3F + M \rightarrow MCl_2 + CCl_2FCCl_2F \text{ (Propellant 112)} \qquad [1]$$

$$2CCl_3F + 2H_2O + 2M \rightarrow MCl_2 + M(OH)_2 + 2CHCl_2F \qquad [2]$$

$$\text{(Propellant 21)}$$

$$CCl_3F + H_2O \rightarrow CCl_2FOH + HCl$$

$$\downarrow$$

$$\underset{\underset{Cl}{|}}{\overset{\overset{F}{|}}{C}}\!\!\equiv\!\!O + HCl$$

$$\xrightarrow{\quad H_2O \quad} CO_2 + HCl + HF \qquad [3]$$

Although Church and Mayer were unable to identify any of the reaction products, they favored reaction [1] on the basis of their rate data, obtained by chloride ion analysis. Subsequently Sanders (35) also stored combinations of Propellant 11, water, and metals for various periods of time. The reaction mixtures were analyzed and found to contain Propellant 21, Propellant 112, acid, and, in some cases,quantities of insoluble solids. The formation of Propellant 112 indicated that reaction [1] had occurred, while the presence of Propellant 21 suggested that [2] had also taken place. Although the acid-forming reaction was not identified, [3] is certainly a logical possibility.

Mixtures of Propellant 11 and water caused considerable corrosion in tin-plate containers at 130°F. Nitromethane was found to have a retarding effect on the corrosion. It is possible that the deactivating effect of nitromethane was due to the formation of a monomolecular layer of this compound on the surface of the metal.

On the basis of our present knowledge, the overall hydrolytic stability of the propellants can be summarized as follows. The most stable are Propellants C-318 and 115. These compounds have been approved by the Food and Drug Administration for use in foods. In order to obtain this approval, it was necessary to show that they have an extraordinary degree of stability in aqueous systems. Hence, they are sufficiently stable to be used in aqueous aerosol systems.

Propellants 12 and 114 are also very stable, except under conditions of high alkalinity and temperature. They can be used for practically all aqueous-based products except those containing appreciable concentrations of caustic, such as oven cleaners.

TABLE 6.9. HYDROLYSIS RATES OF THE FLUOROCARBONS (33)

Fluorocarbon	Water Alone	1% Sodium Carbonate	Water Steel Strips	10% Sodium Hydroxide at 140°F
Propellant 11	0.005	0.12	19	100
Propellant 12	0.005	0.04	0.8	40
Propellant 114	0.005	0.01	1.4	50
Propellant 21	0.01	330	5.2	Very rapid
Propellant 22	0.01	220	0.14	Very rapid
Fluorocarbon 113	0.005	0.01	50	80

TABLE 6.10.   THERMAL STABILITY OF FLUOROCARBON COMPOUNDS (9)

Fluorocarbon	Decomposition Rate at 400°F in Steel (%/year)	Temperature for First Trace of Decomposition in Quartz (°F)
Propellant 11	2	840
Propellant 12	1	1000
Propellant 114	1	940
Propellant 22	. . .	550
Fluorocarbon 113	6	570

Propellant 11 is stable in water alone but not when metals such as iron is present. The pH also affects the rate of hydrolysis, which increases with increasing alkalinity.

Propellants 21 and 22 are stable in neutral or acidic solutions but are unstable in alkaline solutions. The hydrolysis rates of Propellants 142b and 152a have not been determined. Propellant 152a, however, would be expected to be particularly stable.

Propellants 12, 114, 21, and 22 are stable in the presence of alcohols. Propellant 11 is also stable under most conditions, as is evident from the fact that millions of cans of products such as hair sprays have been packaged and successfully marketed. However, there are some conditions under which Propellant 11 will react with alcohols via a free-radical reaction. This reaction is promoted by very low concentrations of oxygen or the presence of some other free-radical catalyst (36).

The following reactions can occur:

$$CCl_3F + C_2H_5OH \rightarrow CH_3CHO + HCl + CHCl_2F$$

$$CH_3CHO + 2C_2H_5OH \rightarrow CH_3CH(OC_2H_5)_2 + H_2O$$

$$C_2H_5OH + HCl \rightarrow C_2H_5Cl + H_2O$$

The initial reaction results in the formation of acetaldehyde, HCl, and Propellant 21. Subsequent reactions form acetal and ethyl chloride. Acetaldehyde can cause off-odors to develop, and this is one of the first indications that the reaction has taken place. The HCl can cause corrosion of the metal container.

The free-radical reaction has been known to take place in alcohol-based products loaded under warm conditions, where the excessive purging of the air from the container by the propellant resulted in a very low oxygen content. Some concentrates also apparently contain free-radical catalysts. However, even if the reaction does occur, the problem is not serious because nitromethane has been shown to be a very effective inhibitor for this reaction. Stabilized Propellant 11 containing nitromethane as an inhibitor is available commercially (37).

Thermal Stability. The fluorocarbon propellants are so stable at elevated temperatures that this factor can be disregarded as far as most aerosol applications are concerned. This is illustrated by the data in Table 6.10 (9).

Toxicity

There are some properties that are desirable in a propellant, and others that are absolutely essential. A low order of toxicity is one of the latter. Some aerosol pharmaceutical products are designed for oral inhalation, while others, such as hair sprays and room deodorants, can be inhaled inadvertently during use. Many cosmetic and pharmaceutical products are applied directly to the skin. These applications emphasize the requirement of low toxicity for aerosol propellants.

During the period from 1931 to 1941, the Underwriters' Laboratories studied the inhalation toxicities of a number of gases, including the fluorocarbons. As a result of this work, the Laboratories developed a system for classifying gases and vapors according to their toxicity. The compounds are divided into six large groups (Table 6.11). Group 1 includes the most toxic compounds; Group 6, the least toxic. Details of the test used as a basis for the classification of the compounds are given in the Underwriters' Laboratories Reports (38) and have been further summarized (39).

The data in Table 6.11 refer to acute conditions for a comparatively short time. However, the body can withstand higher concentrations of many gases for a short time than it can tolerate over a longer period. Hence threshold limit values (TLV) have been established by the American Conference of Governmental Hygienists for continuous exposure to common industrial chemicals (39). Some of these are listed in Table 6.12.

As mentioned earlier, the initial application of the fluorocarbons was in the refrigeration field, where the compounds were used as gases. For this reason, inhalation studies were emphasized first. However, with the expanding use of the fluorocarbons into the aerosol industry, it became necessary to consider their toxicities when applied by other methods. A number of studies on the acute oral, skin, and eye toxicities

TABLE 6.11.   UNDERWRITERS' LABORATORIES CLASSIFICATION OF COMPARA-
TIVE LIFE HAZARD OF GASES AND VAPORS (38)

Group	Definition	Examples
1	Gases or vapors which in concentrations of the order of ½ to 1% for durations of exposure of the order of 5 min are lethal or produce serious injury.	Sulfur dioxide
2	Gases or vapors which in concentrations of the order of ½ to 1% for durations of exposure of the order of ½ hour are lethal or produce serious injury.	Ammonia Methyl bromide
3	Gases or vapors which in concentrations of the order of 2 to 2½% for durations of exposure of the order of 1 hour are lethal or produce serious injury.	Carbon tetrachloride Chloroform Methyl formate
4	Gases or vapors which in concentrations of the order of 2 to 2½% for durations of exposure of the order of 2 hours are lethal or produce serious injury.	Dichlorethylene Methyl chloride Ethyl bromide
Between 4 and 5	Appear to classify as somewhat less toxic than Group 4. Much less toxic than Group 4 but somewhat more toxic than Group 5.	Methylene chloride Ethyl chloride Propellant 113
5a	Gases or vapors much less toxic than Group 4 but more toxic than Group 6.	Propellant 11 Propellant 22 Carbon dioxide
5b	Gases or vapors which available data indicate would classify as either Group 5a or Group 6.	Ethane Propane Butane
6	Gases or vapors which in concentration up to at least about 20% by volume for durations of exposure of the order of 2 hours do not appear to produce injury.	Propellant 12 Propellant 114 Propellant 13B1

of the propellants have been carried out (40, 41). These again demon-strated that, as a class, the fluorocarbons possess a low order of toxicity by several routes of bodily contact. As Clayton (40, 41) has pointed out, there is no evidence that the fluorocarbon propellants in aerosol products contribute to the toxicity of these products.

There are two major reasons why the fluorocarbon propellants pos-sess low toxicity. One is that the bond distance between the carbon and fluorine is short. This results in high bond energy and a bond that is difficult to rupture. The second reason is that the presence of fluorine in a molecule increases the stability of adjacent C–Cl bonds, making them much more difficult to break. It has been shown that, in many cases, toxicity is inversely related to bond dissociation energies. This is illustrated in Table 6.13, which shows the relationship between bond dissociation energy and toxicity (41).

There are several other situations to be considered with reference to toxicological problems that may arise from the presence of fluorocar-bons. The fluorocarbon vapors are heavier than air; they will settle at the floor in a room or displace air from a mixing kettle or in a tank. Their action in this respect is similar to that of carbon dioxide. A worker entering a tank containing only fluorocarbon propellant will die

TABLE 6.12. THRESHOLD LIMIT VALUES FOR FLUOROCARBONS (39)

Compound	Threshold Limit Value (ppm)
Propellant 11	1000
Propellant 12	1000
Propellant 114	1000
Propellant 152a	. . .
Propellant 142b	. . .
Propellant 21	1000
Propellant 22	1000
Vinyl chloride	500
Methylene chloride	500
Methyl chloroform	350

TABLE 6.13.   BOND DISSOCIATION ENERGY AND TOXICITY (41)

Bond dissociation:	$Cl_3C \rightarrow Cl$	$Cl_2HC \rightarrow Cl$	$Cl_2FC \rightarrow Cl$	$F_3C \rightarrow Cl$	$F_3C \rightarrow F$
Energy (Kcal):	68	72	74.5	83	121
Order of toxicity:	CCl_4	$CHCl_3$	CCl_3F	$CClF_3$	CF_4

of lack of oxygen if he is not provided with an air mask. This, of course, has nothing to do with the toxicity of the propellants, but results only from a lack of air. In a sense, this is equivalent to drowning in water, which is known to have a low order of toxicity.

Another problem can arise when aerosols containing fluorocarbons are sprayed into an open flame or onto a red-hot surface. Under these conditions, decomposition of the fluorocarbon can occur in the presence of moisture to form hydrogen chloride, hydrogen fluoride, and traces of phosgene. As Reed (42) has pointed out, however, the decomposition products that are formed (i.e., hydrogen chloride, etc.) are so irritating that a person cannot remain long enough in an atmosphere with a toxic concentration to sustain any permanent damage. In other words, the irritation level is far lower than the toxic level. However, the possibility must be considered that in a plant or laboratory there could be repeated exposure to the decomposition products below irritation level and that ultimately damage might occur.

One final point should be mentioned. The propellants have low boiling points, and if a product is sprayed onto the skin from such a short distance that liquid propellant reaches the skin, freezing of the tissue can occur. Actually, several aerosol products have been formulated specifically for skin-planning applications in which it is necessary to freeze the surface of the skin. The chances of this occurring with normal aerosol products are slight but should be kept in mind.

Summary

The properties and uses of the various fluorocarbon propellants are summarized below.

Propellant 12

Propellant 12 is the most generally used and the next-to-least expensive of the fluorocarbon propellants. It is nonflammable and has a low order of toxicity. Propellant 12 is very stable and can be used in practically all aerosol formulations except those with appreciable concentrations of caustic. Because of its high vapor pressure, Propellant 12 is normally combined with Propellants 11 and 114, which act as vapor pressure depressants. Propellant 12 is a poor solvent.

Propellant 11

Propellant 11 is the next most commonly used propellant and is the least expensive. It has too high a boiling point and too low a vapor pressure to be an effective propellant by itself and is normally used in combination with Propellant 12. It is nonflammable, low in toxicity, and sufficiently stable in nonaqueous systems to be used in most aerosol products. Although it can react with alcohols, this is not a serious problem and can be solved, when it does occur, by the use of a suitable inhibitor.

Propellant 11 will hydrolyze in products formulated as oil-in-water emulsions and packaged in metal containers. Therefore it cannot be used for such products as window cleaners and starch sprays. However, in water-in-oil emulsions it is sufficiently stable to be used in aqueous-based room deodorants, insecticides, and similar products.

Propellant 11, with a kauri-butanol value of 60, is an excellent solvent and ranks next to Propellant 21, with a kauri-butanol value of 102. Fairly high concentrations of Propellant 11 are sometimes used because of its high solvent power.

Propellant 114

Propellant 114 is another commonly used propellant. It is nonflammable, has a low order of toxicity, and is extremely stable. It can be used in practically all aqueous and nonaqueous systems. Propellant 114 has very little odor and is preferred for aerosol colognes and perfumes. It is somewhat more expensive than either Propellant 12 or 11. Propellant 114 is often used in combination with Propellant 12 in aqueous systems where the use of Propellant 11 would be inadvisable because of instability.

Propellant 22

Propellant 22 is used extensively in the refrigeration industry but very little in aerosols. It has too high a pressure to be used by itself and is somewhat more expensive than Propellant 12, 11, or 114. It is unstable in aqueous alkaline solutions. Combinations of Propellant 22 with 11 produce mixtures with high solvent properties and reasonable pressures. Propellant 22 may find limited use in special applications where these properties are particularly desirable.

Propellant 21

Propellant 21 is used to only a limited extent in aerosols. It is somewhat more expensive than Propellant 12, 11, or 114 but is the best solvent in the fluorocarbon series, with a kauri-butanol value of 102. For this reason, it has been of continuing interest to the pharmaceutical industry, where high solvency for active ingredients is important. It is nonflammable, is low in toxicity, and has a fairly low vapor pressure. Propellant 21 is stable in neutral or acidic aqueous systems but not in alkaline systems.

Propellant 115

Propellant 115 is very stable, has an extremely low order of toxicity, and has been approved by the FDA for use as a food propellant. It has a high vapor pressure and is used as a compressed gas or as a liquefied gas in combination with Propellant C-318, which acts as a vapor pressure depressant. Its application in nonfood aerosols is limited because of its relatively high cost. Propellant 115 is nonflammable.

Propellant 152a

Propellant 152a has a relatively high vapor pressure and costs more than Propellant 12, 11, or 114. This has tended to limit its use in aerosol products. Propellant 152a has some special properties, however, which may result in increased application of this propellant in the future. It has a density of less than 1 and can be used to prepare propellant blends with densities close to the value for water if the proper proportions of other fluorocarbons are employed. Propellant 152a also appears to produce less stable foams in some systems than the more commonly used propellants and has attracted attention because of this property.

Propellant 142b

Propellant 142b is similar to Propellant 152a. It costs more than Propellants 12, 11, or 114 and has a density slightly greater than 1.

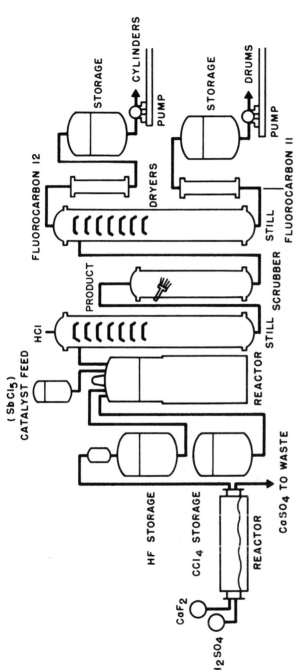

Figure 6.7. Typical flow diagram for Fluorocarbon 12-Fluorocarbon 11 manufacture.

129

Propellant 142b also has the property of producing foams of limited stability in some systems and has been used in quick-breaking foam formulations. It is used to only a limited extent in aerosols.

MANUFACTURING PROCESSES

Most of the commercial processes for preparing the fluorocarbons are based on the Swartz reaction, that is, the substitution or addition of fluorine to a suitable hydrocarbon or chlorinated hydrocarbon, using pentavalent antimony as the catalyst. The starting material (e.g., CCl_4 or $CHCl_3$) and hydrogen fluoride are fed continuously into a reaction vessel containing antimony. A small amount of chlorine is also added to keep the antimony in the pentavalent state.

Propellants 12 and 11 are prepared from carbon tetrachloride:

$$2CCl_4 + 3HF \rightarrow CCl_3F + CCl_2F_2 + 3HCl$$

A typical diagram for these two propellants is shown in Figure 6.7. The mixture of propellants is taken continuously from the top of the reaction vessel into a fractionating column, where by-product hydrogen chloride is removed. The propellants are scrubbed to remove traces of acids and are fractionated. The product is dried with silica or alumina and stored until delivery to the consumer (1).

Propellants 21 and 22 are obtained from chloroform:

$$2CHCl_3 + 3HF \rightarrow CHCl_2F + CHClF_2 + 3HCl$$

The reaction conditions for the preparation of Propellants 12 and 11 or Propellants 21 and 22 can be varied to produce any desired ratio of the products.

Propellant 114 is prepared from perchloroethylene and chlorine:

$$CCl_2{=}CCl_2 + 2Cl_2 + 4HF \rightarrow CClF_2CClF_2 + 4HCl$$

Propellant 152a is prepared from acetylene by the addition of hydrogen fluoride:

$$CH{\equiv}CH + 2HF \rightarrow CH_3CHF_2$$

Propellant 142b is obtained from Propellant 152a by chlorination:

$$CH_3CHF_2 + Cl_2 \rightarrow CH_3CClF_2 + HCl$$

Hydrocarbon Propellants

HISTORY

Hydrocarbons were suggested as aerosol propellants as early as 1933 by Rotheim (43), who disclosed the use of isobutane for spraying lacquers. In 1937 Iddings (44) mentioned ethane, propane, and ethylene as potential propellants for spraying liquid insecticides, mothproofers, and other such products. Goodhue, Fales, and McGovern (45), in an extension of their work on insecticides, evaluated a number of liquefied gases, including butane and propane, as propellants and published their results in 1945.

Goodhue, in 1947, worked on developing methods for preparing and blending commercial quantities of the pure hydrocarbons. During the following year, Shepherd evaluated samples of hydrocarbons from the Phillips Petroleum Company as propellants for various aerosols. In 1950 the Colgate-Palmolive Company investigated the use of P-46, an isobutane/propane blend, as a propellant for aerosol shave creams (46).

GENERAL PROPERTIES

The most common hydrocarbon propellants are propane, isobutane, and n-butane. They are stable and essentially odorless and have boiling points and vapor pressures in about the same range as the fluorocarbon propellants. Their major advantage, in comparison with the latter, is their low cost. The main disadvantage of the hydrocarbon propellants is their flammability; adequate precautions must be observed in transportation, storage, and loading.

The hydrocarbon propellants are used extensively in water-based products such as shaving lathers, window cleaners, starch sprays, room deodorants, and space insecticides. The concentration of propellant in these products usually is sufficiently low so that the hazard due to propellant flammability is minimized. Hydrocarbon propellants are also used in combination with the nonflammable fluorocarbon propellants in a variety of propellant blends.

PHYSICAL PROPERTIES

The physical properties of the hydrocarbon propellants are listed in Table 6.14. Two grades of hydrocarbon propellants are available—an aerosol grade designed for general applications, and a food grade for food and pharmaceutical uses (49).

TABLE 6.14. PHYSICAL PROPERTIES OF THE HYDROCARBON PROPELLANTS (18, 19, 47, 48)

Property	Propane	Isobutane	*n*-Butane
Formula	C_3H_8	C_4H_{10} (iso)	C_4H_{10} (*n*)
Molecular weight	44.1	58.1	58.1
Boiling point (°F)	−43.7	10.9	31.1
Vapor pressure (psig at 70°F)	110.0	30.4	16.5
Liquid density (g/cc at 70°F)	0.50	0.56	0.58
Flammability in air			
Lower limit (vol %)	2.2	1.8	1.9
Upper limit (vol %)	9.5	8.4	8.5
Flash point (°F)	−156	−117	−101

The boiling points of the hydrocarbon propellants vary from -43.7 to 31.1°F, with vapor pressures ranging from 16.5 to 110 psig at 70°F. A propellant with any desired vapor pressure between the two limits can be obtained by using mixtures of the hydrocarbons. The vapor pressures of blends of the propellants over a range of temperatures are illustrated in Figure 6.8 (49). The range of vapor pressures provided by the hydrocarbons corresponds roughly to that of the fluorocarbon propellants.

The hydrocarbon propellants have low molecular weights. Therefore, for the same propellant weight, the hydrocarbons produce a greater volume of gas when vaporized than do the fluorocarbons. The hydrocarbons all have densities less than 1. The fluorocarbon propellants, with their higher densities, permit a greater weight of product to be packaged in a given volume than is possible with the hydrocarbons.

The hydrocarbons are relatively poor solvents, with kauri-butanol values of about 25. They thus fall between Propellant 12 and Fluorocarbon 113 in solvent characteristics on the basis of the kauri-butanol values.

The hydrocarbon propellants are extremely flammable. The potential hazard introduced by the use of flammable propellants must be considered in formulating aerosol products. The degree of safety of petroleum hydrocarbons in aerosols has been discussed by Fiero and Johnsen (50). One of the main problems occurring with the flammable propellants involves their handling and storage by the loader. Local regulations should be checked. Another potential hazard could arise

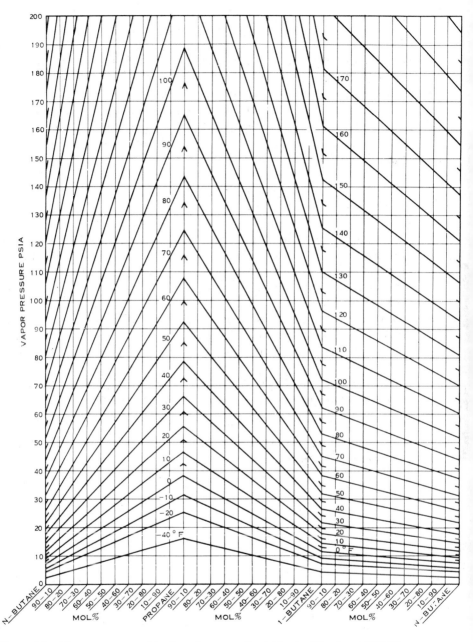

Figure 6.8. Vapor pressures of hydrocarbon propellant blends. (Courtesy of Phillips Petroleum Co.)

from leakage of propellant from aerosol containers in a closed storage area. This aspect has been discussed in detail by Johnsen (51) and Parks (52).

The hydrocarbon propellants have a low order of toxicity and are classified in Group 5b of the Underwriters' Laboratories (Table 6.11). The FDA considers propane and butane as substances generally recognized as safe for use where contact with foods occurs (49). The propellants for such applications must be very pure.

Miscellaneous Liquefied Gas Propellants

The two liquefied gas propellants in this group are vinyl chloride and dimethyl ether. Neither of these compounds is used alone as a propellant in aerosol products in the United States, but both are components of propellant blends. Vinyl chloride is also reported to be under investigation as a propellant component in aerosol products in Japan (53). Vinyl chloride was suggested as an aerosol propellant by Rotheim (43) in 1933 and by Iddings (44) in 1937. The use of dimethyl ether as an aerosol propellant was disclosed even earlier in a patent by Rotheim (54) in 1931.

Vinyl chloride and dimethyl ether are very flammable and cost more than the hydrocarbons. This is one reason why their use as propellants

TABLE 6.15. PHYSICAL PROPERTIES OF VINYL CHLORIDE AND DIMETHYL ETHER (56–59)

Property	Dimethyl Ether	Vinyl Chloride
Formula	CH_3OCH_3	$CH_2{=}CHCl$
Molecular weight	46.1	62.5
Boiling point (°F)	−12.7	7.9
Vapor pressure (psig at 70°F)	63.0	34.0
Flammibility in air		
Lower limit (vol %)	3.4	4.0
Upper limit (vol %)	18.0	22.0
Flash point (°F)	−42	..
Kauri-butanol value	91	58.0
Density (g/cc at 70°F)	0.66	0.91
Toxicity (Underwriters' Laboratories classification)	...	4–5

has been limited. However, they are better solvents than the hydrocarbons and are used in blends with the fluorocarbons as propellants for coating compositions.

Both propellants are quite stable. However, vinyl chloride may polymerize if not properly inhibited. Kubler (55) reported that under certain conditions vinyl chloride was more stable to hydrolysis than Propellant 114.

The physical properties of the two propellants are given in Table 6.15. Vinyl chloride has a vapor pressure close to that of Propellant 152a, while dimethyl ether has a vapor pressure slightly lower than the value for Propellant 12/11 (50:50). On the basis of the kauri-butanol values, dimethyl ether is an excellent solvent and approaches Propellant 21 in this respect. Vinyl chloride has a kauri-butanol value close to that of Propellant 11 and has similar solvent properties. Both vinyl chloride and dimethyl ether have densities less than 1.

Blends of Nonflammable and Flammable Propellants

Blends of nonflammable and flammable propellants were known more than 25 years ago. The development of aerosol insecticides by Goodhue and Sullivan in World War II and the resulting demand for the aerosols led to a projected shortage of Propellant 12. In an attempt to extend the supply, Jones and Scott at the U. S. Bureau of Mines investigated blends of Propellant 12 with propane and butane (46). In 1945 Goodhue et al. (45) compared the efficiencies of aerosol insecticides incorporating blends of Propellant 12 with various hydrocarbons and dimethyl ether.

The fluorocarbon/hydrocarbon blends were developed initially in an effort to provide a propellant mixture that cost less than the fluorocarbons alone but retained most of the nonflammability characteristics of the latter. One of the first blends of any commercial significance was Propellant A, developed by Reed (60) in 1956. Initially, the term "Propellant A" was used by Reed to designate a range of compositions of Propellant 12/11/isobutane blends with varying pressures. However, it has now come to indicate a blend with a composition of Propellant 12/11/isobutane (45:45:10). Johnsen (61) has mentioned that Propellant B, with a composition of Propellant 12/11/propane (47.5:47.5:5), was not accepted to any extent by industry.

Blends of the fluorocarbons with vinyl chloride or dimethyl ether were developed to provide propellant mixtures that were less expensive than the fluorocarbons and had good solvent properties. Both vinyl

TABLE 6.16. PROPERTIES OF SOME COMMON PROPELLANT BLENDS (66–69)

Propellant Blend	Vapor Pressure (psig at 70°F)	Density (g/cc at 70°F)	Flammability	Kauri-Butanol Value
Propellant 12/propane (91:9)	82	1.15	Nonflammable	18–25
Propellant 12/vinyl chloride (80:20)	66	1.21	Nonflammable	25
Propellant 12/vinyl chloride (65:35)	61	1.13	Flammable	33
Propellant P (Propellant 12/11/dimethyl ether) (75:10:15)	62	1.21	Nonflammable	31
Propellant A (Propellant 12/11/isobutane) (45:45:10)	38	1.22	Nonflammable	...

chloride and dimethyl ether are excellent solvents. The physical properties and flammabilities of the fluorocarbon blends containing vinyl chloride and dimethyl ether have been investigated in considerable detail by Scott and his coworkers (62–64) and by Phillips and Holler (65). These blends are used primarily in coating compositions where their superior solvent properties compensate for the fact that they are somewhat more expensive than blends with hydrocarbons. The properties of some common blends are given in Table 6.16, and their uses and properties are summarized below.

Propellant 12/Propane (91:9)

This mixture has been used to some extent in paints and other coating compositions. It does not form an explosive mixture with air when completely vaporized but will form flammable fractions during evaporation. The first fraction contains the highest proportion of propane and is the most flammable. The blend has a fairly high vapor pressure, but the solvents in paints and lacquers act as vapor depressants.

Propellant 12/Vinyl Chloride Blends

Vinyl chloride is a much better solvent than Propellant 12 and is less expensive. Blends of this type are commonly used in paints and lacquers. The Propellant 12/vinyl chloride (65:35) blend is quite flammable and forms explosive mixtures with air. The Propellant 12/vinyl chloride (80:20) blend is also used to some extent. It does not form explosive mixtures with air but will form flammable fractions during evaporation.

Propellant A

One of the first commercial blends, this has about the same vapor pressure as Propellant 12/11 (50:50) and is used in some products as a substitute for this propellant mixture because it is slightly less expensive. Propellant A is a somewhat poorer solvent than Propellant 12/11 (50:50), and occasionally the direct substitution of Propellant A for the 12/11 (50:50) blend has caused problems for this reason. Propellant A does not form explosive mixtures with air but will form flammable fractions during evaporation.

Propellant P

Propellant P is used as a paint propellant. Dimethyl ether is an excellent solvent and increases the solvent properties of Propellant 12/11 mixtures. The blend forms explosive mixtures with air.

COMPRESSED GAS PROPELLANTS

Introduction

The compressed gas propellants—carbon dioxide, nitrogen, and nitrous oxide—are essentially odorless, tasteless, low in toxicity, and inexpensive. Their pressures change relatively little with temperature in comparison to the liquefied gases. These properties have made them of interest to the aerosol industry.

According to Teter (70), carbon dioxide was used for aerosol beverages and foods as far back as 1869. The use of nitrous oxide for charging various liquids was disclosed in a patent in 1870. Nitrous oxide and carbon dioxide are used extensively today in beverages and foods such as whipped cream. In the aerosol industry, however, the use of compressed gases is minor compared to that of the liquefied gas propellants and generally is confined to products for which a coarse spray is acceptable.

In the decade from 1950 to 1960, the use of nitrogen as a compressed gas propellant was greeted at first with considerable enthusiasm by the aerosol industry (71–74). Nitrogen was inexpensive, low in toxicity, and stable. However, the disadvantages of nitrogen soon became apparent, and the enthusiasm declined rapidly. The unpopularity of nitrogen results mostly from the fact that it is almost completely insoluble in the liquid phase of aerosols. This introduces a number of problems.

TABLE 6.17. PHYSICAL PROPERTIES OF THE COMPRESSED GAS PROPEL-
LANTS (76, 77)

Property	Carbon Dioxide	Nitrous Oxide	Nitrogen
Formula	CO_2	N_2O	N_2
Molecular weight	44.0	44.0	28.0
Boiling point (°F)	-109^a	-127	-320
Vapor pressure (psig at 70°F)	837	720	477^b
Flammability	Nonflammable	Nonflammable	Nonflammable
Toxicity (Underwriters' Laboratories Classification)	5	...	6

[a] Sublimes.
[b] At critical point, -233°F.

Today nitrogen is used very little and, as Johnsen has said, "... has
fallen out of vogue as an aerosol propellant" (75). Carbon dioxide and
nitrous oxide are much more soluble in liquids than nitrogen is and
have been accepted to a much greater extent.

Physical Properties

Some physical properties of the compressed gases are given in Table
6.17.

Advantages and Disadvantages

SOLUBILITY

Nitrogen is practically insoluble in liquids and therefore concentrates
in the vapor phase in aerosol products. As the product is discharged,
the volume of the vapor phase increases, with a consequent decrease in
the pressure exerted by the nitrogen. This causes a noticeable change
in spray characteristics. In addition, since practically all of the nitrogen is in the vapor phase, the initial volume of the vapor phase has to
be considerably larger than is required with a conventional liquefied
gas propellant in order to have sufficient nitrogen to discharge the
product completely. These are some of the disadvantages that caused
the decline, mentioned above, in the use of nitrogen. The same also

TABLE 6.18.   SOLUBILITIES[a] OF GASES IN ORGANIC LIQUIDS (78)

Liquid	Nitrous Oxide	Nitrogen	Carbon Dioxide
Water	0.6	0.016	0.82
Acetone	5.3	0.15	6.3
Ethyl alcohol	2.8	0.14	2.6
Amyl acetate	5.9	0.15	4.1

[a] Solubility in volumes of gas per volume of liquid at 25°C and 760 mm.

applies, although to a somewhat lesser extent, to carbon dioxide and nitrous oxide. One advantage of nitrogen was the fact that, since it was insoluble in the liquid phase, it did not alter the characteristics of the concentrate.

As mentioned above, carbon dioxide and nitrous oxide are much more soluble in liquids than nitrogen is. This difference is shown by the data in Table 6.18.

The solubilities of carbon dioxide and nitrous oxide in a variety of solvents at a pressure of 100 psig have been determined by Johnsen and Haase (75). The solubilities of compressed gases in fluorocarbon propellants have been reported by Dantzler, Holler, and Smith (79); some of the data from their paper are given in Table 6.19.

One of the major problems experienced with the compressed gas aerosols has been misuse by the consumer. Whenever the container is accidentally inverted and the valve actuated, the vapor phase is discharged instead of the liquid phase. Since most of the propellant is concentrated in the vapor phase, the propellant is lost and the rest of

TABLE 6.19.   SOLUBILITIES[a] OF COMPRESSED GASES
IN FLUOROCARBON PROPELLANTS (75, 79)

Propellant	Carbon Dioxide	Nitrous Oxide
114	12.1	13.0
12	3.05	4.9
11	10.5	16.6

[a] Solubility is expressed as mole per cent at a pressure of 90 psig and a temperature of 70°F.

the product cannot be obtained from the container. This problem, which was particularly severe with nitrogen aerosols because of the insolubility of nitrogen in the liquid phase, is alleviated somewhat with nitrous oxide and carbon dioxide because of their increased solubility in the liquid phase. The gas dissolved in the liquid phase provides somewhat of a reservoir of propellant in case the aerosols are accidentally inverted and discharged. The effect of misuse on product performance with aerosols formulated with compressed gases has been investigated by Hinn (74) and Webster (80).

Spray Characteristics and Uses

Since the compressed gases are soluble to only a limited extent in the liquid phase of aerosols, coarse sprays are produced. Mechanical breakup actuators are necessary. In most cases, the pressure drops noticeably during use, partially because the gas dissolved in the liquid phase is not released fast enough to maintain equilibrium between the liquid and vapor phases. This can result in a noticeable change in spray characteristics. Anthony (81) has reported that the drop in pressure of compressed gas systems during spraying can be minimized by the use of methylene chloride and methyl chloroform (1,1,1-trichloroethane) as the solvents in the liquid phase. These solvents have comparatively high solubility for nitrous oxide.

At the present time, a number of products for which the application requires only a coarse spray, are pressurized with compressed gases. Products of this type include window cleaners, furniture polishes, and

TABLE 6.20. PRESSURE CHANGES OF LIQUEFIED
AND COMPRESSED GASES WITH TEMPERATURE (8)

| Temperature (°F) | Pressure (psig) | |
	Compressed Gas	Propellant 12[a]
70	70.2	70.2
32	64.1	30.1
0	59.1	9.2

[a] Propellant 12 obviously would not be a very effective propellant at 0°F.

starch sprays. Two products now on the market, windshield deicers and tire traction sprays, perform satisfactorily only with compressed gases as propellants. Both of these products are kept in automobiles during the winter and must operate at low temperatures. Compressed gases are ideal for products of this type because their pressures change relatively little with temperature as compared to liquefied gases. This is shown in Table 6.20.

SOLVENTS

Introduction

The majority of aerosol products contain conventional solvents in addition to the liquefied gas propellants. These solvents perform a variety of functions in aerosols. In the first place, they act as cosolvents for the active ingredients and propellants. The liquefied gas propellants generally are poor solvents, and most of the active ingredients used in aerosol products are insoluble in the propellants. A cosolvent usually is necessary in order to obtain a solution of the active ingredients in the liquid phase of the aerosol.

Another very important function of solvents is to help modify the discharge properties of the aerosol. The liquefied gas propellants by themselves flash into vapor almost immediately after discharge, and this would produce very fine and dry particles of the active ingredients if no other solvents were present. Solvents evaporate much more slowly in air than do the propellants. By proper choice of the solvent and propellant or by adjustment of the propellant solvent ratio, it is possible to obtain a spray with almost any desired characteristic.

A third function of solvents is to reduce the vapor pressure of propellents. Quite often, for economic or other reasons, it is desirable to use a propellant that by itself has too high a vapor pressure. If the solvents are miscible with the propellant, they will reduce its vapor pressure. The extent to which the vapor pressure of the propellant is reduced depends on such factors as the concentration and molecular weight of the solvent and its vapor pressure.

Finally, in most aerosols formulated with fluorocarbon propellants, the addition of the solvent reduces the cost of the product. Most solvents are less expensive than the fluorocarbon propellants.

There is such a wide variety of aerosol products that a corresponding diversity of solvents used therein is to be expected. Included are alcohols, halogenated compounds, esters, ketones, aliphatic hydrocarbons, and aromatic hydrocarbons. Some of these are widely used, whereas other find only limited application. Because of the number of solvents utilized, it would be inappropriate in this chapter to list and discuss the properties of all of them. These properties are given in many reference books and other sources. Instead, the discussion is limited to the two classes of solvents most commonly used in aerosols, halogenated solvents and alcohols.

Halogenated Hydrocarbon Solvents

The solvents in this group are methylene chloride, methyl chloroform (1,1,1-trichloroethane), and Fluorocarbon 113. Methylene chloride is the most common solvent and has been used in such products as hair sprays, insecticides, room deodorants, mothproofers, artificial snows, paints, oven cleaners, and paint removers (82). Methyl chloroform is used to a lesser extent in hair sprays and household products. Fluorocarbon 113 finds limited application in degreasers for electronic equipment and in some household products.

GENERAL PROPERTIES

Some of the physical properties of the solvents are listed in Table 6.21. Propellant 11 has been included for comparison purposes, since methylene chloride and methyl chloroform are sometimes substituted for Propellant 11 in aerosol formulations. The compounds in Table 6.21 are listed in increasing order of their boiling points.

Solvent Properties

Methylene chloride and methyl chloroform are similar in their solvent properties, judging by their kauri-butanol values, and both are much better solvents than Propellant 11 or 113. Their solubility parameters show that they will attack elastomers and plastics in sufficiently high concentrations (30). Their effect on the elastomeric and plastic components in valves therefore must be considered. Concentrations as high as 10 to 15% apparently have been used in aerosol formulations without difficulty. Methylene chloride-resistant valves are available from most valve manufacturers.

Toxicity

Both methylene chloride and methyl chloroform have fairly low orders of toxicity, as indicated by the Underwriters' Laboratories Classification and their threshold limit values. However, they are definitely more toxic than Propellant 11. Since absorption of these compounds

TABLE 6.21. PROPERTIES OF SOME HALOGENATED SOLVENTS (83–88)

	Propellant 11	Methylene Chloride	Propellant 113	Methyl Chloroform
Formula	CCl_3F	CH_2Cl_2	CCl_2FCClF_2	CH_3CCl_3
Molecular weight	137.4	84.9	187.4	133.4
Boiling point (°F)	74.8	103.6	117.6	165.2
Liquid density (g/cc at 70°F)	1.48	1.32	1.57	1.34 (68°F)
Kauri-butanol value	60	136	31	124
Solubility parameter	7.5	9.5	7.2	8.4
Flammability	Nonflammable	Nonflammable	Nonflammable	Nonflammable
Toxicity (Underwriters' Laboratories classification)	5a	4–5	4–5	4–5[a]
Threshold limit value (ppm)	1000	500	1000	350

[a] Estimated.

TABLE 6.22.   HYDROLYSIS RATES IN WATER[a]

Compound	Saturation Pressure at 122°F in the Presence of Steel
Methylene chloride	55
Fluorocarbon 113	40
Propellant 11	28
Propellant 12	10
Propellant 114	3

[a] Grams of compound hydrolyzed per liter of water per year.

through the skin may produce the same effects as inhalation, therefore excessive contact with the skin should be avoided. Contact of the vapor with the eyes is particularly undesirable. Propellant 113 appears to be somewhat less toxic than either methylene chloride or methyl chloroform, judging by the TLV values.

STABILITY

Both methylene chloride and methyl chloroform are unstable in the presence of oxygen, heat, light, and metals. Most commercial products contain stabilizers. Methylene chloride is somewhat more susceptible to hydrolysis than is Propellant 11 (84) or Propellant 113, as shown by the hydrolysis rates in Table 6.22. Several other propellants have been included for comparison.

According to Kubler (82), methyl chloroform has about the same hydrolytic stability as Propellant 11. Care should be taken to remove excess water from methyl chloroform, however, since the wet compound can cause corrosion even at room temperature (85).

REACTION WITH ALCOHOLS

The free-radical reaction between Propellant 11 and ethyl alcohol, which leads to the formation of acetaldehyde, hydrogen chloride, and Propellant 21, was discussed earlier. Similar experiments with methylene chloride indicate that the reaction does not take place with this compound. Propellant 113, however, reacts with ethyl alcohol via a free-radical reaction, similarly to Propellant 11.*

Archer (89) investigated the reactivities of various systems containing combinations of methyl chloroform, ethyl alcohol, iron, tin, and water. Under some conditions, there was a reaction between methyl chloroform and ethyl alcohol which gave the reduced chlorohydrocarbon, 1,1-dichloroethane, and the saturated chlorinated dimer. Traces of acetaldehyde were also found. However, Archer concluded from his work that the use of methyl chloroform and ethyl alcohol mixtures in tin-plate containers should not produce a serious corrosion problem as long as the concentration of water is not above 0.1%.

Alcohols

Ethyl alcohol is the most commonly used organic solvent in aerosol formulations. It is a component of hair sprays, deodorants, some antiperspirants, and a variety of other cosmetic and pharmaceutical products. Ethyl alcohol is low in toxicity and relatively inexpensive and is the best cosolvent for the liquefied gas propellants.

There are several points to remember, however, about the use of ethyl alcohol in aerosols. The free-radical reaction that can occur with Propellant 11 or 113 has been discussed. A second point is that the vapor pressures of mixtures of ethyl alcohol with the liquefied gas propellants are higher than the calculated values, because ethyl alcohol molecules are associated in solution and the actual average molecular weight of ethyl alcohol is higher than is indicated by the monomeric formula C_2H_5OH. This has been shown by Flanner and Risacher (90), who have published the vapor pressures of a variety of propellant/solvent blends.

The spray from alcohol/liquefied gas propellant mixtures can be quite cold, and this chilling effect has been investigated by Broderick and Flanner (91) and Dunne (92). Broderick and Flanner reported that use of the following could reduce the chilling effect of propellant/ethyl alcohol mixtures:

1. High or low concentrations of alcohol.
2. A mechanical breakup actuator.
3. Vapor tap valves.
4. Propellants 12 and 114.
5. A high ratio of Propellant 12 to 114.

* Unpublished data, E. I. du Pont de Nemours & Company, Freon® Products Division Laboratory.

REFERENCES

1. R. C. Downing, Simple Organic Compounds Containing Fluorine, in J. S. Sconce, (Ed.), *Chlorine, Its Manufacture, Properties and Uses*, American Chemical Society Monograph 154, Reinhold Publishing Corporation, New York, 1962, p. 864.

2. S. Young, *J. Chem. Soc.*, **39,** (1881), 489.

3. R. C. Downing, History of the Organic Fluorine Industry, in Kirk-Othmer, *Encyclopedia of Chemical Technology*, Vol. 9, 2nd ed., John Wiley and Sons, New York, 1966, p. 704.

4. T. Midgley, Jr., *Ind. Eng. Chem., 29* (1937), 239.

5. T. Midgely, and A. L. Henne, *Ind. Eng. Chem.,* **22,** (1930), 542.

6. L. D. Goodhue and W. N. Sullivan, U. S. Patent 2,321,023.

7. *American Standard Number Designation of Refrigerants*, ASHRAE Standard 34-57 and ASA Standard B79.1, 1960.

8. E. I. du Pont de Nemours and Company, *Vapor Pressure and Liquid Density of Freon® Propellants,* Freon® Aerosol Report FA-22, April 1970.

9. E. I. du Pont de Nemours and Company, *Properties and Applications of the Freon®* Fluorinated Hydrocarbons, Freon® Technical Bulletin B-2, March 1971.

10. E. I. du Pont de Nemours and Company, *Propellant Solutions Containing Freon® 22 Chlorodifluoromethane,* Freon® Aerosol Report FA-7, December 1960.

11. E. I. du Pont de Nemours and Company, *Thermodynamic Properties of Freon® 115 Monochloropentafluoroethane,* Freon® Technical Bulletin T-115, April 1959.

12. E. I. du Pont de Nemours and Company, *Properties of 1,1-Difluoroethane (DFE) CH_3CHF_2,* Freon® Technical Bulletin D-62, April 1960.

13. E. I. du Pont de Nemours and Company, *Food Propellant Freon® 115,* Freon® Aerosol Report A-63, January 1967.

14. E. I. du Pont de Nemours and Company, *Properties of FC-142b, 1-Chloro-1,1-Difluoroethane CH_3CCLF_2,* Freon® Technical Bulletin D-60, April 1960.

15. E. I. du Pont de Nemours and Company, *Physical Properties of Freon® C-318 Perfluorocyclobutane,* Freon® Technical Bulletin B-18B, February 1963.

16. E. I. du Pont de Nemours and Company, *Freon® 21 Aerosol Solvent and Propellant,* Freon® Aerosol Report FA-28, April 1968.

17. A. Herzka, *International Encyclopaedia of Pressurized Packaging (Aerosols)*, Pergamon Press, New York, 1966.

18. P. A. Sanders, *Principles of Aerosol Technology*, Van Nostrand Reinhold Company, New York, 1970.

19. H. R. Shepherd, *Aerosols: Science and Technology*, Interscience Publishers, New York, 1960.

20. E. I. du Pont de Nemours and Company, *Surface Tension of the Freon® Compounds,* Freon® Technical Bulletin D-17, January 1967.

21. E. O. Haenni, et al., *Ind. Eng. Chem., 51* (1959), 685.

22. H. F. Coward and G. W. Jones, Limits of Flammability of Gases and Vapors, *U. S. Bur. Mines Bull.* 503, 1952.

23. *Genetron Aerosol Propellants*, General Chemical Division, Allied Chemical and Dye Corporation, 1955.

24. E. I. du Pont de Nemours and Company, *Kauri-Butanol Number of Freon® Propellants and Other Solvents,* Freon® Aerosol Report FA-3, October 1969.

25. E. I. du Pont de Nemours and Company, *Solvent Properties Comparison Chart,* Freon® Aerosol Report FA-26, May 1968.

26. J. Hildebrand and R. Scott, *The Solubility of Nonelectrolytes,* 3rd ed., Reinhold Publishing Corporation, New York, 1949.

27. H. Burrell, *Offic. Dig. Fed. Paint Varnish Prod. Clubs, 27* (1955), 726.

28. H. Burrell, *Offic. Dig. Fed. Paint Varnish Prod. Clubs,* (1957), 1159.

29. E. I. du Pont de Nemours and Company, *Solubility Parameters,* Freon® Technical Bulletin D-73, October 1969.

30. E. I. du Pont de Nemours and Company, *Effect of Freon® Compounds on Plastics,* Freon® Technical Bulletin B-41, May 1966.

31. J. A. Brown, *Soap Chem. Spec.,* **36,** (March 1960), 87.

32. H. M. Parmlee and R. C. Downing, *Chem. Spec. Mfr. Assoc. Proc. Ann. Meet.,* 1950, p. 45 (Freon® Aerosol Report A-19).

33. E. I. du Pont de Nemours and Company, *Hydrolysis of Freon® Refrigerants in Aqueous Systems,* Freon® Technical Report D-74.

34. J. M. Church and J. H. Mayer, *J. Chem. Eng. Data, 6,* **3,** (July 1961), 449.

35. P. A. Sanders, *Soap Chem. Spec.,* **41,** (December 1965), 117.

36. P. A. Sanders, *Soap Chem. Spec.,* **36,** (July 1960), 95.

37. F. A. Bower and L. J. Long, *Soap Chem. Spec.,* **37,** (August 1961), 127.
    1961, p. 53.

38. Underwriters' Laboratories Report MH-2256 (1931), MH-2375 (1933), MH-2630 (1935), MY-3134 (1940), MY-3072 (1941), MH-3135 (1940).

39. E. I. du Pont de Nemours and Company, *The Freon® Compounds and Safety,* Freon® Technical Bulletin S-16.

40. J. Wesley Clayton, Jr., *Fluorine Chem. Rev., 1,* **2,** (1967), 197.

41. J. Wesley Clayton, Jr., Toxicity, in P. A. Sanders (Ed.), *Principles of Aerosol Technology,* Van Nostrand Reinhold Company, New York, 1970.

42. F. T. Reed, *Amer. Perfum. Cosmet., 77,* (October 1962), 48.

43. E. Rotheim, U.S. Patent 1,892,750 (1933).

44. C. Iddings, U.S. Patent 2,070,167 (1937).

45. L. C. Goodhue, J. H. Fales, and E. R. McGovern, *Soap Sanit. Chem., 21,* (April 1945), 123.

46. G. F. Ford, Private communication.

47. H. Franz and G. F. Ford, Hydrocarbon Propellants, in A. Herzka (Ed.), *International Encyclopaedia of Pressurized Packaging (Aerosols),* Pergamon Press, New York, 1966.

48. *Thermodynamic Properties of Refrigerants,* ASHRAE, 1969.

49. *Phillips Hydrocarbon Aerosol Propellants,* Phillips Petroleum Company, 1969.

50. G. W. Fiero, and M. A. Johnsen, *Aerosol Age,* Part I, **14,** (September 1969), 22; Part II, **14,** (October 1969), 59; Part III, **14,** (November 1969), 48.

51. M. A. Johnson, *Aerosol Age, 9,* (July 1964), 18.

52. G. Parks, *Aerosol Age, 7,* (May 1962), 31.

53. H. Iosake, *Aerosol Age,* **3,** (February 1958), 22.

54. E. Rotheim, U.S. Patent 1,800,156 (1931).

55. H. Kubler, *Aerosol Age,* **3,** (September 1958), 26.

56. *The Isotron Blends,* Isotron Technical Information Files, Vol. 2, No. 1, Pennwalt Chemicals Company.

57. *Vinyl Chloride Monomer,* Technical Bulletin, Dow Chemical Company, 1954.

58. P. Dyson, *J. Soc. Cosmet. Chem.,* **14,** (1963), 402.

59. *Matheson Gas Data Book,* The Matheson Company, 1965.

60. W. H. Reed, *Soap Chem. Spec.,* **32,** (May 1956), 197.

61. M. A. Johnsen, *Aerosol Age,* **8,** (May 1963), 29.

62. R. J. Scott and R. R. Terrill, *Soap Chem. Spec.,* **38,** 2 (February 1962), 130 and **38,** (january 1962) 142; *Aerosol Age,* **7,** (January 1962), 18.

63. R. J. Scott and R. R. Terrill, *Soap Chem. Spec.,* **38,** (February 1962), 130 and **38,** (January 1962) 142; *Aerosol Age,* **7,** (January 1962), 18.

64. R. J. Scott and R. C. Werner, *Soap Chem. Spec.,* **42,** (December 1966), 190.

65. T. W. Phillips and F. C. Holler, *Chem. Spec. Mfr. Assoc. Proc. Mid-Year Meet.,* (1936), p. 46.

66. *Genetron/Vinyl Chloride Aerosol Propellants,* Genetron Product Information Bulletin, General Chemical Division, Allied Chemical Company.

67. *New Ucon Paint Propellant,* Technical Bulletin, Union Carbide Chemical Company.

68. *Ucon Propellant A,* Technical Bulletin, Union Carbide Chemical Company.

69. T. M. Hartley, *Drug Cosmet. Ind.,* **91,** (December 1962), 707.

70. C. A. Teter, *Aerosol Age,* **1,** (August 1966), 16.

71. J. Kalish, *Drug Cosmet. Ind.,* **81,** (October 1957), 441.

72. A. R. Marks, *Aerosol Age,* **3,** (December 1958), 64.

73. F. A. Mina, *Drug Cosmet. Ind.,* **82,** (March 1958), 321.

74. J. S. Hinn, *Chem. Spec. Mfr. Assoc. Proc. Mid-Year Meet.,* 1961, p. 53.

75. M. A. Johnsen and F. D. Haase, *Soap Chem. Spec.,* Part I **41,** 7 (July 1965), 73, Part II, **41,** (August 1965), 93.

76. F. T. Reed, Propellants, in H. R. Shepherd (Ed.), *Aerosols: Science and Technology,* Interscience Publishers, New York, 1960.

77. R. C. Webster, Compressed Gases, in A. Herzka (Ed.), *International Encyclopaedia of Pressurized Packaging (Aerosols),* Pergamon Press, New York, 1966.

78. R. C. Webster J. S. Hinn, and P. A. Lychalk, *Chem. Spec. Mfr. Assoc. Proc. Mid-Year Meet.,* 1963, p. 39.

79. E. M. Dantzler, F. C. Holler, and P. T. Smith, *Soap Chem. Spec.,* **41,** (January 1965), 125.

80. R. C. Webster, *Aerosol Age,* **6,** (June 1961), 20.

81. T. Anthony, *Aerosol Age,* **12,** (September 1967), 31.

82. H. Kubler, *Aerosol Age,* **5,** (February 1960), 49.

83. H. Kubler, Chlorinated Hydrocarbons, in A. Herzka, (Ed.), *International Encyclopaedia of Pressurized Packaging (Aerosols),* Pergamon Press, New York, 1966.

84. E. I. du Pont de Nemours and Company, *Properties of Methylene Chloride for Aerosol Applications*, Freon® Technical Bulletin AM-1, November 1969.

85. *Du Pont Chlorinated Hycrocarbon Solvents*, Technical Bulletin, Electrochemicals Department, E. I. du Pont de Nemours and Company, 1966.

86. *1,1,1-Trichloroethane, Stabilized (Methyl Chloroform)*, Du Pont Technical Information, Electrochemicals Department, E. I. du Pont de Nemours and Company, 1966.

87. A. Herzka and J. Pickthall, *Pressurized Packaging (Aerosols)*, 2nd Ed., Butterworths Scientific Publications, London, 1961.

88. A. E. Schober, *Soap Chem. Spec.*, **34**, (August 1958), 65.

89. W. L. Archer, *Aerosol Age*, **12**, (August 1967), 16.

90. L. T. Flanner and R. L. Risacher, *Soap Chem. Spec.*, **40**, 1 (January 1964), 108.

91. G. F. Broderick and L. T. Flanner, *Soap Chem. Spec.*, **41**, (June 1965), 98.

92. T. F. Dunne, *Aerosol Age*, Part I, **4**, (May 1959) 36; Part II, **4**, (December 1952), 52.

STEEL AEROSOL CONTAINERS

Before an aerosol package is marketed, a great deal of thought and effort has to be expended. Beginning with a product concept, the steps required in getting a product to the market include formulation, choice of container style and decoration, materials, method of fabrication, and so on. Specifications, compatibility testing procedures, and, finally, filling criteria must also be established. These steps are very much interrelated, however, and all must be considered during the completion of each step to minimize the total effort required in getting a product to market.

PRODUCT CONCEPT

Product concept is completed when the requirements are satisfied that the product has application, is practical, and has good market and profit potential in view of existing and/or anticipated competition. Before these requirements are satisfied, however, all of the other steps have to be considered to determine profitability, since most affect market potential and competitiveness. The point that we are trying to make is that the container must be an early consideration, and container suppliers should be consulted in the product concept stage for maximum efficiency.

PRODUCT FORMULATION

Product formulation can be defined in a number of ways, including "that which is filled into the container" or "that which exists inside the

151

container." The total contents of the container include the materials of construction of container and valve and the entrapped air, as well as the propellant and product. All of these materials influence one another, and the influence of each is further affected by age and temperature. It should also be mentioned that additives not originally considered may become necessary, such as inhibitors, chelating agents, preservatives, antioxidants, buffers, emulsifiers, and antifoaming agents.

CHOICE OF CONTAINER STYLE AND DECORATION

Selection of container style and decoration is very much a marketing decision. It should be made, however, with the advice and consent of the package engineering and quality-control groups. There can be conflicts between the structural materials required for a particular container style and the materials needed for product compatibility. Certain styles or lithography decorations may encourage or enhance the presence of blemishes, scratches, or dirt incurred during fabrication, conveying, shipping, or whatever handling is necessary, whereas other designs reduce the occurrence of such esthetic defects or tend to hide them. Container styles may also conflict with the desired fill and the desired filling line efficiency. The following description of styles, together with the fabrication methods and construction materials required, should clarify these considerations.

Stylized Containers

Stylized containers are those that have been significantly changed from their originally fabricated, straight-wall cylinder shape. If the modified shape is to be a full circle, the soldered three-piece can cannot be considered. The maximum diameter of a stylized can may exceed the diameter of the double seams and require special conveying and packaging to avoid scratching and denting. The stretching of the body and the internal organic coating may require the addition of a sprayed coating for corrosive or metal-sensitive products. Stretched tin plate loses reflectivity, affecting the brilliance of transparent lithography.

Embossed Containers

Such containers, having an emboss depth of 0.008 in., should perform inside and outside essentially the same as a straight-wall can. Embossing is generally restricted to three-piece cans, since it is performed after lithographing but before fabrication when the plate is in the flat.

Necked-In Containers

These units, having outside double-seam diameter equal to or less than body outside diameter, are prone to lithography abrasion during conveying and shipping. There is no effect of necking-in on the internal organic coating. A necked-in top double seam permits an uninterrupted straight-line contour of body and overcap. Also, the smaller-diameter ends on a necked-in can permit higher internal pressures, since the strength of a can is inversely proportional to the end diameter.

Straight-Wall Containers

These are the more commonly used and least expensive containers. Straight-wall, as well as necked-in, cans may be soldered, cemented, welded, or drawn and wall-ironed.

The types of steel plate currently available for the various types of body fabrication are as follows:

Fabrication	Components	Tin Plate	Tin-Free Steel
Soldered	3-piece	X	
Cemented	3-piece		X
Welded: Conoweld	3-piece		X
Welded: Soudronic	3-piece	X	
Drawn and wall ironed	2-piece	X	

Lithography

Lithography is an extremely imprtant part of any package. It is used, in conjunction with an attractive overcap, to persuade the consumer to select the package in question from among the many competitive labels surrounding it on the market shelf. Every imaginable color is available in both opaque and transparent lithogrphy inks. Half-tone printing can exactly duplicate the most beautiful of color photographs. Texturized finishing varnishes can give the container a surface appearance and feel resembling fabric or leather over all or part of the lithography surface. Since the registration of many colors, especialy in half-tone design, is critical, plate decorated in the flat, before fabrication, produces the highest-quality decoration.

CHOICE OF CONTAINER MATERIALS

These materials include steel and the various metallic and organic materials that protect surfaces and provide seals.

Steel

Steel is the subject of this chapter and, therefore, the first material to be discussed. Steel has many desirable qualities that make it an attractive packaging material. It is readily available, strong, and relatively inexpensive. It also is magnetic, a quality that assists in conveying during manufacture and filling and, equally important, in separating steel containers from solid waste for recycling. Steel also is a reactive material and readily oxidizes. Therefore surface protection of steel by the use of other metals, chemical passivation, organic coatings, or a combination of these is required.

Alloys

Alloys of iron are numerous, and scores of steels exist, each suiting a specific need. Three types of steel may be found in aerosol cans, but one type, identified as MR, is by far the most widely used for this purpose. The other steels of potential use in aerosol cans are identified as MC and D. Elemental analyses of these alloys are given in Table 7.1.

Alloying elements found in can-making steel include the following:

Carbon:	Imparts hardness, reduces ductility.
Manganese:	Contributes to strength and hardness, reduces ductility.
Silicon:	Contributes to brittleness and reduces corrosion resistance.
Phosphorus:	Contributes to hardness, reduces ductility and corrosion resistance.
Sulfur:	Reduces ductility and is detrimental to surface quality.
Copper:	Reduces corrosion resistance.

TABLE 7.1.   STEEL-PLATE CHEMISTRY

| Constituent | Type of Plate | | |
	MR (%)	MC (%)	D (%)
Carbon	0.13 max.	0.13 max.	0.13 max.
Manganese	0.60 max.	0.60 max.	0.60 max.
Silicon	0.01 max.	0.01 max.	0.02 max.
Phosphorus	0.02 max.	0.02–0.15	0.02 max.
Sulfur	0.05 max.	0.05 max.	0.05 max.
Copper	0.20 max.	0.20 max.	0.20 max.

BASE BOX

This is the standard unit of area used in the steel and steel container industries. A base box is the area of 112 sheets of plate measuring 14 in. × 20 in. or 31,360 in.²(1). This odd area was determined in the day of manual conveying, when 112 sheets of medium-size plate was considered a "package" for convenient handling.

BASE WEIGHT

Base weight refers to the weight of steel per base box. The weight of tin per base box is over and above the base weight. For example, 112

TABLE 7.2.   STEEL-PLATE BASE WEIGHT TABLE

Weight (lb/base box)	Theoretical Thickness (in.)
70	0.0077
75	0.0083
80	0.0088
85	0.0094
90	0.0099
95	0.0105
100	0.0110
107	0.0118
112	0.0123
118	0.0130
128	0.0141
135	0.0149

sheets measuring 14 × 20 in. (a base box) of 100-lb plate would weigh 100 lb. If the plate were also No. 25 (0.25-lb) electrolytic tin plate, the total weight would be 100.25 lb. This example also illustrates how little tin is used to make tin plate. If specific gravity differences are ignored, the steel thickness is 400 times that of tin. However, the tin is applied to both sides of the steel, and therefore No. 25 100-lb plate has a tin surface 1/800 as thick as the steel core.

The relationship of base weight to thickness is given in Table 7.2. The thickness of plate of known base weight is calculated by multiplying the base weight by 0.00011; the result is the thickness in inches (1).

Surface Protection

To avoid atmospheric corrosion some surface protection is required. Such protection can be achieved by means of metal plating, chemical surface treatment, or organic coating.

TIN PLATING

This was the predominant method in use through the 1960s. The strength requirements of aerosols, especially for the side seam, which is the weakest area of the can body, dictate that solder flow throughout the side seam be complete, that is, be apparent throughout the length inside the can, and be relatively free of voids. Metal plating of the steel with tin, therefore, provides steel protection, good solderability, and an attractiveness that is quite often utilized to enhance lithographed decoration.

Electrolytic Tin Plating

This is the process predominantly used to tin-plate steel since World War II. The electrolytic method produces a more consistent coating than the older hot-dipped method, a more efficient use of tin, and a more economical product for end uses that require less tin coating than the minimum of 1.25 lb per base box available with hot dipping. Electrolytic tin plate is readily available in tin coating weights of 1.35, 1.00, 0.75, 0.50, 0.25, and 0.10 lb per base box.

Differentially Coated Tin Plate

This type of plating was introduced in the 1960s. It reduces the cost of plate, since a heavy tin coating is required on one side but not on the other. A surface marking is incorporated on all differentially

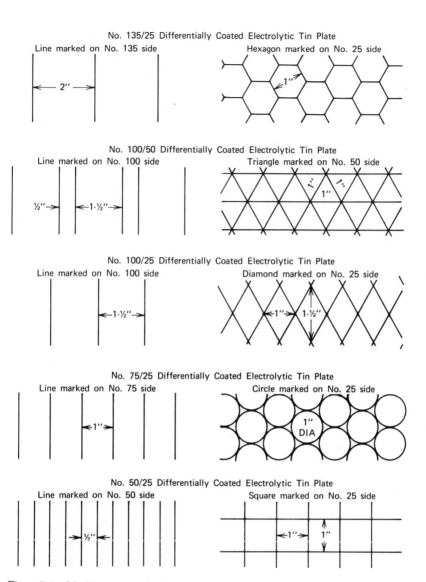

Figure 7.1. Marking system for electrolytic tin plate (2).

coated tin plate to identify the combination used. The identifying marking is placed on the surface destined to be the can interior of the can. Figure 7.1 illustrates the identifying markings used on differentially coated electrolytic tin plate.

Hot-Dip Tin Plate

Hot-dip tin plate is steel coated by passing through a bath of molten tin. Tin weights from 1.25 lb per base box and higher were available. The hot-dip process, however, was inefficient, expensive, less consistent in tin coating weight than electrolytic plating, and restricted the use of lithography inks to the opaque type, since the grainy surface hindered the appearance of transparent inks.

Other Metal Plating

This includes the lead coating of steel known as Terne plate, which is of limited availability, and little, if any, was ever used in aerosol can manufacture. Zinc-coated steel or galvanized plate is not used in the aerosol can industry. Research in the application by vapor deposition of other metals to steel, such as aluminum, has been conducted recently, and such methods may become available in the near future.

Side-Seam Considerations and Bonding

Chemically treated steel is used in the fabrication of aerosol containers, using the welding or cementing methods of side-seam bonding. Although this steel is not tin plated, it is produced in a steel plant's tin mill and therefore was given the name tin-free steel or TFS. For aerosol use, TFS is surface treated chemically to prevent corrosion caused by handling or atmospheric conditions. The most common chemical surface treatment in use today is an electrolytically deposited layer of chrome-chrome oxide.

Methods of side-seam bonding currently in use include soldering, cementing, and welding. Solder has very low tensile strength, and therfore the typical side-seam construction found in food cans is not used in the aersol industry. The low tensile strength of the solder would permit the solder to creep and the side seam to unfold. Instead aerosol cans employ an interrupted lock and lap seam construction. The lock portion resembles that found in food cans, is four layers of plate thick, and holds the shape of the can during soldering. The lap portion is two layers thick and covers a sizable area of overlap. The thin layer of solder in the lap portion is under a shearing force, and solder has high shear resistance. Therefore the lock and lap seam provides sufficient strength so that soldered cans can meet 2P and 2Q specifications. The lap area on soldered aerosol side seams can be found both inside and outside the can; the inside lap is stronger.

SOLDER

Solder used in aerosol cans is of two basic types, lead and tin. Each of these may be obtained in specially alloyed states to provide extra strength for withstanding higher than normal internal pressures.

Lead

This is the most commonly used solder. It has fairly good creep resistance to normal pressure ranges and is less expensive to use than tin solder. Its composition is 98% lead and 2% tin. The small amount of tin enhances its ability to wet the surface of tin plate. A tin content beyond 7% would begin to reduce the creep resistance of lead solder. Some tin content increase may result because of pickup from the tin plate, but not enough to effect performance.

Lead solder for high-pressure systems is a slightly modified 2/98 solder. Silver is used to replace 1/2% of the tin content, resulting in a solder with exceptionally good creep resistance. Obviously the addition of silver increases the cost of the container slightly.

Tin

Solder having a composition of 100% tin has poor creep resistance and is seldom used. It also is significantly more expensive than lead. Tin solder for high-pressure systems is alloyed with either silver or antimony; 1/2% silver or 2% antimony provides strength similar to that of high-pressure lead solder. The cost of high-pressure tin solder is essentially the same as that of 100% tin, and therefore high-pressure tin solder has all but eliminated the demand for 100% tin.

Test packing that includes the use of high-pressure storage at 120°F and above should incorporate the use of high-temperature solders even if 2/98 or 100% tin is satisfactory commercially. Many a can has ruptured before the completion of accelerated testing. To prevent lost time but primarily to avoid accidents, one should consider high-pressure solder when testing. It must also be borne in mind that test cans are not always hot tanked before being stored; therefore a defective can requiring nothing more than a little agitation to explode may be sitting on the shelf.

CEMENT

Since the sealing of aerosol side seams with cement is a relatively new venture, the materials that may evolve as aerosol cements are unpredictable. Currently nylon is under test. Nylon has performed with excellent results as the cement in cans for beer and carbonated beverages, the internal pressures of which closely approach those of aerosols.

WELDED

Welded side seams provide the strongest bond. The welded area is 1.7 to 2 times thicker than the cylinder wall and is composed of the same material fused together. It, therefore, is the strongest area of the can.

ORGANIC PROTECTIVE COATINGS

Coatings of this type are used for a number of reasons.

1. To minimize metallic ion pickup that may cause bleaching of colored ingredients, the adverse formation of color by reacting with ingredients, the degradation of perfumes, or possible skin irritation in the users of pharmaceuticals, or may affect the solubility of potentially unstable solutions or alter in any other manner the efficacy of the product.

2. To minimize the surface of the cathodic area, thus lowering the driving force of potential corrosion and reducing the corrosion rate. The result is increased shelf life.

3. To minimize the surface of the anodic area sufficiently so that adhering products of corrosion can effectively protect any exposed metal that remains. It should be mentioned here that minimizing the anodic surface area sometimes reduces shelf life by concentrating a given reaction rate at a point of metal exposure that is not protected by its own reaction product.

4. To minimize reaction product formation when the potential exists to form large, insoluble particulate matter capable of valve clogging.

5. To provide surface protection from atmospheric corrosion during the period between manufacture and filling only.

6. To provide lubrication in the fabrication of a metal having a surface with a high coefficient of friction.

Types of Organic Coatings

The materials commonly used in the organic coating of aerosols include epoxies, vinyls, phenolics, organosols, alkyds, and acrylics.

EPOXY

Epoxy protective coatings are synthetic resins formed by the condensation reaction of epichlorohydrin and bisphenol and dissolved in esters, ketones, and special chlorinated solvents (2). Curing requires heat to vaporize the solvent and to complete polymerization. Epoxy films fabricate well, have good adhesion, and have excellent resistance to certain solvents and chemical solutions in addition to acids and moderate alkalies.

VINYL

Protective coatings made of vinyl are synthetically formed by the polymerization of vinyl acetate and vinyl chloride. Vinyls do not cure, but heat is usually employed to accelerate solvent vaporization. Vinyls fabricate well and produce tough, hard films with good chemical resistance, although they have poor resistance to certain solvents, especially ketones. Boiling water tends to "fog" or "blush" vinyl films, thus restricting their use to containers not exposed to high temperatures. Metal contact may darken and high dry heat (above 400°F) can char vinyls; hence they should be used predominantly in the top coat of two-coat systems and not in too close proximity to side seams, where temperatures may reach 700°F during soldering or higher during welding. Spray application after aerosol body formation is completed is the ideal method of taking advantage of the properties of vinyl.

PHENOLIC

Phenolic coatings are synthetic resins and are the polymerization products of phenol and formaldehyde. The properties of the phenolic resin or condensate can be developed by controlling the reaction conditions and interrupting the reaction at a point where the resin is still soluble (2). Curing requires heat to vaporize the solvent and to complete polymerization. Phenolic resins are thermosetting and are very inert to chemicals and solvents. Phenolics, however, are very brittle and therefore fabricate poorly. They also have poor resistance to strong alkalies.

Because of their poor fabrication resistance, phenolics are seldom used for aerosol end units without first being modified with epoxy or vinyl resins to improve their fabrication characteristics while retaining their excellent solvent and chemical resistance.

ALKYD

Alkyd protective coatings are ester polymers or polyester resins formed by the reaction of dibasic acids, such as phthalic anhydride, malic anhydride, and fumaric acid, wth polyhydric alcohols, such as mannitol, sorbitol, glycerol, and glycol. Curing is accomplished by heat to vaporize the solvent and to polymerize and oxidize the resin. Alkyds are thermosetting resins, are insoluble when cured, and form a very durable coating. The most frequent application of alkyds, however, is for outside container coatings.

ORGANOSOL

Organosol coatings used in steel aerosols are a blend of polymers that may include epoxy, phenolic, acrylic, and others, with the predominant polymer being polyvinyl chloride (PVC). Organosols differ from plastisols in that they contain solvent, and from the more common protective coatings in that they do not dissolve in the solvent. Organosols are dispersions, and therefore much higher solids are possible without the normal increase in viscosity expected with solutions of polymers. Dispersions permit thicker films. Cross lingking occurs between polymers during curing, providing improved physical properties. Pigment may also be incorporated to enhance the physical properties. Heavy pigmented organosol coatings, therefore, provide a degree of protection against corrosive products that the more common coatings do not. Organosols having a large percentage of PVC have excellent fabrication resistance but unfortunately are as heat sensitive as vinyls.

ACRYLIC

Acrylic coatings of the thermoplastic type are sometimes used to provide the necessary lubricity for tin-free steel, which has a surface with a high coefficient of friction. Without such a coating to replace the missing tin plate, mobility through the fabrication process would be sluggish and tool wear would be excessive. The acrylic coating is insoluble in alkali and many solvents, making it ideal for oven cleaner and paint aerosol cans, where tin is unnecessary or undesirable. Thermosetting acrylics are seldom used for the internal coating, but are used externally with lithography as a base or finishing varnish.

Application of Coating

Organic coatings may be applied to steel or tin plate predominantly by the roll coating of flat metal stock for the manufacture of three-piece aerosol cans. The spraying of fabricated bodies of three-piece cans is sometimes employed over a roll coat when low metal exposure is a requirement. Two-piece steel cans that require coating must be sprayed with organic coating after fabrication.

When roll-coating flat stock is to be used in the fabrication of can bodies, the coating must have registration to provide a bare metal surface for the portion that forms either soldered or welded seams. The roll that applies the coating, therefore, must be cut in a pattern that corresponds to the irregularities of the type of side seam to be used. Since some of this bare surface may be exposed after body fabrication, it is frequently necessary to repair the exposed area with a side stripe, that is, a coating sprayed over the side-seam area during the period of forming and sealing. The more popular coatings, such as epoxies and phenolics, should be placed close enough so that they come into contact with the solder as it flows into and through the side seam, or up to the area that is to be surface cleaned for welding. Vinyls, however, are so heat sensitive that their margins must be located far enough from the side-seam area that the heat used to seal the body does not burn or char the vinyl. Since vinyls are used only as top coats, the wide vinyl margin does not leave exposed plate, and the side stripe should have a spray pattern wide enough to provide the second coat between the vinyl margins.

Side-Seam Stripes

Side stripes can be sprayed on the side seam either before or after the soldering or welding takes place. Since solders containing predominantly tin are applied at around 475°F, and those containing predominantly lead at around 700°F, and since welded seams exceed 1000°F, the striping material applied before heat sealing must be proportionately heat resistant and capable of proper cure under such conditions. Obviously a different material is required for each application. These materials may differ only in the solvents employed, or in the absence of solvents, not necessarily in the resin used, when applied before heat sealing. If applied after the side seam is heat sealed, however, a different resin is usually employed in each application unless additional heat is incorporated to provide cure.

When side stripes are applied before side-seam sealing, the heat generated in sealing is also sufficient to cure or polymerize the organic stripe. When side stripes are applied after the sealing of the side seam, the residual heat is sufficient only to drive off solvents and is inadequate to polymerize. Therefore pre-heat-sealing stripes are epoxy or phenolic types, and post-heat-sealing stripes are vinyl types of protective coatings. On the surface, it would appear that pre-heat-sealing stripes would be preferable, and, admittedly, polymerized coatings have advantages over nonpolymerized ones. Post-heat-sealing stripes can be applied in a much heavier film weight, however, and where vinyl-type side stripes are compatible, the heavier film weight is an advantage.

Welded seams are striped after welding and do not require rapid cool down, such as that required by solder to set. Stripes on welded seams are, therefore, polymerized by maintaining high temperatures at the side seam for a short duration after application.

MANUFACTURE OF THE AEROSOL CONTAINER

Container manufacture begins with steel plate, which is shipped from the mill as a bundle of precut sheets or as plate wound in a coil. The steel has definite specifications that depend on its end use. They include:

Chemical analysis	Surface Treatment
Thickness	Width (precut and coil)
Temper	Length (Precut only)

Coiled plate usually is cut before subsequent operations. Plate to be converted into ends is scroll cut to avoid waste. The dimensions of the cut sheet are dependent on the container size and range up to 40 in. in either width or length. Organic coating protection and lithography, if required, are carried out before further reduction of sheet size.

Fabrication of Ends

Ends are formed from plate that is further scroll cut to a width of approximately 1.75 times the cut-edge diameter of the disk required to form the end. From this sheet usually two disks at a time are punched out and formed. They are then curled and compound lined, and the finished ends sent to the container assembly area. Bottoms are usually formed by a single punch into a die. Domes or specially adapted ends may require two to ten die punch operations.

USE OF SEALING COMPOUNDS

"Compound" is the term used for the rubber-base material applied to end units and put under compression in the double seam to afford a seal. Compound is usually composed of synthetic rubber, resin, and filler, and occasionally plasticizers or other modifying ingredients. Compounds must be compatible with the aerosol contents, and they must be FDA approved when used with foods and some pharmaceuticals.

Whenever test packs are made, a portion of the cans should be weighed initially and then at various stages of the evaluation to determine weight loss. Should excessive weight loss take place, the can manufacturer should be consulted for a determination of cause. If a defective seal is responsible, the extent of the defect must be determined and the test repeated if necessary. If incompatibility exists, it is necessary to change the compound or reformulate the product and then retest. If permeation is the cause and is due to the nature of the particular formulation one may wish to consider:

Substitution of the permeable ingredient.

Increased concentration of the permeable ingredient to compensate for the permeation.

Adjustment of the declared label weight.

Some of the stronger solvents, such as methylene chloride, have the property of permeating compounds and gaskets. Such weight loss is usually very slow, but over a year or two it can be significant. Some solvents cause compound or gasket degradation during permeation, and hence their permeation rate is not constant but accelerates. Thorough evaluation of such weight loss is, therefore, necessary.

Bodies

Bodies are formed from plate that is slit and trimmed into the blank size required. Different amounts of steel are needed to make the same size can by different fabrication processes.

FOR TWO-PIECE STEEL CANS

Two-piece steel cans have their bodies formed by first drawing flat sheet into a cup and ironing out the wall beyond the desired height and then trimming back to the desired height. The end contour is established in the drawing operation. If the integral end is the dome, it has a neck that must be perforated and curled to form the opening for valve attachment.

FOR THREE-PIECE SOLDERED STEEL CANS

Three-piece steel soldered cans have their bodies formed by first notching and folding the body blank edges, after which the blank is wrapped around a mandrel that expands to interlock the folded edges and is then hammered tight to form the side seam. After edge folding but before mandrel wrapping, flux is introduced to clean the tin-plate surface for proper solder wetting. The cylinder is then preheated in the side-seam area and passed parallel over a roll rotating in a bath of solder, placing an excess of solder at the outside side-seam fold. The solder flows into the seam by capillary action. After soldering, the side seam is kept hot until the excess solder is wiped off and then is rapidly air cooled to a temperature low enough to cause solidification of the solder.

FOR THREE-PIECE WELDED STEEL CANS (CONOWELD)

These cans have their bodies formed by first abrasively removing the chemically treated surface in the area to be welded, then wrapping around a mandrel and spot welding to maintain proper metal overlap while in a cylindrical shape, and finally forge welding continuously along the entire length of the side seam. This provides a seam of approximately 1.7 times the plate thickness, having a surface easily repaired with a side stripe.

FOR THREE-PIECE WELDED STEEL CANS (SOUDRONIC)

These cans have their bodies formed by rolling the body blank around a small cylinder to preset roundness to less than the final diameter. The rounded blank is supported at the side-cut edges by a Z-bar that positions the edges in the proper overlap position as the blank moves through the welding process. The weld is nonforge, therefore, a higher-temperature is required. The temperature reaches the molten state of the steel permitting the welding of tin plate. However, the entire overlap area is not bonded as in the Conoweld process. This results in plate surface area within the side seam overlap that is difficult to organically protect. It also provides a channel into the double seam that is difficult to plug with compound.

THREE-PIECE CEMENTED STEEL CANS

Three-piece cemented steel cans first undergo a cement application to one edge of the body blank. The body blank is then passed through burners, heating the cement and both side edges. The blank is then wrapped around a mandrel, and the seam bumped by a refrigerated hammer, thus settling the cement.

Flanging

Flanging of each open end occurs after cylinder formation by any of the processes described above. The flange maintains cylinder roundness during conveying, provides ready end engagement, and permits easy and consistent body hook formation during double seaming.

Double Seaming

Double seaming, the operation that attaches an end unit to a flanged body cylinder, can be performed in a variety of ways. The most common method is with the cylinder axis vertical and with the cylinder and end rotating together. The cylinder is conveyed into a double seamer and positioned on the base plate of one of probably six stations. The end is conveyed to a position above the cylinder. The base plate rises, the cylinder engages the end, and all rise into the seaming chuck, which contacts the end in the countersink area (Figures 7.2 and 7.3). Compression of few hundred pounds is created to assist in forming the flange into a body hook and to provide sufficient friction for the mechanically rotated chuck to rotate end, cylinder, and base plate without slippage. Two seaming rolls move one at a time into contact with the curl of the rotating end, as shown in Figure 7.2. The first-operation seaming roll continues the curling of the end until the cut edge is well under the flange and the flange is bent downward, as seen in Figures 7.2 and 7.3a. The seam is now formed. The second-operation roll moves in and rolls the double seam tight; see Figures 7.2 and 7.3b. The areas of the first-and second-operation seam that must adhere to specific dimensions are shown in Figure 7.3. The dimensions depend on the type of seam (shape) and the material thickness.

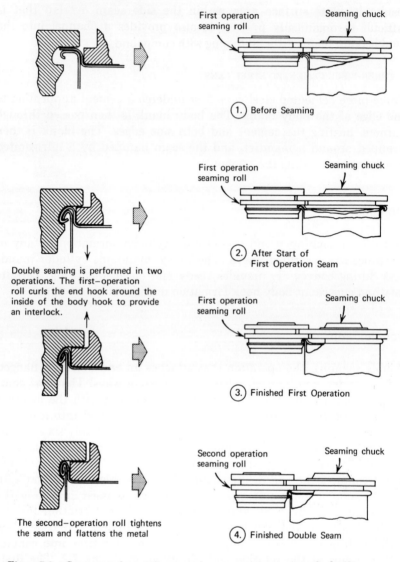

Double seaming is performed in two operations. The first-operation roll curls the end hook around the inside of the body hook to provide an interlock.

The second-operation roll tightens the seam and flattens the metal

1. Before Seaming

2. After Start of First Operation Seam

3. Finished First Operation

4. Finished Double Seam

Figure 7.2.   Sequence of operations seaming a can end onto a can body (3).

COMPARTMENTED AEROSOLS

These were introduced in the late 1950s, and today many different types exist. They were developed for the following reasons:

1. To separate product from propellant.
2. To separate product ingredients, such as catalysts or heating agents, unmixable before application.

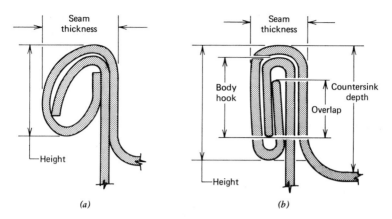

Figure 7.3.   Formation of double seam. (a) First operation; (b) second operation.

3. To provide a container for previously unpackageable corrosive products.

4. To provide a convenient dispenser for viscous products.

Types of compartmented aerosols include piston, bag-in-can, siphon, and codispensing systems. Containers used for the first three systems are discussed in detail. No special containers are required for codispensing systems.

Piston Type

The piston type of double-compartment can, developed and commercialized in the late 1950s, is more closely associated with drawn and wall-ironed aluminum aerosols. It is included in this chapter on steel aerosol containers, however, because all known double-compartment innovations can be coupled with two-and some three-piece steel containers.

Although few drawn steel aerosol containers are marketed today, the possibility exists that more will appear. Also the new bonding techniques for steel open the way to possible use of pistons in welded and cemented, as well as drawn steel containers. The piston container is restricted to the more viscous type of products that prevent propellant bypass of the piston. The product should be filled under pressure or pressurized after filling to force it down the side walls past the piston and thus create a propellant–product interface in th propellant chamber, rather than in the product chamber, to minimize propellant absorption and contamination of the product. Propellant or product permeation is not as acute a problem with the plastic piston as it is in some other types of double-compartment systems. The piston cans commercially available today are charged through a bottom grommet, which is then sealed with a rubber plug by means of specially designed equipment.

Bag-in-Can Type

This method of separating product from proepllant has been worked on, and patents have been issued for it, since the inception of the aerosol. Until the late 1960s, however, no technique of this type was a commercial success. The main problems causing earlier failures were that plastic bags collapsed above product level, separating product from valve, and metal bags formed rigid wrinkles, preventing total collapse and, therefore, total product removal. Today specially constructed and anchored plastic bags eliminate seal-off and either undergo controlled collapse or compensate for uncontrolled collapse, permitting 95 to 98% removal of product.

The bag-in-can method of separating product from propellant offers the ability to dispense a product of any viscosity as long as it will flow. Viscous semisolids, for example, can be dispensed as readily as an alcohol solution. One must remember, however, that the wall thickness of the plastic bag must be thinner than that of a common nonaerosol plastic bottle and that its surrface area is large, permitting potential permeation problems. It should also be borne in mind that the plastic bag-in-can aerosol is in an atmosphere of propellant and not of air. Experience with nonaerosol plastic containers, therefore, cannot serve as the sole basis for predicting product performance in such a double-compartment aerosol. The propellant may modify the ability of the plastic to resist permeation of various product components, and vice versa.

ADVANTAGES

One distinct advantage of the bag-in-can aerosol is that product components that readily permeate plastic bottles are controlled in the double-compartment aerosol, since permeating components are trapped in the propellant chamber and permeation eventually stops because of back pressure. Products can be formulated to take advantage of this fact.

PERMEATION

If the effect of the product on the plastic bag is ignored, propellant permeation generally can be classified as follows:

Least permeable:	Propellant 318
	Propellant 318/115
	Propellant 12
	Isobutane
Moderately permeable:	Propellant 12/114
Most permeable:	Compressed gases
	Propellant 12/11
	Propellant 11

Propellants can be incorporated into the products packed in bag-in-can aerosols to assist spray characteristics or to produce a latent foam. They must be soluble, however, and the vapor pressure differential across the plastic wall must be adequate to prevent the formation of a head space in the product compartment—otherwise, popping results during dispensing.

Bag-in-can aerosols, like piston aerosols, are charged through a bottom grommet and plugged, or through a special valve placed in the grommet by the container manufacturer.

Siphon Systems

Siphon-type aerosols, such as the Preval® and Inovair® systems, although adaptable to steel containers, are neither dependent on nor restricted to steel or other metal containers. This type really should be classified as parts of a double container, one placed within the other, rather than as a double-compartment single container. Therefore siphon systems should not be discussed at length in this chapter, although worthy of mention in the context of double-compartment aerosols.

The primary function of such a device is to cause, through siphon action, the dispensing of product from an attached chamber as the result of dispensing of propellant from the Preval® or Inovair® propellant chamber. Such a device eliminates the need for pressure-resistant product compartments and allows the use of any type of container to hold the product. Product viscosity, however, must be low.

CONTAINER SPECIFICATIONS

Physical

These specifications include size, capacity, strength, and any other attributes that now exist or may come into being. Pressure-relief mechanisms may become one such factor. External decoration and container shape are also physical specifications but have already been discussed.

SIZE IDENTIFICATION

This is accomplished by the external dimensions of double seam to double seam. The system of numbers represents inches and sixteenths of inchs. The first group of numbers describes the diameter; the second group, the height. Within a group of numbers the last two numbers represent sixteenths, and the numbers preceeding the last two represent inches. For example, a 211 × 604 is a can having a diameter of 2 11/16 in. and a height of 6 4/16 in. The height is not the overall height from the top of the dome to the bottm, but the height from the top of the double seam that seals he dome to the body, to the bottom of the double seam that seals the bottom to the body.

CAPACITY

The capacities of the various three-piece aerosol cans are listed in Table 7.3. Although the figures are not necessarily exact since slight variations may exist between aerosol manufacturers, they do serve as a guide.

CONTAINER STRENGTH

The strength of an aerosol can is dependent on time and temperature, as well as pressure. The weak areas of a can are the seams—top, bottom, and side. End seams (top and bottom) fail by unfolding. Before unfolding takes place, the end must first invert or buckle. Inversion or buckling occurs when the countersink area, that is, the recessed area of the end that contacts the seaming chuck, pops out. The chuck positions the end on the body and backs up the force of double seaming. Buckle resistance is a function of metal thickness, hardness, countersink depth, and end diameter. Buckle resistance increases when metal thickness increases, hardness increases, countersink deepens, and diameter diminishes. The minimum pressure at which an end should buckle is 140 psig at 130°F, in order to satisfy the Department of Transportation (DOT) requirements of no distortion for the nonspec-

TABLE 7.3. OVERFLOW CAPACITY OF THREE-PIECE AEROSOL CANS WITHOUT VALVE

Can Size	Fluid Ounces	Cubic Inches	Cubic Centimeters
202 × 200	3.4	6.0	98.3
202 × 214	5.0	9.0	147.5
202 × 314	6.8	12.2	199.9
202 × 406	7.7	13.9	227.8
202 × 509	9.9	17.8	291.7
202 × 700	12.5	22.5	368.7
202 × 708	13.4	24.2	396.6
207½ × 413	11.6	20.9	342.1
207½ × 509	13.4	24.1	395.5
207½ × 605	15.2	27.4	449.0
207½ × 701	17.1	30.8	504.7
211 × 413	13.7	24.7	404.7
211 × 510	16.1	29.0	473.6
211 × 604	17.8	32.1	526.0
211 × 612	19.2	34.6	568.6
211 × 713	22.3	40.2	658.7
300 × 709	26.9	48.5	795.0

ification aerosol can. However, in practice, buckle resistance will be found to be higher.

Unfortunately, differences do exist with regard to the type, temper, and thickness of steel used for aerosol domes and bottoms. These differences are the result of other differences in end design, fabricating equipment, and the end use of the aerosol can. Difficulty is encountered, therefore, when expected minimum buckle resistance is listed as a function of end diameter only. The following figures are, therefore, only a guide to what may be expected to be the minimum buckle of the ends on the majority of cans made.

Diameter	Minimum Buckle Resistance (psig)
202	190
207 1/2	180
211	170
300	160

Buckling is somewhat dependent on temperature. It takes possibly 10 psi less pressure to buckle ends at 130°F than at room temperature. Concave bottoms usually buckle at higher pressures than domes. The minimum bursting pressure of the same ends would be approximately 40 to 60 psi above their buckle levels. In DOT specification cans, changes in the design of the end or in the steel used may be required if the nonspecification properties do not also satisfy 2P and 2Q.

Side-seam strength is more dependent than buckle resistance on temperature and very dependent on time. For a short time and at room temperature, any solder will hold up to the 350 psi level before the side seam will burst. A temperature increase, however, will decrease solder tensile strength and shear resistance. These two terms are usually combined and called creep resistance. Side-seam strength is also dependent on the side-seam construction and the solder composition.

Since side-seam construction varies between manufacturers, and some solders do also, the following information is offered as a guide; it is not the result of testing. At a temperature of 130°F and a pressure of about 130 psig, the following times to failure by solder creep may be expected:

Solder	Side Seam	Time to Failure
100% Sn	Outside tab	Days
2% Sn/98% Pb	Outside tab	Weeks
100% Sn	Inside tab	Weeks
2% Sn/98% Pb	Inside tab	Months
High-pressure Pb solders	Inside tab	Year
High-pressure Sn solders	Inside tab	Months

If a can at 130°F is undisturbed at the time of failure, such failure is not violent; there is just a gradual relaxing of the seam, resulting from solder creep, until the contents vent.

PRESSURE-RELIEF MECHANISMS

Pressure-relief systems for aerosol cans have constituted a frustrating problem for the industry for some time. In the case of incineration of empty aerosol cans in household garbage, the prevention of violent bursts at very high temperature is unquestionably desirable. The incorporation of a pressure- or temperature-sensitive device to vent internal pressure under such circumstances becomes a source of concern, however, lest it function elsewhere with filled aerosol cans. The thought of prematurely vented paint and the material-damage potential, or of oven cleaner and the material-damage and human-hazard potential, or of flammables and the fires that they may start or the controllable fires that they may put out of control, worries the industry. Nevertheless, a number of different approaches have been investigated.

Fusible metal plugs have been tried and found to result in premature can failure due to corrosion caused by the dissimilar metal or failure to melt quickly due to heat dissipation by the contents. Plastic plugs have been tried and found to result in premature failure due to product attack of the plastic. A safety valve has been considered, but its effect on the retail price appears prohibitive.

Currently, attention is centered on two methods of accomplishing the same end result. One is a device that punctures the dome as, in reacting to internal pressure, it inverts or buckles. The other is a built-in weakness in the end that results in a venting action when the end buckles. The main shortcoming of these two methods in that violent rupture can occur through the side seam of soldered cans because of solder melting before internal pressures build up sufficiently to buckle the ends in a significant percentage of cans. This can be compensated for by the use of welded or two-piece drawn and wall-ironed metal aerosols. However, the industry's capacity is in the soldered can.

Actually, a practical need for a pressure-relief mechanism has not yet been demonstrated. It is true that such demonstrations can be made by intentional misuse of the container. But can the need for such a mechanism be convincingly established in view of the disadvantageous side effects? Certainly the industry will continue to investigate all aspects of this question.

Material

The materials used for aerosol cans for various product categories constitute a very difficult subject. Two formulations of a single type of product may be almost identical, but their compatibilities with a given container may differ drastically. Therefore some important general considerations in regard to materials are listed in Table 7.4 and are offered as a starting point. Several other important considerations, such as air removal, side-stripe application, physical effects on organic coatings, and product changes have been omitted from Table 7.4.

All products should have economic maximum air removal by means of vacuum crimping or purging after filling until proof is established that this procedure is not necessary. All organically protected cans should have side-seam striping until this is demonstrated to be unnec-

TABLE 7.4. CONTAINER MATERIAL SPECIFICATIONS

Product	Organic Protection	Solder	On Finding:	Correct with or Change to:
Water base pH 10–14	None	Lead	Head-space rust	Ammonia, morpholine, or equivalent
Water base pH 8–11	Single coat	Lead	Tin or lead corrosion	Increased air removal or tin-free steel
			Steel corrosion	Increased pH and/or inhibitors and elimination of organic protection
Water or alcohol base pH 5–10	Single coat	Lead	Underfilm detinning	Double-coat organic protection
			Solder attack	Tin solder
			No corrosion	No organic protection
Water or alcohol base pH 4–8	Double coat	Lead	Steel corrosion	Inhibitors or formula change
Alcohol base pH 3–5	Spray lined	Lead	Steel corrosion	Formula change
Detergent	Double coat	Lead	Steel corrosion	Formula change or spray-lined protection
Water-in-oil emulsion	None	Lead	Instability	Stability
Anhydrous	None	Lead	Corrosion	Single-coat organic protection or moisture reduction

essary, or solder exposure is proved to be beneficial. Lead solders provide container integrity regardless of content, but all products cannot tolerate lead. Therefore lead solder should be replaced by tin solder or solderless cans as the product dictates. Undesirable organic coating physical changes are not uncommon, and sometimes are not matters of serious concern. To eliminate such changes, however, the first approach is to try another type of coating or coating system. It is not uncommon to have a different system, with two or three components, on each of the several cans.

The subject of product changes is too broad for discussion here and is best solved by collaborating with informative authorities. In fact, such persons should be consulted in regard to any problems involving product–container incompatibility.

COMPATIBILITY TESTING

Testing is required because the aerosol can affects and is affected by its contents. Whether compatibility exists, therfore, is cause for concern, and the period of compatibility or shelf life and methods for determining this are important considerations with each new product.

Shelf Testing

Shelf testing or test packing is still the only reliable method currently available for determining the future performance of aerosol packages. Yet a shelf test has the shortcoming that it tells us only of performance under a single set of conditions. The ingredients of a product may be familiar materials, known individually to exist compatibly in an aerosol can. However, when they are blended together, compatibility does not necessarily continue to exist. Compatibility does not mean absolute inactivity—total inertness. Compatibility does mean that, if chemical activity exists, it does not affect the container or the end use of the product. Everything has impurities, and impurities influence compatibility.

If a shelf life of 2 years is desired, the shelf testing should last for at least 2 years. It is not necessary that 2 years' testing be completed before marketing commences, but that 2 years' testing be done. The shelf testing required before marketing is dependent on the product and the type of marketing to be done initially. Some products, such as shave creams, may be marketed simultaneously with the initiation of shelf tests; others, such as an antiperspirant solution, may require years of shelf study before even test marketing can be considered. Basic knowledge, experience, and legal considerations—all these influence timing.

The extent of the test pack, the variables, and their size and number should be determined by consolidating all the available advice of knowledgeable and experienced people and fitting it to the available facilities, manpower, and priorities. Products, facilities, and priorities vary considerably. Each product's testing should be conducted according to its own merits and not on the basis of some inflexible schedule. It is very important that more than one shelf test of a new product be made. If the first test was laboratory prepared, a scond formal shelf-life study should be conducted on a sampling of the first commercial run. Preferably the first test should be made on the line that will be used commercially. Subsequent formula modifications or supplier changes make new shelf studies mandatory. Each commercial run should be sampled and placed in storage, not for formal shelf tests, but for periodic checks and comparisons.

Absolutely everything involved varies. Product ingredients vary, impurities vary, nominally identical materials from different suppliers vary, containers vary, filling conditions vary, and marketing conditions vary. If, for example, a product has a potential to detin, the commercial performance must be studied to determine rate and degree. If the findings are unacceptable, further action with regard to product, container, or both is essential. A single laboratory test pack cannot always provide this information.

Responsibility for deciding whether the package specification or specifications satisfactorily permit the product to perform according to label claims throughout the required shelf life should be in the hands of one person. This person logically should be an employee of the organization that stands to suffer most should container or product fail. It follows, therefore, that the marketer should conduct the shelf studies if he has the facilities or else supervise the shelf studies wherever the facilities exist. By all means, the marketer should use the talents and abilities of all potential suppliers in setting up and evaluating the studies whenever and however possible, but he should control the testing, make the decisions, and closely follow commercial performance to guarantee success. The marketer should also control or have final approval over any subsequent changes in purchasing and filling, whether this is done inside or outside his own organization.

The quantity of cans that should be put under test depends on many factors. If several product variables are involved, all should be tested in the container representing the specifications that are expected to be successful. If many container specifications are to be considered, containers of all types should contain the formulation most likely to be marketed. Putting all possible formulations into containers representing all possible specifications, however, should be avoided. This is nothing more than a waste of manpower, materials, and possibly production time. If a product is not expected to exhibit a potential to perforate a container through pin-point iron corrosion, 24 cans per formula per container variable is sufficient. If perforations are expected, at least 100 cans should be added to each test pack. When "cuttings" or product–container evaluations are made, a minimum of two cans should be examined. Should the two cans differ in performance, it is desirable to examine a third can and proceed according to the majority. If a subsequent evaluation also produces a difference in performance between two cans, the best policy is to discontinue the test, make necessary changes, and then start again.

Storage Conditions

Storage conditions are an important consideration when conducting shelf studies, and they should reflect known or anticipated shipping, warehousing, and marketing conditions. Refrigerated storage must be included when applicable. Room-temperature storage is absolutely essential and serves as a basis of correlation for all elevated-temperature studies. Storage at 100°F, or thereabout, is very desirable because this condition frequently exists commercially for short periods and generally is equivalent to storage at about double room temperature.

High-temperature storage in the 110 to 140°F range has no value initially, but it should not be discouraged. If correlation exists between room-temperature and high-temperature storage, it can be an aid in subsequent tests of the same product after modifications have been made. The test pack should be placed in a hot water bath prior to storage. This will avoid injury from defective cans, should they exist.

Examination and Evaluation

Examinations of test packs should be made on a schedule consistent with time available, knowledge of past performance of same or similar formulations, and knowledge and experience of the examiner. If one has unlimited time, a feel for the product–container performance, and experience, the following schedule is suggested:

Temperature	1 Month	3 Months	6 Months	12 Months	24 Months
70°F		X	X	X	X
100°F	X	X	X	X	
130°F	X				

Evaluation of test packs involves two concerns: how long the can will hold the product, and how long the product will properly serve its purpose. When the can is affected by the product, the product in turn may be affected. Detinning of exposed tin plate, however slight, does not affect the can's ability to hold the product. However, a small amount of metallic salt in, for example, a topical antiseptic spray may cause irritation; therefore a can with large areas of exposed tin represents an unsatisfactory specification. Such a test pack should be immediately discontinued and remade to a different can specification. Slight discoloration of product due to physical or chemical changes caused by metal or protective coatings may affect neither container nor product. Knowledge of product alone or container alone makes total evaluation difficult. To be efficient, test pack evaluators should have experience or easy access to the experienced. Conducting test pack examinations is costly, and when a pack or a variable no longer provides information, it should cease to exist.

Physical changes in the container occur frequently and sometimes signal potential trouble. Such changes usually involve the protective organic coating. It is common for product components to permeate this coating. Such permeation is selective, and the rate is variable. As the result of permeation, the organic coating may soften, loosen, lose color, or discolor. Perfume components and materials capable of plasticizing often cause softening of the organic coating. If softening takes place, the permeation resistance of the organic coating to other materials will decrease.

Permeation of the product can sometimes result in reduction or loss of organic coating adhesion. If the propellant also permeates an organic coating, it can exaggerate a decrease in adhesion by expanding under the organic coating when the can is punctured; therefore, when rapid depressurization takes place before examination, blistering will occur. Under pressure, however, blistering would not exist. Alkaline ingredients can yellow phenolic components in coatings, and strong solvents can leach some of this coloration into the product.

Epoxys and vinyls are colorless materials and frequently are dyed for identification. The dyes are soluble in some solvents and may be leached into the prduct. These conditions can be overcome by changing the type of organic coating, incorporating a double coat of two dissimilar organic materials, or eliminating the dye. Frequently such physical changes cause no difficulty and can be toleraed. For peace of mind, however, it is suggested that, while tolerating such conditions, other specifications be tested in the hope of improving container compatibility.

Corrosion

Corrosion is chemical change. In an aerosol can, corrosion is the destruction of the metallic materials of construction. Corrosion generally is undesirable, except when it is sacrificial to steel without further deleterious effect. Corrosion is the result of oxidation-reduction reactions involving electron exchange and most frequently involving current flow, whether galvanic or not. Chemical corrosion usually would not be expected to cause container failure, but it can produce undesirable product or functional effects. Electrochemical corrosion is the type capable of causing the pitting of steel and subsequently perforating the container, permitting leakage.

Iron is a fairly reactive metal. The steel aerosol—whether two-piece or three-piece, whether tin plated or tin-free, whether soldered, cemented, or welded—is basically a piece of iron and should be expected to respond to its environment as iron would. Iron frequently is affected by tin and lead when electrically coupled, since these metals are below iron in the electromotive series and, more often than not, are cathodic to iron. Can-making steel is slightly more passive than iron, and as a result tin and lead are occasionally sacrifically anodic. On rare occasions aluminum and stainless steel are introduced for valve cups. Aluminum would be expected to be anodic, and stainless steel cathodic, to steel.

FACTORS AFFECTING CORROSION

Electrolytes

Electrolytes in solution and in contact with the container wall are required before electrochemical corrosion can occur. Water solutions and oil-in-water emulsions are obvious potential sources of corrosion problems. Anyhdrous solutions, especially anhydrous alcohol solutions, contain some moisture and should also be assumed to have the potential to corrode. Organic solvents with a low capacity to dissolve water are usually trouble-free; but when all the moisture contributed by all the ingredients exceeds the capacity of the solvent to dissolve water, a water phase, although small, settle out in the bottom of the can or on the can wall and corrosion should be expected.

Water-in-oil emulsions can also present problems. If the emulsion forms after filling, the oil phase should be filled first. If the emulsion is not stable, the water phase by itself must be noncorrosive. If the water phase is corrosive, the emulsion must be stable under all the environmental conditions that the package and its contents will experience.

Alkalies

Alkalies will attack tin, lead aluminum, and organic coatings. Solutions with pH values above 10 are best packed in the absence of these materials, that is, in steel alone. If they are packed in soldered tin plate, air removal and soldering quality must be maximized. Lead solder is preferred over tin.

Acids

Acids, acidic salts, and hydrolyzable organics are difficult to hold and require maximum protection in the form of organic coatings on the metal surface, minimum head-space air, and preferably inhibiters to reduce corrosion potential.

Impurities

Impurities may be present in any ingredient. Their presence can frequently catalyze or inhibit corrosion. Copper impurities of a few parts per million will cause rapid perforation of steel in an acid solution that otherwise may be passive.

Air

The presence of air in an aerosol container is very often critical. If head-space air could be essentially eliminated, products having pH values from 3 to 14 would be considerably easier to deal with, even without organic coating protection of metal. Aerosol cans with a 1-in. opening and large head-space requirement prevent efficient air removal on an economic basis. Commercial aerosol filling equipment when operating efficiently will remove approximately 70% of head-space air at speeds of 300 cans per minute or less. However, this evacuation precedes propellant filling. After propellant filling, the final head space may actually have compressed air, that is, more air than would exist under atmospheric pressure. Therefore this is one of the difficulties involved in satisfactorily packaging strong acids capable of perforating and strong alkalies capable of detinning and attacking lead solder.

Nevertheless, the removal of air is helpful, and the investigation of better ways to accomplish air removal from aerosols should be encouraged. Occasionally the effect of head-space air can be minimized by the use of low-density propellant that will float on the concentrate, insulating the effect of air on product corrosivity.

MECHANICS OF CORROSION

These can vary as ingredients and conditions vary and therefore are difficult to describe. Nevertheless, many facts that may be helpful in understanding corrosion can be mentioned here.

Steel corrodes, or oxidizes, at the anode when corrosion is electrochemical. The anode is usually found where metal exposure exists—it is ordinarily in the lower portion of the can furthest from the head-space air, and, where a choice exists, in an area more worked during fabrication than other areas. The area that best satisfies these three conditions is the side seam of the can body where it meets the bottom double seam. If the anode covers a large area, pitting of the steel may be found in the bottom seam round the entire circumference of the can, and possibly one third of the way up the side seam. Any damaged area exposing steel, away from the seams but in the same general area, would be expected to also undergo corrosion. The bottom end is not electrically insulated from the can body and therefore would also have steel anodic areas. They usually are in the double seam. The bottom end is one third to one half thicker than the body and hence is less prone to perforate first.

Steel corrosion that is widespread will usually show evidence of its activity by producing orange or brown discoloration. The amount of discoloration may be proportional to the amount of steel corroded, but it has no bearing on the degree to which corrosion is penetrating the plate and eventually perforating. Some products are capable of perforating a can very cleanly; there is only a single point of steel attack— no detinning around the perforation, no obvious corrosion product, no V-shaped excavation, just a tunnel. Such perforations are difficult to find even with the aid of a microscope, and fortunately corrosion of this type is rare.

Occasionally steel corrosion may occur all over the can, or in the head space only. Such corrosion can be electrochemical and therefore capable of perforating, but the chance that it will not perforate is better than with bottom-area corrosion.

Avoiding steel corrosion is possible. Obviously changes have to be made in either the can or the product, or in both. There are many ways to avoid steel corrosion; and as products differ, the practical approaches also differ. Chemical corrosion of steel can be minimized by increased tin coating, increased organic protective coating, increased product or vapor pH, or decreased head-space air.

Electrochemical corrosion of steel requires a more thorough investigation of causes and effects before a logical approach to eliminate the corrosion can be taken. The anodic area is usually obvious; the cathodic is not. Both of these areas should be defined, and the relationship of surface area to reaction rate determined. If the surface area of the cathode is proportional to the rate of reaction, the cathodic area must be reduced. Of course, the method used to reduce the cathodic area probably will also reduce the anodic area. If reducing the cathodic area does not lower the reaction rate, the reduced anodic area may perforate sooner because of the restricted surface to corrode. Therefore an off-hand, apparently logical solution may prove to be just the opposite.

The elimination of tin or lead may eliminate the cathode. The introduction of tin or lead may provide a sacrificial anode. Of course, in using a sacrificial anode the reaction rate of this substituted corrosive action must be determined if tin or lead pickup by the product is detrimental. Preferably, the corrosion rate with the sacrificial anode will be minimal, and the corrosion product insoluble, so as to minimize product effects.

In electrochemical corrosion, chemical changes are occurring at both the cathode and the anode. If these products of corrosion are not physically or chemically removed from the electrode surface, they will insulate the surface and reduce the corrosion rate. This phenomenon is called polarization (4).

How does one go about determining cathodic areas, rates of corrosion, polarization effects, and everything else that is necessary to solve corrosion problems? In the past, the vast majority of these problems have been solved by trial and error through test packing. The test pack includes a number of container variables and the variables cover comparisons of lead versus tin solder, tin plate versus tin-free steel, and organically coated versus uncoated plate, plus variations in the thickness of tin and organic coatings, quite a bit can be learned. First of all, before any test pack is made, someone with knowledge of aerosols is involved in the selection of container specifications. This helps to set the stage initially for the development of proper information. The interpretation of developed information by and large involves nothing more than common sense, and the greater the interpreter's experience, the earlier he may make judgments with confidence.

Cathodic areas, galvanic action, polarization, and all the other phenomena explainable in scientific terms have been recognized and utilized, even when not understood completely or the knowledge synthesized scientifically. Container evaluation has been an art, and for a time it will continue to be an art. However, science is making inroads into this field, and such intervention should be welcomed.

If corrosion is electrochemical, one can define anodic areas, cathodic areas, potential to corrode, polarization effects, and the expected corrosion rate by electronic means. Such information, however, is developed at a point in time. The natural effects of aging or the cumulative effects of container–product interaction may eventually modify significantly the corrosion pattern of a product. Test pack evaluation, therefore, cannot be replaced by corrosivity tests. Corrosivity testing can assist in the initial selection of container specifications to be test packed, however, and can be used to confirm conclusions derived from test pack evaluations.

Corrosivity tests require the assistance of an electrolyte. Anhydrous solutions, oil-out emulsins, and viscous water solutions do not permit current flow and cannot be corrosivity tested by electronic means. Detergents also frequently defy the corrosivity tester. Therefore the test pack remains the basis for decisions, but time and effort may be reduced by electronic assistance.

CONTAINER CONSIDERATIONS DURING PACKAGING

The activities involved in the packaging operation include incoming can inspection, product and propellant fill, container seal, and container testing. All of these operations are critical in maintaining consumer acceptance and satisfaction.

Incoming Inspection

Incoming inspection of cans is a good practice. Production interruptions are minimized when incoming materials are sampled and irregularities eliminated or compensated for before use. When cans are sampled, they should be checked for compliance with material and dimensional specifications. In addition, it may be advisable to match components, such as valve cup and overcap fit to the can. The can weight should be checked; variation from shipment to shipment can be significant. Running a sample through the bottom-end ink coder will serve to verify height adjustment and end surface lubricant and ink adhesion compatibility. It is essential to check the overall can height and compare it to the normal crimper setting. The final check is for esthetics.

Container Fill

Fill of container is dependent on a number of factors. Nonsoluble compressed gasses require about 50% head space; soluble compressed gasses, less. The volume of compressed gas needed to properly dispense the last of the contents determines the amount of head space. The liquefied propellants should have about 15% head space; this amount should allow sufficient space for liquid expansion and the compression of entrapped air at 130°F. If, however, considerable propellant (50% or so) is introduced after evacuation, additional head space may be required to avoid excess pressure caused by air compression. If little propellant is introduced after evacuation, less than 15% head space is possible. Hydrocarbons have a higher coefficient of expansion than fluorocarbons but permit greater solubility of air than the latter.

Fill, therefore, becomes a function of the total contents of the container and cannot be preordained. The fill, however, should be justifiably a safe one and nothing less, since less than a safe fill may be considered to represent deceptive packaging.

Valve Crimping

The crimping of a valve-mounting cup into the 1-in. opening of an aerosol can began in the early 1950s. Earlier valve assembly was done by soldering. In those days of obviously limited experience few crimp leakage problems occurred. At that time aerosol products were either solvent-based insecticides and paints, water/soap solutions, or alcohol/shellac solutions. These products either assisted the gasket to seal by swelling it, or complemented the gasket by being capable of self-sealing. Also, in those early days the curl of the can's 1-in. opening was of closed construction.

In the early 1960s the under-the-cap method of propellant filling became popular. The closed curl method of construction gave way to the open curl, which provided rapid venting of entrapped propellant under the curl before hot-tank testing. Some latitude in the crimping operation was lost because of the open curl's potential for collapsing. A collapsed open curl or an improperly formed closed curl permits propellant, while gassing, to enter under the curl but permits very slow venting. Such a can appears to be a crimp leaker in the hot tank because of the propellant slowly venting from under the curl.

As a result of the open curl and its inherent lower resistance to crimper abuse, as well as the current extensive use of non-self-sealing alcohol/fluorocarbon formulations, which also have the ability to shrink some types of valve and can gaskets, crimp leakage is a frequent nuisance and an occasional catastrophe. Because of the variables that exist in curl strength and crimper adjustment, some organizations packing alcohol/fluorocarbon products are increasing their crimping latitude by incorporating closed 1-in. curl construction or by using a special valve-mounting cup gasket, or both.

CRIMP SPECIFICATIONS

Crimp specifications vary according to can type. The three-piece steel can having a 1-in. opening should be crimped as follows:

1. Minimum axial load required.
2. Crimp depth of 0.185 $\pm$ 0.005 in.
3. Crimp diameter of 1.070 $\pm$ 0.005 in.
4. Good torque resistance.

The specifications listed above must be considered only as a starting point. Cups, cans, and equipment vary, and crimp dimension changes must be made to compensate. For example, curl thickness should be 0.130 + 0.006 in., but if a particular shipment of cans has a nominal thickness of 0.125 in., a crimp depth of 0.180 $\pm$ 0.005 in. may be in order.

In addition to external measuring, cross-sectioned crimps should be periodically examined. Since metal spring-back occurs during cross sectioning, the dome should be cast in plaster or plastic and then cut so that the actual position is maintained. The seal in a crimped aerosol can is not the vertical compression of gasket on the horizontal surface, but the horizontal compression of gasket on the vertical surface. Crimping and crimp evaluation is an art. Until one masters this art, crimp decisions should not be made without knowledgeable help. A wrong decision may not become known for several months. In the meantime a large number of cans may be filled and possibly adversely affected by this wrong decision.

Torque testing is done by very few organizations. The torque test provides a good indication of crimp quality. When the result indicates that something is wrong, an investigation should be conducted. All too often blind adjustments are made, not to improve the crimp, but to improve the torque reading. Such an approach can further adversely affect quality while improving the torque reading.

Leak Testing

Hot-tank testing of aerosol cans is required for the purpose of detecting defective containers; mass-produced goods are not 100% perfect. The container contents should be heated to 130°F, or heated in a manner that will produce an internal pressure equal to the equilibrium pressure at 130°F.

With some products, such as a water-base formulation with hydrocarbon propellant, nonequilibrium conditions may exist and higher than normal pressures may result. This happens when hydrocarbons float on the heavier concentrate and heat independently and rapidly to the hot-tank temperature, without the benefit of any pressure depression that may exist under equilibrium conditions. When pressures are in excess of what should normally exist at 130°F, container distortion may take place.

Since hot tanks detect defective containers through the generation of pressure, it follows that some defective cans will explode in the hot tank. Hot tanks should be well covered, therefore, to protect employees, especially the hot-tank inspector.

Defects found in the hot tank are not restricted to containers. Defective filling practices and faulty crimping practices may also be revealed in the hot tank. On the other hand, all defects, whether they involve container, fill, or crimp, do not show up in the hot tank. Defects have to be gross to be detected, and lesser imperfections can pass unnoticed. Faulty hot-tanking practices also may either cause defects or permit detectable defects to pass unnoticed.

The most common defects found in the hot tank are container distortion, such as dome elongation or end inversion, leakage due to malformed seams, or bursting caused by misformed seams. Malformed and misformed seams are obvious defects. The cause of inverted ends are not as apparent. Occasionally false crimp leakers or "phantom leakers" appear to be obvious crimp leakers but actually involve only the slow release of curl-entrapped propellant.

Buckles or inverted ends can be the result of excessive fill, insufficient can evacuation or purge, excessive hot-tank water temperature, excessive time exposure to hot tank, incorrect propellant blend, nonequilibrium pressure generation, improper end formation, low-baseweight end metal, or low hardness of end mtal. More often than not, a combination of these conditions causes buckles. If a pressure-relief mechanism triggered by buckling is employed, all of the above conditions should be closely monitored when potentially dangerous contents are filled.

When crimp leakers or what appear to be crimp leakers are detected in the hot tank, a thorough examination should be conducted to determine the cause of crimp bubbling. As mentioned previously, an aerosol dome with an open curl configuration that has closed during crimping will appear to be a leaker if charged by an under-the-cap filler, as propellant escapes in the hot tank from its entrapment under the collapsed curl. The cause is either improper dome specification or improper crimper adjustment and should be remedied immediately. A collapsed curl may not permit true leakage immediately, but has the potential to do so eventually. If the aerosol dome has a closed curl configuration, the closure is probably not complete, and an inspection should be made to determine whether the resultant crimp has the potential to permit leakage in time. Probably it does not, but the possibility exits. When what appears to be a crimp leaker is found to actually be one, it is essential to stop the filling line and correct the condition. Crimp leakage detectable in the hot tank is usually the result of a maladjusted crimper.

REFERENCES

1. *Steel Products Manual: Tin Mill Products*, American Iron and Steel Institute, July 1968.
2. *Can Making Materials and Methods Manual*, Continental Can Company.
3. *Top Double Seaming Manual*, Continental Can Company.
4. H. T. Johnson, *Chem. Spec. Mfr. Assoc. Proc. 52nd Mid-Year Meet.* May 1966, p. 47.

GLASS CONTAINERS

The use of glass containers for holding products under pressure is not a new application. Glass aerosol containers technically can be said to have come into being with the first soda water or beer bottle. These products utilized a compressed gas such as carbon dioxide and a pressed-on, pressure-tight closure. However, if a valve were to be crimped onto the bottle in place of the conventional closure, the beer or soda would be dispensed as a stream or spray. This development can be considered to be a forerunner of the present-day uncoated glass aerosol container.

Ethyl chloride, a local anesthetic, was marketed in a glass ampoule for many years before the development of the aerosol bottle (1). This compound has a low boiling point and when held in the warmth of the hands will form a spray if dispensed through a suitable valve. A moth-proofing substance has been packaged in a glass container using a three-phase system.

Historically, the glass industry has almost always been in a position to manufacture a container suitable to withstand internal pressure. The development of suitable propellant and valve systems by the aerosol industry made possible the safe packaging of many products in glass containers and thereby created a new market for such containers. This, together with the ever-present technology to make pressure containers, resulted in the development and marketing of workable aerosol units.

Obviously many modifications were necessary in glass manufacturing in order to adapt existing technology to the production of an aerosol container. The proper size and shape of the finish (the part of the bottle to which the valve is crimped) and particularly the ratio of glass weight to capacity had to be determined.

Bottles, which can withstand a greater amount of internal pressure than the conventional three-piece fabricated tin-plated steel container, lagged far behind cans for the packaging of aerosol products. Even though an aerosol bottle can withstand greater internal pressure than the conventional can, glass is brittle and very likely to break if dropped. This, coupled with the fact that a liquefied gas is inside the container, led marketers to believe that glass aerosol bottles were unsafe to use under existing pressures of 35 to 40 psig at 70°F and as much as 85 to 90% of propellant. Under these conditions of use, glass is potentially hazardous, since, if the bottles were dropped and the glass broken, the combination of the high prssure and the high concentration of liquefied gas would result in a constant regeneration of the pressure, resulting in the glass being shattered with tremendous velocity.

In 1953 Mina (2) developed an "ultra-low-pressure" system which, together with the "three-phase" system, made possible the widespread use of glass aerosol containers. The ultra-low-pressure system is based on the use of relatively small proportions of propellant. The propellant is dissolved in the product, which generally contains ethyl alcohol. The pressure of this system is less than 15 psig, and as the droplets of product are released, the relatively small quantity of propellant rapidly boils off, leaving behind a fine mist of product. By varying the quantities of alcohol, water, and propellant, it is possible to obtain different spraying characteristics.

Under these conditions it was observed that, if a bottle were dropped, the force with which the broken glass shattered was less hazardous than that prevailing when a soda bottle shattered. Marketers' objections started to disappear, resulting in the introduction of many products (3-5).

The ability to place a vinyl coating around a bottle also furthered the use of glass containers for pressurized products (6). The coating served to contain the glass if the bottle shattered. Also, single or double vinyl coatings permit the use of higher pressures within the container, as well as larger containers and shapes that are not practical for an uncoated bottle. In these cases, pressures of 25 to 35 psig at 70°F can be used.

From all of this there emerged the uncoated and the plastic-coated glass aerosol. Both of these containers offer the marketer a more attractive method of packaging aerosol products than was ever available before. Seven advantages of the use of glass are the following:

1. Inertness.
2. Ability to be easily sterilized if necessary.
3. Lack of compatibility problem.
4. Ability to be made light resistant.

5. Relatively low cost.
6. Esthetic appeal and attractive appearance.
7. Ability to be molded into a variety of sizes and shapes.

TYPES OF GLASS

There are several types of glass. Ordinary sand-soda-lime glass used for common bottles and windows is remarkable for its suitability over a wide range of conditions. However, changes and additions to the common formula produce glass specially designed for limited purposes and for use when cost is not a primary concern.

Pyrex glass, especially valuable for its resistance to thermal shock, has a very high silica and boron oxide content.

Borosilicate glass, used in manufacturing containers for pharmaceuticals, is more chemically resistant than ordinary container glass. It is high in boron and potassium oxide content.

Optical glass has a more diversified composition, containing higher percentages of barium, lead, potassium, and zinc oxides.

Although all of the above types of glass can be used for the production of aerosols, because of economics it is practical to use only the sand-soda-lime glass. Fortunately, this type of glass has proved to be satisfactory for this purpose.

Common glass for containers has the following composition:

Silica:	71%
Alumina:	2%
Calcia:	10%
Soda:	14%

The remainder consists of fluxes, fining agents, and decolorizers and or coloring agents.

MANUFACTURE OF GLASS

Container glass is made by melting "batch," which is a mixture of several ingredients; the major ones are sand, soda ash, burnt lime, and syenite, The sand used in glass manufacturing must be very pure with a negligible iron content. Only a few deposits in the United States are of sufficient size and adequate purity to be economically satisfactory for glass manufacturing. Notable among these are deposits in northern Illinois.

Soda ash has, until recently, been largely manufactured from salt. In the past few years, however, the mining of natural soda ash deposits in Wyoming has supplied an increasing amount of this product to the glass industry.

Burnt lime is made by heating a high-grade limestone in kilns to obtain a maximum calcium and magnesium content. Sources suitable for glass manufacturing are located in northwest Ohio.

Nepheline syenite, a gray crystalline igneous rock compound of feldspar and other minerals, is a naturally occurring mineral mined and crushed in the central area of Ontario, Canada.

Slag and saltcake are minor ingredients used in container glass. Slag, a by-product of steel manufacturing, is responsible for maintaining the quality of the glass within certain limits. Saltcake is both a manufacturing by-product and a naturally occuring mineral; both sources supply the glass industry.

Sand alone makes an extremely durable glass, but its melting temperature is very high and the cost of an all-sand glass would be prohibitive. Soda ash, possessing a relatively low melting point, is added to the batch. The prime purpose of soda ash is to dissolve sand, as water dissolves salt.

The resulting product of sand and soda ash is a glass that is readily attacked by water. This is remedied by the addition of lime or limestone. Syenite is added as a source of alumina, which increases the chemical durability of the glass. Slag and saltcake serve largely as melting and refining agents. In addition, slag is important to the color of amber glass.

Materials used in much smaller (even minute) amounts are the decolorizers. These are commonly selemium and cobalt for flint glass, iron sulfide for amber, and iron chromite for emerald green.

All batch materials must meet rigid chemical and physical specifications. Harmful contamination must be avoided throughout the mining, manufacturing, and shipping processes, When batch materials are put into the melting furnace as a dry mixture, the various ingredients should be similar in grain size so that segregation is avoided in the handling process. Large percentages of materials that are either too coarse or too fine may cause melting problems within the furnace or melter.

Nearly all raw materials are shipped by rail, and it is at the unloading area that the glassmaker first becomes physically involved with the care and quality of his product. Each material must be put into its proper bin or silo at the batch house.

The several batch materials are weighed individually to close scale tolerances. Cullet, which is broken glass from the rejected ware, is also added to the batch. This tends, however, to lower the melting temperature. All of these materials are mixed together in a large mixer for 1 to 3 min, and then the batch is conveyed by various methods to the batch chargers at the glass melting furnace.

The tank or glass furnace is basically a large covered container of molten glass, generally rectangular, made of refractories that have a high melting temperature and are only minimally attacked by glass. The batch is melted by projecting gas or oil flames out over the glass surface. Temperatures range from 2600 to 2800°F. Lower temperatures are desirable to prolong the furnace life. Several physical and chemical reactions take place at these temperatures. Water evaporates; gaseous components formed during the process escape; and there is mutual solution of the liquid phases, resulting in a homogeneous molten glass. Bubbles are removed in the fining area of the furnace. Electrodes inserted below the surface of the glass are sometimes used to provide electric "boosting" to conventional firing methods, and a few small tanks are totally electric. A very narrow temperature range, if adjusted to match the amount of glass being drawn out, does not affect the quality of the glass. Also, the amount of batch being fed into the melter must be closely controlled to the amount of glass being drawn out (called the "pull" on the furnace) in order to maintain the glass level in the tank to within a few hundredths of an inch. Excessive variations in the molten glass level will affect the "gob" weight which is used to mold each container.

MANUFACTURE OF GLASS CONTAINERS

Glass from the tank flows through a small "throat" into the "nose" or refiner and then into the "feeders," which are finger-like extensions to each glass-blowing machine. In the refining and feeder areas, the glass is conditioned and the temperature adjusted to fit the forming requirements of the particular type and size of ware being made on each machine.

Several types of glass machines are available (7), including the following:

1. Owen's bottle machine.
2. Lynch blow-and-blow machine.
3. Individual section machine.
4. Pressing one-stage machine (Figure 8.1).

Figure 8.1. Overview of one-stage pressing machine (Foster-Forbes Glass Co.).

At the front of the feeder is an orifice topped by a tube and plunger apparatus. The hollow tube revolves around the plunger and provides a slight mixing action that modifies slight variations in glass conditions. The glass flows slowly down through the orifice and is given a push by the plunger. A "gob" is cut off by water-cooled shears, and this piece of hot glass drops into the blank mold. The temperature of the feeder, the action of the plunger, and the timing of the shears must be carefully controlled in order for the desired gob weight to be uniformly maintained. The blank mold is designed to be of nearly equal volume to the gob of glass entering the mold. As a result, the glass is a solid mass formed into a particular shape that will behave in a desired manner in the next stage. The finish, that is, the part of the bottle to which the cap fits, is also formed in this initial stage.

The glass is next transferred to the blow mold, where compressed air pushes the glass out against the sides of the mold to form the shape of the finished glass container. The bottle is then taken from the mold and set on a conveyor for movement to the annealing lehr. The proper design of the mold equipment, the timing of the various machine functions, the swabbing of the molds, the amount and location of cooling air, and the weight and temperature of the gob are all necessary and important factors in making a high-quality glass container.

After the glass containers are formed, they travel on a heated conveyor chain and are then loaded automatically into an annealing lehr. The purpose of the glass containers passing through an annealing lehr is to stress-relieve or normalize the glass by subjecting it to a closely controlled heated atmosphere and then gradually cooling it under controlled conditions to room temperature at the exit end of the annealing lehr.

Upon leaving the annealing lehr, the glass containers pass through a variety of automatic inspection equipment on a conveyor called a single line. During this trip through the single line the quality of the glass container is closely checked by highl trained inspectors, as well as the aforementioned automatic inspection equipment, such as plug gaugers and electronic check inspectors.

At this point the inspected containers continue down the single line conveyor to be packed in cases by several methods: automatic case packing, bulk palletizing, or hand packing by trained selector-packers.

MANUFACTURING PLASTIC-COATED AEROSOL BOTTLES

The plastic coating of glass aerosols received its greatest impetus early in the development of the market for aerosols. At that time marketers and fillers were intensely concerned about various imaginary hazards presented by glass for aerosol use. Consequently, a coating that would not diminish the esthetic properties of the container and that assured fragment retention upon breakage was considered imperative for safety.

The coating used is a plasticized, high-molecular-weight polyvinyl chloride resin applied from a plastisol system. (A plastisol is a dispersion of colloidal-size resin in a plasticizer.) Coating thickness on the order of 0.010 to 0.100 in. is easily applied from such a system. Normally, the coating on an aerosol bottle will be from about 0.035 to 0.055 in. thick, depending on the size of the bottle and the surface area to be covered.

The coating is applied by a hot-dip process, wherein the bottle is preheated and then dipped into a fluid plastisol. The plastisol forms a gel structure that has little physical integrity, conforming essentially to the geometry of the bottle. After the gel structure has been formed, the bottle must go through a heat cycle to fuse the resin and to develop desirable physical properties such as tensile strength, tear resistance, abrasion resistance, and impact-absorbing capability. Both the dipping and fusing cycles must be carefully controlled to produce coatings of optimum thickness and physical properties.

Three major coating systems are in use at present. All of these systems use a polyvinyl chloride resin in a plastisol, which acts as a film former. They differ only in the end product produced. These coating systems are as follows.

NONBONDED VENTED COATINGS

These coatings require a bottle finish different from the normal one. During manufacture of the glass, a groove is placed directly under the lip of the bottle. In the coating process, this groove fills with plastisol, forming a thickened ring. This thickened ring, in combination with the aerosol valve ferrule, holds the plastic and the glass together. Pierced or drilled vent holes in the plastic coating act to relieve the pressure if the bottle is broken.

BONDED COATINGS

This system uses a bottle with a conventional finish. An adhesive primer is applied to the glass before the cycle during which the coating is applied. Upon fusing, the coating is firmly adhered to the glass. If such a bottle breaks, the energies contained within the package are not immediately released. The coating remains intact for an indeterminate period of time, even though the bottle is obviously broken. This condition continues to exist until plastic fatigue allows a small splinter of glass to penetrate the plastic jacket. At this time the energies will dissipate through the small hole that has been formed. The effect can be compared to breaking a piece of safety glass.

LAMINATED COATINGS

This coating system differs from the first two types in that two coatings of differing physical properties are applied to the bottle and are fused together to form a true laminate that is virtually impossible to separate. Because these coatings are so different in physical properties, a synergistic increase of the most desirable property (tear strength) results from the lamination. This enables the coating to resist pressures higher than those normally used for single coatings, bonded or unbonded. This system is not normally bonded to the glass; therefore the special finish described above must be used and the coating must be vented, preferably with a drilled hole.

In addition to fragment retention the plastic coating performs the following functions:

1. Preserves the glass surface from weathering.

2. Protects the surface from abrasion and contusions resulting from bottle-to-bottle contact, or contact with surfaces denser and harder than glass. (These two conditions are known to diminish measurably the internal-pressure-retention properties of the pristine containers.)

3. Increases impact resistance. Impacting forces on the order of 4 to 5 times greater than those required to break an uncoated bottle are needed to break a plastic-coated bottle, because of the impact-absorbing property of the coating.

Plastic coatings can be formulated to produce high gloss or low gloss; they can be textured or smooth, water-white or colored, transparent, translucent, or opaque in virtually any color imaginable. This capability opens an entirely new dimension in package design.

QUALITY CONTROL

The following is a general outline of testing and inspection procedures for all types of glass containers.

RAW MATERIAL ANALYSES

Both chemical and physical analyses are made on each carload of raw material.

BATCH SEGREGATION

A survey is made occasionally by outside firms to check the batching process and to ensure that it is operating satisfactorily.

ALKALINITY TEST

The alkalinity of the glass is checked periodically.

DAILY GLASS ANALYSIS

This covers the following:

1. Petrographic inspection of stones and foreign matter.
2. Density.
3. Cord.
4. Seeds.
5. Flint color.
6. Amber-light transmittance.

CHEMICAL ANALYSIS

Samples are submitted at regular intervals for complete glass analysis.

Hot End Inspection

FOREMAN AND MACHINE OPERATORS

Front end personnel are responsible for the quality of ware produced. Considerable time is spent in using gauges and making visual inspections.

LEHR MEN

Lehr men are directly responsible for the operation of the lehrs under their jurisdiction. Their careful attention to their duties assures good annealing at all times.

Inspection Department

INSPECTORS

Bottles are weighed, inspected, measured, and gauged on an hourly basis before they reach the selector. Capacity checks are frequently made.

TESTERS

At this point, ware is tested and checked for thermal shock, pressure strength, vertical load strength, and glass distribution.

SELECTORS

One hundred per cent of the bottles are inspected and placed in shipping cartons.

PACKED-WARE INSPECTION

All packed ware is inspected on an hourly basis. The quantity of ware inspected is determined by the number of pieces per hour and by the required A.Q.L.

REPACK DEPARTMENT

All ware that fails to pass the packed-ware inspection is routed through the repack department, where it is carefully reinspected.

Specifications Department

A thorough inspection of body dimensions, finish dimensions, and volume is made by this department at regular intervals to ensure that containers are within specifications at all points.

Other Functions of Quality Control

The entire system of inspecting and testing is under the direct supervision of management and the quality-control department. This department is responsible for making all decisions regarding defect limits, testing, and inspection procedures.

In addition to the testing procedures already mentioned, the following tests are performed for uncoated and coated glass aerosols:

THERMAL SHOCK

Glass bottles are placed in a water bath at a controlled temperature and then quickly immersed in a bath with a temperature 130°F above the original temperature. This is shown in Figure 8.2.

Figure 8.2.   Thermal shock tester (Foster-Forbes Glass Co.).

HYDROSTATIC PRESSURE

The bottle is completely filled with water and then subjected to hydrostatic pressure until the breaking point is attained, as shown in Figure 8.3.

VERTICAL LOAD TEST TO THE BREAKING POINT

Downward pressure is exerted on the bottle, and the point of breaking is noted.

IMPACT TEST

This is performed with the use of the Preston impact testing machine, model 400, manufactured by American Glass Research, Butler, Pennsylvania. This instrument provides reproducible blows for testing the impact resistance of bottles. The equipment is illustrated in Figure 8.4. In use, the glass bottle to be tested is placed in position on the bottle support and against the back stop. The instrument is adjusted to give the exact point of impact. The severity of the blow can be adjusted from 0 to 22 in.-lb. of energy. The hammer is then released and strikes the glass.

Figure 8.3. Increment pressure tester (Foster-Forbes Glass Co.).

Figure 8.4. Impact tester (Foster-Forbes Glass Co.).

DESIGN CONSIDERATIONS FOR GLASS CONTAINERS

Glass aerosols are designed to be as safe and as attractive as possible. The cosmetic industry, being particularly sensitive to esthetics, has become the prime user of glass aerosols.

In designing uncoated glass aerosols, an attempt is made to avoid sharp angles and edges. Uniform glass distribution affords strength. A round container, for example, is stronger than a square or oval one because the circular shape lends itself to uniform distribution. The question arises as to how much strength is required and how one draws the line between required strength and esthetics. The glass companies involved in manufacturing glass aerosol containers have established the testing procedures outlined above to determine the strength of the container. One should always keep in mind the importance of not permitting the customers' constant demand for new shapes, and the glass companies' natural desire to sell merchandise, to compromise the requirements of safety.

To enhance beauty, the bottles can be molded with various surface designs. A smooth container may be silk screened. The bottles may be sprayed in various colors with ceramic paint. A variety of closures, generally designed in such a manner so as to cover the fill level, are available for aerosol packages.

It is of considerable importance to determine the required capacity before designing the container. The glass companies generally develop glass bottles to fluid ounce capacity, but aerosols are labeled by weight. Therefore the density of the formulations should always be transmitted to the glass company so that the desired weight can be determined and converted to fluid ounce capacity. This also makes it possible to arrive at the approximate fill level, and the customer can decide to leave it exposed or to conceal it by labeling, ceramic decoration, or special closure.

Plastic-coated glass bottles gain safety and color through the same process. In addition, the coating permits use of more radical shapes and larger sizes. An opaque coating automatically eliminates the necessity of disguising the fill level. Silk screening may be applied over the coating for additional decoration and labeling requirements.

REGULATIONS COVERING GLASS CONTAINERS

Virtually no government regulations cover glass aerosols at the present time. Most of the existing regulations or ground rules have been self imposed by the glass industry. This speaks well for industry integrity. For example, all of the tests and inspection procedures outlined above were developed by the glass manufacturers. For uncoated glass aerosol bottles the size has been restricted to a maximum of 4-fl-oz capacity and a limitation of 15 to 18 psig at 70°F for a standard cologne formulation. Products utilizing 8 to 10% propellant would be satisfactory up to a pressure of 25 psig at 70°F.

Practically speaking, a head space of about 15% at 70°F is generally required. Existing regulations state that the contents should not fill the

TABLE 8.1. DETERMINATION OF MEAN BOTTLE CAPACITY"

Propellant 12/114 (10:90)   50% weight
Ethanol, anhydrous   50% weight

	Weight of Desired Fill (avoir oz)			
	1	2	3	4
Fill (g)	28.35	56.70	85.05	113.40
Filling tolerance (+3%) (g)	0.85	1.70	2.55	3.40
Excess fill (g)	0.80	1.00	1.00	1.00
Total fill (g)	30.00	59.40	88.60	117.80
Density at 70°F of propellant/ ethanol mixture	1.02	1.02	1.02	1.02
Volume of fill (ml) at 70°F	29.41	58.24	86.86	115.50
Valve displacement (ml)	1.4	1.5	1.5	1.5
Total volume (ml) at 70°F	30.81	59.74	88.36	117.00
Density at 130°F of propellant/ ethanol mixture	0.985	0.985	0.985	0.985
Volume of fill (ml) at 130°F	30.46	60.30	89.95	119.59
Valve displacement (ml)	1.4	1.5	1.5	1.5
Total volume (ml) at 130°F	31.86	61.80	91.45	121.09
Bottle tolerance (fl oz)	$-\frac{3}{64}$ (0.0469)	$-\frac{1}{16}$ (0.0625)	$-\frac{1}{16}$ (0.0625)	$-\frac{5}{64}$ (0.0781)
Bottle tolerance (ml)	1.387	1.848	1.848	2.310
Mean bottle capacity (ml)	33.25	63.65	93.30	123.40

container at 130°F. It is known that in going from 70 to 130°F the propellant and solvents (ethyl alcohol in the case of perfumes and colognes) will expand. The degree of expansion is dependent on the specific solvent and propellant used. For perfumes and colognes, the density of the product is about 1. This is, however, not always true, since the density of the mixture is dependent on the amounts of alcohol and propellant present. Generally, as the propellant content decreases, the density of the system also decreases, so that 2 oz by weight of product will occupy a volume greater than 2 fl oz. By using an average value of 15% for the head space, sufficient safety is built in to ensure that the contents do not completely fill the container at elevated temperatures.

During the past several years, increased emphasis has been placed on decreasing this head space to a minimum.

Present regulations state that the quantity of product packaged in an aerosol container must be stated on the label in terms of net weight and not volume. Before this regulation was imposed, it was common practice to give this quantity in terms of volume measure. Perfumes and colognes were labeled as containing 1 fl oz, 2 fl oz, and so forth. With the change in regulations, the weight appears on the container as 1.8 oz, 2.75 oz, and so on. Once the density is determined. the equivalent weight from the volume can be quickly calculated.

It is important to calculate the "mean bottle capacity" of various sizes of aerosol containers. If the densities of different mixtures of

TABLE 8.2.   DETERMINATION OF MEAN BOTTLE CAPACITY"
Propellant 12/114 (10/90)   40% weight
Ethanol, anhydrous   60% weight

	Weight of Desired Fill (avoir oz)			
	1	2	3	4
Fill (g)	28.35	56.70	85.05	113.40
Filling tolerance (+3%)	0.85	1.70	2.55	3.40
Excess fill (g)	0.80	1.00	1.00	1.00
Total fill (g)	30.00	59.40	88.60	117.80
Density at 70°F of propellant/ ethanol mixture	0.968	0.968	0.968	0.968
Volume of fill (ml) at 70°F	30.99	61.36	91.53	121.70
Valve displacement (ml)	1.4	1.5	1.5	1.5
Total volume (ml) at 70°F	32.39	62.86	93.03	123.20
Density at 130°F of propellant/ ethanol mixture	0.956	0.956	0.956	0.956
Volume of fill (ml) at 130°F	31.38	62.13	92.68	123.22
Valve displacement (ml)	1.4	1.5	1.5	1.5
Total volume (ml) at 130°F	32.78	63.63	94.18	124.72
Bottle tolerance (fl oz)	$-\frac{3}{64}$ (0.0469)	$-\frac{1}{16}$ (0.0625)	$-\frac{1}{16}$ (0.0625)	$-\frac{5}{64}$ (0.0781)
Bottle tolerance (ml)	1.387	1.848	1.848	2.310
Mean bottle capacity (ml)	34.17	64.48	96.03	127.03

propellant and alcohol are known, the mean bottle capacity can be determined after taking into account filling tolerances, volume occupied by the valve, and tolerances for glass bottles. Table 8.1 and 8.2 illustrate this for 1-, 2-, 3-, and 4-oz fills, using two typical aerosol formulations for perfumes and colognes. The assumption is made that the filling tolerance is +3%, an excess fill is used (0.80 to 1.00 g), and the valve displacement is from 1.4 to 1.5 ml. The final value represents the maximum amount of liquid which can be filled into the container; that is, if the glass aerosol container is made using this value as the "mean bottle capacity" under the worse conditions, the contents will reach this volume at 130°F. In other words, this value minus the total volume at 70°F will give the head space.

REFERENCES

1. H. R. Shepherd, *Aerosols: Science and Technology*, Interscience Publishers, New York, 1961, p. 2.

2. F. A. Mina, Dispensing Products by Internally Produced Ultra-low Pressure, *Chem. Spec. Mfr. Assoc. Proc. 40th Ann. Meet.*, December 1953.

3. T. Herst, Uncoated Glass Aerosol Containers, *Aerosol Age,*, **9,** 5 (May 1964), 26.

4. F. A. Mina, Glass Aerosols for Antiseptics, *Soap Chem. Spec.*, **31,** 10 (October 1955), 163.

5. F. A. Mina, Glass Aerosols for Pharmaceuticals, *Drug Cosmet. Ind.*, **75,** 5 (November 1954), 625.

6. A. R. Marks, Glass Containers for Pressurized Products, *Drug Cosmet. Ind.*, **84,** 4 (April 1959), 451.

7. F. V. Tooley, *Handbook of Glass Manufacture*, Vol. 1, Ogden Company, New York, 1953.

8. W. M. Robertson, The Evolution of Coated Glass Aerosols, *Aerosol Age*, **4,** 4 (April 1960), 26.

9. A. R. Marks and E. Budzilek Plastic Coated Push Button Containers, *J. Soc. Cosmet. Chem.*, **8,** (1957), 386.

ALUMINUM CONTAINERS

Aluminum has found widespread use as a material for the manufacture of aerosol containers. Although pure aluminum is an extremely reactive metal and is attacked by both acidic and alkaline substances, the formation of a continuous film of aluminum oxide produces a high degree of chemical inertness. This film is easily formed in the presence of oxygen and water vapor, and it is for this reason that aluminum is resistant to atmospheric corrosion.

ADVANTAGES OF ALUMINUM CONTAINERS

Aluminum aerosol containers can be more costly than comparably sized tin-plate containers but offer a number of advantages, among them the following:

1. Bursting strengths are higher, particularly in the one-piece aerosols since they are seamless. There is no danger of leakage of the product.

2. Wrap-around lithography is possible. Without a seam to contend with, the aluminum can be decorated in its entirety, resulting in a more attractive appearance. In addition, colors and printed designs can be altered without major line changeover, provided that the can size remains the same. This advantage permits long production runs in combination by color or by decorative image, which, in turn, offers consumers long-run benefits for short-run items without the investment of preprinted stock.

3. The natural "printability" or printing characteristics of aluminum permit a more sophisticated degree of lithography, as well as silk screening and other decorative finishes that are difficult to achieve on tin-plate containers.

4. There is freedom from external rusting, which is an important consideration for products intended for use in the bathroom, kitchen, or other high-humidity areas.

5. A wide variety of special decorative treatments can be used, including brushed textured effects, gold dye sprays for iridescence, burnished finishes, and special wedding band and cap closures for a flush, straight-line appearance that has been widely employed for cosmetic products.

6. There is no danger that corrosion will take place on the side-seam stripe and around the ends of the container where the top and bottom portions have been seamed in place, as occurs in tin-plate. These areas have been difficult to protect, and since aluminum containers can be made seamless, this problem does not exist.

7. Shipping weights are reduced since aluminum is considerably lighter than tin-plated steel and glass. For example, a conventional three-piece, tin-plated steel 16-oz container weighs about 105 g, whereas an 18-oz aluminum container weighs about 60 g and a 35-oz aluminum container about 115 g. These weights will vary somewhat, depending on internal coatings and external lithography, although these will be about the same in both cases. Thicker-walled containers will be slightly heavier.

8. Products that show a preferential attack on steel may not have this effect on aluminum.

9. Conventional coatings can be applied to aluminum containers as well as to tin-plated steel containers.

MANUFACTURE OF ALUMINUM AEROSOL CONTAINERS

Size and Type

Aluminum aerosol containers are produced in either one- or two-piece forms. The one-piece containers are extruded from aluminum slugs as single units. In the manufacture of two-piece units, a flat circle of metal is "drawn" into a cuplike shape and then stretched to a desired height by a process called "ironing." The top is then attached by a seamer. If the container initially is made with integral sides and top, the bottom is rolled on by a seamer.

By far the most popular of the aluminum aerosol containers is the one-piece unit currently offered by four American manufacturers and several foreign suppliers. Although one-piece, extruded aluminum containers are made in sizes up to 45 oz, the bulk of domestic production is concentrated on capacities ranging from 1/4 oz to 8 to 10 oz. In this size range, the differential in price between aluminum products and their tin-plate counterparts is substantially more competitive than in the larger sizes. In the case of the smaller-capacity containers—1/4 to 3 oz—the only practical package available is the one-piece aluminum aerosol container.

The development of this container was pioneered by the Peerless Tube Company. For several years the company had been extruding aluminum condenser cans for radios. However, with the advent of transistorized electronic components, Peerless soon found itself burdened with expensive processing equipment and an obsolete product line. In turning to the manufacture of aerosol containers, the company found the basic manufacturing techniques to be very similar. During its first year of operation in the aerosol container business, Peerless turned out slightly less than 500,000 units. In the past 15 years, however, the aerosol operation has expanded many fold and has become a basic and highly important part of the company's overall business.

Today one-piece, extruded aluminum aerosol containers are manufactured in a fluid capacity range from 1/4 to 10 ounces. Other domestic manufacturers include American Can Company, Victor Metal Products, Virjune Manufacturing Company, Risdon Manufacturing, Emson Manufacturing, Aluminum Company of America, Impact Container, Continental Can, Tubing Seal Cap Company, Apache Container Corporation, and J. L. Clark Aluminum General Corporation.

Pressure Considerations

Since these one-piece aluminum containers are seamless, they are capable of withstanding rather high pressures, depending on the thickness of the container wall. Pressures in excess of 500 psig can be safely withstood by certain aluminum containers. However, Tariff 19 of the Department of Transportation, under which aerosol pressure requirements are set, states that all aerosol containers must meet a minimum pressure performance of 140 psi at 130°F without bursting. For 2P specification containers the pressure performance required is 140 to 160 psi at 130°F, and for 2Q specification containers 160 to 180 psi at the same temperature is required without bursting. Although most products can be dispensed effectively with a low-pressure fill of around 35 psi, some, including feminine hygiene deodorants and butane lighter fluid, require pressures in the 2P and 2Q pressure ranges. One-piece aluminum containers meet all of these pressure requirements.

Production of Slugs

The manufacture of the one-piece aluminum aerosol container begins with an 1100 series aluminum ingot which has a purity content rating of 99.7%. The ingot is melted in a gas-fired furnace at a temperature of approximately 1300°F. The molten aluminum feeds through a trough onto a casting wheel, which produces a continuous slab or strip as shown in Figure 9.1. The latter progresses through a series of rollers that reduce the strip to the desired thickness in relation to the length of the container to be produced. For example, in the production of nominal 6-oz containers which measure overall 5 5/16 in., the continuous strip is rolled to a final thickness of 0.410 in. This is shown in Figure 9.2. After the final rolling reduction, the slab moves through lubricators and into a blanking press, where 6 to 12 dies (depending on the diameter of the container being produced) blank out round aluminum slugs. For example, the nominal 6-oz container, with an outside diameter of 1 1/2 in., requires a slug 0.410 in. in thickness and 1.485 in. in diameter and allows the use of only a six-die blank. In blanking smaller containers, more dies per blank can be employed.

The blanked slab is then guillotined at the end of the blanking press into manageable portions of scrap, which are conveyed directly back to

Figure 9.1.   Furnace, rolling mill, slugs (Peerless Tube Co.).

Figure 9.2. Rolling mill—reduction of slab for required slug thickness (gauge) (Peer-less Tube Co.).

the melting furnace (Figure 9.3). The blanked slugs drop through the press and are conveyed to an annealing oven to soften them for the extrusion process, after which they are transferred to a large tumbling barrel, where the rough edges are smoothed. The annealed and smoothed slugs move through a final lubricating bath and another tumbling barrel in order to remove excess lubricant, and they are then ready for production into aerosol containers.

Production of Aluminum Containers and Finishing (1, 2, 3)

EXTRUSION

The annealed and lubricated aluminum slugs are loaded into an automatic feed hopper serving a 200-ton impact extrusion press located at

Figure 9.3.   Blanking press scrap return (Peerless Tube Co.).

the head of an aerosol container production line. Both flat-bed and upright presses can be employed. As the shank hits the slug, which is in the die, with pressures of approximately 100 tons, the aluminum flows up the side of the shank to form the side walls and bottom of the container. The thickness of the slug, therefore, determines the length of the extrusion.

A mechanical stripper removes the extruded container from the press, and a belt transfers it to a trimmer, which cuts the side walls to the desired length, as shown in Figures 9.4 and 9.5. From there, the trimmed containers move into a necking machine, where, through the use of progressive dies, the open end of the container is reduced to a standard Chemical Specialties Manufacturers Association opening, which, depending on the diameter of the container, is either 1 in. or 20 mm. The necked containers are next conveyed through a three-phase trichlorethylene bath for degreasing.

INTERNAL COATING

If required, the clean containers are sprayed internally with an epoxy lining for product compatibility, after which they travel though a preheater oven for curing. Figure 9.6 illustrates the spraying operation. The original epoxy coating is a clear liquid, but upon curing takes on a golden or yellow-brown hue. Coverage of the internally lined containers is spot-checked via the standard copper sulfate testing process. The copper sulfate solution attacks voids in the coverage, producing a noticeable rusty coloration in these areas. Containers with insufficient coverage of the internal lining will be rejected by the quality-control department.

EXTERNAL COATING

The containers now receive an external coat of epoxy paint (Figure 9.7). Any color may be applied, and application is accomplished either by

Figure 9.4.   Press for making aerosol shell and trimmer (Peerless Tube Co.).

automatic spray guns or roller coating. After another trip through a curing oven for drying, the base-coated containers are conveyed to an offset lithographic printing press for final decoration.

The most common offset equipment employed is the highly economical Rutherford or cylindrical type, which can print up to four colors. At each of the four-color stations (see Figure 9.8) a series of rollers picks up and evens out the inks, transferring them to the zinc lithographic plates made from the customer's original art work. A separate plate is used for each color and is attached to the printing cylinders at each color station. The printing cylinder butts against a large drum, on the face of which is fixed a thin rubber blanket. As the drum and printing blanket move in a counterclockwise direction, the blanket picks up the printing inks from the zinc lithographic plates. All of the ink pickup from the four plates is minutely registered so that no color mixes with another or lies on top of it on the blanket. After picking up the fourth color, the blanket "kiss" prints on the aerosol container, which is conveyed past the blanket on a turning spindle. All four colors are transferred onto the container instantaneously. The printed con-

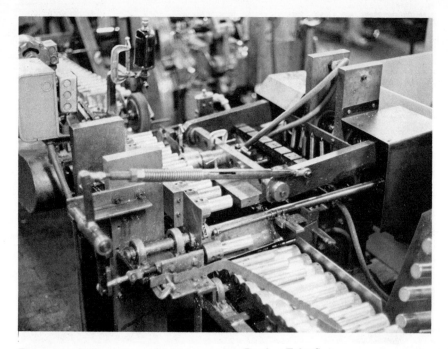

Figure 9.5.   Roll-over for making 1-in. opening (Peerless Tube Co.).

Figure 9.6.   Spraying internal linings (Peerless Tube Co.).

tainers are again ovendried and finally sprayed with a clear epoxy lacquer to protect both the lithography and the base coat.

During the painting and decoration process, the containers are inspected against base-coat color standards and for accuracy of printing registration. After final inspection at the end of the line, the containers are packed in boxes with layers of polyethylene sheets between or in special partitioned boxes as shown in Figure 9.9.

Other Finishing Techniques

Although the foregoing review of the manufacturing process describes the most common form of container decoration, offset lithography, there are, of course, other highly effective and attractive techniques that can be employed.

SILK SCREENING

Silk screening involves a nylon mesh screen that is treated chemically and photographically cut in a process similar to photoengraving. This process results in a stencil that will allow ink to be forced through the open design area in the screen mesh by a squeegee. Silk screening makes available a wide range of high-quality and opaque colors but is somewhat more expensive than offset lithographic decoration.

THERMIAGE TRANSFER

Thermiage transfer involves printing a design on a release-coated paper. The process makes possible five-color rotogravure printing and

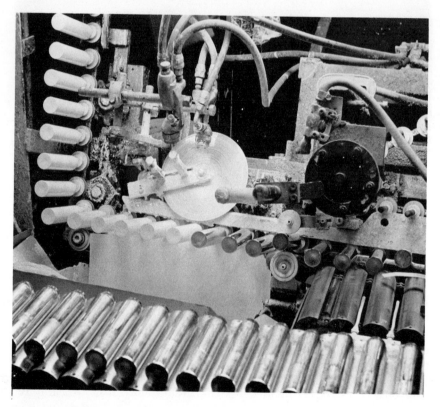

Figure 9.7.   Spraying base coat (Peerless Tube Co.).

Figure 9.8.   Offset lithographic press for up to four-color decoration and lacquer station for overcoat of printing inks (Peerless Tube Co.).

the utilization of half-tone and process art. Transfer of the design to the container is accomplished on a special machine that preheats the carrier and then presses it against the container with a heated roller.

BRUSHED FINISHES

By means of various types of abrasive wheels, brushed finishes produce a wide range of special textures on the raw metal of the container itself. Generally, after texturing, the containers are sprayed with a clear epoxy for protection against scratches and degradation from certain products.

Wedding band necks and matching caps can be used to give the container a complete straight-line appearance. Both bands and caps are made by progressively drawing aluminum sheet stock under the

Figure 9.9.   Final packing of containers (Peerless Tube Co.).

eyelet process. They can be buffed, gold or silver dye-lacquered, textured, or painted.

Caps for aluminum aerosol containers are available in an almost infinite variety of styles and shapes. Although metals, as indicated above, are employed to some extent, by far the greater number of caps used are made of plastic. In many cases, the type of valve used dictates the style of cap to be employed, but in any case the style range is sufficiently broad to meet nearly any esthetic requirement.

FILLING OF ALUMINUM CONTAINERS

The aluminum aerosol is normally filled in one of three ways: cold fill, undercap fill, or pressure fill. These methods are discussed in detail in Chapter 12.

Pressure Fill

Pressure fill is the most common method employed. After the product concentrate is introduced into the container, the valve is inserted and crimped. The propellant is then pressured into the container through the stem of the valve. Most of the aerosol valve crimping performed in the United States and Canada is done with a 3/32-in.-diameter foot with a six-finger collet.

After being filled and valve-crimped or sealed, containers are conveyed through a water bath to test for leakage. Tested and approved containers are coded and dated with special ink and are then ready for marketing.

Although the valve unit includes a more or less uniform dip tube and mounting cap, the valve itself is available in many different forms. A satisfactory aerosol package requires that the valve and the formulation be selected to complement each other. In addition to a wide variety of the more common spray valves, there are valves to dispense foaming products such as shampoos and shaving creams, metering valves that deliver measured quantities of product each time they are actuated for items such as mouth or nasal sprays, food and paste valves, and paint-and powder-dispensing valves. Among special-purpose valves are those for fire extinguishers, hermetically sealed valves, and transfer valves. Equally extensive is the range of push buttons or actuators available to match the container in esthetics and utility.

Formulation Aspects

Aluminum is an extremely reactive metal and because of its amphoteric nature is attacked by both alkaline and acidic substances. However, the formation of a continuous film of aluminum oxide produces a high degree of chemical inertness. This film is easily reformed in the presence of oxygen and water vapor, and it is for this reason that aluminum is resistant to atmospheric corrosion. Whereas in most aerosol tin-plated steel containers the air is evacuated to prevent corrosion, it is possible that this practice may hasten the breakdown of an aluminum container. It is for this reason that all formulations must be thoroughly tested before they are finally packaged in an aluminum container. For some products there may be no need for any further treatment of the aluminum surface, while other products may require an organic coating of an enamel, vinyl, phenolic, or epoxy-type system. The choice of system will depend on the product to be packaged in the aerosol container. As many as four coats of an epoxy-type enamel have been used. This coating has excellent flexibility and adhesive properties. Generally, a double-coated system will suffice (5).

For fragrances, a second enamel system of a phenolic-base coating is required. These coatings are generally odorless, tasteless, and nonabsorptive, although not as flexible as other types. This deficiency can be modified by the addition of suitable plasticizers. Vinyl top coats seem to work very well for water-base formulations.

Many different types of resins have been investigated for possible application for products that may be extremely damaging to their containers. Some of these have shown promising results.

Although aluminum is extremely unreactive, it does react with ethyl alcohol as follows (6,7):

$$2Al + 6C_2H_5OH \rightarrow 2(C_2H_5O)_3Al + 3H_2$$

Experience has shown that this reaction can be inhibited by the presence of small amounts of water. However, large amounts of water, especially in the presence of entrapped air, will hasten the corrosion.

Aluminum containers have found widespread use for many different aerosol products, including pharmaceuticals, cosmetics, paints, and specialty items. They have been used because of flexibility in size, chemical inertness, lightweight, and/or esthetic appeal.

REFERENCES

1. *A Glance at Peerless*, Peerless Tube Company, Bloomfield, N.J., pp. 19-24.
2. "The Peerless Tube Story," *Aerosol Age*, **5**, (July 1960), 21.
3. J. Jackson, Aluminum Containers, *Soap Chem. Spec.*, **36**, (March 1960), 119.
4. P. W. Sherwood, The Aluminum Aerosol, *Drug Cosmet. Ind.*, **91**, (1962), 440.
5. G. F. Hanna, New Coating Technology for Aluminum Aerosol Cans, *Aerosol Age*, **22**, (February 1961), 29.
6. J. J. Sciarra, Pharmaceutical Aerosols, Chapter 21 in L. Lachman, H. A. Lieberman, and J. L. Kanig (Eds.), *The Theory and Practice of Industrial Pharmacy*, Lea and Febiger, Philadelphia, 1970, pp. 621-622.
7. Y. Yamamoto, M. T. Tsurumaru, and T. Kurihara, An Attempt to Evaluate the Corrosion of Aluminum Aerosol Containers Packed with Ethyl Alcohol-Chlorofluorocarbon Systems, *Amer. Perfum. Cosmet.*, **86**, (November 1971), 41.

PLASTIC CONTAINERS

HISTORY

Plastics have long been used for inner coatings of metal aerosol containers and for outer coatings of glass aerosols. Most aerosol valves are now
made in whole or in part of plastic materials, and dip tubes are almost
invariably of plastic, as are most cover caps.

But the all-plastic container has only recently emerged as a serious
contender for the aerosol container market. Price trends, new production technology, and ecological pressures have combined to make the
plastic aerosol a commercial reality. The first large-volume production
of an all-plastic aerosol occurred in the United States in 1971 with
injection-molded (and ultrasonically sealed bases) thermoplastic acetal
copolymer 1-oz (purse-size) containers for hair spray.

Numerous attempts have been made in the past 13 years to produce
an economical plastic aerosol container. As early as 1960, a two-part
aerosol container, molded of acetal plastic for a spray-mist cologne for
teenagers, was marketed by a leading personal product manufacturer.
This attempt and several subsequent ones at all-plastic aerosols were
short-lived, however, and failed to live up to their early promise primarily because of technical limitations in producing containers economically from the available plastic resins.

Acetal, nylon, and polypropylene plastics have been the ones most
seriously considered for an aerosol container, but melamine, phenolic,
and even high-density polyethylene resins have also come in for a
share of attention. Polypropylene enjoyed a brief popularity in England for dispensing perfume and shave cream but is currently without
commerical application. Nylon and acetal aerosols have also been
produced briefly in Europe, Japan, and South Africa, without outstanding commercial success (1). Problems have included fabrication
limitations, permeation, and unwanted odor (2).

223

Recent developments in plastics resin technology and engineering design have combined to bring about a significant change in the economics of plastic aerosols. Inflation and the continuously rising prices of metals and other container materials have also helped to place plastics in a more price-competitive position. Whereas tin-plate, aluminum, and glass prices have risen steadily in the past decade, the prices of most engineering (predictable-performance) plastics have declined or have stayed at the same level.

Nevertheless, it has been the advances in injection molding, extrusion blow molding, injection blow molding, ultrasonic welding, and other plastic-forming techniques that have provided the impetus for a breakthrough. New concepts in plastic valve and body design developed through a systems approach to the basic problems have also helped to bring about this rapid change. A direct result is the aforementioned purse-size hair spray container molded of acetal copolymer (3). Aside from basic economies in fabricating this 1-oz plastic aerosol container, via injection molding and ultrasonic sealing, cost advantages have also been realized because of a new and patented valve system that is integrated with the container body. The plastic and rubber valve components are easily assembled into place before the container is filled. All parts of this container are injection molded in multicavity molds. The base of the container is joined to the molded body by ultrasonic sealing after cold filling of the product on high-speed automatic machines. Larger-size containers, up to 6 oz, can also be produced by means of this procedure at competitive prices. Figure 10.1 illustrates one such container.

This new economic advantage of the commercial 1-oz unit gives the plastic container more than just a toehold in a growing market for purse-size containers that in 1972 reached about 100 million units per year in the United States alone. The small touchup spray, on-the-go deodorant samplers, other small personal product aerosols, and vending machine items are expected to boost this market to about 500 million units by 1975.

Although plastics are nothing new to the aerosol industry, they have remained, until recently, secondary materials used in protective coatings and valve parts. The resistance of most plastics to corrosion made them the answer to some of the early difficulties with tin-plate corrosion when more highly acidic and alkaline formulations such as antiperspirants and oven cleaners were introduced. The protective linings used in metal cans are derived largely from the thermosetting group of plastics, such as epoxies, phenolics, and polyesters. These three dimensional cross-linked polymers make durable coatings but are

Figure 10.1. Injection-molded acetal copolymer containers with ultransonic sealed base and integral-molded valve mechanism. (Courtesy of Celanese Plastics Co.)

relatively difficult to fabricate into container shapes at high production rates. In the early stages of plastic container development initial success was achieved with melamine resins as material for a complete aerosol container.

Thermoplastic resins, however, are more easily formed into containers by a variety of processes, including injection molding, injection blow molding, blow molding, and forging. Low-and high-density polyethylene, polypropylene, vinyls, and styrenes have been successfully used in the packing industry. Unfortunately, none of these has the strength necessary for use as a container material for pressurized aerosol products.

A new group of high-strength thermoplastics, headed by nylon, acetal, and polypropylene, bridged the strength gap to provide a range of tensile and impact strengths sufficient to satisfy aerosol requirements. These new rigid thermoplastics have enough strength to compete seriously with metals and even to replace them, in many industrial applications. They have tensile strengths upwards of 12,000 psi and, when reinforced with glass fibers, as high as 30,000 psi.

SUITABLE PLASTICS

For use as an aerosol container, a requirement equally important to high tensile strength for a plastic material is creep resistance. This is the property of a material that enables it to resist deformation by a continuously applied stress over a considerable period of time. Good creep resistance is particularly necessary for aerosol products that have high-pressure, relatively long shelf life and are stored in warm environments for extended periods. Accelerated flexural creep data comparing various plastics at 180°F and 500 psi stress are shown graphically in Figure 10.2. It will be noted that several plastics, specifically polycarbonate, nylon, and acetal copolymer, have relatively flat flexural creep curves. These good creep-resistant materials have been termed "engineering resins." The newly introduced thermoplastic polyester materials, for example, polybutylene terephthalate (PBT), also fit into this category although flexural creep resistance at 180°F is marginal.

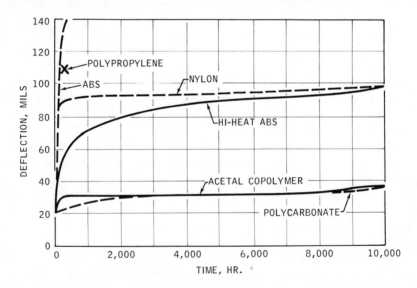

Figure 10.2. Flexural creep for various plastics at 180°F and 500 psi stress (Molded flexural bar 5″ × ½″ × ⅛″).

Nylon (or polyamide resins) attracted considerable attention at an early stage in the development of the plastic container because of its excellent strength characteristics. Nylon was tried as early as 1954 in an all-plastic aerosol container (4). Despite many years of research and testing, though, no nylon aerosol container has been produced in any significant volume. The major deterrent to commercial success is its critical melting point range, which tends to limit container size and also makes processing difficult and containers expensive. Also, the ability of water and other solvents to cause nylon to swell and distort makes precision molding to exact tolerances difficult. In addition, many perfume and cologne formulations cause the nylon containers to become stained on their exterior surfaces. Nylon, like many other plastics, is extremely permeable to many gases, but particularly to water vapor and the low-molecular-weight alcohols frequently used in aerosol formulations (5).

Polyacetal plastics, particularly the acetal copolymer,

$$HO-\overset{\overset{\displaystyle H}{|}}{\underset{\underset{\displaystyle H}{|}}{C}}-\overset{\overset{\displaystyle H}{|}}{\underset{\underset{\displaystyle H}{|}}{C}}-O-\overset{\overset{\displaystyle H}{|}}{\underset{\underset{\displaystyle H}{|}}{C}}-O-\overset{\overset{\displaystyle H}{|}}{\underset{\underset{\displaystyle H}{|}}{C}}-O\cdots\overset{\overset{\displaystyle H}{|}}{\underset{\underset{\displaystyle H}{|}}{C}}-\overset{\overset{\displaystyle H}{|}}{\underset{\underset{\displaystyle H}{|}}{C}}-O-\overset{\overset{\displaystyle H}{|}}{\underset{\underset{\displaystyle H}{|}}{C}}-O-\overset{\overset{\displaystyle H}{|}}{\underset{\underset{\displaystyle H}{|}}{C}}-\overset{\overset{\displaystyle H}{|}}{\underset{\underset{\displaystyle H}{|}}{C}}-OH$$

have thus far demonstrated the best combination of properties to fulfill the requirements for an aerosol container (6). Soon after the first acetal was introduced in early 1959, attempts were made to mold an aerosol container. As mentioned earlier, by late 1960, an acetal homopolymer aerosol container containing a spray-mist line of cologne for teenagers was placed on the market (7). The tear-drop-shaped container was injection molded in two halves that were joined by spin welding. Although this acetal aerosol container and several others were actually produced and sold commercially, none lasted on the commercial scene for more than a few months before being withdrawn. Again, container fabrication limitations were chiefly responsible for these early failures.

Aside from the fabrication limitations, odor problems continued to plague the early, most promising, acetal resins despite intensive and persistent developmental efforts. Reformulation of the chemistry of the acetal copolymers in the mid-1960s resulted in the development of a series of proprietary compositions which eliminated the extractable fugitive-odor components (8).

It is important to note that the chemical nature of plastic resins makes them suspect as unwanted odor contributors to the delicate fragrances of cosmetic and pharmaceutical products. New generations of plastic resin candidates for aerosol container materials must satisfy this often overlooked but critical requirement for this application.

TABLE 10.1. ACETAL COPOLYMER BLOW-MOLDING-GRADE RESIN

Property	Units	ASTM Test Method	Typical Value
Specific gravity	g/cc	D792-60T	1.410
Melt index	g/10 min	D2133	1.0
Tensile strength at yield	psi	D638-61T	8800
Elongation at break	%	D638-61T	40
Flexural modulus	psi	D638-61T	335,000
Flexural strength at 5% strain	psi	D790-61	11,800
Notched Izod impact strength	ft-lb/in. of notch	D256-56	1.0
Hardness, Rockwell		D785-62	M84
Heat distortion			
At 66 psi	°F	D648-56	318
At 264 psi	°F	D648-56	205
Vicat softening point	°F	D1525-65T	327
Melting point	°F	D2117-64	334

Figure 10.3. Plastic aerosol containers provide safety because of few lightweight parts that result, should breakage occur. (Courtesy of Celanese Plastics Co.)

Special higher-melt acetal copolymers have also been tailored for the specific requirements of the aerosol industry, especially for containers to be formed by blow-molding techniques. Now odor-free acetal copolymer resins have been made available in both injection- and extrusion-molding grades. Typical physical property data for these resins are listed in Table 10.1.

SAFETY AND ECOLOGICAL CONSIDERATIONS

Impact strength has become increasingly important to a more and more safety-conscious consuming public. Pollution and the environment have also become broad public concerns that affect packaging and waste disposal. The new plastic aerosol containers provide solutions to the problems of improved product safety and disposability. Not only do plastic containers offer much greater impact safety, but also they have an inherent safety factor in a fire.

Figure 10.4. Safe release of pressurized contents of plastic aerosol container through molten pinhole created by heat of fire. (Courtesy of Celanese Plastics Co.)

Unlike glass containers, plastic containers do not shatter at impact; rather, they break into two or three parts. When under pressure, of course, the parts may fly about. However, since they are light in weight, these fugitive parts are of minimal danger.

Figure 10.3 illustrates the typical break pattern of a plastic aerosol container. Only three pieces were formed when this container was deliberately destructed via impact failure.

The safety of plastic containers is equally noteworthy in incendiary environments. This is attributed to their relatively low melt points. The heat from a fire is sufficient to locally melt a section of the plastic aerosol container wall and thereby allow the pressurized product to be released (Figure 10.4). In an incinerator or open fire, plastic containers first soften and then expand, pinhole, and release the internal pressure safely before it can build up to explosive violence, as occurs with metal or glass containers.

Pollution problems can be greatly reduced by the use of plastics in containers. Most plastics used for this purpose burn completely and harmlessly under efficient combustion conditions. When they burn, little ash remains (9).

CONTAINER DESIGN

Impact Considerations

Impact strength is one of the three major factors that determine the wall thickness and, hence, the weight specification of a plastic aerosol container. Permeation resistance and pressure retention are the other two factors. All other criteria will be described before illustrating how the container specifications are established.

Once the optimum container dimensions, including shape, size, and wall thickness, have been fixed by permeation and long-term pressure considerations, utilizing established relationships, the minimum weight and wall thickness necessary to provide a safe container that will resist breakage in normal handling accidents must be determined. This is usually done empirically by dropping filled containers of varying wall thicknesses and thickness distributions ("plastic programmed containers" can have areas of heavier wall thicknesses on their most vulnerable sections). Usually a mounted steel plate is used for the "floor," and the containers are dropped from a collapsible, variable-height platform (10). This testing procedure also serves as an excellent quality-control tool for production targets.

Impact requirements for the containers are established by preselecting heights that are considered to be practically safe and then setting acceptable limits for breakage. Containers breaking prematurely do not have sufficient wall thickness (a design problem) or are highly stressed (because of improper molding conditions). Once it is established that the containers to be tested contain minimal stresses, the wall thickness can be increased until impact results are satisfactory. In this way the minimum wall thickness specification design for impact considerations is established. Production containers satisfying the minimum wall thickness but failing the impact standard are suspect as having been improperly molded. It is important to note that plastic containers can be easily designed to have greater impact resistance than glass or vinyl-coated glass.

Next, it must be determined whether or not the thickness dictated by impact considerations is sufficient to provide the container with adequate permeation resistance and pressure retention.

Permeability and Chemical Resistance Considerations

All plastics are permeable in varying degrees to most vapors, liquids, and gases. Acetal copolymers are no exception, but they are extremely resistant to permeation by many of the propellants and ingredients used in aerosol products. Fortunately, however, permeation is predictable, as is loss through valves on metal and glass containers. It can be compensated for by careful formulation and overfill of ingredients and /or possible use of barrier coatings for critical products (11).

Much of the permeation information compiled for acetal copolymers has resulted from three container compatibility studies (12).

1. Two-year testing of 60 common aerosol product ingredients.

2. One-year testing of aerosol formulations selected from nine personal product categories.

3. Three-year testing of customer formulations submitted for compatibility determinations.

Table 10.2 lists permeation factors generated over 2 years for some ingredients commonly used in aerosol products. The data are presented in the form of permeation rates or factors.

Permeation rate is a common tool for expressing product transfer through a substrate. In this manner, a normalized value is obtained that can be used to calculate the permeation for a vessel of any size and any wall thickness (13).

TABLE 10.2.   EQUILIBRIUM PERMEATION FACTORS FOR
COMMON AEROSOL INGREDIENTS IN ACETAL COPOLY-
MER CONTAINERS

Ingredient	P [grams/mil/day/100 sq. in.][a]
Methanol, 95%	−1.7
Ethanol, 90%	−0.9
Isopropanol, 95%	−0.3
Mineral oil	0.0
Water	−3.6
Propellant 11	+0.2
Propellant 12	+0.2

[a] Permeation at room temperature.

The permeation rate equation has proved quite useful in predicting
product loss:

$$P = \frac{L \times M \times 100 \text{ in.}^2}{d \times a} \qquad [1]$$

where P = permeation rate [grams/mil/day/100 sq. in.], L = loss of
product (g), M = average wall thickness (mils), d = time of container
exposure (days), and A = surface area of container (in.²).

The permeation rates (factors) shown in Table 10.2 are used to judge
the relative permeability through the container. The factor can also
be used in the transposed equation to determine average wall thickness, as will be seen later in [4]. Permeation predicts only average
(and not minimum) thickness because of the average nature of this
phenomenon.

A negative value indicates that the ingredient has permeated out of
the container. Similarly, a positive rate signifies a package weight
gain. This is explained by the fact that moisture has been imbibed
into the container from the storage environment. This phenomenon
occurs when a deliquescent (hygroscopic) product such as anhydrous
ethanol is packaged in the container. Here more moisture is permeating
into the container than there is loss of the deliquescent product out
of it.

After the deliquescent contents of the package have been fully saturated with water, permeation of the product out of the container is the
principal mechanism taking place. Thus, in time, there is a change in

the rate of permeation of a deliquescent product from positive to negative. The time to reach permeation equilibrium (i.e., the stage where the permeation factor levels off) is a function of the temperature and humidity of the storage environment (13).

The majority of aerosol propellants, including the flurinated hydrocarbons, hydrocarbons, and compressed gases, can be packaged in

TABLE 10.3.   PERMEATION OF CERTAIN PRODUCT FORMULATIONS THROUGH ACETAL COPOLYMER AEROSOL CONTAINERS

No. 1: Sun-Tan Oil Preparation

Concentrate	*% by Weight*
Glyceryl *p*-aminobenzoate	2.00
Mineral oil	9.00
Propylene glycol monolaurate	5.00
Isopropyl myristate	14.00
Isopropyl alcohol, 99%	36.50
Dipropylene glycol	33.00
Perfume International Flavors and Fragrances No. 4689-G	0.50
	100.00

Fill	*%*
Concentrate	50
Propellant 12/11 (40:60)	50
	100

Pressure: 30 psig
Container: 3-oz Utility Round
Valve: 20 mm with buna N gasketing
Fill level: 65 g (2.3 oz avoir)

Test Results

Component	Weight Loss (%)		
	75°F (12 months)	100°F (6 months)	120°F (3 months)
Container	+0.8	+0.7	−1.2
Valve	+0.4	−1.2	−1.1
Total	+1.2	−0.5	−2.3

TABLE 10.3. CONTINUED

No. 2: PVP-Based Hair Sprays

Concentrate	% by Weight
PVP/VA E-535 (50% solution)	20.00
Lanethyl[a]	2.50
Dimethyl phthalate	0.50
Silicone fluid 555[b]	1.25
Perfume	1.25
Ethyl alcohol, SDA-40, anhydrous	74.50
	100.00

Fill	%
Concentrate	20
Propellant 12/11 (35:65)	80
	100

Pressure: 33 psig
Container: 3-oz Utility Round
Valve: 20 mm with buna N gasketing
Fill level: 85 g (3 oz avoir)

Test Results

	Weight Loss (%)		
Component	75°F (12 months)	100°F (6 months)	120°F (3 months)
Container	−0.9	−4.6	−6.2
Valve	−1.3	−1.9	−1.4
Total	−2.2	−6.5	−7.6
Discoloration	None	None	None
3-Month Container strength	Acceptable	Acceptable	Acceptable

[a] Croda, Inc.
[b] Dow Corning.

TABLE 10.3. CONTINUED

No. 3: Personal Deodorant

Concentrate	% by Weight
Glygerin	1.00
Hexachlorophene	0.50
Isopropyl myristate	1.00
Dipropylene glycol	4.50
Perfume	2.00
Ethyl alcohol, SDA-40, anhydrous	91.00
	100.00

Fill	%
Concentrate	70
Propellant 12/114 (40:60)	30
	100

Pressure: 33 psig
Container: 3-oz Utility Round
Valve: 20 mm with buna N gasketing
Fill level: 70 g (2.6 oz avoir)

Test Results

	Weight Loss (%)		
Component	75°F (12 months)	100°F (6 months)	120°F (3 months)
Container	+2.6	+0.3	−2.0
Valve	−0.4	−0.3	−1.0
Total	+2.2	0.0	−3.0

acetal copolymer containers. It should also be noted that acetal copolymers are quite stable to Propellant 11 in aqueous media. Since the small amount of acid generated in the hydrolysis reaction that occurs with Propellant 11 in aqueous media in the presence of metal corrodes metal cans, use of this propellant in uncoated metal cans is precluded. Instead, one must use Propellant 114, which is approximately 3 times more expensive than Propellant 11, or internally coated metal containers.

Similarly, certain base-type chemical products, such as oven cleaners, can be packaged in acetal copolymer containers.

Hydrocarbons, such as butane and isobutane, exhibit little or no permeation through plastic containers made from acetal copolymers. Butane cigarette lighter bodies in acetal have been marketed since 1964 in the United States and Europe.

In order to explore the performance of actual products in acetal copolymer aerosol containers, typical formulations were placed on test (12). The test formulations included hair sprays, baby products, sun tan preparations, antiperspirants, personal deodorants, disinfectants, insect repellents, room deodorants, and topical pharmaceuticals. One-year test data indicate, on the basis of what is considered to be favorable permeation behavior, that products in each of these categories can be packaged in acetal copolymer containers. Three typical formulations and their permeation results, as measured in acetal copolymer containers under accelerated and long-term testing, are listed in Table 10.3.

PERMEATION REDUCTION BY MEANS OF BARRIER COATING SYSTEMS

In areas in which it is determined that the permeation level is excessive, barrier coating systems have been developed. For acetal containers, the barrier is a two-coat system composed of a specific primer and a polyvinylidene chloride-based topcoat. The internal coatings are applied automatically via hot, airless, lance-spray techniques. Subsequent oven baking ensures the proper curing of the primer and solvent removal. With very delicate fragrances, where it is desirable to maintain an extremely inert interior finish, thermosetting epoxies may be applied over the primer coat. The cost to apply the coating is minimal, approximating that for coating metal cans. External coatings using the same primer coat barrier systems have been almost as successful.

OTHER PERMEATION REDUCTION METHODS

Other techniques to minimize or reduce permeation are also available. One effective method is to overwrap the plastic aerosol package with a tight skin-type wrap. Shrink wrapping using polyolefins has been found to significantly reduce the permeation of ingredients during shelf life.

Another system receiving attention is double injection molding. A low-cost, low-permeable plastic such as polyethylene can be injection molded inside the higher-melting engineering thermoplastic. The combination acts similarly to film laminates used for such applications as "boil-in-the-bag" and luncheon meat wrapping.

Pressure Retention Considerations

The third consideration in determining the required plastic aerosol wall thickness is pressure retention. All materials, when under sufficient stress, undergo a phenomenon called *creep*. Plastic materials creep at lower pressures and temperatures than do metals. Engineering thermoplastics (e.g., acetals, polyamides, and polycarbonates) resist creep forces much more effectively than conventional resins such as polyethylene, vinyl, and styrene.

In an aerosol container, the internal pressures exert forces inside the container, causing a tendency toward a spherical shape. Hence container design is especially important for pressure vessels. More material must be placed in the sections of the container that are, because of design, the weakest ones. The amount of extra material needed can be calculated for engineering resins because of their predictable nature. With knowledge of the expected life of the container and its probable environment, the proper calculations can be made. For cylindrical

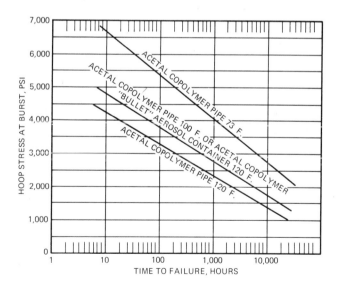

Figure 10.5. Hoop stress versus time to burst for acetal copolymer. Pipe and blow molded bullet containers.

containers the calculations are simple; plastic "pipe" or cylinder formulas apply (14). The problem becomes complex, however, as the container shape deviates from the cylinder.

The calculations for cylindrical shapes can be illustrated as follows. First, the hoop stress or the force acting inside the container on the side walls must be determined. For the plastic resin being considered, a series of hoop stress relationships over a practical temperature range is established by deliberately failing containers with varying pressures and temperatures. Thus, for acetal copolymer containers, the middle curve in Figure 10.5 was established for the "standard" bullet shape. The curve approximates the curves measured for true cylinderical pipe, which are also shown in Figure 10.5. It deviates from the 120°F curve for pipe on the "safe" side, however, because of the inherent reinforcement due to the pressure effects on the top and bottom walls of the container.

For cylindrical containers the hoop stress is determined by the Barlow pipe formula (14):

$$S = \frac{P(D - t_m)}{2t_m} \tag{2}$$

where S = hoop stress, P = pressure, D = container diameter, and t_m = minimum wall thickness.

Also, the minimum wall thickness can be calculated from the same equation presented in this manner:

$$t_m = \frac{P \times D}{2S + P} \tag{3}$$

where t_m = minimum wall thickness, P = pressure, D = Diameter of container, and S = hoop stress.

To illustrate how the minimum thickness is determined, the following example is offered.

1. A 3-oz (Figure 10.6) cylindrical aerosol container has the following specifications:

Shape   = "bullet"
Diameter   = 1.58 in.
Surface area = 20.4 in.^{2}
Practical fill = 80 g of 1.1-specific-gravity product

2. The product to be packaged is water based, and 5% per year product loss is the maximum tolerable for this item.

3. The product pressure is 40 psig.

4. No failures occurred in 5-ft random drop of 30 containers.

Figure 10.6. Standard 3-oz "bullet"-shaped plastic aerosol container used for permeability measurements. (Courtesy of Celanese Plastics Co.)

Permeation thickness determination

$$t = \frac{P \times d \times a}{L \times 100} \qquad [4]$$

where t = average wall thickness, P = permeation factor [-3.6 grams/mil/day/100 in.^{2} for water at 73°F], d = storage life of container (days), and a = surface area of container (in.^{2}).

For the "bullet," using 5%/year water loss as stated above,

$$t = \frac{-3.6 \times 365 \text{ days} \times 20.4}{-5.0 \times 100} = 0.054 \text{ in.}$$

Pressure thickness determination

Similarly, the *minimum* wall thickness requirement for pressure retention for cylindrical containers can be calculated by using the Barlow equation [3] for hoop stress in cylindrical pipe. In this equation S will be set at 2000 psi because it predicts (see Figure 10.4) a container shelf life of 1 year at 120°F or 20 years at 73°F.

To determine the *minimum* wall thickness requirement for pressure retention, assuming a product pressurized with Propellant 12/11 (50:50) in a 100°F environment (72 psig), we use:

$$t_m = \frac{72 \times 1.58}{2(2000) + 72} = 0.028 \text{ in.}$$

In blow and injection blow molding it is difficult to maintain exact uniform wall thickness. Most molders should hold ±0.010-in. tolerance in the wall of a container, so the minimum thickness is useful in predicting average thickness and container weight specifications.

Figure 10.7. Typical commercially available acetal copolymer aerosol containers. Container sizes range from 0.5 to 3 oz. (Courtesy of Celanese Plastics Co.)

Figure 10.8.   Photo of commercialized room freshener. (Courtesy of Celanese Plastics Co.)

Thus, in this example, permeation considerations call for an average wall of 0.054 in., but for pressure retention 0.028 in. is sufficient. Since the specification is determined by the factor with the larger value, the container must have a minimum wall thickness of 0.054 in. Now, by trial and error, starting with a 0.054-in. minimum for the particular container shape, the wall thickness is incrementally built up until it satisfactorily passes the impact criteria established. In this manner the container specification is established.

DESIGN AND FABRICATION

Thermoplastics are, as a rule, easily formed into a wide variety of shapes, as illustrated by the various commercially available containers shown in Figures 10.7 and 10.8. Plastic materials release the aerosol container from the cylindrical "beer can" shape that has so long dominated retail aerosol containers. Shapes as exciting as those shown in Figure 10.9 were suggested years ago by leading designers upon learning about the feasibility of manufacturing aerosol containers.

In addition to the multitude of shape designs, the plastic aerosol offers even greater variety in molded-in color, textured surface effects, and printed or decorated surfaces.

Design guidelines for achieving optimum performance, economy, and safety dictate the use of curved side walls, oval and rounded shapes, and embossed but not deeply molded surface effects. Radical departures in design should be evaluated for cost versus performance.

Figure 10.9. Design freedom provided by plastics aerosol is illustrated by these basic shapes molded of acetal copolymer. (Courtesy of Celanese Plastics Co.)

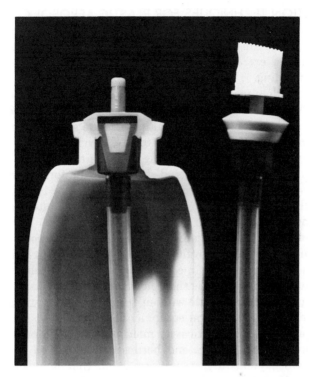

Figure 10.10. New development with plastic aerosol container suggests departure from standard metal skirt valve. New concept depicts valve with skirt molded from plastic resin identical with that used for container to allow efficient ultrasonic sealing. Cross section on left illustrates excellent ultrasonic seal which is readily obtained. (Courtesy of Celanese Plastics Co.)

Plastic containers have the inherent flexibility of being readily formed with any type of neck configuration. This means that any of the standard openings can be molded on these containers. Hence existing valves and crimping equipment can be used in sealing plastic aerosol containers.

Additionally, caps, valves, and neck finishes can be designed for maximum simplicity and easier adaptability to plastic molding and plastic sealing techniques. For example, valves with plastic "bodies" identical or similar in composition to the containers have easily been sealed in place ultrasonically, as shown in Figure 10.10.

PRODUCTION TECHNIQUES FOR PLASTIC AEROSOL CONTAINERS

Currently, there are three practical methods of molding "plastic" aerosol containers: extrusion blow molding, injection blow molding, and injection molding.

Extrusion blow molding is made feasible by the high-melt-strength plastics. Acetal polymers are among the new thermoplastics that can be blown in this manner while making a bottle strong enough to withstand aerosol pressures and environments. Several different processing systems are adaptable to fast-cycling, multiunit production. Control of wall thickness by parison (extruded molten tube formed in the blow-molding operation) programming makes it possible to "beef up" the sections that take the greatest stress and to distribute material efficiently and economically for optimum results.

Injection blow molding combines the dimensional control of injection molding with the design flexibility of blow molding. Critical areas, such as neck finishes, are injection molded to close tolerances, along with a molten parison that is then blown up to bottle shape. This process produces a finished bottle without trim, uses less material and floor space, and maximizes capital equipment use.

Injection molding in multicavity molds permits high rates of production at low unit cost. Open-end bottles can be sealed by ultrasonic welding, which yields a fast, uniform bond that is clean and easily automated. High-pressure hermetic welds are obtained in a fraction of a second. By using newly devised joint designs, bond strengths equal to 100% of the tensile strength of the plastic are obtained.

Plastic aerosol containers can be decorated by a number of common plastic decoration techniques. Flame treatment (surface treatment) of the object is not required before decoration. However, as with all plastics, surface treatment improves the adhesion of the decorative material (e.g., ink, glue, foil) to the polymer.

The established decoration techniques evaluated to date include printing, either silk screen or offset, transfer label, decal, or ink and paper labeling and hot stamping.

REFERENCES

1. South African Moulder Makes First Oval-Shaped All-Plastic Container, *Plastics, Paint Rubber*, March/April 1967.

2. M. J. Kakos and M. K. Black, Personal Product Formulations for Plastic Aerosols, *Amer. Perfum. Cosmet.*, **82,** 11 (November 1967), 57.

3. Plastics Breathe New Life Into Aerosols, *Plastics World,* **29** (July 1971).

4. Robert H. Abplanalp, U.S. Patent 2,799,435 (June 9, 1954).

5. R. H. Abplanalp and J. Pizzurro, Nylon Opens a New Era in Aerosol Containers, *Aerosol Age,* **2,** 6 (June 1957), 37.

6. H. Doll, H. E. Roger, and K. D. Voigt, Hostaform, A New Plastic for Packaging, *Verpack.-Rundsch.,* No. 3 (1965), 222–234.

7. Plastic Containers Revived, *Aerosol Age,* **12,** 8 (August 1967), 27.

8. Michael J. Kakos, U.S. Patent 3,484,399 (Dec. 16, 1962).

9. S. J., Barker and M. . Price, *Polyacetals,* ILIFF, Boos, London, 1970.

10. ASTM D-2463-70, *Annual ASTM Standards,* 1972.

11. Maurice J. Gifford and Oscar E. Selferth, U. S. Patent 3,438,788 (Apr. 15, 1969).

12. *Celcon Aerosol Container Manual,* Celanese Plastics Company, Newark, N.J., 1967.

13. C. E. Rogers, Permeability and Chemical Resistance, Chapter 9 in Eric Baer (Ed.), *Engineering Designs for Plastics,* Reinhold Publishing Corporation, New York, 1964.

14. Theodore R. Olive, Pipe Fittings and Valves, Chapter 39 in Charles L. Mantell, (Ed.), *Engineering Materials Handbook,* 1st ed., McGraw-Hill Book Company, New York,1958.

AEROSOL VALVES

The valve is an essential element in the makeup of the total aerosol package. For this reason the industry has developed a wide variety of styles and materials of construction in order to meet the requirements of all the packages to be marketed.

We define the aerosol valve as a device that is used to seal the aerosol container and to permit controlled discharge of the contents in a fashion desired by the user.

HISTORY

The history of aerosol valves as given by Beard (1) and Harris and Platt (2) shows that the concept originated as early as 1862 but that inexpensive and efficient models were not produced until the late 1940s.

With the rapid post-World War II growth of aerosol products, many new developments in aerosol valves occurred. Valves were engineered to deliver the product from the aerosol pressure package as a space spray, residual spray, foam, or powder from either the classical two-phase solvent system or the three-phase water system and, more recently, from codispensing systems.

TYPES OF VALVES

Valves are referred to as either continuous spray or metered. Continuous-spray valves operate in a simple "on" or "off" manner and are in reality two-way valves which permits either the addition of product into the container, as in pressure filling of propellant, or the withdrawal of product during use of the aerosol package. Continuous-spray valves are used for liquids, emulsions, powders, paints, and food products.

Metered valves are designed to permit the uniform delivery of a premeasured quantity of product from the aerosol container. Metered valves represent only a small portion of the aerosol industry valve requirements and are used primarily when a controlled dosage is required, as in certain pharmaceutical applications or for dispensing perfumes and colognes. Metered valves, when actuated, can be thought of as simultaneously discharging a premeasured quantity of product and sealing off the aerosol container from further delivery. These are one-way valves, since product can be expelled from the container but cannot be added to it. There are metered valves, however, which are designed to allow for pressure filling of the propellant.

Valves are further subdivided according to the product categories for which they are intended. Valves for aerosol cans are mounted in standard 1-in. metal cups; valves for aerosol bottles and metal containers with neck finishes of 20 mm or less are mounted in metal ferrules that are affixed to the containers by outside closure methods.

Aerosol valves are available for liquefied propellant systems (fluorocarbons or hydrocarbons) and for compressed gas systems (nitrogen, carbon dioxide, nitrous oxide). Further groupings for sprays, foams, streams, or powders are made, and as a result the aerosol formulator has available an exceptionally wide range of valves and variables.

The actuator or button is considered an integral part of the aerosol valve, and it is essential for proper functioning of the package that the proper actuator be selected.

VALVE COMPONENTS

The normal aerosol valve has seven basic parts (Figure 11.1).

1. Stem.
2. Stem gasket.
3. Spring.
4. Mounting cup with associated seal gasket.
5. Body or housing.
6. Dip tube.
7. Actuator.

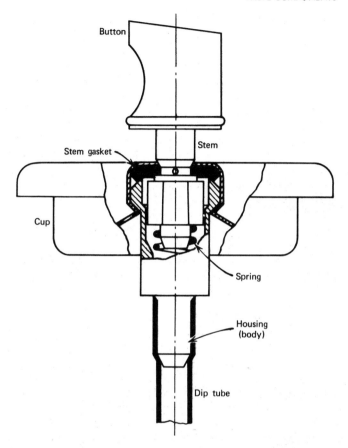

Figure 11.1.   Aerosol valve assembly (Precison Valve Corp.).

Although some valve designs add to these basic components, combine the function of several into a single part, or do without one part
for a specific application, a knowledge of the function of each of these
parts is essential to the understanding of aerosol valves.

Stem

The stem provides a metering orifice or orifices and sealing surfaces
for the stem gasket. It has attachment points at the top for the actuator and at the bottom for the spring. Generally, stem orifices are avail-

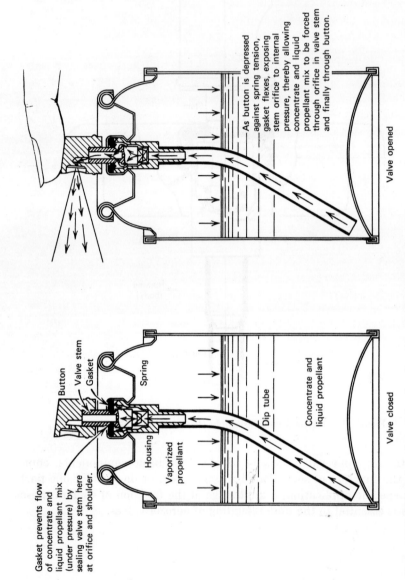

Gasket prevents flow of concentrate and liquid propellant mix (under pressure) by sealing valve stem here at orifice and shoulder.

Button
Valve stem
Gasket
Spring
Housing
Vaporized propellant
Dip tube
Concentrate and liquid propellant

Valve closed

As button is depressed against spring tension, gasket flexes, exposing stem orifice to internal pressure, thereby allowing concentrate and liquid propellant mix to be forced through orifice in valve stem and finally through button.

Valve opened

Figure 11.2. Operation of aerosol valve (Precision Valve Corp.).

able from 0.013-in. diameter to the equivalent of an 1/8-in. diameter, and usually one orifice is present, although stems with as many as four orifices are standard. The orifices are round or rectangular. Stems are commonly made of plastics; nylon, acetal, and polyester are popular, but other plastics, as well as metal, especially brass, are used. In the plastic stem the orifice is molded, whereas metal stems have the orifice drilled or pierced. Orifices may have traces of plastic flashing or rough metal edges that will interfere with the smooth operation of the valve.

The hollow core of the stem provides the first expansion chamber of the valve. Upon actuation, product from the container is forced through the stem orifice and the liquid is allowed to expand in the hollow core (Figure 11.2). From this point the product is forced through the actuator, where further expansion takes place.

Valves that have the stem molded into the actuator are available. The protruding section of the stem is slotted to resemble a rectangular orifice, and when this is inserted into the valve pedestal it becomes a metering orifice.

Stem Gasket

The stem gasket seals the stem and the mounting cup in the stem area. Almost any elastomer can be used for the gasket, but the most popular by far are neoprene and buna N rubber in 72 and 55 durometer (Shore A). The stem gaskets are made by die cutting from flat sheets or by cutting from rubber tubing.

When the valve is in the open position, the product is forced between the stem gasket and the stem and exits through the stem open position, the product is forced between the stem gasket and the stem and exits through the stem orifice. Thus it is of major importance that this gasket not swell, distort, or shrink when exposed to the product. Testing of the gaskets in the solvent constituents of the formula is recommended in order to observe any physical changes that may take place. The gasket should not leach out any materials such as colors or plasticizers. Plasticizers used in the rubber-manufacturing process can be considered a major source of objection in perfume and fragrance aerosol products. Metered valves used for mouth fresheners must be exhaustively tested with the flavors used in the package, since even very slight traces of plasticizers may leave an objectionable aftertaste.

The stem gasket–stem seal is one of the key items to be investigated in weight loss studies. The reader is referred to Chapter 13 for the specific test method for seepage rate. Weight loss should be determined whenever a formulation or propellant change is made, and actuated as well as unactuated samples should be included.

Valve Spring

The valve spring assists in closing the valve and helps to hold the stem upright. Although 302 stainless steel has been the mainstay of the industry for years, the special problems associated with a few new products have led to the development of springs made of 316 stainless and other more resistant materials. As a passing point of interest, 302 SS, which is not normally magnetic, becomes significantly more magnetic in the wire and spring-making process. This attribute can sometimes be used to differentiate the material from other grades of stainless steel.

Mounting Cup

The valve mounting cup closes the can and holds the valve together. It also is used as an attachment point for spouts, actuators, and caps. Tin plate, aluminum, and stainless steel are the most popular materials for cups. Tin plate is often provided with coatings of epoxy, vinyl, phenolics, or organisols. Some manufacturers are experimenting with urethane and acrylic coatings, but these have not yet found major market applications.

The mounting cup is available for crimping in the standard 1-in.-opening metal aerosol can. The complete valve assembly is sealed to the can curl by inside crimping of the mounting cup. Tin-plated steel cups are normally 0.011 in. thick, while aluminum cups may be as thick as 0.025 in. Different cup thicknesses require adjustments in the crimping collet. Mounting cups are marked with an ink to indicate the dip tube alignment in relation to the marking. This feature is used to orient the actuator either in the same direction as the dip tube curvature or in a 180-degree orientation. Orientation is especially important in assuring that the dip tube is always in the liquid phase during actuation. Conventional mounting cups are referred to as flat but are also offered in a concical shape. The conical cup is recommended when product impingement on the cups during actuation may occur, such as is commonly encountered with water-based sprays.

Mounting-Cup Gasket

The mounting-cup gasket provides the can-to-valve seal and must be properly compressed between the valve cup and he can bead. Cut rubber gaskets are used to create this sealing in special applications, usually involving high internal pressures. All-glass bottle valves use cut rubber gaskets for the glass-to-valve seal. The cup-to-valve seal, where weight loss occurs because of seepage, is an important check point.

Darex GK-45, a solvent-base latex rubber, is most popular in the United States for flowed in gaskets. It is not available in some world locations, however, and Darex 1564 or other water-based red compounds are used in Europe.

Valve Body or Housing

The valve body or housing serves to hold the valve in place and is also a base for the return spring. It supplies a point for dip tube attachment and provides metering orifices, a main metering orifice, and a vapor tap if desired. The valve body is usually made of plastic similar to that used for the stem. Main orifice sizes vary from 0.012 in. in diameter to over 0.25 in., while vapor taps generally range from 0.012 to 0.045 in. Sanders (3) and Beard (1) discuss the advantages and disadvantages of vapor taps.

The valve body is the metering chamber in metered valves. In such a valve the volume of the body is important, since the quantity of product discharged per actuation depends on the volume of the product stored in the valve body. Metered valves are closely inspected to maintain a narrow range of variation in the capacity of the valve body. Since the valve body is immersed in the concentrate and propellant, it should not contain plasticizers or residual materials that can be extracted into the product. If the valve body is colored, the color should not discolor the contents of the aerosol.

Vapor tap bodies provide for the draining of the valve body and dip tube when the aerosol is in the nonuse position. If suspended or precipatated solids remain in the valve body because of this draining, the valve body may become plugged upon prolonged standing.

Dip Tube

The dip tube of the valve performs the rather obvious function of drawing the material from the bottom of the can, but it can also be used as a metering device. Dip tubes are commonly made of specially compounded polyolefins, as well as nylon and other thermoplastics. Although dip tubing is available with inside diameters from 0.016 to almost 0.5 in., an I.D. of about 1/8 in. is by far the most popular. Jumbo dip tubes with diameters of 1/4 in. or more are commonly employed to provide a reservoir of product for limited inverted use.

Dip tubes are subject to the solvent and stress cracking effect of the propellants and the concentrate. Most common is dip tube elongation or "growth" of the dip tube. If this elongation is excessive, the tube may touch the container bottom and seal. This is especially true of short, flat-bottom containers. Tubes are available with a "notch cut" which, it is claimed, minimizes this effect. Dip tubes must be firmly attached to the valve body if the valve is pressure filled; loose-fitting dip tubes may be blown off the valve body during pressure filling.

Actuator

The actuator on top of the valve is the prime determinant of the form in which the product will be dispensed—as a foam, spray droplet, or stream. The term "actuator" here includes the common sprayhead button, spouts, overcap actuators, and a myriad of other forms. The actuator contains the final expansion chamber and the final orifice through which the product is forced. To a very large extent this fitment controls the appearance of the product. Actuators are either of one-piece or two-piece construction. In the two-piece actuator, a small disk or plug is inserted into the discharge port. The orifice is in the disk or plug, and behind the insert can be a swirl chamber which provides the mechanical breakup feature or a straight-through chamber. Other means for mechanical breakup are available in one-piece actuators with internal slotted swirl chambers that connect to the discharge ports. By varying the size and taper of the orifice, the spray is established. Water-based products are normally discharged as sprays by the use of mechanical breakup actuators. Propellants such as the compressed gases require mechanical breakup actuators in order to achieve a spray.

Foams and products dispensed as liquid streams or syrups use actuators with large, straight-through discharge orifices.

VALVE SELECTION

The process of the technical selection of the proper combination of valve components to assist in producing an effective product consists primarily of materials selection to provide product compatibility and component design selection to attain the desired delivery characteristics.

The materials selection should take into account the following:

1. Corrosion.
2. Component shrinkage or swell.
3. Odor effects on product.
4. Color changes in product.
5. Extraction of material from components.

Compatibility Studies

PROGRAM

Because of the multitude of actions and interactions that can occur between valve components and various product formulations, the materials selection process is largely an empirical one and must be based on a comprehensive testing program. A good program includes storage and testing with regard to the following:

1. A range of temperatures from 0°F to at least 120°F.
2. Humidity ranges from 20% RH to 80% RH.
3. Various can storage attitudes—upright, side, and inverted.
4. Cycling of containers in both environmental conditions and attitudes, with simulation of actual shipping and storage conditions.
5. Use of the package after storage in the fashion the ultimate consumer will use it.
6. Continued storage under use conditions (repeat actuation).

At regular intervals in the testing program the aerosol package should be disassembled and examined with the following points in mind:

1. Product changes in odor, color, viscosity, and so forth as a result of exposure to valve components. Especially important is the formation of unwanted precipitates.
2. Gasket changes in dimension, color, hardness, and other signs of deterioration.
3. Mounting-cup evidence of corrosion or coating deterioration, especially a tendency for the coating to flake off.
4. Spring corrosion or other deterioration.
5. Stem and body cracking, swelling, or other signs of deterioration.

TABLE 11.1. LINEAR SWELL OF GASKET MATERIALS (%)

Aerosol Material	Buna		Neoprene
	55 Durometer	72 Durometer	
Propellants			
P-11		−0.6	11
P-12	−4	−3	0
P-11/12 (50:50)	−3.5	−3	3
P-114	−2	−3	−1.4
Vinyl chloride	15	14	13
Isobutane	−4	−3	0
Propane	−4.5	−4	−0.8
A-46 (isobutane/propane, 84:16)	−4	−3	0
Alcohols			
Methanol	0.3	2.0	0.3–11
Ethynol	−2	−2	4
Isopropyl alcohol	−0.3	−1.5	0
Butyl alcohol		−1	−1
Dipropylene glycol	−3	−2	−0.5
Halogenated Solvents			
Methylene chloride	47	38	25
Carbon tetrachloride	8	6	33
Ethylene chloride	47	38	29
Trichlorethylene	26	24	36
Perchlorethylene	2	4	32
Chlorothene	23	17	28
Aromatics			
Toluene	17	15	27
Benzene	21	19	30
Xylene	12	12	33
o-Dichlorobenzene	35	31	38
Eugenol		27	27
Ketones			
Acetone	29	24	4
MEK	37	28	14
Methyl amyl ketone	31	23	25
Esters			
Ethyl acetate	19	18	13
Butyl acetate		16	21
Methyl salycilate		25	25
Isopropyl myristate	−3	−3	20
Isopropyl palmatate	−4.5	−3	13

TABLE 11.1.   CONTINUED

Aerosol Material	Buna		Neoprene
	55 Durometer	72 Durometer	
Others			
Kerosene	−3	−2	4
Mineral oil	−3	−2	1
Water	0	0	7
Water/ethanol (50:50)	0	1	13

6. Dip tube swelling and cracking. Dip tube elongation must be taken into account in adjusting dip tube length for the selected container.

7. Actuator cracking, loosening, or clogging.

INTERPRETATION OF RESULTS

Caution should be used in the interpretation of corrosion results. It is known that the effects of oxygen and water content can be insidious, in that product packed in a laboratory may be quite different from product packed on a production line in the amount of oxygen or water included. Small differences in the amounts of these materials can produce drastically different corrosion effects.

Caution should also be exercised in interpreting high-termperature tests commonly used to accelerate aging effects. Aerosols are not commonly exposed for long periods of time to temperatures over about 115°F; however, many laboratories test at 120 and 140°F to accelerate aging. It should be remembered, first, that, thermodynamically, a first-order chemical reaction doubles in rate for each 10°C increase in temperature; thus the acceleration can be significant. Second, at the higher temperatures and longer duration of accelerated storage tests certain reactions may occur that would never take place under actual storage and use conditions. Experience has shown that a package which survives 2 months at 140°F will have acceptable shelf life; one which does not, requires further study.

One prime purpose of simulated storage and use testing is to discover any tendency of the product to develop leakage or clogging. In these areas proper gasket material selection is critical. Most valve constructions will tolerate gasket dimensional changes from 5% shrinkage to 25 or 30% swell. Table 11.1 gives approximate swell dimensions for typi-

TABLE 11.2. CAUSES AND REMEDIES OF COMMON VALVE DEFECTS[a]

Defect	Cause	Check:	Remedy
1. Valve does not spray.	Propellant not added.	propellant filling machine. propellant cylinder; it may be empty or shut off.	Adjust propellant filling machine. Introduce weight control on filling line to detect and reject unfilled dispensers.
	Propellant leaked away.	valve for damage and/or inadequate sealing. for powder or dirt in formula holding gasket open.	Adjust sealing machine. Change to valve which can accommodate powder or reformulate (for powder formulas) or filter solutions and clean dispensers.
		container for perforation, seam leakage, etc.	Reject obvious leakers by passing filled dispensers through a hot-water bath.
	Valve gasket swollen and blocked because of influence of solvent. *For Solvent-Based Formulas:*	dissemble valve and control for swelling.	Use other valve gasket or other solvent system.
	Solid particle in housing, stem, or actuator orifice(s).	propellant line for metal particles (e.g., rust) from cylinder. concentrate for insolubles or dirt. formula in glass for possible precipitation on filling or slow crystal growth on storage. dispensers for dirt or insects.	Introduce propellant filter on filling line. Filter concentrate. Reformulate if necessary. Introduce dispenser air-jet cleaner on filling-line.

For Powder-Based Formulas:

	Check for	Remedy
Blockage of housing, stem, actuator orifice, or dip tube.	particle size of powder (i.e., maximum 200 mesh for regular-shaped particles or maximum 325 mesh for needle-shaped particles).	Reduce particle size of powder. Use special powder valve. If necessary, change to another type of powder valve. Reformulate.
	for clumping or hard packing of powder on standing (i.e., difficult to redisperse on moderate shaking).	Use suitable suspending agent.
	for undesirable crystal growth in powder (use glass control dispensers).	Change solvent or/and propellant system.
	ease of redispersion on shaking.	Reformulate if necessary. Clearly advise: "Shake before use," on consumer instructions.
2. Valve leaks from gasket.		
Gasket shrunken because of solvent or/and propellant influence.	influence of individual components in formulation upon gasket.	Change gasket material or solvent and/or propellant system.
Gasket held open by trapped dirt, insect, or precipitate from the concentrate or dispenser.	the filled formulation for hard precipitates, crystal growth, dirt, or insects.	Filter concentrate. Filter propellant. Introduce dispenser air-jet cleaner on filling line. Reformulate.
Damage to surrounding valve cup during valve sealing or propellant filling processes.	sealing machine and/or propellant filling machine.	Adjust appropriate machine.
Absence of gasket.	other valves from same batch by random test.	If more than an isolated case, contact valve supplier.

TABLE 11.2. CONTINUED

Defect	Cause	Check:	Remedy
3. Valve leaks from loose grommet or flowed-in-gasket.	Swage diameter incorrect.	with swage diameter meter.	Adjust sealing machine.
	Swage depth too great, therefore seal too loose.	with swage depth meter.	Adjust sealing machine.
	Loose grommet absent/fallen off.	before applying to dispenser.	*Immediately:* introduce visual inspection before use. *Long-term:* change to flowed-in gaskets.
	Loose grommet slipped from position.	as above.	As above.
	Flowed-in-gasket dried out.	storage conditions. stock rotation of valves.	Improve storage conditions. Ensure old stock used up in order of delivery.
	Powder from formulation (if applicable) or dirt between valve and dispenser.	powder filling machine or valve and dispenser storage conditions.	Adjust powder hopper, or introduce device for wiping powder from dispenser 1-in. aperture. Improve valve/disperser storage conditions.
4. Valve leaks through stem or/and sprayhead.	See valve leaks from gasket (above).	as above.	See above.
	Spring (or flexible rubber plastic) is too weak or is damaged during filling process.	filling speed and pressure if filled through the valve.	Adjust filling machine.
	Flexible rubber/plastic seal not sealing or only slowly sealing.	effect of formulation on flexible rubber or plastic valve component.	Reformulate or change valve type.

Symptom	Cause	Check	Remedy
5. When normally operated, only propellant sprays from valve.	Spring too weak to shut off rapidly or completely.	shut-off speed of valve.	Ask valve supplier for stronger spring.
	Valve supplier forgot to assemble spring (where normally supplied).	random samples when accepting delivery.	Return valves to supplier.
	Concentrate not added.	concentrate filling machine and concentrate line.	Adjust concentrate filling machine. Introduce weight control on filling line to detect and reject unfilled or in adequately filled dispensers.
	Dip tube fallen off because: through-valve filling speed too high.	for blown-off dip tube.	Use lower filling speed, or ask supplier for valve capable fo higher filling speed, or use under-cap filling machine with existing valve.
	stress cracking.	for split in dip tube.	Reformulate or change dip tube.
	dip tube or housing blocked but propellant still sprays through vapor phase tap.	for dirt, insects, precipitation, etc., in solvent-based formulation.	Filter solutions and/or propellants, clean dispensers, and control formulation in glass.
		for packing, clump formation, difficulty of redispersion, incorrect usage etc., in powder-based formulation.	Control formulation in glass and if necessary reformulate.
	dip tube too long and pressed firmly against bottom but propellant still sprays through vapor phase tap.	whether initial dip-tube length is suitable for dispenser.	Shorten dip tube to suitable length.
		whether dip-tube growth occurs after filling.	

TABLE 11.2. CONTINUED

Defect	Cause	Check	Remedy
6. Sprayhead falls off.	Sprayhead/stem tolerances too great (loose fit).	tolerance of incoming valves.	Ensure that tolerances for components are adequate and maintained.
	Solvent effect.	effect of solvent influence on sprayhead and/or stem.	Change solvent system or stem/sprayhead material.
7. Sprayhead splits.	Sprayhead/stem tolerances too great (tight fit).	tolerances of incoming valves.	Ensure that tolerances for components are adequate and maintained.
	Stress cracking.	whether this is the fault of solvent, propellant, or one of the other formulation components.	Reformulate or change to sprayhead of another material or possibly of another design.
8. Metering valve sprays continuously.	Assembly fault.	random samples when accepting delivery.	Contact valve supplier if this occurs excessively.
9. Metering valve does not spray, or it gives a faulty dosage.	Incompatibility of one or more valve components with formulation.	for change in dimensions of valve components due to swelling or shrinkage.	Reformulate or contact valve supplier for alternative.

a Reprinted from the *Aerosol Troubleshooting Charts* by S. C. Elvin and P. L. F. Pos, Aerosol Application Laboratories, Naarden International.

cal buna and neoprene gaskets in a wide range of commonly used aerosol materials. Harris and Platt (2) suggest that gasket swell for mixtures of liquids can be estimated by weighting the gasket swell in each of the pure components by the percentage of the component in the total solution. In the event that seepage or clogging is found in the stability testing phase, Table 11.2 may be of assistance in evaluating and finding the cause of the problem.

Delivery Characteristics

Proper valve selection can have beneficial effects on the delivery characteristics of a given product, especially if it is to be dispensed as a spray. Some of the spray characteristics that can be significantly altered by valve and actuator selection are the following:

1. Delivery rate (g/sec).
2. Spray pattern.
   a. Shape (circular, fan, etc.).
   b. Description (solid, doughnut, or other).
   c. Dimensions.
   d. Particle size and distribution.
3. Velocity of spray.
4. Throw or extent of spray.
5. Temperature of spray.
6. Flammability (flame extension, etc.).
7. Sound volume and tone.
8. Foaming, fallout of particles, and product buildup on actuator.

From a fluid flow viewpoint, the typical aerosol valve is made up of a series of channels, orifices, and expansion chambers. The prime metering points avaible for manipulation are, in order of flow, as follows:

1. Dip tube. Use of capillary-size dip tubing can be important in regard to metering.
2. Main housing orifice.
3. Vapor tap orifice in the housing. Use of a vapor tap will usually result in a warmer, drier spray with reduced delivery rate, narrower pattern, and finer particle size.
4. Stem orifice.
5. Actuator orifice or orifices. Usually the actuator orifices have the greatest effect on the delivery characteristics listed above.

Most studies of spray characteristics start with the determination of an orifice combination from the five variables just given that will provide the desired product delivery rate. The variables can then be manipulated to optimize the other spray characteristics desired. Fulton, Yeomans, and Rogers (4) have studied the effect of various orifice combinations in conjunction with expansion chambers. In particular, the proper balancing of expansion chamber size and orifice combinations appears to affect particle size and particle size distribution. It has also been found that flame extension is affected to a significant degree by particle size distribution. Some valve manufacturers offer variations in expansion chamber size in the housing, stem, and button; in fact, some aerosol buttons can be viewed as having two or three expansion chambers separated by controllable orifices.

The use of mechanical-breakup-style buttons described in references 1, 2, and 3 usually results in finer particle size and wider patterns as a result of the swirling action imparted to the stream of liquid as it exits from the button orifice.

In addition to the mechanical breakup feature some manufacturers offer buttons with tapered orifices. In general, a forward taper, which is larger at the exit end, gives a more narrow pattern, while a reverse taper with the smallest portion at the exit generally produces a wider spray pattern in low-propellant aerosols.

Stem Height

Another consideration in valve selection is stem height. The Chemical Specialties Manufacturers Association defines stem height as the vertical distance between the top rim of the mounting cup and the top of the stem. This dimension becomes of great importance when a spout, overcap actuator, or other device is attached to the aerosol package.

Most such spouts and other devices are designed to work properly only if the stem height is within a relatively narrow range, usually $\pm$ 0.025 in. Problems arise because the final stem height is a variable function of the can clinch operation, the propellant pressure, and the hot-tank conditions.

For a better understanding of this consideration we define two other dimensions and describe the effects of the above variables on stem height. Stem height is associated with two other dimensions as follows (Figure 11.3):

Stem height (SH) - stem extension (SE) = pedestal height (PH)

where *stem height* is the vertical distance from the top rim of the mounting cup to the top of the stem, *stem extension* is the vertical dis-

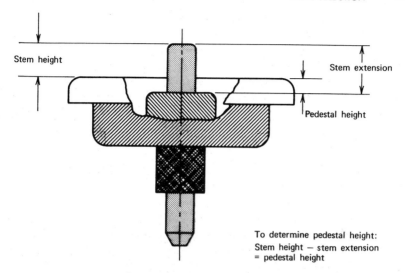

To determine pedestal height:
Stem height − stem extension
= pedestal height

Figure 11.3.   Some aerosol valve dimensions (Precision Valve Corp.).

tance from the top of the valve pedestal to the top of the stem, and
pedestal height is the vertical distance from the top rim of the mounting cup to the top of the pedestal. In this connotation a negative pedestal height means that the pedestal is below the rim of the cup.

Valve manufacturers take pains to deliver "raw valves" with controlled stem heights. In the filling process this stem height changes by a
factor called the "stem height rise" to a new dimension termed the
"finished can stem height." Thus SH (raw valve) + SH rise = SH
(finished can)

Stem height rise is made up of four factors.

1. The clinching operation of swaging the valve into the can spreads
the metal of the cup and results in a rise in stem height and pedestal
height. A rough estimate shows that a 1.070-in. clinch width causes an
increase in stem height of about 0.040 in. This figure may vary as a
result of mounting-cup design and clinch collet configuration.

2. In pressure filling, the pedestal may be depressed by the head
and by the propellant, resulting in a decrease in stem height. This
decrease is normally less than 0.010 in.

3. As the propellant pressure is exerted on the underside of the
mounting cup, the pedestal is pushed upward, resulting in an increase
in stem height. This increase is normally in the range of 0.005 to 0.015
in. but is directly affected by the propellant pressure, especially in the

hot tank. Hot-tank temperature can therefore be used at times to help control stem heights. Deformation of the cup in the hot bath is particularly noticeable with aluminum valve cups. Also some increase in stem extension can be noted after the can is pressurized (usually <0.004 in.).

4. As the can cools from the hot-bath temperature, the pedestal height and stem height are slightly reduced, usually by about 0.004 in.

The sum of all these effects is the above-defined "stem height rise," which too is a function of formulation and filling conditions. In order to provide some flexibility to account for the different values of stem height rise encountered in different products, some valve suppliers offer variations in cup configurations (different raw-valve pedestal heights), such as conical or flat cups. In addition, various stem lengths (stem extensions) are offered. Thus good valve selection involves first the determination of stem height rise under production conditions and then the proper selection of cup pedestal height and stem extension to attain a finished can stem height suitable for the chosen spout or the overcap actuator design.

Codispensing Valve Selection

Codispensing valves are described by Johnsen and Dorland (5) and by Sanders (3). The codispensing system is the first package concept with aerosol convenience which offers the formulator the opportunity to effect a chemical reaction at the time the product is dispensed. This capability opens new horizons for aerosol product formulation.

Along with this advance come increased problems of valve selection. In addition to the considerations in regard to a normal valve discussed above, four additional areas become of concern.

1. Ratio of the two phases present. This is usually controlled by orifice size selection in the valve or actuator, but it is important to test to be sure that the ratio does not change beyond reasonable bounds as the product is used.

2. Permeation of one phase through the internal container and valve components. This must not be of such magnitude as to affect the product characteristics.

3. Phase mixing. The various codispensing valves available differ widely in design regarding where in the system the two phases are mixed to form the final product. Some mix internally, some mix externally, and some mix internally in such a way that the final product is trapped inside the valve. This means that valve selection now involves

knowledge of the effects of three different chemical systems on the valve components.

4. *Safety*. If inadvertent mixing of the two components can cause rapid or extreme increases in temperature or pressure, all elements of the total package must be selected to avoid any hazards.

REFERENCES

1. W. C. Beard, Valves, Chapter 5 in H. R. Shepherd (Ed.), *Aerosols: Science and Technology*, Interscience Publishers, New York, 1961, pp. 119–211.

2. R. C. Harris, and N. E. Platt, Valves, Chapter 6 in A. Herzka (Ed.), *International Encyclopaedia of Pressurized Packaging (Aerosols)*, Pergamon Press, New York, 1966, pp. 78–118.

3. Paul A. Sanders, *Principles of Aerosol Technology*, Van Nostrand Reinhold Company, New York, 1973, pp. 46–53.

4. R. A. Fulton, A. Yeomans, and E. Rogers, *Chem. Spec. Mfr. Assoc. Proc. 36th Mid-Year Meet.*, June 1950, p. 51.

5. M. Johnsen and W. Dorland, *Aerosol Handbook*, Wayne Dorland Company, Caldwell, N. J., 1971, pp. 159–214.

MANUFACTURE AND PACKAGING OF AEROSOLS

The term "manufacture" is applied to both the preparation of the various mixtures to be dispensed from aerosol containers and the cofilling of these products and their propellants into suitable aerosol containers. The products can be made by simply mixing two liquids or by effecting the reactions of both organic and inorganic materials. The filling operation is done by two basic methods that can be considered as "cold filling" and "pressure filling." In simple terms, in the cold-filling method, the product to be filled and the propellant are chilled to low temperatures so that the propellant can be added as a liquid at room temperature. The pressure-filling method allows the filling of the product at room temperature (or elevated temperature if required), and the propellant, which is handled as a liquid under pressure, is injected into the packaged product by some mechanical means. Most products can be filled by either one method or the other.

Public acceptance of the aerosol container as a convenient form of packaging has widened the scope of aerosol manufacturing. Originally, aerosols were available mostly as insecticides, mothproofers, and room deodorants. The list of aerosols now includes such consumer items as paints and coatings, household cleaners and waxes, foods, pharmaceuticals, cosmetics and toiletries, and many other categories. These products are processed as liquids, creams, solids, emulsions, and suspensions. They are filled into aerosol containers (Figure 12.1) fabricatd from metal, glass, or plastic and range in size from 1/4 oz to about 32 oz. Such great diversity in product types and aerosol systems has created the need for developing new and highly sophisticated manufacturing processes.

Figure 12.1.   Cosmetic, perfume, and powder filling line (Peterson/Puritan, Inc.).

Historically, the industry manufactured most aerosols by the cold-filling method on equipment usually modified from other industries. Products were mostly of a low viscosity, such as solvent-based insecticides and room sprays. These products are well suited for cold filling at reasonably high speeds and extended production.

As aerosol requirements increased, however, it became obvious that cold filling was an expensive method. Today its popularity has waned, chiefly for three reasons.

1. The high investment cost required for expensive and elaborate refrigeration systems.

2. The high operating costs due to power requirements and propellant losses.

3. The lack of flexibility, since neither water-based systems nor hydrocarbon propellants can be cold filled.

As new products were introduced, propellant filling came to be handled more and more on "pressure fillers," which can be installed at a lower investment cost. The chief advantage of pressure filling lies in the fact that many more formulations can be pressure packed since chilling of the product is no longer a packaging specification. With the "chilling" restriction removed as one of the aerosol manufacturing conditions, an increasing number of "self-fillers" began to market their products in aerosol packages. This is in contrast to the early aerosol manufacturing plants, which were operated mostly by "contract fillers."

With the tremendous market growth which occurred, it was obvious that many new aerosol valves and mechanisms would appear. Valve orifice restrictions placed a severe impediment on pressure filling. Formulations using a large quantity of propellant or large aerosol containers slowed pressure filling considerably. As a result, simple in-line filling was abandoned in favor of rotary pressure fillers, which have as many as thirty separate gassing heads. Rapid filling was again possible, but the new pressure fillers were extremely complicated and costly to operate.

The best method to date for the addition of propellant to the product is the "under-the-cap" pressure filler. This latest filling machine operates by forcing the propellant under the valve mounting cup into the container. The operation is performed just before valve crimping.

As more new products were marketed, a need for lower-cost materials developed, and hydrocarbon propellants were substituted in part for the more costly flurocarbons. Hydrocarbon propellants permit the formulation of many water-based aerosol systems and other products. In manufacturing, the use of hydrocarbons as propellants has meant that propellant filling must now be conducted in "explosion-proof" areas.

As mentioned above, the growth in aerosol production has brought self-fillers (marketers) into the manufacture of aerosols. These captive fillers have been responsible in large part for the steady improvement of the basic aerosol methods in terms of both operating costs and operating efficiency.

In view of the many product areas in which the aerosol package is found, one must consider aerosol manufacturing in relation to six categories.

1. Cosmetics and toiletries.
2. Foods.
3. Pharmaceuticals.
4. Household products.
5. Paints and coatings.
6. Industrial products.

This classification may be further divided into five systems.

1. Solvent or anhydrous liquids.
2. Water-based liquids.
3. Creams and emulsions.
4. Powders.
5. Suspensions and slurries.

Any treatment of the above-mentioned classes must include product preparation, material handling, propellant and product filling, and the manufacturing controls required during production.

The subject matter in this chapter is presented under the following headings:

- Plant design.
- Processing, mixing, and storage.
- Product filling.
- Propellant filling.
- Crimping.
- Auxiliary packaging equipment.
- Product Quality Control.

The two major areas are processing and mixing, and filling and packaging. These areas have many specific aspects unique to aerosol manufacture.

PLANT DESIGN

A manufacturing facility should be designed with prime consideration for the nature of the manufacturing process that will be carried on in the area. Such a facility may be thought of as one big machine that contains all the smaller machines, which are so integrated in the plant layout as to provide a coordinated flow of marketable production. The aerosol plant may be arranged according to two fundamental methods.

One method is to group together all machines of a similar kind in one area. All work to be done on a particular machine must be sent to that area. This method is referred to as the "process layout" method. Manufacturing of this kind is usually intermittent and on an order-to-order basis. Process layout is used when the product cannot be standardized and the production volume is low. This method of plant design permits flexibility in the manufacturing sequence, and the investment in machines and equipment is less since each product produced does not require a complete production line. The handling and labor costs in this form of plant are high, however, since the product must go from one department to another for each assembly operation.

When specific products are made on a continuous and mass-produced basis, the "product layout" method of production is used in the plant design. The product layout method arranges the manufacturing equipment according to the operations that will be performed during the manufacturing cycle. The material flow is continuous from start to finish for each particular product. If more than one product is made on a large scale, each product has its own equipment and product layout. A typical illustration of this method is a plant designed to produce hair sprays on a high-volume continuous basis. This method of manufacturing is also referred to as the straight-line or direct method. Here the design and layout of the plant are concerned with the characteristics of the product to be manufactured, and it is necessary that all of the various components used in production be standardized and variations minimal.

Lines of this type are installed to produce a certain number of units per hour, and each machine employed in the line must be selected with this objective in mind. Inefficient line utilization with resulting high operating costs will occur if the production of one machine is greater than the capacity of the manufacturing line as a whole.

A most important criterion in the selection of a layout design for the manufacturing area must always be the cost of equipment. The objective for the design engineer should be to deliver the best product at the lowest cost to the purchaser.

In the early stages of plant design, it is essential to define the critical factors influencing the overall operation. Consideration must be given to the geographical area in which the plant is located. Pembroke (1) cites the following factors to consider in selecting an aerosol plant location:

- Availability of an adequate labor force.
- Availability of land for future expansion.
- Proximity to routes of transportation.
- Proximity to major customers.
- Proximity to major suppliers.
- Availability of abundant and inexpensive power.
- State and local regulations governing the manufacture of aerosols.

With these points defined, selection of the physical size and design of the plant can proceed smoothly. The general areas of warehousing and storage, receiving and shipping, processing and mixing, filling and packaging, production control, and material handling are defined and their space requirements allocated.

The types of products that will be manufactured dictate the manner in which the plant will be outfitted. One special aspect to consider is plant sanitation. This area once was considered to be applicable mainly to the food and pharmaceutical industries, but now must be considered when manufacturing cosmetics and household products as well.

Burg (2) describes the sanitary design requirements that must be taken into account when processing food into aerosols. Included are the following:

Plant site location. Water supply, solid waste disposal, liquid waste disposal, sanitary sewage disposal, and absence of potential harborages for insects and rodents in the surroundings must be considered.

Floors and drainage. Special materials to resist erosion and provide proper drainage should be used.

Walls and ceiling. Walls should be glazed, since paint peels in high-humidity areas.

Ventilation. Ventilation must be adequate to remove odors and maintain a supply of clean, filtered air. Air conditioning is highly recommended, with the air temperature ideally maintained at 68 to 78°F and the relative humidity between 30 and 70%.

Lighting. Proper lighting is essential for the health, safety, and efficiency of the workers. All electrical outlets, connections, and junctions should be vapor and moisture proof as well as grounded.

Materials of construction. Processing, storage, and filling equipment should be of stainless steel 304L and/or 316L. All wetted surfaces should have a No. 4 polish. Where flexible tubing is required, clear plastic rated for food service should be used.

Safety must be the uppermost consideration in designing an aerosol facility. Included here are the following factors:

- The placement of storage tanks for propellants, both fluorocarbon and hydrocarbon, and the installation of monitoring devices to report explosive levels from flammable vapors immediately as detected.
- The installation of sprinkler systems throughout the manufacturing plant.
- Provisions for emergency power generators to be used during power failures to provide electricity for critical operations (i.e., the refrigeration of tanks storing concentrates and propellants).
- Installation of explosion-proof motors, lights, and equipment where flammable materials are mixed or filled.

The final factor to be considered in plant design is provision in the layout for a pilot-plant manufacturing operation. The pilot plant is equipped with the same filling apparatus that is used in full-scale production and should be operated in a joint manner with the development department to establish meaningful manufacturing specifications. A well-equipped pilot plant can provide numerous answers about process capability, filling speeds, and pitfalls in processing and will also disclose whether all the package components perform as planned for in the product concept. In operation, the pilot plant should use, whenever practicable, the same chemicals, propellants, and packaging components that will be utilized during full-scale production. Units made from the pilot plant will then represent actual production conditions, and any samples used for product performance tests, stability tests, clinical tests, and consumer tests are excellent approximations of commercial production.

A properly designed pilot plant provides the manufacturing department with vital information and the training necessary for full-scale production.

PROCESSING, MIXING, AND STORAGE

The preparation of the product which is dispensed in the aerosol package requires the same methods and standards that are used in good manufacturing practices for nonaerosol products. Aerosol products are produced generally in two operations. After manufacture, the concentrate must be brought to the filling operation, where the propellant is added and the finished product prepared.

Aerosol products can be generally classified into groups that require special processing methods. Such a grouping includes the following:

1. Solutions.
   a. Solvent-based.
   b. Water-based.
2. Emulsions and Suspensions.
3. Powders.

The maintaining of stringent manufacturing standards during processing of the concentrate is imperative, since errors made during mixing may not be detected until the addition of the propellant and the crimping of the valve to the finished container. Whenever possible, the aerosol concentrate is tested before the addition of the propellant. Good manufacturing practice dictates that all chemicals be arranged and properly identified for use in manufacture. Processing equipment must be kept clean and should be sanitized before manufacture of the concentrate is begun. Generally, the processing department is responsible for four functions.

1. Raw material storage.
2. Compounding and mixing.
3. Bulk storage of finished concentrates.
4. Disposal of scrap concentrates.

Many aerosols are based on water/solvent systems, and accordingly of prime importance is the treatment of the raw water that is added to a suitable solvent for preparation of the aerosol concentrates. The trace levels of dissolved salts present in untreated water have a marked effect on the stability of the aerosol container and in certain cases can cause container failure and product breakdown. Raw water should be filtered to remove suspended solids, which are usually of organic nature. The water is then processed through demineralizers to remove metallic ions. Both cationic and anionic exchange units are necessary in delivering demineralized water to the processing tanks. Demineralized water should be piped in chemically inert plastic, which prevents contamination by metallic ions. Treated water should not be stored for prolonged periods lest it become contaminated by metallic ions and air-borne microorganisms.

When concentrates are prepared in the anhydrous state, the raw materials must be protected from atmospheric moisture during storage and use. Bulk storage tanks should have air driers which remove moisture from atmospheric air before it enters the tank. The vessels holding the finished concentrates are sealed, and a blanket of nitrogen is maintained in the head space to prevent moisture pickup.

Where filling of a product is continuously carried out, automatic batching may be preferred over manual batch making. In automatic systems, the raw materials are added to the mixing kettle through metering devices. Liquids are pumped through recording meters, which are preset to deliver the proper quantity of each material. Dry materials are preweighed into hoppers that feed automatically into the mixing kettle. More recent developments in concentrate mixing include processing by computer and programmed techniques.

The preparation of emulsions and suspensions usually requires a mixing area which is supplied with steam and cooling water. Emulsions are normally prepared by the addition of one phase to another phase. Although emulsion manufacture is not the subject of this chapter, it should be mentioned that emulsions [usually either oil in water (o/w) or water in oil (w/o)] have critical mixing times and addition rates. Kettles in which emulsions are prepared should be equipped with variable-speed mixers of suitable design. Most often, mechanical equipment providing high-shear mixing is required. The dispersion is effected through the use of three different types of high-shear equipment.

1. Rotor-stator machines, such as the colloid mill.
2. Pressurized fluid machines, such as the homogenizer.
3. Vibrational machines, such as the ultrasonic acoustical emulsifier.

Liquid concentrates that have been approved for filling should be filtered before being added to the container. Filters provide for the efficient removal of any suspended materials that may be present in the concentrate. These small particles, if packaged, can obstruct product flow through the valve and cause plugging of the aerosol unit. Viscous concentrates should be passed through strainers to remove large bits of solids.

Powders used in aerosol sprays are processed in ribbon or tumble blenders. When more than one powder is blended, the particle sizes of the various powders should be similar to each other to assure a homogenous mixture. Prolonged blending times can reduce the particle size of powders, and too small a particle size may create a pasty mass after propellant addition. Powders selected for use in aerosols should be free from lint and fine threads, as the presence of these foreign materials can cause obstruction in the aerosol valve, resulting in nonoperative packages.

Processing equipment should be constructed from materials that provide an easily cleaned system. The materials of construction should be nonreactive with the chemicals or concentrates handled in the system. In most cases, tanks, pumps, heat exchangers, and piping are constructed of stainless steel 304 or 316. Equipment for the processing of foods must conform to local and federal regulations. Food tanks, pumps, and equipment must be so designed as to permit sanitizing and cleaning between batches and during downtime. Where foods are mixed, it is required that the area and equipment be tested for microbial contamination. This is true also for cosmetic and pharmaceutical aerosols.

Propellant storage and handling are major considerations in the manufacture of aerosols. Propellants are classified as fluorocarbons, hydrocarbons, or compressed gases. Each type is handled and stored differently and will be considered separately.

Figure 12.2. Bulk storage of propellants (Barr-Stalfort Co.).

Propellants can be purchased in bulk quantities (i.e., tank trucks, tank cars, or cylinders). Storage and handling of cylinders are the same for all propellants, but flammable propellants must be stored in special areas. Fluorocarbons are available in 1-ton cylinders, whereas cylinders for hydrocarbons rarely exceed 150 lb. Compressed gases are available in standard-size cylinders that are rated in cubic feet of gas per cylinder.

Bulk storage of propellants (Figure 12.2) can be a major factor in the economical operation of the aerosol filling plant, and such facilities, when properly installed, provide an inexpensive method of handling propellants. The storage tank should be of sufficient size to accommodate tank truck shipments of 4000 gal. Bulk tanks range in size from 2000 to 6000 gal. The storage tank is constructed of steel, and the design is based on the vapor pressure of the propellant to be stored. When Propellant 11 is stored, the tank design pressure should be at least 30 psig. For Propellant 12, the minimum design pressure for the tank should be 200 psig. All tanks are equipped with pressure-relief valves that will alleviate excess pressure should it occur.

Pumping apparatus is of the positive-displacement type, and selection depends on four specific requirements: (a) flow rate, (b) delivery pressure, (c) cost, and (d) reliability.

Flammable propellants require larger bulk storage tanks, since their liquid densities are approximately 4.5 lb/gal versus 11.7 lb/gal for fluorocarbons. Tanks of 18,000-gal storage capacity are often recommended. Tanks for hydrocarbon storage have a minimum design pressure of 250 psig for propane service and 125 psig for butane. All piping is designed for 250 psig minimum, with relief valves installed to protect against rupture from expansion of a liquid-full system.

Specific storage tank and unloading facilities are recommended by the National Fire Protection Association Standards, Pamphlet 58.

All bulk storage and handling installations must conform to state and local ordinances and to insurance company regulations.

PRODUCT FILLING

The filling of aerosol products is one of the more complex manufacturing processes. Aerosols are filled as clear, one-phase solutions, powders, dispersions, or emulsions. The product being filled is called the concentrate or base and so is differentiated from the propellant. In aerosol filling, the concentrate or base is normally added to the container before the addition of the propellant. As a result, aerosols are filled by a two-stage system.

Cosmetics, pharmaceuticals, foods, paints, and insecticides are specific product categories and accordingly require filling procedures designed to fit their quality protocols. Filling of the concentrate can be done by the "cold-fill" or refrigerated method and the "pressure-fill" or ambient-temperature method. In the cold-fill method, the concentrate is chilled to extremely low temperatures between −40 and 0°F. These low temperatures must not alter the product when it returns to room-temperature equilibrium. The cold-fill process excludes the filling of water-based concentrates such as starches, window cleaners, room deodorants, and spray waxes. Similarly, emulsions (shaving cream) and viscous products are not cold filled. As mentioned earlier, pressure filling is the more versatile procedure for the manufacture of aerosol products and has thus become the predominate method used.

The filling process cannot begin until the concentrate has been approved for packaging by the quality-control department. The approved concentrate is transferred into holding tanks which supply the product fillers. These tanks may be equipped for low-temperature cooling if the cold-fill method is used. An alternative to a chilling tank is an external heat exchanger through which the concentrate is circulated and chilled. When external heat exchangers are installed, however, caution must be exercised to prevent extremely low temperatures at the cooling interface. This condition can cause ingredients to freeze out within the heat exchanger, possibly resulting in a loss of active ingredients. Perfume solutions, when chilled harshly, may lose certain constituents by solidification. These solids will not redissolve at room temperature in the alcohol/propellant mixture. In these cases, the fragrance impact from the finished aerosol unit is appreciably altered from the standard.

Storage tanks are provided with pumps, filters, screens, and return lines from the filling machine. All tanks and lines must be accessible for washing and cleaning. Concentrates that separate during storage, such as emulsions, require that slowly turning mixers be mounted in the storage vessels.

Filling areas for foods or sterile products should be specifically designed to provide for the critical stages of storage and filling. The need for aseptic filling mandates an aerial environment free or virtually free from microorganisms. Floors and walls should be uncluttered, and all equipment designed for easy dismantling and sterilization. The entire filling area must be cleaned and disinfected with sanitizer solutions which are rinsed through all tanks, lines, and filters. The aseptic area should be maintained under a positive pressure by utilizing a system of double-door airlocks to minimize the entry of untreated air, which carries microorganisms. Harris (3) has stated that in food and pharmaceutical filling, "The air inside and outside of plants, the building, raw materials, packaging components, and machinery are all sources of contamination."

Some food products are filled at elevated temperatures (180 to 200°F) and the hot product is used to self-sterilize the internal components of the package. After the sterilization cycle, the aerosol containers are chilled and the internal temperature is brought to 70 to 80°F. Where hot packing cannot be accomplished, the valves and containers must be sterilized before use. Valve sterilization can be achieved by radiation or with ethylene oxide. Dry heat will ruin the valve by causing dimensional changes of the plastic components. Metal containers are sterilized by autoclaving and dry heat, or, depending on the product, the container may be washed with a sanitizing solution and rinsed with filtered or pyrogen-free sterile water. Glass containers are washed as above and if uncoated can be treated by dry heat or by autoclaving.

Nonfood liquid filling can be classified as solvent based (anhydrous) and water based. The first category includes colognes, hair sprays, room sprays, and personal deodorants, as well as insecticides and paints. As a product class, alcohol-solvent systems are filled on either vacuum fillers or piston fillers (Figure 12.3). Selection of a filler depends on filling rate, size of container, and filling accuracy. Colognes or perfumes filled into clear glass containers must be filtered to a sparkling clarity, and no precipitate should be present. Clarity is achieved by filtration after chilling to 20°F and passing the solution through acceptable filter pads. Because of toxicity considerations, filter pads containing asbestos should not be used. After filtration, coloring is added if required. To achieve uniformity of color, dyes are always added after filtration.

Presant (4) reports a method for filling uncoated glass bottles which, he claims, overcomes the problem of precipitation due to the poor solvent nature of the fluorinated propellants. In this procedure, the aging of the perfume solution is carried out in the presence of Propellant 12 and/or 114 at low temperatures and in pressure tanks. This single-stage liquid solution is then chilled, filtered, and filled. One-stage fill-

Figure 12.3.   Piston-type product filler (Barr-Stalfort Co.).

Figure 12.4.   Accuflow filler (Peterson/Puritan, Inc.).

ing is a technique said to provide the additional advantage of filling accuracy, especially in small packages of under 1 oz. (Figure 12.4).

Solvent-based products are filled on aerosol lines that can operate as either cold or pressure fill. When the concentrate is filled through a refrigerated machine, moisture tends to condense on the filling nozzles. Such moisture can collect and drip into the concentrate being filled. This moisture may create a stability problem in the finished aerosol when a corrosive or chemical action results. Excess moisture condensation is prevented in the filling area by enclosing the product filler in a low-humidity room. Anhydrous specifications are easier to maintain if the concentrate is filled at room temperature and the propellant is then pressure filled. Metered valve systems, how ever, are seldom filled on pressure-fill equipment.

Single phase water based concentrates are filled at ambient temperatures because it is impossible to cool these solutions to low temperatures before filling. When the water-based product is added at room temperature, the propellant must be filled by pressure through the valve or under the cap.

When filling water-based products, removal of oxygen from the product is a particularly important operation and diminishes the corrosive effect of oxygen on the metal container. Lychalk and Webster (5) have reported the effect of oxygen removal in aerosol starch systems. Their study concluded that a low oxygen content provides a noncorrosive aerosol system with superior shelf stability.

Oxygen can be removed from the liquid concentrate by injecting nitrogen under pressure into the liquid concentrate. The nitrogen displaces the oxygen and forces it out of the solution. The head space in the storage vessel is also maintained under a blanket of nitrogen. Oxygen that is contained in the head space of the aerosol container is removed by vacuum crimping or purging. By combining both methods, an effective control of oxygen in the container can be maintained. Trider (6) has published a report on nitrogen sparging of concentrates for aerosol systems.

Most water-based concentrates have a tendency to foam during filling and this necessitates that product fillers operate with a minimum of foaming in order to prevent foam from entering the vacuum lines during vacuum crimping. Persistent foams, such as those encountered in shave creams, are best filled at elevated temperatures between 110 and 120°F. This method provides a concentrate with a minimum of foam. Filling water-based concentrates at elevated temperatures has the additional advantage of requiring a shorter heating time in the hot tank.

Aerosol powders are filled as dry powders by means of an auger-type filler. Alternatively, the powder is slurried with a small amount of propellant (usually Propellant 11) and filled as a suspension.

Dry filling of powder is considerably slower than filling the powder as a suspension, but it affords closer weight control of the powder phase and reduces the chance of picking up moisture at the filling nozzle. Moisture pickup is considered one of the major factors that cause clumping or agglomeration of talcum powders in aerosol systems. Powders are filled with a minimum of mixing, thus avoiding excessive air entrapment in the product. Interstitial air, when present in the finished aerosol, will give high pressures.

In a multicompartment container, the product must fill each compartment completely to the top and without any voids or pockets of air, which will cause spattering during use. In the filling of creams or emulsions, the product should not be overworked by excessive pumping. Such a condition can result in product instability and phase separation.

Codispensing aerosol systems require that the inner system or phase be filled and then attached to the valve body. The outer system or phase is filled into the aerosol container, and the valve assembly then inserted into the container. Propellant is added by pressure filling through the valve or under the cap. These multicompartment systems have not achieved any degree of automation and are expensive to fill and package from a labor consideration.

PROPELLANT FILLING

Propellant filling constitutes the core of aerosol manufacturing, and the addition of the propellant to the container is a unique filling process. As mentioned at the beginning of this chapter, propellants are loaded into aerosol containers by two methods: pressure filling and cold filling. Compressed gases (nitrogen, carbon dioxide, and nitrous oxide) are loaded as gases and must, therefore, always be pressure filled. Hydrocarbons and fluorocarbons are propellants that are liquids at room temperature because of cooling or pressure and are loaded into the containers as liquids.

The cold-filling method requires the propellant be liquefied by cooling to a temperature below its boiling point. The concentrate must also be cooled to a temperature of about $0 \pm 10°F$. The cooled concentrate is metered into the open container at the first filling stage. At the next stage, the propellant, cooled to a temperature of about -50 to

$-20°F$, is metered into the open container. If the concentrate is not sufficiently cold, the propellant boils violently when added to the container, resulting in excessive propellant losses and an unbalance of the propellant/concentrate ratio. After the propellant is added, valves are placed on the containers and crimped. The containers are heated in a hot-water bath, as required by governmental regulations, and observed for leakage and distortion.

Cold filling provides a method for the production of aerosols at a fairly fast rate. Large quantities of propellant can be added rapidly to the containers, a feature that is advantageous for large-size units or formulations containing high concentrations of propellant. In this process, purging of air is very efficient since enough propellant vaporizes during filling and before crimping to force any remaining air out of the container.

Cold filling is limited to products whose concentrates can be chilled to the required temperature without any harmful results.

Ingredients in certain concentrates will precipitate when chilled, water-based formulations will freeze, and other solvent systems may become extremely viscous. Each formula must be checked by cold-fill laboratory procedures to determine the cooling temperature required for both the concentrate and the propellant.

Since the propellant is added at very low temperatures, moisture from the air freezes on the cold-filling nozzles. These icy nozzles will occasionally drop small particles of ice into the containers, and the resulting excessive moisture can cause stability problems in aerosols, such as gelling, precipitation, and container corrosion.

Pressure filling began as the alternative method for concentrates that could not be chilled. The concentrate is added to the container at room temperature, and the valve placed into the container and crimped. Before crimping, air is removed by purging with a small quantity of propellant or by vacuum. The crimped container is then pressurized by forcing the propellant through the valve. In pressure filling, the liquid propellant is metered into the container under sufficient pressure by a positive-displacement piston pump. If the propellant is added at extremely high filling pressures, valve damage can result. The gassing head can be designed to allow pressure filling of the valve with the spray button on. This eliminates the additional step of placing buttons on the containers.

Pressure filling has the advantage of being capable of handling any type of concentrate without expensive refrigeration. Also, it provides extremely accurate fill volumes of propellant with resulting fill weight accuracy. Propellant losses are slight, since there is no propellant boil-

Figure 12.5.   Under-the-cap propellant filler (Kartridg-Pak Co.).

off as in cold filling. Both nonflammable and flammable propellants can be safely loaded. Since the filling is done at room temperature, moisture contamination is kept at a minimum.

A disadvantage to conventional pressure filling is the slow filling speed, due to the fact that the propellant is forced through small openings in the valve. As mentioned earlier, this was overcome by building rotary pressure fillers with multiple gassing heads—up to 36 heads per machine. However, these machines are expensive, and one defective gassing head may force the filling process to stop for repairs.

Recently a propellant filling method referred to as under-the-cap filling (Figure 12.5) has been introduced. The under-cap machine handles extremely large propellant fills and valves that cannot be rapidly pressure filled by conventional methods. The machine performs three operations. The container with concentrate and valve has the air removed, the propellant added, and the valve crimped during the cycle. This is accomplished through one filling head, which contains the vacuum line, the propellant line, and the crimping collet. The propellant is added to the container by being forced between the valve cup and the can curl. After the propellant is added, the valve is seated on the can and the crimping collet crimps the valve. Under-the-cap filling is suitable for almost all concentrates and liquefied propellants. Filling speeds of 20 cans per minute per head are reported, and filling speeds of over 300 cpm on large containers are not unusual (7).

One disadvantage claimed for this system is propellant loss. The filling head has a relatively large head space, which is subject to propellant loss during each cycle. Losses as high as 3 cc of propellant per cycle have been reported, but with an additional unit called a propellant reclaimer these losses are allegedly cut in half. It should be noted that, although this loss is constant, it is of small consequence when the propellant fill is large.

Gases such as nitrogen, carbon dioxide, and nitrous oxide are always charged into the containers on pressure gassers. Nitrogen, which is virtually insoluble, can be charged into the container by a simple gassing head, and carbon dioxide and nitrous oxide are added using a gasser-shaker. These propellants are used mostly for food products but are also employed in engine starters, deicers, furniture polishes, residual insect sprays, and other specialties. Gasser-shakers add the gas to the container in one simple operation. The gas is injected through the valve at pressures of about 150 psig while the can is mechanically shaken. In this manner, the shaking cycle achieves solubility equilibrium in a few seconds. When organic solvents are present, the solvent system can be saturated with the compressed gas before filling. This is done on a piece of equipment called a saturator and has been described by Anthony (8).

Johnsen and Haase (9) list twelve factors that influence compressed gas filling.

1. The solubility of the gas in the liquid.
2. The amount of liquid.
3. The volume of liquid in the can.
4. The viscosity of the liquid.
5. The temperature of the liquid.
6. The nature of the compressed gas.
7. The use of purge or vacuum crimp versus nonpurge conditions.

8. The size and design of valve orifices.
9. The pressure setting of the gassing head.
10. The pressure drops during gassing.
11. The gassing time.
12. The violence of shaking.

CRIMPING

The assembly of the aerosol valve and the container is the most critical mechanical operation in the production of an aerosol unit. An improperly applied crimp is certain to result in a leaking, nonoperating, defective aerosol container. Another aspect of crimping is the valve configuration after the crimping operation. Distortion of the valve during crimping can prevent the application of the overcap.

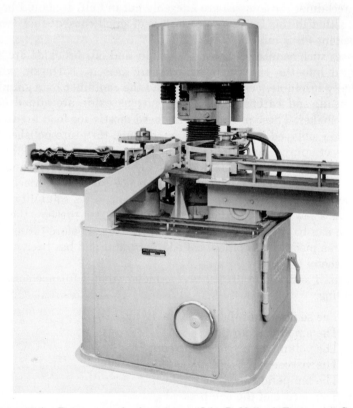

Figure 12.6.   Rotary aerosol valve crimper (John R. Nalbach Engineering Co.).

Valves are affixed to aerosol containers (Figure 12.6) by either an inside or an outside crimp. For the standard 1-in.-opening aerosol container the inside-crimp method is used. Glass bottles and containers with openings of 20 mm or smaller have the valves affixed by the outside-crimp method. When the outside method is used, the valve can be attached to the container by a roll seamer. This machine affixes the valve by rolling the ferrule onto the container in a manner similar to ampoule sealing.

After the valve is inserted into the aerosol can, it should be squarely placed or seated in such a way that the valve cup is properly placed in relation to the can curl. This step minimizes crooked crimps and deformed valves. The crimping collet is normally set to give a crimp diameter of 1.065 ± 0.005 in. to 1.070 ± 0.005 in. on a standard valve cup of 0.011-in.-thick metal. A valve cup of thicker metal will require a decrease in the crimp diameter equal to twice the difference between the metal thicknesses of the standard valve cup and the one in question. Aluminum valve cups are supplied in a thicker metal and can range up to 0.024 in. Crimp depth is another critical dimension. This is measured as the distance from the top of the valve cup to the center of the crimp. The normally accepted standard for this dimension is 0.190 ± 0.010 in. This dimension will vary with the thickness of the sealing gasket and the diameter of the can curl. After crimping, the valve cup should be wrapped around the can curl at the crimp line. An acceptable crimp has the sealing gasket compressed and the valve cup held tight to the can curl with a metal-to-metal seal.

Brinkley (10) lists four variables that affect the crimp depth.

1. The valve cup thickness.
2. The valve gasket compound, which varies in thickness and durometer reading.
3. The crimping collet, which is available in both 1/32- and 3/64-in.-radius tips.
4. Can curls that vary from those specified by CSMA standards.

Crimps can be checked by torque testing as a quality control procedure. This is a convenient method to use during production. Torque specifications will vary with the crimp depth and crimp diameter but in most instances will range from 40 to 60 in.-lb.

Distorted or crooked crimps can cause misfitting of the overcap and improper actuation of the valve. If too much of the metal in the valve cup is drawn during crimping, the valve pedestal will be too high. This distortion usually causes actuation when the overcaps are placed on the containers. One-piece actuator–overcaps will not seal tightly since the valve stem is too high for a proper snap or friction fit. Crooked crimps can result from a number of factors, such as the following:

1. The container may be crooked.
2. Worn crimping machines may no longer be square with the base, and the crimping head will tilt when pressure is applied.
3. On the undercap crimper the O-ring cap seal can become partially distorted and make crooked crimps.

Assembly of the aerosol valve and a glass container is done by external crimping or seaming. Variables in glass container crimping include the following:

1. Bottle variation (height and neck finish).
2. Thickness of plastic coating on bottle.
3. Plastic overrun on neck.
4. Valve gasket thickness.
5. Ferrule material.
6. Ferrule length.

Since the valve crimper is set to a predetermined dimension, large variations of the neck finish can result in loose crimps (leakers) or broken necks during valve crimping. Roll seamers that have spring-loaded rollers will contour the valve ferrule to the neck finish, and therefore this method of closure can accommodate a greater variation of neck dimensions.

Just as in the can, the valve gasket creates a pressure-tight seal between the valve and the glass bottle. Proper hold-down pressure to compress the gasket and correct shape of the collet insert or valve nest are of prime importance in achieving a nonleaking crimp. Most valve ferrules are fabricated of aluminum, which is easily distorted by incorrect pressure or collet insert. Fishman (11) has stated that an improperly shaped valve nest will hold the valve on the extreme outside edge. In this case, when the collet closes, the valve, which is held down solely by the outer edge, will balloon. The end result may be a valve leaker at the seal.

Some valve crimpers operate at speeds of over 200 cpm and are available with an air-removal feature. Such units are known as vacuum crimpers and are installed on pressure-filling lines. Vacuums can be operated from 15 in. to 24 in. of Hg. When air in the head space is detrimental to product stability, vacuum crimping offers the most consistent air-removal method.

AUXILIARY PACKAGING EQUIPMENT

The need for auxiliary equipment in the filling of aerosols has developed as production speeds have accelerated. Uniformity of the components used in the aerosol package is the most rigid requirement for the proper operation of auxiliary equipment. A listing of auxiliary equipment, with a description of the function of each item in the aerosol line, follows.

Container Cleaners

This equipment is employed mostly for glass bottles and in cases where lint or dust may be present in the empty containers. Cleaning is done just before filling. Container cleaning is considered mandatory in the pharmaceutical and food industries.

The empty containers are air cleaned by spouts or jets that inject a stream of clean dry air into the container, thus blowing out lint or dust. Some air cleaners invert the containers during the cleaning cycle, allowing heavy particles of dirt to drop out of the containers by gravity

Figure 12.7.   View of container unscrambler (Barr-Stalfort Co.):

Unscramblers and Depalletizers

These pieces of machinery are designed to maintain a steady flow of empty containers to the filling line with a minimum of handling. Unscramblers (Figure 12.7) have a capacity of up to 300 cpm. Depalletizers are designed to automatically discharge empty aerosol containers from palletized loads to an unscrambler, which feeds onto the filling machine conveyor.

Unscramblers can be either rotary or straight-line tables. With a rotary table unscrambler, containers from cartons which are inverted on the feed table are pushed onto the revolving table or disk and are then automatically discharged individually onto the conveyor. Straight-line unscramblers feed the containers onto the conveyor at timed intervals, which are programmed in accordance with the filling speed of the packaging line. The containers are pushed intermittently onto the conveyor belt.

Valve Placers

These machines (Figure 12.8) can automatically select, orient, and place valves with dip tubes into aerosol cans. Valve inserters with capacities of up to 360 valves per minute are available. The valves are sorted by either a mechanical or a magnetic type of sorter. The magnetic sorter is limited in use to valves with ferrous metal cups. Valve sorters are available to handle the normal 1-in. mounting cup and also the smaller ferrules which come in millimeter sizes.

In operation, the valves are dumped into a large holding bin, which feeds them into the sorter. The sorter unscrambles the valves and feeds them down a chute or escapement. Each time that a container moves into position, a valve is dropped into the opening of the container. Newer machines have guides that straighten the dip tube curvature as the valve is inserted into the container and the valve cup is accurately seated. This last feature is very important on high-speed filling lines where the container, after valve placement, enters the undercap propellant filler.

Both types of valve sorters operate by churning or agitating a large number of valves in order to supply the escapement with an overfeed of valves. This factor may cause variations in valve dimensions, configuation, or structural stability. Wonneman (12) states that aerosol valve handling presents a challenge to future computer-monitored and automatically controlled operations of aerosol filling that can be solved

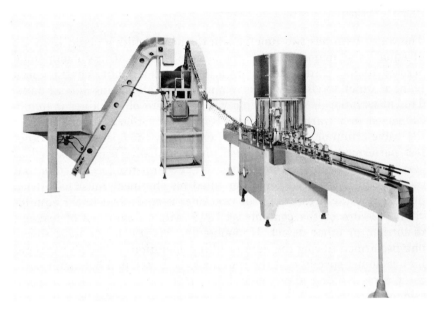

Figure 12.8. Automatic valve placer (PMC Industries).

only by mutual cooperation between valve manufacturers and fillers. A poorly seated valve can cause a line jam and costly downtime for repairs. Valve placers can also handle valves with or without the standard-type activator button.

Pressure Testers

Pressure testers are essential for products that are pressurized with compressed gases. Pressure testing each unit from the gasser is the only manner by which one can determine whether the unit has been properly pressurized.

A pressure testers operates similarly to a compressed gas pressure filler but with the accept-reject limits set slightly over the minimum-maximum acceptable limits for pressure. The testers are designed not to introduce gas into the container or to draw product up into the valve stem. This is extremely important for food products that are packaged with nitrogen or other compressed gases. Any product drawn up into the valve stem could be subject to microbial contamination during warehousing and storage.

Water Baths

The water bath has two functions in the aerosol filling plant. Both are of equal (and major) importance for the success of any commercial aerosol product. First, the water bath serves as a quality-control check point at which to visually observe and remove all leaking aerosol units. This observation is quite simple when the aerosol container is totally immersed in a warm-water bath. Rapid evaluation of bubbling from the valve crimp or can seams can be made by the operator. The other test performed by the water bath involves quality control as well as serving as a regulatory compliance test. Federal regulations state that each completed aerosol container filled for shipment must have been heated until the pressure in the container is equivalent to the equilibrium pressure of the contents at 130 F without evidence of leakage, distortion, or other defect. Undoubtedly, this test is the most severe one performed during the aerosol filling operation. Although the regulations do not specify that the containers be tested in a hot-water bath, most aerosol filling plants consolidate leak testing with the elevated-temperature test.

Hot-water test tanks are of the "straight-through" type. The length of the tank is between 20 and 40 ft, and the containers travel on guide rails up to four abreast. The water tanks are constructed of stainless steel, as are the steel chains on which the containers travel. These steel chains ride on a magnetic plate, so that the containers are held in place as they pass through the tank. Handling of glass and nonferrous containers requires the use of grippers or "pucks" that hold the aerosol containers below the water as they travel through the leak detector tank.

All tanks should be fitted with a water overflow pipe to remove any dirt and grease which accumulate during the operation. Leaking units, when encountered, will also contaminate the hot water with the leaking material, and hence a system that maintains a clean water bath is a necessity. The water temperature is maintained and regulated by either circulating the water through an external heat exchanger or placing steam coils on the bottom of the leak detector tank. The water temperature in the tank depends on the length of time that the container is in the bath, the size of the container, and the temperature of the container before entering the tank. To achieve the desired internal temperature, the variable condition adjusted is retention time of the container in the hot tank. By attaching a small maximum-reading thermometer to the dip tube of the valve on a test container, the maximum internal temperature generated in the test sample can be determined by opening the container and recording the temperature. Another method to measure the internal temperature is by using a thermocouple, which is inserted in the container. Klisch (13) has proposed a method that records the maximum pressure in the container while in the hot-water tank.

During the operation of the water bath, it is recommended that a small quantity of a corrosion inhibitor be added to the water. This aids in preventing the formation of rust on the containers and valves. In addition, a nonfoaming surfactant should be added to the water to solubilize any grease and oil found on the containers. As the containers leave the hot tank, jets of dry air should be blown over them, with special attention to the mounting cups. This step removes small droplets or beads of water found clinging to the containers. The purpose is the prevention of rust and corrosion.

Appropriate protective covers and shields are installed on the water tanks, since the internal pressures generated within the aerosol containers are quite high. This protective shield must be sufficient to restrain any exploding unit from leaving the test tank. Personnel assigned to remove defective units from the test tank must wear face shields and such protective clothing as is necessary.

Check Weighers

Check weighers are installed to ensure that the weight of each container leaving the production line is within the specified limits. These units are usually placed in the filling line just after the containers have been leak tested. Check weighers monitor 100% of the production by passing each container over a weighing platform or other sensing device. The check weigher is divided into three lanes: one for underweight, one for overweight, and one for correct weight. These units check the gross weight of the container; when many units are rejected, a quick check must be made of the concentrate and propellant fillers.

Actuator Button Placer

This equipment (Figure 12.9) is utilized when valves have been affixed to the containers without actuators. The machine can be an in-line or rotary chuck type of placer. Rotary machines operate at up to 360 cmp and can be designed to automatically spray-test all units. Nonspraying units are rejected by the machine.

Figure 12.9.   Actuator button placer on valves (John R. Nalbach Engineering Co.).

Overcap Placers

Protective overcaps are applied automatically by machines that are either in-line or rotary cappers. When the overcap also contains the actuator, capping machines can orient the actuator cap and container and thus align the spray direction with the dip tube curvature.

Coding

Coding machines selected for aerosol lines should be versatile enough to code the different container sizes and the various curved surfaces. Normally, the code is placed on the bottom of the aerosol containers. The curvature of this bottom surface, however, presents two major difficulties: (a) the number of digits per line of code is limited, and (b) the number of lines in the code is limited. Coding ink should dry rapidly and adhere to the metal, which may have a thin film of oil or water. Flexographic inks are most commonly used. A more uniform and legible code is obtained when the container is stabilized with a top hold-down belt during the coding cycle.

Glass bottles present an extremely varied bottom surface, and in many instances a hand-applied code is used. A method of coding glass containers by stamping onto the metal valve ferrule is available. The code is stamped into the side of the ferrule after the valve is affixed to the aerosol container.

PRODUCTION QUALITY CONTROL

The production of aerosols requires different standards and specifications from those used for nonaerosol units. The specifications for processing, filling of the special quality-control program during manufacture. Such a program can be classified into three general areas.

1. Ingredients.
    a. Concentrate.
       1. Mixing.
       2. Storage.
    b. Propellants—receiving and storage.
2. Filling,
    a. Concentrate.
    b. Propellant.
    c. Crimping.
3. Components.
    a. Containers.
    b. Valves.
    c. Actuators.

These areas are critical and must be monitored by a sampling program that will reveal deficiencies either before or during manufacture.

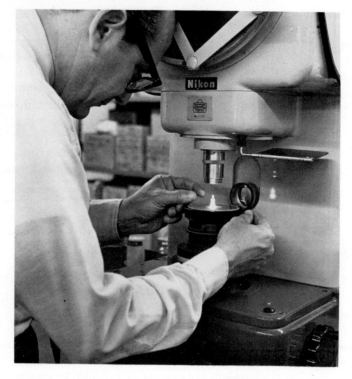

Figure 12.10.   Microscopic examination of valve body (Barr-Stalfort Co.).

Sampling of chemicals, propellants, and package components (Figure 12.10) for physical and chemical inspection before they are used will determine conformance or degree of conformance in regard to the following characteristics:

1. Chemistry.
2. Physical properties.
3. Dimension or shape.
4. Function or performance
5. Component compatability.

Inspection of materials before manufacture provides a positive control against the production of scrap. This is especially important for aerosol packages, since defective aerosols have little if any salvage value after the valve is crimped to the container.

During manufacture, a quality-control program will ensure that required properties are being maintained within tolerance limits. This program will forecast or detect trends away from the limits, so that adjustments can be made on the filling machines, crimpers, and temperature controls.

The production data sheet developed by the quality-control department for the manufacturing group should indicate the following:

1. Handling requirements for the concentrate.
2. Fill weights and range.
3. Filling temperatures and range.
4. Propellant pressure.
5. Net weight and range.
6. Moisture limits.
7. Crimp dimensions.
8. Spray rate.
9. Valve and actuator dimensions.
10. Container specifications.

All materials brought to the aerosol filling lines should be checked at regular intervals to prevent any misuse of components.

The inspection of finished goods may be based on a sampling plan which operates on the basis of probability that the quality of a given lot is typified in samples of proper size taken from that lot. Thus a statistical sampling of the lot will show whether defects are present in excessive amounts; if so, the lot is rejected. Sampling procedures are found in Military Standard Mil-STD-1-5D and can be modified for use in aerosol manufacture.

REFERENCES

1. J., Pembroke, Plant Design, *Che. Spec. Mfr. Assoc. Proc. 55th Ann. Meet.*, December 1968, p. 47.
2. R. Burg, Sanitation Aspects of Aerosol Packaging of Foods, *Aerosol Age,* **10,** 12 (December 1965), 66.
3. R. C. Harris, Pharmaceuticals and Foods—Sterile Packing, *Chem. Spec. Mfr. Assoc. Proc. 54th Mid-Year Meet.*, May 1968, p. 34.
4. F. Presant, Preventing the Formation of Precipitates in Aerosol Colognes, *Aerosol Age,* **9,** 8 (August 1964), 18.
5. P. A. Lychalk and R. C. Webster, The Importance of Total Oxygen Removal from Water-Based Aerosol Systems, *Aerosol Age,* **8,** 3 (March 1963), 33.
6. R. M. Trider, Nitrogen Sparing of Aerosol Products, *Chem. Spec. Mfr. Assoc. Proc. 52nd Mid-Year Meet.*, May 1966, p. 29.

7. P. A. Schuler, Under Cap Filling at S. C. Johnson & Son, Inc., *Chem. Spec. Mfr. Assoc. Proc. 51st Mid-Year Meet.*, May 1965, p.

8. T. Anthony, New Technology Brightens Profit Outlook for Aerosols, *Soap Chem. Spec.*, **43**, 12 (December 1967), 185.

9. M., Jonhson and F. Haase, Technical Aspects of Gasser-Shaker Injection of Non-condensable Gases, *Chem. Spec. Mfr. Assoc. Proc. 51st Mid-Year Meet.*, May 1965, p.

10. E., Brinkley, Crimping Aerosol Cans, *Aerosol Age,* **12**, 12 (December 1967), 24.

11. M. Fishman, Crimping Glass Bottle Aerosols, *Aerosol Age,* **12**, (January 1967), 33.

12. F. F. Wonneman, The Handling and Insertion of Aerosol Valves, *Aerosol Age,* **8**, 10 (October 1963), 80.

13. S. Klisch, A Method of Evaluating Water Bath Effectiveness in Aerosol Line Production, *Aerosol Age,* **9**, 9 (September 1964), 16.

TESTING OF AEROSOL PRODUCTS*

The testing of aerosol products is an important aspect of aerosol science and technology when one considers the complexity of the aerosol system. A rather comprehensive testing program is required during all stages leading to the formulation and development of a new product. In addition, it is only through the use of a complete quality-control system, which includes many test procedures, that the quality of the final product can be maintained (1). Finally, various tests must be performed on aerosols in order to determine compliance with federal, state, and local governmental regulations in regard, for example, to pressure, net weight, and flammability.

The tests used specifically for aerosol products range from a simple determination of the vapor pressure within the container to a determination of many other physicochemical and chemical characteristics, including pH, viscosity, density, qualitative and quantitative analyses of active ingredients and propellants, and moisture determinations. In addition, good manufacturing practice dictates that microbiological determinations should be made on the product so that the total count of bacteria as well as the nature of any bacteria present can be ascertained.

SPECIFIC AEROSOL TESTING PROCEDURES

Various test procedures have been developed by the Chemical Specialties Manufacturers Association. Details of the procedures have been covered elsewhere (2, 3).

* The author would like to thank the publisher of *Aerosol Age*, in which portions of this text have previously been published, for permission to use this material.

Vapor Pressure (4)

Vapor pressure within the aerosol container is important from many standpoints. Vapor pressure is related to the spraying characteristics of the product and is also an indication of the nature and the amount of propellant used. Pressure readings can also be used to detect air present in the container.

Pressure can be determined accurately by means of a special "can-piercing pressure device" which is available commercially. This device

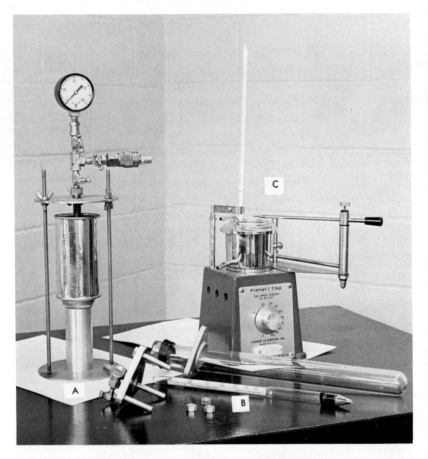

Figure 13.1.   Some aerosol testing equipment: *A* vapor pressure determination; *B* density apparatus; *C* flash point determination by Tag open cup.

pierces the container and allows the pressure to be measured accurately with a pressure gauge. After the gauge has been standardized, the test can is fitted into the special apparatus and pierced. The entire apparatus, including the test container, is then immersed in a water bath that is set at a given temperature (generally 70°F). The can is held at this temperature until equilibrium has been attained; this takes generally from 30 to 60 min, depending on the size of the can. The container is shaken several times during this heating period, and then the gauge is prepressurized with an inert gas, such as nitrogen or air, to within 5 psi of the estimated pressure. The purpose of prepressurizing is to decrease any error due to the gas present in the apparatus. The valve connected to the can is slightly opened, and the needle on the gauge is observed. Depending on the movement of the needle, the prepressurizing procedure is repeated by increasing or decreasing the prepressurizing pressure by 2 psig. Then the procedure is repeated until there is little or no movement of the needle. At this point, the valve is completely opened and the reading recorded. Three cans of product are tested in this manner, and the average reading is taken as the final pressure. This equipment is illustrated in *A* of Figure 13.1.

This method gives an accurate measure of vapor pressure and should serve as the basis for determining final specifications for each product. However, it is time consuming, destructive to the sample, and limited to metal containers which can be punctured.

Modifications of this method can be used for glass and plastic containers, as well as for a rapid check on the pressure within a metal container. Pressure gauges are used (either of the prepressurizing or regular type) which have been fitted with an adapter that corresponds to the valve stem. A fairly accurate result can be obtained by immersing the samples to be tested in a water bath held at the required temperature until the contents attain the temperature specified. Then the containers are removed, and the pressure is measured by placing the gauge over the valve stem and actuating the valve. After this reading is recorded, the containers are shaken and placed in the water bath for 1 min. and another reading is taken. This procedure is then repeated for a third time. The three readings will decrease in a fairly linear manner, and the pressure reading is determined by extrapolating these readings to the *o*th reading. The same procedure can be used for metal containers. Unfortunately the accuracy of the results is limited by the way in which the test was conducted. Figures 13.2 and 13.3 show several of these gauges.

Depending on whether or not a dip tube is present, the pressure in the container can be measured with the container in either the upright or the inverted position. It is always best to position the container so that the pressure of the vapor phase is measured. This prevents contamination of the gauge with the product concentrate. Care should be taken to clear the dip tube of any product that may be present. When metered and vapor tap valves are used, the pressure must be determined by the "can-puncturing method." In the case of the vapor tap valve, the modified method can be used, although care must be taken since the product will contaminate the gauge. This cannot be avoided unless the gauges are fitted with a proper protector.

Figure 13.2.   System for measurement of pressure as contents of aerosol product are discharged (Builders Products, Inc.).

Figure 13.3. Vapor pressure gauge and adapters (Builders Products, Inc.).

Density (5–7)

Density has been defined as "the weight of a given volume of a gas, liquid, or solid." It is usually expressed as grams per milliliter or pounds per cubic foot. When density is expressed in terms of the metric system (g/ml), it is numerically equal to the specific gravity.

Usually, the liquid formula for a given aerosol product is expressed in terms of weight; however, when certain ingredients are used, they may be actually measured in terms of volume. Density is also of considerable importance in the storage of liquid propellants, as well as in setting the metering devices on various filling lines. Since density is a physical constant, it can also serve as a guide to the identification of propellants. It has additional value in determining the composition of various propellant blends.

Among the various external factors that will affect the density value of a propellant, temperature must be considered. Generally, density will decrease with an increase in temperature.

Since most of the propellants used in aerosols are of the liquefied gas type, the density of the vapor must also be considered. Vapor density is influenced by changes not only in temperature but also in pressure. In the case of vapor density, there is an increase with a rise in temperature.

Several methods can be used to determine the density of various propellants. They are based on buoyancy principles (hydrometric) and determination of the weight of the liquid occupying a known volume (pycnometric).

HYDROMETRIC METHOD

This method is based on Archimedes' principle, which states that the buoyant effect of a substance immersed in a liquid is directly proportional to the weight of the liquid displaced. A conventional hydrometer used in conjunction with a glass pressure tube can serve for this determination. The liquid propellant is added to the pressure tube containing the hydrometer. The entire apparatus is immersed in a water bath and allowed to attain the desired temperature. The density can then be read directly from the hydrometer. This method is relatively simple and, with most liquids, presents no problems. Since, however, with aerosols a liquefied gas is involved, the tube that is used must be capable of maintaining a gastight seal and withstanding the vapor pressure of the liquefied gas. Such tubes are available and can be fitted with a metal flange and gasket to maintain a pressure seal.

The hydrometer should have a range from 1.000 to 1.600 for fluorocarbons and from about 0.400 to 0.800 for hydrocarbons. When considering blends of propellants, the range must be extended to include the density of each pure component. Generally a hydrometer with a wide range is first used to determine the approximate density and then a narrow-range hydrometer to determine the actual density. For most determinations, the uncorrected reading can serve as the value for the density. However, an error is introduced because of the size of the hydrometer. The appropriate correction can be found by using the hydrometer to determine the density of a compound of known density, such as water, Propellant 11, or carbon tetrachloride. The average difference between the actual density and the determined density is taken as the correction and is deducted from the actual reading. This value must be determined for each hydrometer but remains constant once the instrument is calibrated.

Provision must be made to evacuate the tube of all air before the addition of propellant. This is generally accomplished by fitting the top of the tube with a Hoke valve, which is connected to a vacuum pump.

The tube is evacuated to less than 5 mm Hg or flushed with the liquid to be determined. The sample propellant is then added to the tube (containing the hydrometer) through the Hoke valve. Sufficient liquid propellant is added to ensure that the hydrometer floats at least half way up the tube. The propellant source is then removed, and the tube placed in the water bath until equilibrium is attained. The density is then read.

Since the glass tube is filled with a liquefied gas under pressure, there is always the danger that the glass will break. Therefore it is essential that safety precautions be taken when carrying out this procedure. The use of safety glasses, face shields, wire guards, leather gloves, and other protective items is absolutely essential. This apparatus is illustrated in *B* of Figure 13.1.

PYCNOMETRIC METHOD

A pycnometer is a device that is capable of keeping constant the volume of a given material. The weight of this volume of substance can be determined easily by weighing.

For nonliquefied gases, a glass-stoppered tube or flask is generally used as a pycnometer. However, in the case of liquefied gases, the pycnometer must not only be capable of withstanding pressure but also be closed to prevent loss of propellant.

One such pycnometer consists of a high-pressure aerosol container of approximately 500-ml capacity, which has been fitted with a bleed valve at one end and a graduated Saran tube and needle valve at the other. The apparatus is evacuated and weighed. It is then filled with distilled water through the bleed valve until the liquid level is visible in the Saran tube. The apparatus is placed in a water bath set at the desired temperature; when equilibrium has been attained, the volume is read directly. The apparatus is then weighed, and the volume can be determined from the weight of water and its known density.

To determine the density of the propellant, the apparatus is evacuated to less than 2 mm pressure and weighed. The apparatus is then chilled and the sample propellant added through the needle valve until the liquid level can be seen in the Saran tube. After equilibrium has been attained by immersion in a water bath, the volume is read. The entire apparatus is again weighed. The density can be calculated from the following equation:

$$\frac{\text{Weight apparatus full} - \text{weight evacuated}}{\text{Total volume}} = \text{density}$$

Caution must be exercised to ensure that all of the air has been evacuated from the tube. Any residual amount can be minimized by evacuating to a very low pressure or by flushing the pycnometer with the liquid propellant before weighing.

BEADS OF KNOWN DENSITY

Another method, which has been proposed by Reed (8), involves the use of beads of known density. These beads are placed in the solution, and by noting the position of the beads in the solution, one can determine the density of the product rather quickly. However, the values obtained in this way may not be as accurate as the results of other methods. Reed's method is suitable for quality-control purposes. The temperature must also be controlled through the use of a constant-temperature water bath.

Determination of Volatile Contents

Aerosol products consist essentially of active ingredients dissolved, suspended, or emulsified in solvents and propellants. The nature of the propellant, as well as the quantity present, determines to a great extent the performance of the aerosol product. Too little propellant may result in a spray with little, if any, drying power and inability to discharge the complete contents of the container. Too much propellant, on the other hand, results in an exceptionally dry spray and one that may have excessive pressure.

In order to detect these changes when present, methods have been developed that are capable of determining the volatile and nonvolatile contents of aerosol products. The simplest procedure which can be used consists of weighing the complete aerosol container, followed by the spraying of a sample of the contents into a tared evaporating dish of suitable size. The aerosol container is then reweighed, and the difference in weight represents the weight of the sample. The sample in the evaporating dish is allowed to vaporize, and after a given period of time the dish is reweighed in order to determine the nonvolatile residue. The difference between the sample weight and the nonvolatile residue weight represents the volatile components. The percentage of a volatile component can be calculated as follows:

$$\frac{\text{Weight of volatile component}}{\text{Sample weight}} \times 100 = \% \text{ of volatile component}$$

Although the method is relatively simple, it is not without error and therefore must be modified in order to produce satisfactory results that can be duplicated. Caution must be exercised in spraying the sample of aerosol product. Some of the product will be lost because of air-borne particles, as well as some remaining on the outside of the valve actuator. If the propellant is allowed to vaporize uncontrolled, there is danger of loss of ingredients because of rapid vaporization. In addition, the temperature, as well as the length of time the sample is exposed to the atmosphere, will affect the results. In order to prevent many of these errors as well as to be able to obtain consistant results, a modified method is available. A vacuum distillation procedure has been used for this purpose.

VACUUM DISTILLATION METHOD

This method can be used for formulations that do not contain volatile active ingredients and may contain suspended solids. The procedure consists of filling a sampling device or "thief" with a sample of the aerosol and weighing the entire device. The sample is added to a tared vacuum flask that has been fitted with a one-hole rubber stopper containing a capillary tube extending to the bottom of the flask. The thief is then reweighed to obtain the sample weight. The flask is placed in a water bath at 120°F, and the propellant is allowed to boil off through the side arm. Vacuum is applied gently for about 30 min, followed by gentle swirling. This procedure is repeated until the difference in successive weighings is slight. From these figures one can determine the percentage of volatile and of nonvolatile residue present. By keeping the vacuum and temperature constant and similar for both the known and the unknown samples, suitable results can be obtained. This also will yield results that are consistent and can be duplicated.

DENSIMETRIC METHOD

Another method which can be used to determine volatile/nonvolatile ratio is a densimetric analysis. This method is dependent on the relationship that, under isothermal conditions, the density of an aerosol product is almost a linear function of the volatile and nonvolatile portions. This method is especially applicable to aerosol products that do not contain suspended solids. It consists of determining the density of the sample at approximately −20°F, using a hydrometer. In order to correlate this value to volatile material in the formulation, standards must be utilized. Samples containing about 1% more and 1% less volatile material than the unknown must be treated in a similar manner. Since this is a linear relationship, a plot of the density versus the percentage of volatile material will give a straight line. The volatile content of the unknown can then be determined from this plot, and the results calculated from the following equation:

$$K \text{ [density (standard)} - \text{density (unknown)]} + \% \text{ volatile (standard)}$$
$$= \% \text{ volatile (unknown)}$$

The value of K is obtained from a plot of density versus per cent volatile content and represents the slope of the line.

Since this is a relatively simple procedure that can be easily carried out, it is useful for routine in-plant control procedures. Its degree of accuracy is dependent on the accuracy with which temperature is controlled, as well as the accuracy of the hydrometer.

Discharge or Delivery Rate (9)

It is important to determine the amount of material that can be discharged from an aerosol container per unit time. This can generally be accomplished by allowing the test cans to reach a given temperature

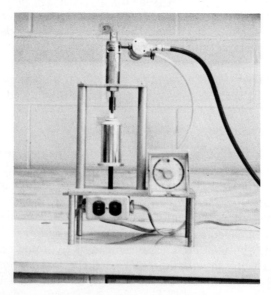

Figure 13.4.   Apparatus for determination of delivery or discharge rate.

and then weighing the containers. The valve is then actuated for 10 sec, and the container reweighed. The difference in weight represents the amount of product discharged per 10 sec, from which the delivery per second can be calculated. This value can be used as a check on the valve orifices, since the delivery is affected by orifice size. Fig. 4 illustrates apparatus useful in determining this value.

Another method developed for pharmaceutical aerosols is dependent on collecting the amount sprayed and then assaying for the active ingredient (10). For example, in the case of pharmaceuticals, a general test is carried out on at least four containers. Each container is weighed and placed in a constant-temperature water bath until the internal temperature reaches $25 \pm 1°C$. The containers are then taken from the water bath, and the excess water is removed by blotting with a paper towel. Each container is actuated for exactly 5 sec and then reweighed. The containers are again placed in the water bath, and the test procedure is repeated 3 times for each container. The average delivery rate is calculated in terms of grams per second for each container.

The delivery rate for dexamethasone aerosol is given in terms of amount of active ingredient (dexamethasone) delivered per second and not in terms of total amount of product. Since a vapor tap valve is used for this product, the test container, after reaching a temperature of 25°C, is inverted into a beaker containing alcohol. The valve is actuated by pushing the container against the bottom of the beaker for exactly 10 sec. The amount of dexamethasone dispensed is then determined by a spectrophotometric assay.

The delivery rate for dexamethasone determined in this manner should be not less than 0.060 and not more than 0.090 mg/sec. This method is important when dispensing potent materials and represents a way in which dosage can be controlled. This test is of particular interest to the Food and Drug Administration when evaluating New Drug Applications.

Spray Pattern (11)

Spray patterns are useful in determining the performance of a specific formulation–valve combination. In addition, the test can serve as a check on the batch-to-batch uniformity of aerosols. The spray pattern should remain the same, provided that the same formulation and valve are used. A permanent record can be made of the spray pattern and saved for future reference. A given amount of material is sprayed onto a specially coated paper or onto alcohol-sensitive paper, depending on the nature of the product. A spot is impinged at each point touched by the aerosol spray. This produces a record of the spray, which should be retained with the batch card.

Various methods have been developed for determining and evaluating spray patterns. The simplest method consists of spraying the finished product onto some type of absorbent paper such as filter paper. In order to obtain reproducible results, the distance separating the container from the target must be kept constant. Additionally, the amount of material hitting the target area must be controlled and kept constant. If too much material is allowed to come into contact with the target area, the spray pattern becomes obscure and clouded. Products containing colored materials can be evaluated by this method since they will form a visible picture on the filter paper.

In order to overcome the limitations of this method, Root (12) suggested coating a piece of paper with a mixture of dye and talc, which is brushed onto the paper. The dye must be a water-soluble one for aqueous- and alcohol-based products; for oil-soluble bases, an oil-soluble dye is required. When the spray comes into contact with the dye on the paper, the dye will dissolve and make a recording on the paper. Gunn-Smith and Platt (13) suggested that the dye be added to the container to be tested and sprayed directly onto absorbent paper. In this manner, the spray can be directly recorded. Dixon (14) suggested that a glass plate be coated with magnesium oxide; when the spray comes into contact with the coated plate, the force of the particles will disturb the magnesium oxide film, forming an impression which can then be related to the spray pattern. Other methods that are available include spraying into a beam of light and observing the shape of the spray, and spraying the aerosol and photographing the actual spray by means of high-speed photography.

In order to control the amount of spray coming into contact with the target area, various devices have been developed. Most of these devices make use of some type of shutter, which controls the amount of material hitting the absorbent paper. The aerosol product is sprayed continuously, and only when the shutter opening is exposed does product impinge upon the surface of the paper. By controlling the size of the orifice and the time of exposure, this device can be adjusted to accommodate all types of spray patterns.

The CSMA method for spray pattern determination consists of brushing onto a sheet of paper a mixture of 5% dye in talc. The dye used must be soluble in the spray. Gentian violet is the customary choice for alcohol and water-based aerosols. Oil-soluble dyes are used for oil-soluble sprays such as insecticides. In order to control the amount of spray on the paper, the paper is attached to a rotating disk containing a cut-out section. To this section is fitted a sliding shutter. The spray is allowed to come into contact with the paper only once; it must be removed before the second revolution. By adjusting the speed of the motor (rpm) and the width of the opening, the amount of material deposited can be varied. It can be calculated from the equation:

$$\frac{SD}{2\pi r\,R} = \text{grams deposited}$$

where S = slit width at circumference (cm), D = delivery rate of valve (g/sec), R = rate of revolution of disk (rps), and r = radius of disk (cm).

As the spray particles hit the paper, they dissolve the dye. The dye is absorbed by the paper and produces a spot the size of which is directly related to the size of the spray particles. The apparatus used for this purpose is shown in Figure 13.5.

Flammability

Flammability can arise in an aerosol product from one or both of two causes: the nature of the propellant, and/or the nature of the active ingredients and solvents used.

Figure 13.5.   Spray pattern apparatus.

The fluorinated hydrocarbons are classified as nonflammable and as such do not form flammable mixtures with air. The compressed gases such as nitrogen, carbon dioxide, and nitrous oxide are also nonflammable and do not present any flammability hazards. However, the pure hydrocarbons (e.g., propane, butane, and isobutane) present a flammability problem since under certain conditions they will form explosive mixtures with air. For example, from 2.3 to 7.3 vol % in air of propane will form an explosive mixture.

The presence of a solvent may affect the flammability of an aerosol product. Such products as room deodorants, hair sprays, and perfumes are rendered flammable by the presence of solvents like ethyl alcohol, acetone, toluene, and benzene. However, in many instances the presence of the nonflammable fluorocarbon propellant decreases the flammability of the solvent. An outstanding example of this phenomenon is provided by a mixture of ethylene oxide and Propellant 12. The addition of as little as 10% Propellent 12 to ethylene oxide will decrease the flammability of the latter to the point where the mixture can be classified as nonflammable. This same phenomenon has been shown to take place with mixtures of fluorocarbon propellants and hydrocarbons.

In order to determine to what extent aerosol products are flammable, several tests have been devised and are utilized in the development of aerosol products. These tests include the flame projection and the modified tag open-cup tests.

FLAME PROJECTION TEST

This test is conducted by using a ruler measured off in inches and a lighted candle. The contents of the aerosol can are sprayed and directed toward the candle. The extension of the flame is measured and becomes an indication of the flammability. This test is indicative of the actual performance of the aerosol when it is sprayed in use. Depending on the length of the flame extension, as well as the flash point, products are classified as combustible, flammable, or extremely flammable. These classifications are indicated in Chapter 23.

MODIFIED TAG OPEN-CUP TEST

This test is carried out using a standard Tagliabue open-cup tester, Tag gas-flame testing burner, flash test thermometer, and heat source (C of Figure 13.1). The propellant is allowed to escape, and the remaining liquid transferred to the Tag open cup. A thermometer is placed in the solution, which is then heated at a rate of about 2°F/min. This is continued until the solution flashes completely across the top of the cup, the solution reaches a temperature of 20°F, or the product has evaporated. This provision may be slightly different in several States or cities in the United States.

Leak Testing

Loss of volatile components from an aerosol container can occur as follows:

1. Through various seams of the container.
2. Around the valve crimp.
3. Through the valve itself.

All aerosol products will lose propellant upon standing for long periods of time. Since the seal between the container and the valve is made of either rubber or a gasket compound, small amounts of propellant will escape. The same holds true, but to a lesser extent, for the various seams in the container. Flanner (16) indicates the amount of material that can be lost because of leakage through the valve.

SEEPAGE

A procedure has been developed by which the normal seepage of an aerosol product (2) can be determined. The test containers are immersed in a water bath, and a small test tube (about 5-ml capacity) is filled with water and inverted over the valve stem. If seepage through the flowed-in gasket or cut gasket of the valve cup is to be determined, a similar test tube with a specially made wide opening is placed over the valve cup. The containers and tubes are allowed to stand for 48 hours, after which the amount of water displaced in the tubes is measured. After taking into account the diffusion of gases and the solubility of the gas in water, the seepage rate per year can be calculated.

Another procedure has been developed to determine the weight loss of pharmaceutical aerosols (15). This test is designed to give the leakage rate of each of 12 containers of the product. After recording the date and time the test is started, each of the 12 containers is weighed to the nearest milligram. The test containers are then allowed to stand upright for at least 3 days at room temperature. After this period, the date and time are again recorded and the containers reweighed. The leakage rate is calculated from the equation:

$$\text{Leakage rate/year} = (W_1 - W_2)\left(\frac{365}{T}\right) \tag{24}$$

where W_1 and W_2 are the weights of the container before and after the test period, respectively, and T is the time period of the test in hours.

The containers are acceptable if the average leakage rate is not more than 3.5% of the labeled weight per year and if none of the containers has a leakage rate greater than 5%. If the latter standard is not met, the test is repeated on an additional 24 containers. The lot is then acceptable provided that not more than 2 of the 36 test containers leak more than 5% and none of the containers leaks more than 7% of the labeled weight per year. This places a limit on the leakage of pharmaceutical aerosols. Although the first part of the test places this limit at an average of 3.5%, it is possible during the second part of the test to accept a lot of containers with a slightly higher leakage rate.

LEAKAGE

In most instances, small leaks can be detected and generally are indicative of faulty equipment (valve crimper) in need of adjustment or possibly of containers that do not meet specifications (16).

The simplest way to detect leakage from an aerosol product is by completely immersing the container and valve stem in a hot-water bath. By heating the contents to about 130°F, the pressure is increased sufficiently to cause the gas to escape through any possible openings. Leakage is detected by bubbles which appear at the surface of the water.

Although this is a relatively simple procedure and is performed routinely in production, it does not detect comparatively small leaks. Depending on the size of the container and the need for accurate control of the propellant/concentrate ratio, it may be desirable to utilize more elaborate and accurate equipment for the detection of small leaks.

One such type of leak detector depends on the difference in thermal conductivity between untreated air and air containing a halogenated compound. A platinum wire is heated to about 900°C, and the instrument is balanced for the conductivity of normal air. A tube is used to sample the air around the leak and to bring it into contact with the heated wire. If a halogenated compound is used, the instrument will register a change. This instrument is extremely accurate and is capable of detecting leaks of about 0.01 oz (0.3 g) of propellant per year.

In use, a probe is placed around the area of possible leakage. The instrument responds instantly and returns to its zero reading within 1 sec. Since the instrument depends on the presence of a halogen compound, any propellant that may be in the air during the filling operation can decrease the sensitivity of the instrument as well as produce false results. In order to minimize this source of error, the containers to be tested must be placed in a sealed area free from halogen compounds. After a given period of time, a sample of air is withdrawn and tested for the presence of halogens. The containers can be tested individually or collectively. If the test is positive and it is necessary to identify the leaking container, then, each container must be tested individually.

This leak detector can also serve as an in-process control and be used to test each container. However, the unit must be completely enclosed to prevent contamination with the propellant vapors normally present during the filling operation. The product to be tested comes into contact with the leak detector, and if this is then connected to an automatic kick-out control, the leaking containers will be rejected. Aerosol containers that have several potential points of leakage can be passed into a hooded area which has a constant supply of fresh air. Since the containers spend sufficient time in the hood, the presence of halogen compounds can be easily detected.

Moisture Determination (17)

The presence of water can be detrimental in an aerosol product that has been specifically formulated as anhydrous. Insecticides, hair sprays, antiperspirants and aerosol coatings are several examples of this type of system.

When no special precautions have been taken to prevent contamination with water, it is possible that the amount of water picked up by the system will be sufficient to cause corrosion of the container or valve, as well as decomposition of the concentrate. Relatively small quantities of water can bring about corrosion or increase the opportunity for it to take place.

There are several ways in which moisture can enter the filled aerosol container. The greatest amount may be present in the concentrate. In many instances the ingredients used may contain traces or even large quantities of moisture, and unless they are treated to remove this moisture they may introduce serious corrosion problems. The specifications regarding the allowable moisture content of propellants are extremely rigid, and it is very unlikely that the propellant will be a source of moisture. The method of filling aerosol products can also serve to introduce moisture. The cold-filling method is generally considered to present more of a problem as to moisture contamination than the pressure method. In the cold method, there is a danger that the moisture condensing on and around the refrigeration equipment will fall into the aerosol container as the product concentrate and propellant are being added. On extremely humid days this can result in a significant increase in moisture. In addition, air trapped in the container can also contribute to the overall moisture content of the finished product.

Since the concentrate is generally responsible for the introduction of a significant amount of moisture into the aerosol container, the handling, as well as the preparation, of the concentrate should be carefully controlled. Proper precautions must be taken to ensure the anhydrous condition of the concentrate. Heating an ingredient in an oven at 105°C for about 2 to 3 hours is generally the easiest way to remove moisture; however, this method is limited to nonvolatile ingredients and to those not decomposed at this temperature. Many solvents can be passed through drying tubes containing desiccants such as silica gel, calcium chloride, activated alumina, and calcium sulfate. These materials can be easily regenerated after absorption of moisture by placing them in an oven and heating to 120 to 300°C, depending on the desiccant. Molecular sieves are generally useful for removing moisture from polar liquids, alcohols, ketones, and aldehydes. Since water is also polar in nature, the separation can best be accomplished by means of molecular sieves.

When moisture content is critical and traces of moisture can be detrimental to the success of the aerosol product, additional precautions should be taken during the filling operation. Ideally the entire filling procedure should be carried out in a humidity-controlled room, but in many instances this is not possible or practical. However, the propellants can very easily be passed through a desiccant which has been fitted into the propellant line feeding the propellant filler, and the concentrate can be handled in the same manner. All containers can be flushed with dry air. In this way, the overall moisture content of the finished aerosol package should be relatively low.

Various methods have been utilized for the analysis of moisture in aerosol packages. Since the aerosol is a pressurized package, many of the conventional methods (e.g., oven drying) cannot be utilized for this purpose. Of the methods used, gas chromatography and the Karl

Fischer titration have proved successful. The Karl Fischer titration has shown the greatest application for aerosols.

KARL FISCHER METHOD

The method depends on the reduction of iodine by sulfur dioxide in the presence of water and anhydrous pyridine in a solvent of anhydrous methanol. The reaction is said to occur in two steps, which can be shown as follows:

$$H_2O + I_2 + SO_2 + 3 \left\langle \bigcirc \right\rangle N \rightarrow 2 \left\langle \bigcirc \right\rangle N{\overset{I}{\underset{H}{\diagdown}}} + \left\langle \bigcirc \right\rangle N{\overset{SO_2}{\underset{O}{\diagdown}}} \qquad [1]$$

$$\left\langle \bigcirc \right\rangle N{\overset{SO}{\underset{O}{\diagdown}}} + CH_3OH \rightarrow \left\langle \bigcirc \right\rangle N{\overset{H}{\underset{SO_4CH_3}{\diagdown}}} \qquad [2]$$

The first reaction is responsible for the stoichiometry, since it can be seen that 1 mole of water reacts with 1 mole of iodine from the Karl Fischer reagent. At the completion of this reaction, there is a color change in the reagent from a deep brownish red to a straw yellow-brown color. However, since this end point is difficult to detect and, in many cases, is masked by highly colored solutions, an electrometric end point can be used by noting when the current across an electrode circuit does not drop below a set value for a given period of time. This is based on the fact that the iodine in the reagent will depolarize the cathode, resulting in a current flow between the electrodes. At the end point, a minimum excess of free iodine maintains this depolarization and the flow of current for a given period of time.

The aerosol sample is allowed to vaporize through a dry methanol solution. The Karl Fischer reagent is then added as needed. Depending on the volume of reagent used, the amount of water in the sample can be accurately determined in the range below 20 ppm.

The titration is generally carried out in a moisture-free atmosphere and in a closed system to prevent contamination with atmospheric moisture. The Karl Fischer reagent decomposes upon standing, and therefore its titer must be determined before the actual analysis. The sample is treated in a similar manner, and the amount of water can be calculated from the equation:

$$\frac{\text{Karl Fischer titer} \times \text{Karl Fischer reagent (ml)}}{\text{Sample weight (mg)}} \times 100 = \% \text{ water}$$

Parts per million of water can be obtained by

$$\frac{\text{Karl Fischer titer} \times \text{Karl Fischer reagent (ml)}}{\text{Sample weight (mg)}} \times 10^6 = \text{ppm water}$$

This method is advantageous in that it yields rapid, accurate results. In addition, it can be used with highly colored solution. However, the apparatus is rather difficult to handle, since precautions must be taken against contamination with atmospheric moisture. Several types of apparatus have been suggested for this purpose.

OTHER METHODS

Another method used to detect moisture in aerosol products, especially propellants, involves the use of an electrolytic moisture analyzer. This instrument is dependent on the measurement of current due to the electrolysis of water after it is absorbed in a film of phosphoric acid. It is advantageous in that it can detect any traces of moisture that may be present in propellants. In this method, water from a vaporized sample is absorbed in a phosphoric acid film. A direct-current voltage is applied across two platinum electrodes that are in contact with the acid film. By Faraday's law, the current flowing between the electrodes is related to the amount of water present. This method has been used for the analysis of moisture in pure propellants but not in aerosol products to date.

Another moisture meter makes use of two hygroscopically coated quartz-crystal oscillators. Changes in frequency due to moisture being alternately sorbed and desorbed are measured and related to moisture in parts per million. The range of some of these analyzers is from 0 to 25,000 ppm. Since they are designed primarily for use with gases, their application may be limited to the propellant phase only and not extend to the entire aerosol product. Many of these instruments are utilized in the refrigerant industry, where the presence of moisture in the fluorocarbon is extremely critical. These meters are relatively simple to use and produce rapid results. They are suitable for both laboratory use and in-process applications.

Propellant and Other Volatile Ingredient Analysis (Gas Chromatography)

The propellant is an extremely important part of an aerosol formulation, as it not only plays a role in developing the proper dispensing characterisitcs (particle size distribution, foam, spray, etc.) but also ensures that all of the contents can be emptied from the container. Selection of the wrong propellant or propellant combination can be detrimental to the performance, stability, and safety of the product. For example, the replacement of Propellant 114 with Propellant 11 in an aerosol containing water can result in severe corrosion problems. The addition of a hydrocarbon, dimethyl ether, vinyl chloride, or other flammable component may result in a product having a different flash point and flame extension from the product containing Propellant 11 in place of any of these components. Vapor pressure alone is insufficient to identify the composition of the propellant blend, as many combinations will give the same vapor pressure. The use of tertiary systems further increases the number of combinations. The same criticism applies to density determinations. Although both of these characteristics are useful, they should be employed along with some other test to definitely determine the identity as well as the composition of the pure propellant or propellant blend.

Gas-liquid chromatography provides a procedure that is easy to perform. Routine analysis of a propellant system can be accomplished within 5 to 30 min, depending on the complexity of the system. In gas chromatography, one phase is packed into the column and is referred to as the static or stationary phase. The other is the mobile phase. Since propellants can easily revert to the gaseous phase (part of the mobile phase in gas chromatography), they do not present any problem to the analyst.

In introducing a sample into the instrument, sampling chambers or gas microsyringes are used. Generally, the sample is introduced as a liquid through a gastight syringe or by means of special gas sampling valves. The liquid is then carried through heated columns by means of a carrier gas, such as air, helium, or nitrogen. Ingredients that are volatile or that will volatilize at the temperatures used can be separated and identified by this procedure. Previously, the columns have been packed with the stationary phase. The material used must be selected with care in order to ensure accurate results. Kieselguhr impregnated with dinonyl phthalate, Celite impregnated with di-n-octyl phthalate or Carbowax 60, alumina, and firebrick impregnated with butyl phthalate are a few examples of the materials which can be used. The material selected must be capable of separating the volatile components of the mixture in a reasonable length of time—about 30 min or less; in addition, it must be stable at the operating temperatures and should give a symmetrical peak. Prior experimentation will usually determine the best material for a particular column.

Figure 13.6.   Use of gas chromatography for the analysis of propellants.

Before use, the instrument must be calibrated by passing a known propellant or propellant blend through the column. By noting the positions of the peaks and the area under each of the peaks and comparing these to the corresponding results for an unknown propellant or propellant blend, identification of the unknown can be made. Other methods that can be used to calibrate the instrument include determination of the peaks obtained, using varying quantities of each propellant present in the mixture. The results are plotted on a graph as "milligrams of sample versus height (mm) of peak." This is repeated for each component. The sample is treated in a similar manner, and the peak height determined. By comparing the weight of each component present to the weight of the sample, one can calculate the percentage composition. A typical gas chromatograph used for this purpose is shown in Figure 13.6.

Although gas chromatography can be used to identify propellants and other volatile components, it is helpful if the identity of the components is previously known. This makes selection of the columns easier and facilitates rapid determination of the composition of the propellant blend. By using infrared spectophotometry in combination with gas chromatography, both the identity of components and the composition can be determined. The purity of each component can also be measured. The technique basically involves passing radiation of different wavelengths through a substance and measuring the radiation absorbed or transmitted. This is indicative of the molecular structure of the compound. The sampling techniques used for gaseous materials are different from those for solid or liquid samples.

Gases are handled in long cells, varying in length from 1 to 20 cm or over, with sodium chloride or potassium bromide windows. The more useful region of the infrared spectrum is from 2.5 to 15.0 μ. A propellant may be identified by comparing its spectrum to a known spectrum of the pure compound. If no new bands appear and if all of the known absorption bands are present, the compound is considered to be pure. The instrument can be made to yield quantitative results by calibrating its with a known mixture of propellants and plotting concentration of the component versus log of the ratio of the incident radiation (I_o) to the transmitted radiation (I). The value for I_0 is obtained by drawing a straight line tangent to the spectrum curve at the position where the absorption band is to be determined. The value for I is measured at the point of maximum absorption.

Both gas chromatography and infrared spectrophotometry have been used for research and development, as well as in quality-control procedures. Margolin and Fairchild (18) described a gas chromatographic procedure using two chromatographic columns, which gave good results. These columns were packed with diethyl hexyl sebacate, hexamethyldisilizane on Chromasorb W, and hexadecane on Chromasorb W. In 1967, Silverman, Marcis, and Schmidt (19) reported the use of infrared spectrophotometry for the rapid analysis of propellants. They were successful in identifying Propellant A (45% Propellant 12, 45% Propellant 11, and 10% isobutane), Propellant 12/11/isobutane (37:55:8), and Propellant 12/11/isobutane (33:62:5). They noted that these gas mixtures alone give a very strong peak at 3.5 μ and again at 6.75 μ. In the presence of 35% ethanol, the alcohol peak at 3.5 μ is masked by the isobutane.

Other procedures are described in the literature for the analysis of propellants by gas chromatography and infrared spectrophotometry. The combination of these two methods provides a rapid and accurate means whereby the identities and percentage compositions of propellants and other volatile components may be determined.

Net Weight Determinations

During the past several years a great deal of attention has been given to methods for the "determination of the net weight" of aerosol containers. Although this is an important consideration for all products, aerosol and nonaerosol alike, it is especially important for the former. Since the aerosol container is a sealed unit, it is impossible for the consumer to remove traces of the product that remain in the container after all of the propellant has been utilized. There has been a great deal of controversy regarding the interpretation of the statement of contents in terms of net weight which must appear on each container. This has been interpreted by some as the amount of material actually put into the container, and by others as the amount of material that can actually be utilized. The difference between these two values is dependent on many variables, including viscosity, temperature, degree of shaking, and manner in which the contents are dispensed.

Many regulatory agencies are involved as to the net weight of the contents. For example, the Federal Food, Drug, and Cosmetic Act states that the net weight must appear on the container for all pressurized products under FDA jurisdiction.

Net weight has been defined by various groups; for nonaerosols, it is the amount that can be obtained from the opened container. With the exception of the small amount of a viscous material that may adhere to the walls of the container, most of the contents can be expelled. Whereas with nonaerosols this is a relatively simple procedure, aerosols present additional considerations since these containers are sealed and cannot be opened in a manner similar to nonaerosol containers.

Since the amount of aerosol product remaining behind in the container is dependent on several factors, a standard method for the determination of net weight should be utilized. Several test methods are available. The procedure to be used is dependent primarily on the nature of the product, especially its viscosity. The methods are divided by product use into the following categories:

> Foam aerosols (nonfoods).
> Low-viscosity aerosol products.
> High-viscosity aerosol products.
> Food aerosols.

Specific procedures have been developed for each category of aerosol product (20).

Particle or Droplet Size Distribution

Depending on the size range of the various particles found in an aerosol spray, several methods can be used for the measurement of particle size. The techniques may be complicated by the properties of the particles, as well as the difficulty encountered in obtaining a representative sample of the spray. The term "particle size" generally refers to solid particles; "droplet size," to liquid particles. In many instances, however, these terms have been used interchangeably.

For medicinal aerosols, the sizes of both liquid and solid particles may be involved from the time the particles leave the nozzle of the aerosol package to the time they are deposited onto the desired surface of the respiratory tract. These solid particles, which must be finely subdivided initially, are generally in the range from 1 to 10 μ and, in most cases, from 3 to 4 μ in diameter. Other aerosols may produce particles as large as 75 to 100 μ.

MICROSCOPY

Both the optical and the electron microscope have been used to measure particle size. The droplets or particles are collected on slides, and their sizes are measured using a microscope. Through use of a calibrated ocular and a grid, the sizes of the particles, as well as the number of particles in each size range, can be determined. Depending on the size of the particles, different magnifications of the microscope are required. For particles in the range of 0.5 to 10 μ, it becomes difficult to obtain sufficient magnification for accurate measurement. To obtain an accuracy within 1%, a relatively large number of particles must be counted; some workers in this area have suggested as many as 10,000. For materials where the droplet will change because of impact upon the slide or evaporation, the slide can be coated with a substance such as magnesium oxide, and the crater produced by contact of the particle or droplet with the coating on the slide is measured and related to the size of the particle.

Various workers have reported on the use of the electron microscope to obtain greater magnification and resolution of the particles. It has been possible to measure particles having a mean diameter of 0.22 μ. However, the same problems involving collection of the particles are present. Although the electron microscope enables one to measure accurately the sizes of particles, the significance of the results must still be determined.

The sampling of an aerosol spray can present many problems. In many instances elaborate equipment is required. One of the simpler methods consists of spraying an aerosol onto a microscopic slide that has been coated with an inert powder, such as magnesium oxide or charcoal. The droplets will hit the slide, causing an impression. The slide is then placed under a microscope, and the exact diameter is measured by means of a calibrated ocular. A given number of particles (about 200) must be counted and measured in order to obtain the average particle size. In most cases, the particles are measured and then classified, so that the results are expressed in terms of cumulative per cent.

Although the method is a relatively simple one, elaborate precautions must be taken to ensure that a representative sample of the spray is obtained. In order to maximize this probability, the sample is generally collected in a wind tunnel. By adjusting the distance between the aerosol spray and the collecting slide, only particles capable of being carried this distance by the air in the tunnel will be collected. The slide is allowed to rotate in the tunnel in order to further ensure a representative sample.

When the particle hits the surface of the slide, there may be distortion of the droplet. In order to compensate for this, a correction factor that takes into account the spreading of the droplet is applied to each result. The *National Formulary* includes a method which can be used for inhalation aerosols (21). The particle size is determined by applying one measured spray onto a glass microscope slide, which is then examined under a microscope, using a calibrated ocular and 450× magnification. A majority of the magnified particles must be less than 5 μ in diameter, and not more than 10 particles may exceed 10 μ in length, measured along the longest axis. Although this procedure does not give an accurate particle size distribution of the spray, it provides some indication as to the suitability of the product for use.

SEDIMENTATION

Sedimentation sampling consists of the collection of particles which settle under the influence of gravity or other means. Stokes's law has become the standard in determining the rate of settling of particles. The velocity of the particles through a given medium can be expressed as follows:

$$V = \frac{ds}{dt} = \frac{2}{9} \frac{a^2(d_1 - d_2)g}{\text{viscosity}}$$

where a = diameter of particles, g = 980 cm / cm^2 (acceleration due to gravity), d_1 = density of internal phase, and d_2 = density of external phase.

Several assumptions made by the law include (a) that the particles are spherical, (b) that the particles are uniform, (c) that the particles are sufficiently distant from each other so that the movement of one has no influence on another, and (d) that the rate of settling is governed solely by hydrodynamic factors. Since time is needed for the particles to settle, and the time required is related to the size of the particle, the method becomes impractical when the size range is great. When employing a settling chamber, the concentration of the aerosol mixture must be restricted in such a way as to avoid the overlapping of particles, so that an accurate determination can be made. Samples are taken at a constant depth at periodic intervals.

Other sedimentation techniques utilize an automatic-recording sedimentation balance. This balance utilizes a sensitive spring and is used to weigh the particles that have settled on the pan. As the weight on the pan changes, a shutter mechanism intercepts part of the parallel light which has been directed toward a photocell. The resulting change in the photocell current is recorded and related to the weight on the pan. Particles in the range of 2 to 30 μ have been measured by this technique.

Since larger particles will generally fall a longer distance than smaller ones per unit time, by continuously varying the sedimentation depth by means of a synchronous motor, particle size analysis may be achieved. The settling velocity can be measured, and by means of a collimated beam and photocell, an optical tracing can be obtained. This technique has been found to be extremely useful for determining the size of particles in the range of 0.5 to 5 μ.

IMPACTION AND INERTIAL TECHNIQUES

Inertia methods to determine settling rates are employed to speed the rate of particle settling. The cyclone, centrifuge, and the ultracentrifuge are especially helpful in determining the particle size. Various equations are available that relate rate of sedimentation and speed of rotation to the size of the particle.

Figure 13.7.   Cascade Impactor used for particle size distribution studies.

Particle size determination by an impaction technique can be carried out effectively with the Cascade Impactor, illustrated in Figure 13.7. A series of slides is placed in the path of a moving stream of aerosol particles. The particles are deposited on the slides as the gas stream flows past the slides. The instrument is so designed that the larger particles are deposited on the first slides with which the gas stream comes into contact, and the smaller particles are carried along with the gas stream, which passes through a nozzle. This nozzle speeds the flow, and the next smaller particles are deposited on the succeeding slide. The process is repeated until the required number of slides is obtained. This instrument has met with a great deal of success when used for the determination of the particle sizes of various aerosol products. Depending on the size of the opening in each of the different stages of the cascade impactor, particles of about 16 to 0.5 μ can be collected and measured. By using different sizes of stages, particles of other sizes can be collected. In use, the aerosol is sampled through the large opening of the impactor at a constant rate, as shown in Figure 13.8. However, as the stream passes through each of the smaller openings, the velocity of the air stream and, consequently, the velocity of the dispersed particles are increased. This enables the smaller particles to eventually acquire enough momentum to impact on a slide.

LIGHT-SCATTERING TECHNIQUES

A sample of aerosol is dispensed into a chamber, and its rate of flow is regulated. The contents are diluted and mixed as required, and the stream passes through an illuminated region. The scattered light is received by a photomultiplier tube, which connects the light intensity to an electrical impulse. The impulse is then amplified and sorted into

Figure 13.8.   Introduction of sample into Cascade Impactor.

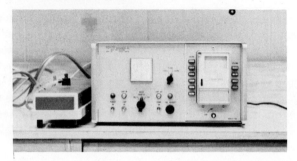

Figure 13.9.   Determination of particle size distribution by use of Royco light-scattering apparatus.

several different size ranges. Counts are registered separately on a set of several electromechanical counters. Since the method is based on the light scattering caused by particles of various sizes, and the particles are "in flight," there is no distortion of the particles due to collection on a slide or plate. However, there is the possibility of change in droplet size as a result of evaporation of solvents and other changes that may take place. This method has been used for determining the particle sizes of various aerosol products. One such unit is shown in Figure 13.9.

Other methods have been investigated for possible application to the determination of particle size of aerosols. One such method makes use of an instrument based on light scattering and sedimentation considerations. Radioactivity, high-speed motion photographs, and photomicrographs have all been investigated in the attempt to develop a method that can be used to determine the particle sizes of different aerosol products.

Microbiological Testing

A most important part of the testing of aerosol products consists of microbiological considerations. Within the past few years a great deal of attention has been given to the subject of "microbiological contamination" of various products. Several instances were detected in which certain food products were heavily contaminated with the *Salmonella* organism; some baby products were found to be contaminated with microorganisms; and, in addition, cosmetics, including creams, lotions, ointments, and eye liners, were found to contain a high concentration of microorganisms. The contamination in several cases was sufficient to warrant a recall of the products from the market place. The methods used to detect microbial contamination of aerosols are similar to those for nonaerosol products. Since the aerosol container is sealed, the presence of anaerobic organisms must always be considered.

Other Test Procedures

A variety of test procedures have been developed for use with specific aerosol products. Several of these test methods are covered elsewhere in this book, along with the discussions of specific aerosol product types. Other methods have been developed for the evaluation of aerosol foams, suspensions, and solutions (22–25).

STABILITY STUDIES (26–31)

Introduction

One of the most important considerations in the formulation of any aerosol is its stability. The stability of aerosols must be considered from several viewpoints. The efficacy of any aerosol is dependent primarily on the nature and the amount of active agent present. Since many of these agents are affected by moisture, atmospheric oxygen, other chemicals, foreign materials, and so on, it is imperative that the effects of these materials be determined during the development program. In addition, the nature of the decomposition products must also be determined, since decomposition not only represents a loss in potency of the active ingredient but also in many instances may produce an effect that is not desired (and, in certain cases, may actually be toxic). Certainly the decomposition of the other ingredients in the formulation must also be considered.

The other stability considerations are directly concerned with the components used to prepare the aerosol. The effect that the container has on the product and, conversely, the effect of the product on the container must be considered. Should the corrosion be severe enough to result in the liberation of hydrogen, followed by pitting and pinholes, the product would escape from the container, rendering the latter completely useless.

The same considerations apply to the valve components. Even slight changes in the various components of the valve may result in an inoperative package. The valve of the aerosol package has several working parts made of different materials, such as natural or synthetic rubber, plastic, polyethylene, and stainless steel. Any of these materials may possibly produce an adverse effect on the product.

A comprehensive stability program should include a study of both the physical and the chemical properties of the aerosol product in question, as well as a critical evaluation of all of the components.

Test Conditions

Since all materials will eventaully "break down," and there is always some deterioration, the main function of a stability program is to determine the shelf life of the aerosol product. This can be determined in a variety of different ways, although the generally accepted method involves storage of samples at room and elevated temperatures. Results obtained from tests conducted on samples stored at ambient temperature are generally the most reliable. However, ambient temperature does not take into account the difference in room temperatures in different parts of the United States and the world. In order to determine behavior at these other temperatures, the containers are stored at various elevated and reduced temperatures.

Tests conducted at room temperature, although most reliable, are time consuming in that it actually takes 1 year to determine each year of shelf life. In many instances, marketing and related considerations make this time interval too long. To avoid these time problems, a stability program may be accelerated; this is generally acomplished by exposing samples of the aerosol to elevated temperatures. The temperatures generally used for this purpose are 37 to 40°C and 55°C. Since heat affects many materials differently, and each material decomposes at a specific rate, it is difficult to correlate results obtained at elevated temperatures with those obtained at room temperature. The rate of breakdown of any ingredient can be determined, experimentally, however, and once this rate is known at different temperatures, the shelf life of a product can be determined with a reasonable degree of accuracy.

A stability program consists of four major steps: (a) design of experiment; (b) preparation of sufficient samples, and storage at desired temperatures for given periods of time; (c) removal of samples periodically and their subsequent analysis and examination; and (d) evaluation of the results. All stability projects should consider these four areas.

Design of the Experiment

The stability and the shelf life can be determined at the same time by a series of experiments. The experiments must be so designed that the desired results will be obtained in a minimum period of time. The stability of aerosol products involves the stability of the concentrate, as well as the stability of the entire package.

Most stability studies are designed to determine the following:

1. Decomposition of active ingredients over a given period of time.
2. Flavor, odor, and color changes that may occur.
3. Development of toxic and irritating decomposition products.
4. Stability of container.
5. Stability of valve (including dip tube).

Since a variety of different materials are used in the construction of the container, valve, and dip tube, it is difficult to determine whether a reaction takes place between the materials and the packaging components. In this regard, a thorough and systematic search of the literature will reveal a wealth of information that will aid in narrowing the number of variables to be studied. In certain instances, a rapid evaluation of several variables can be obtained by preparing a rather small number of samples for each variable, subjecting them to a high temperature (55°C), and noting the effect. However, the results obtained can prove to be misleading, since it is possible that certain reactions will take place at elevated temperatures but not at room temperature.

Preparation of Samples and Storage

A sufficient number of samples must be prepared to allow for testing over a period of at least 1 year. In addition, a representative number of samples must be stored under varying conditions. Generally a minimum of 48 containers must be prepared for each variable. However, the number may vary, depending on the nature of the product and the scope of the stability study. These samples should be prepared in a manner similar to the one in which they will ultimately be produced.

It is wise to introduce as many variables as possible during the early stages of the stability study. In this manner valuable time can be saved in the event that several of the variables prove unsatisfactory. Since all stability studies are based on results obtained over a period of time, and a certain amount of time must of necessity elaspse before results are obtained, the more variables introduced initially, the greater is the opportunity to obtain a satisfactory product. If this were not done and only one set of variables was studied initially, in the event that these proved to be unsatisfactory the study would have to be repeated from the start using a new set of variables, resulting in a loss of valuable time.

Specifically, one should attempt to include, if possible, at least two formulations, two or three container variables, and two or three valve variables. In many instances preliminary evaluations, as well as a search of the literature and available data, will make it possible to reduce the variables to the minimum recommended above. Preliminary studies can be made by placing the various subcomponents, especially the valve, in contact with the solvents used in the formulation. This examination is generally conducted in glass containers.

In order to complete the preparation of samples and to ensure the inclusion of all possible variables, samples of the aerosol product should be packaged in glass containers, so that in the event there is a breakdown of the formulation, one can determine whether the decomposition was due to the effect of the metal container on the product or the product was unstable regardless of the components employed. In addition, these samples will also provide a basis for comparison with results obtained on samples containing the various variables.

Immediately after preparation, several of the important physicochemical constants of the product are determined. These vary, depending on the nature of the product, but may well include vapor pressure, spray rate, pH, density or specific gravity, refractive index, viscosity, total weight, active ingredients, infrared and/or gas chromatography curves, color, and odor. These values are then used for comparison during each evaluation of the product.

Once the samples have been prepared, they are divided into three groups. Each group is stored under one of the following conditions:

Ambient room temperature.
98 to 100°F (36.7 to 37.8°C).
120 or 130°F (49 or 54°C).

Several other samples are stored at refrigerator temperature (45°F or 8°C) to determine the effect of lowered temperatures, and also to have available reference samples with a minimum amount of decomposition.

These samples are stored upright and on their sides, so that the product will come into contact with both the valve mounting cup and the container. When three-piece metal cans are used, the liquid should be in contact with the side seam.

Samples of each variable from each storage condition are periodically removed and examined. The time interval varies, depending on individual preferences. However, a suggested series of intervals would be 1, 2, 3, 6, and 12 months. At each interval approximately one to three containers representing each variable in each group should be examined. This schedule can be modified to suit individual needs.

Application of Chemical Kinetics

Once the samples of aerosols have been prepared and stored under various conditions, a given period of time must elapse in order for any suspected reactions to take place. Most products will show some decomposition or other change over long periods of time, and the problem is to determine the rate at which these changes occur in order to ascertain whether the product is marketable.

Before considering the application of chemical kinetics to an aerosol stability program, it may be advantageous to review briefly some of the principles of chemical kinetics.

PRINCIPLES

When two or more substances are mixed together, many different products are formed, depending on the nature of the substances, any catalysts that may be present, temperature, the presence of oxygen, and other factors. In many instances, by altering the concentrations of reacting materials or the temperature, and by using specific catalysts, the amounts of various products formed may be altered. Of importance in this discussion is that these reactions proceed at a given rate, and eventually an equilibrium is attained. When one of these reactions proceeds at substantially a much higher rate than the other reactions, a large amount of that specific endproduct is produced.

The rate of a reaction is generally expressed as a rate of change of concentration, dc/dt, or any property directly proportional to concentration. This may include changes in amount of active ingredient, pH, color, light absorption, release of gas, or partial pressure. The changes that do occur are measured over a given period of time.

The measurement of these changes can be relatively simple or very difficult, depending on the complexity of the system. In many instances the liberation of a gas and changes in pressure, optical rotation, pH, and other parameters can be easily measured and used as the means for following the decomposition of the product. In other cases, a sample of material must be assayed by chemical or physical means in order to determine its content of active ingredient. Regardless of the method employed, the ultimate purpose is to obtain a relationship between the amount of active materials originally present and the amount remaining after a given period of time.

"The rate of a chemical reaction is proportional to the product of the molar concentration of the reactants, each raised to a power equal to the number of molecules of substances undergoing the reaction." This "law of mass action" serves as the basis for chemical kinetics.

Very simply, this can be shown by the following reaction:

$$CH_3COOH + C_2H_5OH \rightarrow CH_3COOC_2H_5 + HOH$$

The rate of the forward reaction, R_f, is proportional to the concentrations of acetic acid and ethyl alcohol, while the rate of the reverse reaction, R_r, is proportional to the concentrations of ethyl acetate and water. In simple systems only one reaction takes place, whereas in complex systems more than one occurs. The reactions whose rate can be related to the stability of the system, and the one that can be conveniently determined, are chosen for the study.

Various chemical reactions can be classified according to their "order." The order of a reaction may be defined as the influence of the concentrations of the reacting materials on the rate or velocity of the reaction. In simple terms, it is expressed as the "sum of the exponents of the concentration terms that enter into rate expressions."

Depending on the value of the "sum of the exponents," a chemical reaction is said to be first order, second order, third order, zero order, and so forth. Some of the more complex chemical reactions do not fall into any of the above "orders," however, since the rate is dependent on complicated functions of the concentrations and these are assigned fractional orders.

The order of a reaction can be determined only by experimental kinetic studies and cannot be obtained from a knowledge of the balanced chemical equation.

First-Order Kinetics

Mathematically, one can express the rate of a first-order reaction as

$$\frac{dc}{dt} = -kc$$

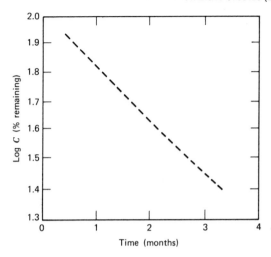

Figure 13.10. Typical plot for first-order reaction.

where c is the concentration of the reacting substance, t is the elapsed time, k is the rate constant, and dc/dt is the rate of change of the concentration of the reactant. In order to make this equation useful in calculating concentrations over a given period of time, it may be integrated to yield:

$$\log c = -\frac{k}{2.303}\,t + \text{constant}$$

If this expression is then evaluated between the limits of two concentrations, C_1 and C_2, and times t_1 and t_2, respectively, one obtains

$$k = \frac{2.303}{t_2 - t_1}\log\frac{C_1}{C_2} \qquad \text{or} \qquad k = \frac{2.303}{t}\log\frac{C_0}{C}$$

where C_0 is the concentration at the beginning of the reaction, when the time is zero, and C is the concentration after time t has elapsed. The type of plot obtained with a first-order reaction is shown in Figure 13.10.

Second-Order Kinetics

When the rate of a reaction is proportional to the concentration of each of two reactants, or to the square of the concentration of one reactant, second-order kinetics are involved. This is shown in Figure 13.11 and can be expressed as

$$\frac{dx}{dt} = k(a - x)(b - x)$$

where x is the number of moles per liter of substances a and b that react in time t. When the concentrations of a and b are equal, the expression becomes

$$\frac{dx}{dt} = k(a - x)^2$$

These expressions can be integrated as follows:

$$k = \frac{2.303}{t(a - b)} \log \frac{b(a - x)}{a(b - x)}$$

Zero-Order Kinetics

In certain instances, the rate of a reaction is unaffected by a change in concentration of the reactants (Figure 13.12). Such a reaction is determined by other limiting factors, such as the amount of a catalyst present or the amount of light absorbed. A reaction of this type can be mathematically expressed as

$$\frac{dc}{dt} = -k \qquad \text{or} \qquad \frac{dx}{dt} = k$$

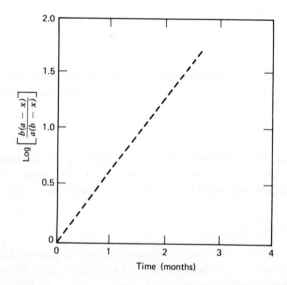

Figure 13.11. Typical plot for second-order reaction.

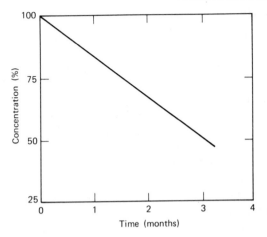

Figure 13.12.   Typical plot for zero-order reaction.

where x is the concentration of the product formed. When integrated, this equation becomes

$$x = kt + \text{constant}$$

The various expressions used to define the order of a reaction are useful in that they can be applied to determine the rate constant. By substituting the experimental values for the concentrations of reactants at any two time intervals into the appropriate equation, the value of k can be ascertained. It can also be evaluated graphically by plotting the proper expression for concentration in log form) versus time. The slope of this line multiplied by -2.303 will be equal to k. The value of k is generally expressed in terms of reciprocal seconds (t^{-1}) or other units of time.

DETERMINATION OF THE ORDER OF A REACTION

Substitution into Formulas

The results obtained are substituted into the equation for a first-, second-, third-, and so forth, order reaction. If the reaction is of one of the above orders, a constant value for k will be obtained. The equation producing the constant value for K is indicative of the order of the reaction. If no equation produces a constant k value, the reaction is more complicated and may be of a fractional order.

Graphical Method

A more convenient method (and one used to a great extent to determine the order of a reaction) involves the plotting of different functions of the concentration against time. If a straight line is produced when log C is plotted against time, the reaction is first order. This has been illustrated in Figure 13.10. The types of plot obtained with second-order and zero-order reactions are shown in Figures 13.11 and 13.12.

Fortunately, the decomposition of most agents can be depicted as first-order reactions. The kinetics of first-order reactions are rather simple and allow for the prediction of stability. In addition, many first- and second-order reactions behave in a similar manner up to about 50% decomposition, so that by restricting the conditions many second-order reactions can be treated as first-order reactions with acceptable results.

FIRST-ORDER REACTIONS

Since the rate or velocity of decomposition is dependent on temperature, samples must be stored at constant temperature, as seen in Figure 13.13.

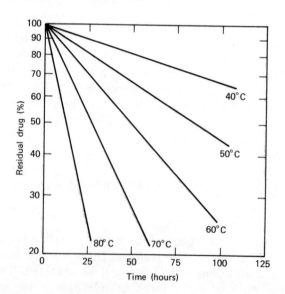

Figure 13.13.   Effect of temperature on rate of first-order reaction. [From L. Lachman, *Amer. Perf. Cosmet.,* **77** (September 1962), 25–38.]

Half-Life

Once a reaction is found to be first order and the value of k is evaluated, other values for time can be inserted into the equation and one may obtain the predicted concentration corresponding to the time. In this manner it is possible to predict the concentration remaining after time (t). From these equations one can derive an expression for "half-life," which is defined as the time necessary for half the material to decompose. Half-life is independent of initial concentration and is extremely useful in expressing the decomposition (decay) of radioactive materials. This parameter is expressed as

$$t_{1/2} = \frac{0.693}{k}$$

Of greater importance to aerosol manufacturers is the time required for the product to decompose a reasonable amount. Many have accepted a 10% figure as being reasonable. The time required for a product to decompose 10% can be expressed as

$$t_{10\%} = \frac{0.104}{k}$$

All that a manufacturer need know is the k value, which can be determined experimentally. From this, the time required for 10% to decompose can be ascertained.

Arrhenius Plot

The rate of the decomposition can be accelerated by using temperatures higher than room temperature. The influence of temperature on the velocity of a reaction is given by

$$k = Se^{-\Delta H_a/RT}$$

where k = specific rate of degradation, R = universal gas constant [1.987 cal/(deg)(mole)], ΔH_a = heat of activation (cal/mole), T = absolute temperature, and S = frequency factor (frequency of encounters between reactant molecules).

This equation can also be derived to yield

$$\log k = \log S - \frac{\Delta H_a}{2.303RT}$$

If certain liberties are taken with this equation, $\log S$ can be neglected and one can still obtain acceptable results. A plot of $\log k$ versus $1/T$ should yield a straight line. The slope of the line is obtained and is used to evaluate ΔH_a:

$$\text{Slope} = -\frac{\Delta H_a}{2.303R}$$

Sometimes ΔH_a is referred to as E_a (activation energy). The y intercept will yield the value of $\log S$. This type of plot is seen in Figure 13.14. By evaluating this equation between temperatures of T_1 and T_2 and between k_1 and k_2, the following equation is obtained:

$$\log \frac{k_2}{k_1} = \frac{E_a}{2.303R} \frac{T_2 - T_1}{T_2 T_1}$$

This equation is valuable in that it can be used to obtain k at any temperature T. This assumes that a straight line exists between the limits of T, an assumption that is not always true but gives acceptable results in most cases. Once the rate of decomposition is known, the stability as well as the shelf life of the product can be ascertained. Through

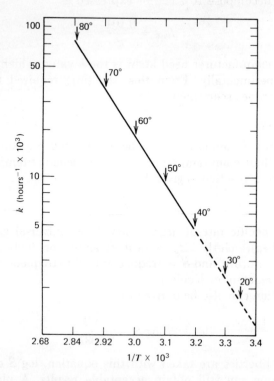

Figure 13.14.   Arrehenius plot of first-order reaction [From L. Lachman, *Amer. Perf. Cosmet.*, **77** (September 1962), 25–38.]

use of accelerated studies one can predict, to a certain extent, what may occur to a product over a longer period of time. However, it should be indicated that accelerated studies may sometimes be misleading and therefore the product should be fully studied before making a final decision.

Analysis of Aerosols

In order to obtain experimental data, the samples that have been stored under varying conditions must be subjected to physical and chemical examination. At periodic intervals samples are removed and examined. The time interval between samplings is largely dependent on the type of product and any prior knowledge of its overall stability. For the most part, samples are taken after 1, 2, 3, 6, and 12 months. This is only a suggested schedule and may vary, depending on specific cases. In many instances additional samples are retained to allow the sampling after 18 months and 2 years. The latter would represent the

Generally 1 year is sufficient for accelerated studies. Products capable of withstanding high temperatures for this length of time usually do not present stability problems. However, each product must be individually evaluated. At least two or three containers of samples stored under the different conditions must be removed during each period. In this manner, it is possible to obtain a more representative sampling. The number of samples selected at each time depends on the initial number of samples prepared, which is generally limited by the size of the storage facilities.

The testing of aerosols must of necessity be concerned with three areas:

1. Concentrate and propellant.
2. Container.
3. Valve.

Evidence of decomposition or deterioration in any of these could result in an ineffective product.

CONCENTRATE AND PROPELLANT

For aerosols, the concentrate is most important since it contains the active ingredients. Depending on the nature of the ingredients, an assay may be necessary in order to determine the amount of active ingredient present. When more than one active ingredient is present, an assay must be run on each ingredient unless it can be shown that the rate of decomposition of one ingredient is indicative of the corresponding rate of the other. In addition to the assay, other tests should include pH, specific gravity, color, odor, refractive index, and any other physical and chemical constants which may give a clue to decomposition. In fact, it is possible that a change in one of the above may be indicative of decomposition of the active ingredients. After these determinations, the results are compared to the original values.

CONTAINER

The contents of the container are removed, and the container is opened and then examined for signs of corrosion. These changes can be detected without too much difficulty, since attack on tin plate is generally visible to the naked eye. In cases of doubt, the container can be flattened and then examined under a microscope. Small pinholes can easily be seen if present. For containers with internal lacquering, further examination is necessary to ensure that the lacquer is not softened or dissolved by the concentrate. Blistering and peeling of the internal liner also may occur. Special attention should be paid to side seams and other seams in the containers, as there is always the danger of attack on these areas. The head space should also be examined to ensure that there is no reaction; if there is, its effect may be noted.

When glass is used for the container, an examination of the container is not needed, since glass does not present the problems found with plated cans. Plastic containers may require special testing in order to determine whether any leaching takes place. There is also the possibility of the active ingredients reacting with the plastic container, and tests should be conducted to ensure that such reactions do not take place. Aluminum and stainless steel, being less reactive, present fewer problems than does tin-plated steel. However, these containers should be fully examined in a manner similar to that used for tin-plate cans.

VALVE

The valve should be examined to ensure that it is functional, that is, will satisfactorily dispense the product and then can be easily turned off. This can readily be determined during the dispensing of the product. The valve cup should be examined for evidence of corrosion. The various valve subcomponents should also be examined to determine whether softening, cracking, elongation, distortion, or the like has occurred. Several of these effects can result in a defective valve that will not operate properly.

Another concern with the valve and dip tube is the possibility of leaching and sorption. These can result in some of the active ingredients being removed from the system, as well as some of the components in the various valve parts being selectively dissolved by the concentrate /propellant system. When this is suspected, special tests must be utilized to check whether it has actually occurred.

After the first month's examination, which should be as complete as possible, one may wish to modify the tests since several variables may no longer be applicable. For example, if the containers stored at 120 to 130°F have shown extreme decomposition during the first month, there is no need to further study containers stored at this temperature. In many instances, however, the results obtained at this high temperature are misleading and require knowledgeable interpretation.

The results of all these tests should be correlated and related to the results obtained with fresh samples of product.

It should be emphasized that the type of program used to determine the stability of the aerosol product is dependent on the type of product, as well as the nature of the ingredients. In many instances, the elaborate testing program described here can be modified to suit specific needs. For example, products that contain only solvents and propellants need not be analyzed as exactly as a pharmaceutical aerosol containing an exact amount of a therapeutically active ingredient and that must be dispensed in an exact dosage.

REFERENCES

1. John J. Sciarra, Quality Control for Pharmaceutical and Cosmetic Aerosol Products, in Murray S. Cooper (Ed.), *Quality Control in the Pharmaceutical Industry*, Vol. 2, Academic Press, New York, 1973, pp. 1–54.

2. *Aerosol Guide*, 6th ed., Chemical Specialties Manufacturers Association, New York, March 1971.

3. M. A. Johnsen, W. E. Dorland, and E. K. Dorland, Chapter 13 in *The Aerosol Handbook*, Wayne E. Dorland Company, Caldwell, N.J., 1972, pp. 377–400.

4. *Method for Measuring Vapor Pressure of Aerosol Products*, Freon® Aerosol Report FA-11, E. I. du Pont de Nemours and Company, Wilmington, Del., 1961.

5. *Determination of Freon Propellent Composition with a Pressure Pycnometer*, Freon® Aerosol Report FA-23, E. I. du Pont de Nemours and Co., Wilmington, Del., 1958.

6. *Vapor Pressure and Liquid Density of Freon Propellants*, Freon® Aerosol Report FA-22, E. I. du Pont de Nemours and Co., Wilmington, Del., 1957.

7. William G. Gorman and Gary D. Hall, Densimetry in Aerosol Products, *Soap Chem. Spec.*, **40** (May 1964), 213.

8. Allan B. Jr. Reed, Measuring the Density of Volatile Liquids, *Aerosol Age,* **9** (November 1964), 20.

9. A Method for the Determination of Discharge Rates from Pressurized Packs, *Soap, Perfum., Cosmet.*, **31**(January 1958), 55.

10. *The National Formulary*, 13th ed., American Pharmaceutical Association, Washington, D.C., 1970, pp. 208, 770.

11. Paul A., Sanders, Evaluating the Spray Properties of Aerosol Products, *Aerosol Age*, 11 (January 1966), 31.

12. M. J. Root, Formulating for Pressure, *J. Soc. Cosmet. Chem.,* **7** (1956), 149.

13. Ronald A., Gunn-Smith and Norman E., Platt, Examining Aerosol Spray Patterns, *Aerosol Age*, **6** (February 1961), 33.

14. K. Dixon, Method for Determining Spray Patterns from Pressurized Packs, *J. Soc. Cosmet. Chem.,* **10**(1959), 220.

15. *The National Formulary*, 13th ed., American Pharmaceutical Association, Washington, D.C., 1970, pp. 770–771.

16. L. Flanner, Aerosol Valve Leakage Can Cause Short Weights, *Aerosol Age,* **11** (March 1966), 26.

17. *Moisture Determination in Freon Fluorocarbons by Karl Fisher Titration*, Freon® Technical Bulletin B-23, E. I. du Pont de Nemours and Co., Wilmington, Del., 1961.

18. I. Margolin and C. M. Fairchild, Gas Chromotographic Method for the Determination of Propellants in Aerosol Cosmetics, Scientific Conference, Toilet Goods Association, 49th Meeting, Americana Hotel, New York, May 9, 1969.

19. P. Silverman, J. Marcis, and C. N. Schmidt, Rapid Analysis of Propellents in Aerosol Products, *Soap. Chem. Spec.,* **43** (February 1967), 78.

20. S. Hasko, Checking Packaged Aerosol Products, *Natl. Bur. Stand. Rept.* 10-379, U.S. Department of Commerce, Washington D.C., Oct. 19, 1970.

21. *The National Formulary*, 13th ed., American Pharmaceutical Association, Washington, D.C., 1970, p. 384.

22. Randall P. Ayer and James A. Thompson, Scanning Electron Microscopy and Other New Approaches to Hair Spray Evaluation, *J. Soc. Cosmet. Chem.,* **23** (1972), 617.

23. K., Vora, L., Augsburger, and R. F. Shangraw, Design and Evaluation of a Pressure Attachment for a Rotational Rheometer, *J. Pharm. Sci.,* **59** (1970) 1012.

24. M. David Richman and R. F. Shangraw, Reological Evaluation of Pressurized Foams, *Aerosol Age,* **11** (July 1966), 28, and (August 1966), 39.

25. Albert Shansky, Evaluation Methods for Some of the Mechanical and Surface Properties of Hair, *Amer. Perfum. Cosmet.,* **83** (March 1968), 25.

26. F. Presant, The Aerosol Stability Program, *Aerosol Technicomment,* **3** 1 (March 1960).

27. E. R. Garrett, Stability Prediction of Drugs and Pharmaceutical Preparations, *J. Pharm. Sci.,* **51** (1962), 811.

28. N., Lordi and M. Scott, Stability Charts, *J. Pharm. Sci.,* **54** (1965), 531.

29. L., Lachman and J., Cooper, A Comprehensive Pharmaceutical Stability Testing Laboratory, Parts I, II, and III, *J. Amer. Pharm. Assoc. Sci. Ed.,* **48** (1959), 226, 233; **49** (1961), 213.

30. J., Cooper and L. Lachman, The Stability Testing of Pharmaceuticals, *Pharm. Weekbl.,* **94** (1959), 457.

31. Fred, Kraus, Accelerated Testing of Shelf Life Parameters, *Soap Cosmet. Chem. Spec.,* **48** (November 1972), 68.

DEODORANTS AND ANTIPERSPIRANTS

BACKGROUND

Personal hygiene can be enhanced by the use of agents that suppress body odors. In recent years the physiology of sweat and the development of body odor have become better understood, and good means of control are available.

Types of Glands

According to skin specialists, there are some 2,300,000 sweat glands located over the entire body surface. There are two types of sweat glands: the eccrine and the apocrine.

ECCRINE SWEAT GLANDS

The eccrine sweat glands occur all over the body. The liquid discharged from these glands is known as sweat or, more elegantly, as perspiration. The perspiration from the eccrine glands consists of about 99% water, with a little salt plus many other chemicals in very small concentration. It is a clear liquid with a pH value of 4.0 to 6.8. The eccrine glands are coiled tubes with a funnel-like opening on the surface of the skin. Eccrine perspiration helps to regulate the body temperature as these glands are the body's thermostat. Thus, when the weather is warm, sweat pours onto the skin. As it evaporates, it cools the skin surface. More than a pint of sweat may be excreted by a healthy individual in a day. Perspiration also responds to emotional stimuli. Eccrine sweat has no odor when freshly discharged and does not develop any significant odor with time. This perspiration results in the wet underarm feel.

APROCRINE SWEAT GLANDS

The apocrine sweat glands are larger than the eccrine glands. The apocrine gland consists of a single duct which proceeds from a coiled glandular mass and empties into a hair follicle. These glands develop during puberty and are present on only a few parts of the body, for example, in the axillae and genital regions. The perspiration from the apocrine glands also consists of water but contains, in addition, milky secretions from the ducts of the glands. It is a turbid-white fluid with a pH value of 5.0 to 6.5. These glands are under the control of adrenergic fibers of the autonomic nervous system. They respond to emotional stimuli but not to heat. The role of the apocrine glands is not fully understood, but is thought to be part of the human being's ancient defensive system against other living species and is also thought to be related to sex stimulation. Apocrine sweat, like eccrine sweat, is odorless when freshly discharged, but an unpleasant odor develops within a short time, especially under favorable conditions of humidity and temperature. The odor is formed when bacteria, which are normal residents of the skin, start acting on the organic waste matter of the apocrine sweat. The decomposed fatty acids and the by-products of fermentation account for most of the body odor that develops.

Deodorant Action

Offensive underarm odor is prevented by the application of antibacterial substances. These materials prevent the bacteria of the skin from reproducing. Therefore the number of bacteria is drastically reduced and is not sufficient for breaking down the organic waste materials from the apocrine glands into odorous components. Gram-positive organisms are responsible for most of the body odor that is produced. There are no known ways of safely preventing the excretion of apocrine sweat.

Deodorant products are now being sold for use on other parts of the body besides the underarm areas, such as on the feet and vaginal area.

Antiperspirant Action

Underarm wetness may be controlled by reducing the amount of eccrine perspiration discharged into the underarm area. Most of the ingredients used in antiperspirants have astringent properties (e.g., aluminum salts). The exact manner of this action is not known. Some skin specialists believe that the astringent materials react with the protein of the skin with the result that the eccrine gland openings are partially blocked (1, 2). Others believe that the antiperspirant substances cause the wall of the eccrine sweat glands to become porous, so that most of the sweat passes out through the walls before reaching the skin surface (3). Several other mechanisms of action are proposed in the literature. The antiperspirant materials do not control the flow from the apocrine glands.

The commonly used antiperspirant ingredients are not effective on the entire population. Furthermore, they are not 100% effective even on individuals who show favorable activity toward antiperspirant chemicals.

GENERAL CONSIDERATIONS

Many product and application improvements have been made since the first deodorant cream appeared on the market in 1888. The most significant improvement has been the aerosol form of dispensing. In 1971 the aerosol deodorant/antiperspirant category, including feminine hygiene sprays, amounted to over $450 million in retail sales. The aerosol form accounted for about 75% of the total deodorant/antiperspirant sales volume.

Advantages of Aerosols

An aerosol spray containing a deodorant and/or antiperspirant substance is more advantageous to use than other product forms, such as roll-ons, creams, and squeeze liquids, for the following reasons:

1. It applies easily and quickly.
2. It may be used by more than one person because the applicator does not touch the underarm area.
3. Usually the product dries very quickly without messiness.

Three possible types of products can be formulated.

1. Deodorant only.
2. Antiperspirant with antibacterial agent added.
3. Antiperspirant without antibacterial agent.

Before elaborating on the specific formulation technology of deodorant and antiperspirant products, many generalities that apply to both types of aerosol products should be mentioned.

Aerosol Components

PROPELLANT SYSTEMS

Usually low-pressure propellant systems of about 20 to 70 psig are used. The propellant systems may consist of the following:

1. Mixtures of Propellant 11 or 114 with Propellant 12, for example, 12/11 (65:35) or 12/114 (50:50).

2. Propellant 12 only.

3. Mixtures of the hydrocarbons butane and isobutane with propane. These mixtures are in disfavor, however, because of their flammability hazard.

4. Blends of fluorocarbons and hydrocarbons, for example, 12/11/ isobutane (45:45:10), also known as Propellant A. The low concentration of isobutane in this blend with fluorocarbons does not yield a flammable propellant blend.

Products placed in evacuated containers with blends containing Propellant 11 and ethyl alcohol may react to produce hydrochloric acid, which may corrode the aerosol container (4, 5). This reaction may be inhibited by oxygen or by using Propellant 11S (patented by E. I. du Pont), which contains 0.3% nitromethane as the inhibitor. Propellants 11 and 11S are not as stable in aluminum containers as they are in tin-plate containers, probably because of their solvent effects on the aluminum oxide film (6).

As previously mentioned, ethyl alcohol is used as a vehicle in many formulations. Anhydrous ethyl alcohol has a pressure-depressing effect on propellants because of mutual solubility. For instance, Propellant 12 has a pressure of 70.2 psig at 70°F. The combination of 65% by weight of anhydrous ethyl alcohol with 35% by weight of Propellant 12 in an aerosol container yields a pressure of about 40 psig at 70°F. The greater the percentage of anhydrous ethyl alcohol, the lower is the resulting pressure. The addition of water minimizes the pressure depressing effect of ethyl alcohol, because the mutual solubility of water and propellant is small. For example, 65% by weight of 190-proof ethyl alcohol with 35% by weight of Propellant 12 in an aerosol container will give a pressure of about 53 psig at 70°F.

The pressure in an aerosol container may be further reduced by evacuation of air during the valve-crimping operation or by cold filling of the propellant, which displaces the air in the head space. For instance, the combination of 65/35% by weight anhydrous alcohol/ Propellant 12 will produce about a 34-psig pressure if a vacuum of 15 in. Hg is used during the valve-crimping operation. Cold filling of the propellant will reduce the pressure even more because more air is removed in the filling operation.

VALVE–ACTUATOR ASSEMBLIES

Two basic types of aerosol valve designs are available for use on aerosol deodorant/antiperspirants. These are vapor-orifice or non-vapor-orifice valves. A vapor-orifice provides the following advantages over a non-vapor-orifice valve:

1. Warmer spray for a more comfortable underarm feeling.
2. Reduced delivery rate to minimize wetness upon application.
3. Finer spray for quicker drying of product.
4. Capability of being used upright or inverted.

The vapor orifice of the valve is exposed to the vaporized propellant in the head space.

When a non-vapor-orifice valve is actuated, only the liquid portion of a two-phase aerosol is emitted. However, as a vapor-orifice valve is actuated, the vaporized propellant in the head space reaches the valve housing, along with the liquid portion that travels up the dip tube. This phenomenon in the valve housing reduces the delivery rate. Also, the vaporized propellant portion atomizes the liquid in a finer and warmer mist. As a container is inverted, the dip tube opening becomes the "vapor orifice," the vapor orifice of the valve becomes the "liquid orifice," and the container continues to operate. No "liquid orifice" is available when a non-vapor-orifice valve is inverted.

Regular (1/8 in.) and capillary dip tubes are available for valves. Capillary dip tubes decrease the flow of liquid product entering the valve housing. Mechanical breakup actuators also provide some atomization of the spray, resulting in an even warmer and finer spray. Directional actuators are widely used on personal deodorants/antiperspirants so that the consumer can easily apply the product to the underarm areas without accidentally spraying his face.

CONTAINERS

Three-piece tin-plate containers are most widely used in the United States. However, two-piece drawn tin-plate, aluminum, and glass containers have also been utilized. Plastic containers are available but have not been widely used. Most metal containers are internally line (e.g., epoxy or vinyl over epoxy) to provide maximum protection of the product over its shelf life.

The Department of Transportation has published regulations pertaining to allowable pressures in aerosol containers (7). These are covered in Chapter 23.

ASSESSMENT OF AEROSOL SPRAY

The following parameters are specific for deodorant/antiperspirant products.

Spray Pattern

This is conventionally characterized by the spray diameter obtained in spraying at a 6-in. distance for 2 sec (usual spray distance and time from an underarm spray). A measurement may easily be made by spraying at the desired distance and time onto a white paper colored with an indelible pencil or other similar means. The diameter of a deodorant/antiperspirant spray should not exceed 3 in.; otherwise, some of the product will fall outside the underarm area. Wider spray patterns, however, may be desirable for deodorant products sprayed on parts of the body other than the underarm.

Delivery Rate

This is usually expressed as the amount of product sprayed per unit time. Delivery rates in excess of 0.7 g/sec usually produce very wet and slow-drying applications if the product has an alcoholic or hydroalcoholic base.

Drippiness

The spray should not be excessively wet when used by the consumer. Spraying for 2 sec from a 6-in. distance onto waxpaper provides a laboratory indication.

Drying time

This will vary with the vehicle, propellant, ratio of vehicle to propellant, valve, and actuator. Most satisfactory alcohol-based products contain about 50% by weight of a fluorocarbon propellant with a vapor-orifice valve and mechanical breakup acutator. These sprays usually dry within a few seconds.

Particle Size

This may be measured by several methods (8, 9). Nearly all of a wet spray with large particles (e.g., low propellant content) will be delivered to the underarm. However, it may be drippy, wet, and slow drying. A fine spray is more pleasant for underarm use.

Coolness

One of the requirements of a deodorant/antiperspirant spray is minimum cooling action on the skin. A simple, mercury-bulb thermometer may be sprayed from a 6-in. distance, and the resulting temperature drop recorded. More sophisticated electronic methods may also be developed to generate more reliable data. Many variables, such as concentrate/propellant ratio, propellant composition (blends of Propellants 12 and 11 are usually cooler than blends of 12 and 114), and valve–actuator assembly (as previously described), significantly influence spray coolness.

EFFICACY

The deodorant and antiperspirant efficacy of a product should be determined to ensure that the vehicle or formulation has not interfered with the active antibacterial or antiperspirant components. Deodorant efficacy is checked by *in vivo* bacterial counts and odor evaluations. Antiperspirant efficacy is best evaluated by *in vivo* perspiration collection methods (10, 11). These tests must be run under carefully controlled conditions so that the results are meaningful. The subjects under evaluation should have refrained from using any deodorant/antiperspirant product for a sufficient period of time before testing. A single application of a product to a "clean" underarm does not show its maximum effectiveness. It may take from three to seven applications to obtain maximum benefits from these products. Many service laboratories are equipped to conduct these tests.

DEODORANTS

Active Ingredients

Some of the most widely used antibacterial components employed in aerosol deodorants are the following:

1. Benzethonium chloride, a quaternary ammonium salt active against skin bacteria. It is soluble in both ethyl alcohol and water. It has both gram-positive and gram-negative activity.

2. Methylbenzethonium chloride, which is similar to benzethonium chloride.

3. Hexachlorophene.* This is active against the gram-positive organisms that reside on the skin and break down apocrine perspiration. Hexachlorophene is substantive to the skin to provide better deodorant protection. It is soluble in ethyl alcohol and insoluble in water.

4. Di- and tribromosalicylanilides. These are also effective antibacterial agents. They are not as soluble in ethyl alcohol as the other materials mentioned above.

5. Zinc phenolsulfonate. This compound is active against gram-positive and gram-negative organisms. It was originally thought to be an antiperspirant agent because of the zinc cation, but this activity is minimal (12).

6. 2,4,4′-Trichloro-2′-hydroxydiphenyl ether. This is active against gram-positive and gram-negative organisms.

The antibacterial agent may often be transferred to the clothing that is in contact with the skin, thus providing deodorant protection to the fabric.

Other Ingredients

EMOLLIENTS

Emollients or protective lubricants may be added in small concentration to the formulation to provide a better cosmetic skin feel. They also leave a thin film on the skin to hold the antibacterial agents. Propylene glycol and isopropyl esters are widely used for this purpose.

PERFUME

Perfume serves four purposes in a deodorant product.

1. Mild antibacterial effect for additional protection.

2. Masking agent for any residual body odor that may still be formed. An ideal perfume for a deodorant has long-lasting fragrance characteristics to prolong this masking quality.

3. Masking agent for the odors contributed by the raw materials in the formulation.

4. Selling point to the consumer.

* See Editor's Note, page 17.

Certain fragrance constituents are avoided by the perfumer because they may react unfavorably with perspiration and cause a complete change in the residual fragrance on the skin.

OTHER ADDITIVES

Film formers (e.g., polyvinylpyrrolidone) and surface-active agents have been added to formulations to increase the product efficacy. However, they are not widely used in present-day formulations. As a matter of fact, some surface-active agents (e.g., phosphate esters) may decrease the effectiveness of certain antibacterial agents (13).

VEHICLE

Ethyl alcohol is widely used as a vehicle or carrying agent for the active ingredients of an aerosol deodorants because of miscibility with propellants. Also, its fast drying qualities are an obvious advantage. Ethyl alcohol has antiseptic qualities of its own. Specially denaturated ethyl alcohol is used; common denaturants for this purpose are brucine, brucine sulfate, quassin, Bitrex, and *tert*-butyl alcohol. For example, SDA-40 designates ethyl alcohol to which 1.5 avoir oz of brucine alkaloid or brucine sulfate or quassin (or 1.5 avoir oz of any combination of two or three of those denaturants) and 1/8 gal of *tert*-butyl alcohol have been added as denaturants to every 100 gal.

Hydroalcoholic vehicles have been used in formulations to maintain lower product cost. However, in this case it is sometimes necessary to incorporate corrosion inhibitors such as benzoates and nitrates if the product is packaged in metal containers (14).

Corrosion and Product Stability

Anhydrous aerosol deodorant formulations are not usually very corrosive. However, certain cautions should be pointed out:

1. Phenolic compounds such as hexachlorophene will discolor in the presence of iron. Three-piece tin-plate cans perform better over a long shelf life if they are internally coated.

2. Quaternary salts may be corrosive because of the halogen anion.

3. Completely anhydrous ethanol formulations should not be packaged in aluminum containers. At least 0.5 to 1.0% water should be present to prevent container perforation. The small amount of water provides a protective oxide film.

Typical Formulations

UNDERARM PREPARATIONS

Underarm Deodorant 1 (Source: Allied Chemical)

Concentrate

	% by Weight
Zinc phenolsulfonate·$8H_2O$	2.72
Hexachlorophene	0.22
Solulan 16 (Amerchol)	0.91
Perfume of choice	q.s.
Ethyl alcohol, SDA-40, anhydrous	96.15
	100.00

Fill

	% by Weight
Above concentrate	55.00
Propellant 12/114 (57:43)	45.00
	100.00

Method of Preparation

Dissolve or mix ingredients in ethyl alcohol. Fill concentrate and propellant by conventional aerosol filling methods.

Underarm Deodorant 2 (Source: Amerchol)

Concentrate

	% by Weight
Acetulan (Amerchol)	1.5
Hexachlorophene	2.5
Dipropylene glycol	20.0
Perfume of choice	2.0
Ethyl alcohol, SDA-40, anhydrous	74.0
	100.0

Fill

	% by Weight
Above concentrate	10.0
Propellant 12/11 (30:70)	90.0
	100.0

Method of Preparation

Dissolve or mix ingredients in ethyl alcohol. Fill concentrate and propellant by conventional aerosol filling methods.

Underarm Deodorant 3
(Source: U.S. Patent 2,995,278)

Concentrate	% by Weight
Antibacterial agent of choice	q.s.
Deionized water	44.0
Solubilized perfume of choice	q.s.
Ethyl alcohol, SDA-40, 190 proof	56.0
	100.00

Fill	% by Weight
Above concentrate	73.00
n-Butane	14.00
Propellant 12/114 (75:25)	13.00
	100.00

Method of Preparation

Dissolve or mix the antibacterial and solubilized perfume in ethyl alcohol. Add the water. Fill concentrate and propellant by conventional hydrocarbon filling methods.

Product must be shaken before use because it is a three-phase aerosol system.

OTHER APPLICATIONS

About 25 to 30% of aerosol deodorant users apply these products to other parts of the body.

Foot Deodorants

Several foot deodorants are available in the market place. Many are very similar to the typical formulations described above. However, the moist and confined conditions of the foot, along with the many crevices around the toes, leave many people prone to fungus growth, particularly athlete's foot (*Tinea pedis*—dermatophytosis). Undecylenic acid, zinc undecylenate, salicylic acid, and boric acid are sometimes added to combat fungus conditions. Aerosol talc formulations are also used (with added medication) to absorb the moisture on the foot.

Feminine Hygiene Sprays

In the recent past, a number of companies have successfully introduced mass-marketed aerosol sprays for use as personal hygiene products. Although sold mostly as feminine hygiene sprays in spray liquid and powder form, similar products are also available in the men's product line.

The sprays are applied to the sensitive genital areas and therefore are formulated with little or no alcohol. A low-boiling, high-pressure propellant and vapor tap valve are used to minimize the chilling effect. Since most of the propellant flash evaporates, very little propellant reaches the target area, reducing the possibility of any irritating effect.

The formulations are based on a blend emollient oil such as isopropyl myristate, a bacteriostat, perfume oil, and propellant; some may. also contain talc or other similar powders.

Feminine Hygiene Spray, Solution Type
(Source: Amerchol)

Concentrate	% by Weight
Antibacterial agent of choice	4.0
Perfume of choice	2.0
Ethyl alcohol, SDA-40, 190 proof	94.0
	100.0

Fill	% by Weight
Above concentrate	5.0
Propellant 12	95.0
	100.0

Method of Preparation

Dissolve or mix ingredients in ethyl alcohol. Fill concentrate and propellant by conventional aerosol filling methods.

Feminine Hygiene Spray, Powder Type
(Source: Givaudan)

Concentrate	% by Weight
Perfume oil	2.00
Hexachlorophene	1.50
Isopropyl myristate	3.50
Talc Lo Micron No. 1 (Whittaker, Clark, & Daniels)	93.00
	100.00

Fill   *% by Weight*

Concentrate   7.50
Propellant 12/114 (40:60)   92.50
                                                ———————
                                                100.00

Body Deodorant Sprays

Several sprays have been developed for use on the entire body. A typical formulation for this type of product follows:

Body Talcum (Source: Allied Chemical)

Concentrate   *% by Weight*

Part A
    Talc   16.6
    Magnesium stearate   2.2
Part B
    Hexachlorophene   0.2
    Isopropyl myristate   0.5
    Arlacel 83 (Atlas Chemical)   0.5
    Perfume of choice   q.s.
    Propellant 11   80.0
                                                ———————
                                                100.0

Fill   *% by Weight*

Above concentrate   55.0
Propellant 12   45.0
                                                ———————
                                                100.0

Method of Preparation

Mix the powders in Part A thoroughly. Dissolve the hexachlorophene in the isopropyl myristate; then add Arlacel 83, perfume, and Propellant 11. Add the Part A powder concentrate to the aerosol container. Add Part B as a liquid concentrate. (Parts A and B may be added as a suspension as long as the mix is stirred continuously while filling.) Fill the Propellant 12 by conventional aerosol filling methods.

A vapor tap valve and a mechanical breakup actuator are recommended to make the spray as warm as possible.

ANTIPERSPIRANTS

History

Aluminum compounds have been considered, almost exclusively, to be the best and safest antiperspirant agents for use in cosmetic products. However, the addition of aluminum compounds to aerosol products can produce three problems.

- Corrosion of metal cans because of the acidic nature of these compounds, particularly when the salt is in solution.
- Clogging of the actuator, leading to inoperative units.
- Deterioration of fragrances due to the reactivity of the aluminum compounds.

Active Ingredients

ALUMINUM SALTS OF STRONG ACIDS

Examples of this classification are aluminum chloride and aluminum sulfate. These materials are very effective antiperspirant agents

Figure 14.1   Right Guard deodorant and antiperspirant (Gillette Co.).

because of their acidic nature and the availability of aluminum to react with skin protein. However, because of their acidic nature they are very corrosive to metal aerosol containers. Stable perfumes are also difficult to formulate. In addition, these salts are irritating to sensitive or abraded skin.

ALUMINUM PHENOLSULFONATE

Aluminum phenolsulfonate has only minimal antiperspirant efficacy because of its covalent structure and therefore has not been used extensively. It is soluble in ethyl alcohol.

A typical formulation using these ingredients is as follows:

Antiperspirant 1 (15)

Concentrate	*% by Weight*
Aluminum phenolsulfonate	10.00
Propylene glycol	7.50
Hexachlorophene	0.2
Perfume of choice	0.35
Versene No. 7 (Dow)	0.1
Water	1.0
Ethyl alcohol, SDA-40, anhydrous	80.85
	100.00

Fill	*% by Weight*
Above concentrate	57.5
Propellant 12/114 (20:80)	42.5
	100.0

Method of Preparation

Dissolve or mix ingredients in ethyl alcohol. Fill concentrate and propellant by conventional aerosol filling methods.

Severe discoloration problems have been encountered with aluminum phenolsulfonate because of the iron that is present in the commercial raw materials.

ALUMINUM CHLORHYDROXIDE COMPLEX

Aluminum chlorhydroxide complex has been the antiperspirant agent of choice for many years in nonaerosol product forms. It is a 5/6 basic aluminum chloride complex with the following empirical formula:

$$[Al_2(OH)_5Cl]y \cdot 2H_2O$$

It has been shown to be an effective antiperspirant agent. Aluminum chlorhydroxide complex has the advantage of being only mildly acidic (the pH of a 15% solution is 4.2). Hence the potential skin irritation usually encountered with aluminum salts of strong acids is minimized.

Aluminum chlorhydroxide complex is soluble in water and insoluble in ethyl alcohol. However, a 50% aqueous solution is miscible in all porportions with ethyl alcohol. This antiperspirant agent is very crystalline when it dries from an aqueous or hydroalcoholic vehicle. Its solutions are also very sticky as they dry, and emollients may be added to minimize the crystalline nature and sticky feel of the material. However, additives must be carefully selected so as not to interfere with the antiperspirant efficacy of the product.

Many different aerosol product forms have been formulated incorporating aluminum chlorhydroxide complex.

Aqueous Solution Diluted with Ethyl Alcohol

The following must be remembered in this regard. Aluminum chlorhydroxide complex is soluble in water and insoluble in ethyl alcohol. However, an aqueous solution of the material is miscible in all proportions with ethyl alcohol. Fluorocarbon propellants are miscible with ethyl alcohol but not with water. Small amounts of water may be tolerated in a hydroalcoholic vehicle with fluorocarbon propellants, depending on the quantity of the latter being used.

Nonalcoholic aerosol formulations would separate into two liquid layers. This separation occurs rapidly, and it would be difficult to dispense a uniform spray. Uniform hydroalcoholic aerosol systems are possible, depending on the concentrations of water and propellant in the system. For example, a concentrate with 8% by weight of water, 8% aluminum chlorhydroxide complex, and 84% anhydrous ethyl alcohol will tolerate about 40% Propellant 12. This mixture produces a wet aerosol spray.

Decreasing the aluminum salt concentration correspondingly lowers the water concentration needed for miscibility with ethyl alcohol. Therefore a concentrate with 3% by weight of water, 3% aluminum salt, and 94% anhydrous ethyl alcohol will tolerate about 60% Propellant 12. The higher concentration of propellant in this combination produces a more acceptable aerosol spray. However, the concentration of active antiperspirant material is extremely low.

A typical formulation of this type is as follows:

Antiperspirant 2 (16)

Concentrate	*% by Weight*
Aluminum chlorhydroxide complex (50% aqueous solution)	10.00
Hexachlorophene	0.30
Tartaric acid, 1% in ethyl alcohol, SDA-40	0.50
Perfume of choice	0.50
Isopropyl myristate	1.00
Dipropylene glycol	4.50
Ethyl alcohol, SDA-40	83.20
	100.00

Fill	*% by Weight*
Above concentrate	70.00
Propellant 12/114 (40:60)	30.00
	100.00

Method of Preparation

Dissolve the hexachlorophene in alcohol and then add the perfume, isopropyl myristate, dipropylene glycol, and tartaric acid solution. Add the aluminum chlorhydroxide solution to this mixture very quickly with agitation. Fill concentrate and propellant by conventional aerosol filling methods.

Belgian Patent 699,773 describes a hydroalcoholic aerosol formulation with aluminum chlorhydroxide complex, using coupling agents to increase the amount of water that can be tolerated. This type of product is sticky to use and corrosive to ordinary metal containers, with a high tendency to cause actuator cloggage.

Hydroalcoholic aerosol antiperspirants have a tendency to gel with time, particularly at elevated temperatures. Also, because of the iron in the complex and the water in the formulation, there is the possibility of discoloration if phenolic antibacterials are used. Chelating agents, such as EDTA and salts, may be added to prevent discoloration.

Impalpable Powder in Suspension

Aluminum chlorhydroxide complex is also available as an impalpable (fine) powder. French Patent 1,494,465 describes the use of this material suspended in an emollient and propellant.

Antiperspirant—Suspension Type 1
(Source: French Patent 1,494,465)

Concentrate	*% by Weight*
Aluminum chlorhydroxide complex (impalpable powder)	2.0
Hexachlorophene	0.1
Bentone 34	0.3
Isopropyl myristate	6.0
Propellant 113	12.6
Propellant 114	65.0
Propellant 12	14.0
	100.0

This product must be shaken before use. The emollient carrier acts to keep the aluminum salt on the skin. The particle size should be 50 μ or less, and the moisture content of the system must be kept to a minimum.

The advantages of this system include nonstickiness, ability of the product to dry, quickly and low corrosion potential since no free water or alcohol is present. The disadvantages are possible agglomeration of the suspension in the aerosol container, misty spray because of the large concentration of propellant present, and the need to shake the product before use.

Other suspending agents, emollient oils, or low-pressure propellants may be substituted in the above formulation. A perfume of choice should also be added.

Antiperspirant—Suspension Type 2
(Source: Croda, Inc.)

Concentrate	*% by Weight*
Aluminum lanolate	6.0
Isopar H solvent (Humble Oil and Refining)	19.0
Isostearic acid	2.0
Isopropyl myristate	40.0
Aluminum chlorhydroxide complex (Impalpable powder)	30.0
Cabosil (Cabot)	2.0
Hexachlorophene	1.0
	100.0

Fill   *% by Weight*

   Above concentrate   10.0
   Propellant 12/114 (40:60)   90.0

                                                    100.0

Method of Preparation

Dissolve the aluminum lanolate in the Isopar and isostearic acid with gentle heating. Add isopropyl myristate and impalpable aluminum salt with adequate agitation. Add Cabosil and hexachlorophene slowly, and then fill concentrate (keep well stirred) and propellant by conventional aerosol filling methods. This product must be shaken before use.

U.S. Patent 3,288,681 describes a similar suspension-type aerosol antiperspirant using a small amount of ethyl alcohol.

Antiperspirant—Suspension Type 3
(Source: U.S. Patent 3,288,681)

Concentrate   *% by Weight*

   Aluminum chlorhydroxide com-
      plex (Impalpable powder)   8.00
   Isopropyl ester   2.00
   Perfume of choice   0.04
   Ethyl alcohol, anhydrous   4.00
   3,4',5-Tribromosalicylanilide   0.08
   Propellant 114   51.46
   Propellant 12   34.42

                                                   100.00

This product must also be shaken before use.

Impalpable Powder Spray

Powder spray antiperspirant formulations have been marketed, but they afford no improvement over the oily suspension form discussed above. Essentially these formulations are based on talc with a small amount of impalpable aluminum chlorohydroxide complex added.

Antiperspirant—Powder Type
(Source: R. T. Vanderbilt Co.)

Concentrate	*% by Weight*
Isopropyl myristate	3.5
Aluminum chlorhydroxide com-plex (impalpable)	7.0
Zinc stearate	10.0
Talc (impalpable)	79.5
	100.0

Fill	*% by Weight*
Above concentrate	8.0
Propellant 12/114 (20:80)	92.0
	100.0

Method of Preparation

Blend all the powders with the isopropyl myristate and a perfume of choice. Micronize the mixture, and fill into aerosol containers. Add the propellant by conventional aerosol filling methods. This product must be shaken before use.

This type of formulation has five disadvantages:
(a) powder flakes or rubs off from the skin; (b) antiperspirant efficacy is only minimal; (c) agglomeration of particles in the aerosol container may occur (the addition of colloidal silica may be beneficial); (d) the spray is misty because of the large concentration of propellant and talc; and (e) the product must be shaken before use.

On the positive side, this system would provide a good foot-care product with deodorant and antiperspirant qualities. It would also be cool and refreshing.

ALCOHOL-SOLUBLE COMPLEXES

As previously mentioned, aluminum chlorhydroxide complex is not soluble in anhydrous ethyl alcohol, and therefore it is not possible to formulate a good uniform (or single-phase) aerosol product. The addition of water helps in the formulation of a uniform system but introduces serious corrosion problems.

In recent years alcohol-soluble complexes have become available to facilitate the formulation of satisfactory uniform aerosol systems.

Aluminum Chlorhydroxyethylate Complex

This complex was one of the first alcohol-soluble complexes to become available. However, it was not successful; the complex was not stable over a long period of time, it was corrosive to metal containers, and hard crystals formed on drying and caused valve–actuator clogging. United States Patent 2,823,169 (Brown and Govett) describes this complex and various other aluminum chloroalcoholates.

Aluminum Chlorhydroxide/Propylene Glycol Complex

This complex has been reported in U.S. Patents 3,359,169 and 3,420,932. The propylene glycol is hydrogen bonded to the aluminum chlorohydroxide complex and therefore becomes soluble in anhydrous ethyl alcohol. As sold, this alcohol-soluble complex contains about 75% by weight of aluminum chlorohydroxide complex, the remainder being propylene glycol. Since the complex dries to a hard crystal, it is necessary to add emollients to soften the film. The proper valve–actuator assembly must also be chosen to minimize clogging.

Because of the presence of relatively high concentrations of this aluminum salt in an aerosol, one commonly encountered difficulty is gelation of aerosol formulations stored at elevated temperatures. However, the addition of lactic acid or certain phosphate esters prevents this solvolytic and/or temperature-related phenomenon (17). These antigelling agents must be carefully selected so as not to increase corrosion or reduce antiperspirant efficacy.

Metal aerosol containers must be carefully screened and shelf tested because this aluminum chlorhydroxide/propylene glycol complex tends to be slightly corrosive. New aerosol container linings have become available because of these compatibility problems. These containers are described in the last paragraph of this chapter.

Antiperspirant—Alcohol-Soluble
(Source: Reheis Chemical)

Concentrate	% by Weight
Aluminum chlorhydroxide propylene glycol complex	18.0
Hexadecyl alcohol	6.0
Polyphenylmethylsiloxane	2.0
Hexachlorophene	0.4
Perfume of choice	0.6
Ethyl alcohol, SDA-40, anhydrous	73.0
	100.0

Fill   *% by Weight*

Above concentrate   50.0
Propellant 12/114 (60:40)   50.0
                                                             ───────
                                                              100.0

Method of Preparation

The aluminum complex is slowly dissolved in the ethyl alcohol. The remaining ingredients are dissolved or mixed in the concentrate. The concentrate and propellant are filled by conventional aerosol filling methods.

Other emollients or antigelling agents may be added to the above formulation. The final product is uniform and need not be shaken.

This type of formulation has the following disadvantages: it is sticky, valve–actuator clogging may occur, and there is a possibility of corrosion and off-odor with extended shelf life.

Other Alcohol-Soluble Complexes

Many other alcohol-soluble complexes have been commercially available or have been developed by consumer-product companies for captive use. They include the following:

a. Aluminum chlorhydroxide polyethylene glycol complex.

b. British Patent 1,113,886 describes a method of producing a uniform aerosol antiperspirant by which aluminum chlorhydroxide complex is dissolved in an amount of substantially anhydrous ethyl alcohol to solubilize the aluminum salt on aging in the aerosol container. Other additives such as lanolin and propylene glycol are included.

c. U.S. Patent 2,872,379 describes "alkoxyaluminum" chlorides that are soluble in ethyl alcohol. However, their aluminum content is low, and their acidity excessively high.

Perfume Selection

The perfume for an aerosol antiperspirant product must be carefully formulated and selected. Aluminum salts are reactive chemicals, and they do react with perfume components to produce off-odors and discoloration. The final formulations should be sufficiently shelf tested at ambient and accelerated temperatures to detect potential problems. This subject is covered in greater detail in Chapter 16.

Antibacterial Properties

The commonly used antiperspirant salts usually have antibacterial properties because they are protein precipitants (18, 19, 20). However, to obtain maximum deodorant protection for at least a 24-hour period it may be desirable to add an antibacterial agent to the formulations.

Special Containers

In order to minimize compatibility problems with certain aerosol antiperspirant formulations, most of the major aerosol container manufacturers have developed special internal linings. For example, vinyl organisol coatings have improved the shelf-life stability of formulations containing alcohol-soluble aluminum complexes. Also, the special coating of the side-stripe area of three-piece tin-plate containers is beneficial. However, one should ensure that the side stripe does not peel away during storage to later cause corrosion, off-odor, and possible valve clogging.

REFERENCES

1. W. B., Shelley and P. N. Horvath, Experimental Miliaria in Man. II: Production of Sweat Retention Anhidrosis and Miliaria Crystallinia by Various Kinds of Injury, *J. Invest. Dermatol.*, **14** (1950), 9.

2. W. B. Shelley, Experimental Miliaria in Man. V: The Effect of Poral Closure on the Secretory Function of the Eccrine Sweat Gland, *J. Invest. Dermatol.*, **22** (1954), 267.

3. C. M. Papa, The Action of Antiperspirants, *J. Soc. Cosmet. Chem.*, **17** (1966), 789.

4. P. A. Sanders, *Chem. Spec. Mfr. Assoc. Proc.*, May 1960, p. 66.

5. F. A. Bower and L. J. Long, *Chem. Spec. Mfr. Assoc. Proc.*, May 1961, p. 57.

6. P. A. Sanders, *Principles of Aerosol Technology*, Van Nostrand Reinhold Company, New York, 1970, p. 129.

7. Agent T. C. George's Tariff No. 19, issued Aug. 6, 1966, published by the Bureau of Explosives, Section 73.306.

8. A Tentative Method for Determination of the Particle Size Distribution of Space Insecticide Aerosols, *Chem. Spec. Mfr. Assoc. Proc.*, December 1956.

9. M. Lefebvre and R. Tregan, *Aerosol Age*, **10** (July-August, 1965), 31.

10. W. G. Fredell and R. R. Read, Antiperspirant—Axillary Method of Determining Effectiveness, *Proc. Sci. Sect. Toilet Goods Assoc.*, **15** (1951), 23; No. 25 (1956), 32.

11. Ouelette, Healy, Jenkins and A. Della Lana, Technique for Perspiration Measurement, *Proc. Sci. Sect. Toilet Goods Assoc.*, **42** (1964), 12.

12. *Schimmel Briefs*, No. 366, September 1965.

13. *Rehydrol ASC Brochure*, Reheis Chemical Company, 1966, p. 15.

14. K. Klausner, Inhibition of Metallic Corrosion in Aerosols, *Amer. Perfum., Aerosol Documentary*, October 1960, p. 53.

15. F. T. Reed, *Chem. Spec. Mfr. Assoc. Proc.*, December 1951, p. 37.

16. V. DiGiacomo *Amer. Perfum.*, **69** (May 1957), 45.

17. *Rehydrol ASC Brochure*, Reheis Chemical Company, 1966, p. 15.

18. I. H. Blank and R. K. Dawes, Antibacterial Activity of Weak Solutions of Aluminum Salts, *A.M.A. Arch. Dermatol.*, **81** (1960), 565.

19. I. H., Blank et al., the Antibacterial Activity of Aluminum Salts, *Proc. Sci. Sect. Toilet Goods Assoc.*, May 1957.

20. N. H. Shehadeh and A. M. Kligman, The Effect of Typical Antibacterial Agents on the Bacterial Flora of the Axilla, *J. Invest. Dermatol.*, **40,** 1 (1963).

COSMETIC FORMULATIONS

Cosmetics are, in general, products that are applied for the purpose of beautifying the face or body. This multifarious product category includes preparations utilized to preserve or alter one's appearance or to cleanse, color, condition, or protect the skin, hair, nails, lips, eyes, or teeth. Cosmetics can therefore be regarded as products that do not physiologically alter a part of the body but rather impart the feel or appearance of effecting an improvement.

Since most cosmetics are applied to the skin, some understanding of the anatomy and physiology of skin is helpful. The skin is the largest organ of the body, weighing approximately 9 lb and occupying an area of about 2 yd^2 or 2592 in.^{2} in the average adult.

The outermost layer of skin is the one mainly involved with cosmetic use. It is a dead layer, composed of horny material loosely connected, and is continuously being removed by friction and bathing. Cosmetic effects are often achieved by first removing the outer dead layer, thereby coming closer to the skin layer that has a vital or living nucleus and is living tissue.

Throughout the body, skin thickness varies from about 0.02 in. on the eyelids to 0.125 to 0.166 in. on the palms and soles. Facial skin is relatively thin, and in any case great care should be given to the composition of the cosmetics applied. Careful consideration of raw materials (separately and in combination) and testing of finished products, particularly as to safety of use, are essential in formulating.

Aerosol cosmetics, although not functionally different from their nonaerosol counterparts, generally offset increased costs and the need for reformulation by providing the following advantages:

1. Ease of use.
2. Convenience.
3. Application control.
4. Esthetically pleasing and hermetically sealed package.

CATEGORIES OF AEROSOL COSMETICS

For purposes of discussion, aerosol cosmetics can be classified into three broad categories: solutions, suspensions, and emulsions. With very few exceptions, every aerosol cosmetic product currently being marketed fits into one of these categories. Specific product formulas and associated problems are discussed in detail later in this chapter. Initially, a general discussion of basic aerosol systems and pertinent terminology is given.

Solutions

The aerosol cologne represents a solution system. This product is composed of an alcoholic solution of perfume oil in which the propellant is soluble. Modification of cologne (e.g., addition of certain amounts of water) can cause a dissolution of the propellant or phasing out. This results in a three-phase system with two discernible liquid layers and a vapor phase. With current advanced technology, however, three-phase systems are rarely employed. Some factors that influence the formulation of a solution system are the following:

1. Pressure requirement for a specific container.
2. Solubility of active ingredients, which sometimes requires solubilization or cosolvents.
3. Addition or deletion of water where a specific corrosion problem exists.
4. Formula balance to prevent soluble solids from depositing in the orifice of the valve and causing leakage or clogging.

Particular attention is given to specific solution and associated problems under the applicable product categories later in this chapter. The solubility and pressure relationships of alcohol and other solvents with propellants are also discussed.

Suspensions

For our purpose, a suspension is a distribution of larger-than-colloidal-size particles of one material throughout another material in which the former is not soluble. An aerosol powder is a suspension of an insoluble solid in the liquid propellant.

Six factors affect the properties of suspensions in aerosol systems.

1. Particle size of the powder to be suspended.
2. Densities of the powder and propellant.
3. Effect of additives such as isopropyl myristate and isopropyl palmitate (used to minimize dusting into the air and to increase adherence of powder to the target area), medicament, or active ingredients, particularly if partially soluble in the propellant.
4. Addition of a surfactant to reduce surface tension and act as an interphase between the solid particles, thereby reducing their tendency to agglomerate.
5. Length of time a uniform suspension is necessary to achieve the requirements of end-use application.
6. Addition of fillers or solid suspension aids such as Cab-o-Sil.

The optimum characteristic of an aerosol suspension varies with the intended end use of the product. Specific additives and their effects on total formulation are discussed in more detail under specific product categories.

Emulsions

An emulsion can be regarded as the mixture produced when one immiscible liquid is dispersed as discreet globules (surrounded by an emulsifying agent film) throughout another liquid. There are two types of emulsions, namely, oil-in-water (o/w), where oil is the internal phase, and water-in-oil (w/o), with water as the internal phase. Two tests are available to determine the type of emulsion.

1. The introduction of an oil-soluble dye, which will remain a discrete drop in an o/w emulsion.
2. The use of a low-voltage conductivity tester, which requires water as the external phase (o/w) to conduct electricity.

The type of emulsion formed is dependent on the relative concentrations of phases and also on the concentration and type of emulsifying agent. Surface tension is the force that causes liquids to assume the least-surface-mass relationship, that is, spherical. Interfacial tension is a similar force occurring when a liquid is in contact with another immiscible liquid. These liquids have greater tendency toward separation, the greater the surface tension. As discussed under emulsifying agents, it is essential to reduce the surface tension of one of the liquids, usually the dispersed or internal phase.

Although many emulsions are based on soap, in one form or another, a great number of synthetic organic surface-active agents are employed as primary or auxiliary emulsifiers. These "surfactants" are materials used as wetting, solubilizing, emulsifying, dispersing, or foaming agents. Generally, a surfactant is classified as anionic, cationic, or nonionic according to its hydrophile (water-loving or polar) group.

The type of emulsion desired (o/w or w/o) and the concentration and characteristic of the oil phase must be considered when choosing a surfactant or blend of surfactants as emulsifiers. Blends of surfactants are often employed, as an individual surfactant usually has a greater affinity for either water or oil. A specific emulsion will require a balance of the hydrophilic (water-loving or polar) and lipophilic (oil-loving or nonpolar) groups. More detailed information on this subject is available in the literature (1).

In addition, the formulator is concerned with the stability of the resultant emulsion and must also consider the following:

1. Particle size of the dispersed phase.
2. Difference in the densities of the two phases.
3. Amounts and characteristics of oils, waxes, and so forth to be emulsified.
4. Potential incompatabilities between raw materials employed.
5. Viscosity, pH, and other characteristics of the end product.

Stability is often a matter of compromise. A cream must be stable in the jar on the shelf but may be required to break readily when used on the skin or hair. The cracking of an emulsion is generally an instability in the form of creamy flocculation, coalescence, or inversion. Among the causes for a broken emulsion are the following:

1. Insufficient particle size reduction of the oil phase can result in coarse oil dispersion. The larger the oil droplets in the internal phase, the greater is the tendency to collect or coalesce. As the droplets become larger, they tend to migrate more readily to the surface. The fat-rich upper layer (in the case of o/w emulsions) creams out on top (e.g., cream separating to the top of unhomogenized milk). In a w/o

emulsion, the internal phase migrates to the bottom in a downward creaming effect.

2. Introduction of bacteria or inadequate preservation can cause emulsions to break.

3. Excessive amounts of electrolytes (salts) will often adversely affect an emulsion.

4. Addition of solvents or substitution of a specific solvent contained in the formula can also cause emulsion breakdown. The addition of 5% or more of alcohol, or the addition of acids, frequently results in emulsion separation.

5. Instability of emulsions can also occur when the product inverts from o/w to w/o or vice versa.

6. Improper preparation can cause emulsion breakdown. Specific causes include:

    a. Air entrapment.

    b. Excessive heating of emulsion phases (or insufficient heat).

    c. Cooling the emulsion too rapidly.

    d. Wrong type of mixing equipment or duration of mixing.

It is necessary for the cosmetics formulator to become familiar with emulsion theory to a degree beyond the scope of this chapter. The literature on the subject should be consulted for a better understanding of emulsion technology (2-4).

Aerosol emulsions are essentially the same as nonpressurized emulsions, with some added formulation considerations. Modifications of emulsions for specific aerosol applications are discussed at greater length under particular product categories in this chapter and others.

Since several of the common aerosol products are essentially soaps, some general information concerning saponification will be useful.

Saponification

Saponification is the interaction between alkalies and fatty acids or natural fats and oils, resulting in the formation of soap. For example, the addition of sodium hydroxide to stearic acid yields sodium stearate (soap).

In formulating soap systems (e.g., shave or shampoo products), the amount of alkali needed for complete reaction will vary with (*a*) the chemical composition of the fatty acid or oil and (*b*) the alkali chosen. The amount of alkali to be employed will, therefore, be calculated by chemical equivalent. To neutralize each pound of stearic acid, triple pressed, would require 0.21 lb of potassium hydroxide, 0.15 lb sodium hydroxide, 0.52 lb triethanolamine, and 0.71 lb sodium borate (borax), assuming each of the alkalies to be 100% pure.

RAW MATERIALS OF COSMETICS

A considerable amount of information has been published in trade literature and books on the vast number of solvents, emulsifiers, and additives employed in cosmetics. The following brief description covers some of the more widely used raw materials, with factors to be considered when employing them in a finished formulation.

Water

It is extremely important that properly distilled or deionized water be used in the preparation of cosmetic products, as the presence of salts and dissolved gases can affect the appearance and consistency of finished products. The presence of electrolytes, calcium and magnesium salts, or iron can adversely affect the stability of the product. Water must, of course, be free of bacteria or mold growth.

Alcohols

Widely used, the alcohols constitute a group of organic compounds varying in physical form from clear liquids to waxy solids. They are used, depending on their properties, as solvents, humectants, emulsifiers, and emollients.

SOLVENT ALCOHOLS

Ethyl alcohol, a colorless liquid with a characteristic odor, is miscible with water and widely used in denatured form as a solvent and antiseptic. Ethyl alcohol is commonly employed in aerosol products (e.g., cologne, hair spray) as either anhydrous (200 proof) or as a 95% (190 proof) grade.

Alcohol is available from several commercial sources and bears a formula designation that indicates the denaturant employed to make it unpotable. Generally, SDA 40-2 containing brucine sulfate as a denaturant has been found to be the least odorous. Each batch should be evaluated to ensure consistent quality.

Isopropyl alcohol, a colorless liquid with a characteristic, slightly acetone-like odor, is miscible with water and is used as a solvent. Although cheaper than ethyl alcohol, it is not as widely used in cosmetic aerosols, primarily because of its odor.

GLYCERIN

Glycerin, a colorless, hygroscopic, clear, neutral, syrupy liquid with a sweet taste, is miscible with water and alcohol. A humectant, glycerin is used in many cosmetic preparations to improve spreading characteristics and preserve consistency.

PROPYLENE GLYCOL

Propylene glycol, a clear, odorless, colorless, hygroscopic viscous liquid, is miscible with water and alcohol. It is used as a solvent and humectant and serves as a substitute for glycerin.

FATTY ALCOHOLS

Cetyl alcohol, $C_{16} H_{33}$ OH, is available as white crystals or plates and is insoluble in water. An emollient and emulsifier, it is used in a variety of creams and lotions to give added emulsion stability and impart a smooth feel to the skin.

Oleyl alcohol, $CH_3[CH_2]_7CH=[CH_2]—CH_2OH$, a clear, slightly yellowish liquid with characteristic odor, is insoluble in water but soluble in alcohol and many oil and wax mixtures. It is an emollient superfatting agent, with good feel on hair and skin.

Esters

Basically, esters are organic salts prepared by reaction of an alcohol with an organic acid. Widely used in cosmetic preparations are the fatty acid esters such as oleic, palmitic, and stearic. They act as emulsifiers, solvents, lubricants, and plasticizers.

Isopropyl myristate, $C_{13}H_{27}COOC_3H_7$, is a clear liquid which, in commercial form, usually contains small quantities of other fatty acid esters (e.g., palmitate). Widely used in bath oils, powders, creams, and lotions as a solvent, extender, auxiliary emulsifier, and penetrant, it imparts a smooth, even, nongreasy feel to skin.

Isopropyl myristate improves the appearance of cosmetic preparations by providing sheen, absence of stickiness or greasiness, and more even spreading. It is an emollient and penetrant superior to other vegetable and mineral oils.

Isopropyl palmitate, $C_{15}H_{31}COOC_3H_7$, is similar to isoproply myristate with like properties.

Other esters that are used as emulsifying agents and are nondrying on skin are glyceryl monostearate, glyceryl monolaurate, and glyceryl monooleate. Also commonly employed as surface-active agents are the esters of sorbitol. These are nonionic emulsifiers exemplified by a series of products such as sorbitan oleate and polyoxyethylene sorbitan monolaurate.

Gums

Prepared synthetically or obtained from natural sources, gums are used primarily to increase viscosity, improve spreading characteristics, and stabilize emulsions. Gums aid in imparting a smooth feeling and can form some degree of protective coating. Water solutions of gums must be preserved. Listed below are a number of gums that can be employed:

Vegetable	Synthetic
Acacia	Methyl cellulose
Tragacanth	Carboxymethyl cellulose
Quince seed	Polyvinyl alcohol
Agar-agar	Ethyl cellulose
Irish moss	

Oils and Fats

Generally, the oils and fats used in cosmetics come from animal, vegetable, and hydrocarbon sources. Employed as emollients, lubricants, and solvents and as aids in forming protective coatings on the skin, they can be divided into the following classes by source:

Animal: cod-liver oil, lard, tallow, and so forth.

Vegetable: corn, sesame, peanut, almond, coconut, and castor oil.

Hydrocarbon: mineral oil, available as fractions obtained by distillation of white oils from crude oil.

Waxes

Waxes are materials that generally melt above body temperature and below the boiling point of water. They are the esters of alcohols other than glycerin and have a characteristic waxy feel.

Beeswax—white (bleached) and yellow: a mixture of cetyl myristate, in aliphatic acid esters and some hydrocarbons. It is used in cold creams in combination with sodium borate to form sodium cerotate, which whitens and stiffens the cream.

Carnauba, obtained from the carnauba palm tree: a hard yellowish or greenish solid. Small quantities are used in creams to give a firmer product.

Paraffin: a mixture of solid hydrocarbons obtained from petroleum, available as a white or colorless mass having a greasy feel and appearing crystalline. Available in a wide range of melting points, paraffin is employed in cosmetics to obtain a firmer product.

Carbowax*—solid polyethylene and polyoxyethylene glycols, available in a variety of molecular weights from liquid to solid. Carbowaxes are water soluble and are useful as nonvolatile unctuous materials with excellent stability. They are used in cosmetics as humectants, lubricants, and plasticizers and for the reduction of tacky feel.

Lanolin: a purified fatlike substance obtained from sheep's wool. More correctly a wool wax as it is free from glycerol, lanolin is available in many grades. It is employed in cosmetics as an emulsifier, a lubricant, and an extremely good emollient. Lanolin and its derivatives have been used extensively in almost all types of cosmetic preparations.

Antioxidant

An antioxidant prevents the absorption of oxygen or takes up oxygen itself. It is used in vegetable oils to retard oxidation or rancidity. Among those used are butylated hydroxyanisole, butylated hydroxytoluene, and propyl gallate.

Preservatives

It is often necessary to add to cosmetic preparations a material that will inhibit mold and bacterial growth. Most preservatives are used at a relatively low level (0.10 to 0.25%), depending on the specific preservative employed and the nature of the base. Some of the chemicals that have been employed as preservatives are methyl p-hyroxybenzoate, propyl p-hydroxybenzoate, formaldehyde, sodium benzoate, alcohol, quaternary ammonium compounds, and Giv-Gard DXN® (6-

---

* Carbowax is the registered trademark of Union Carbide Corporation.

acetoxy-2, 4-dimethyl-*m*-dioxane). There are, of course, many other preservatives, and considerable reference material is available.

Sequestering or Chelating Agents

Trace metal ions can be a problem in preventing deterioration and rancidity and increasing the shelf life of cosmetics. Small amounts of agents can be employed to sequester or bind metals or ions to form water-soluble complexes. Among the materials employed as metal scavengers are the salts or esters of ethylene diamine tetracetic, citric, and phosphoric acids.

Emulsifying Agents

An emulsifying agent is used to attain a mixture of two immiscible liquids, one of which is broken into minute globules each surrounded by a film of the agent and dispersed throughout the other liquid. Emulsification will occur because of one or more of the following:

1. Lowering of surface tension of one of the liquids.
2. Preventing the coalescence or rejoining of the dispersed phase.
3. Forming a film on the globules of the internal phase which reduces the tendency to coalesce.
4. Forming a film at the oil–water interface.
5. Producing a charge on the dispersed globules.

Trade literature and textbooks contain considerable information on emulsion technology. Since some specific treatment of emulsions is given later in this chapter, only a brief description of some emulsifying agents is presented here.

SYNTHETIC

1. Anionic—where a negative ion is the oil-soluble portion, as in fatty acid soaps (e.g., sodium stearate).
2. Cationic—where a positive ion (cation) is the oil-soluble portion, as in quaternary ammonium compounds (e.g., cetyldimethylbenzylammonium chloride.
3. Nonionic—where affinity for water is due to nonionizing polar groups, as in fatty acid esters (e.g., glyceryl monostearate).

NATURAL

1. Carbohydrates from vegetable sources, such as gum tragacanth, alginates, and starches.

2. Agents from animal sources, such as wool fat (lanolin), cholesterol, and lecithin.

COSMETIC PRODUCTS

Fragrance Products: Colognes and Perfumes

Provided that the fragrance has been formulated for aerosol solubility and stability, these are relatively problem-free products to formulate. A typical aerosol cologne formula contains from 2 to 5% by weight of perfume oil based on the total aerosol fill.

Cologne

Concentrate	*% by Weight*
Perfume oil	5.00
Alcohol, SDA-40-2 (95%)	95.00

Method of Preparation

Mix perfume oil and alcohol and age for at least 48 hours. Chill for at least 5 hours at −5 to −10°C. Filter while cold, using a suitable filtering medium.

Fill	*% by Weight*
Cologne concentrate	60.00
Propellant 12/114 (15:85)	40.00

This formula is a guide and can be modified for the achievement of specific spray characteristics or for economic reasons. The approximate pressure of this formula would be 20 psig at 70°F. The formulator should attempt to keep the pressure below 25 psig at 70°F for coated glass containers and 15 psig at 70°F for uncoated glass. Naturally one should attempt to balance the formula to achieve an acceptable spray pattern at a given pressure.

By consulting Figures 15.1 to 15.3, one can quickly pinpoint the vapor pressure area of interest and attain it by either altering the propellant blend and/or changing to anhydrous alcohol. The use of 95% alcohol is preferred, as the small amount of water generally has a better effect on fragrance character. Propellants 12 and 114 are primarily used because of their low odor and their stability. Once the

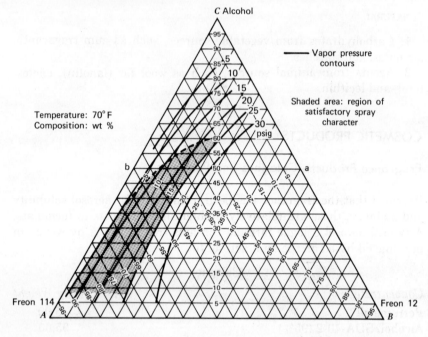

Figure 15.1.   Properties of Freon 12/Freon 114/ethyl alcohol (100%) solutions.

correct pressure and level of fragrance have been established, the spray pattern should be evaluated with the specific aerosol valve desired. A good aerosol cologne represents a proper balance of concentration of perfume oil, alcohol level and type, propellant level and type, and type of aerosol valve, actuator, and container (5, 6).

The preceding discussion deals with colognes intended for glass packaging. Fragrances can also be formulated that will have reasonable stability in lined aluminum containers or in plastic. Aerosol colognes intended for aluminum packages should contain a level of water up to about 1.0% based on the total product.

Aerosol perfume sprays are concentrated (10 to 15%) solutions of fragrance and, where glass packaging is employed, are formulated in a manner similar to colognes.

Perfume

Concentrate	% by Weight
Perfume oil	20.00
Alcohol, 39C (95%)	80.00

Method of Preparation

Age, chill, and filter.

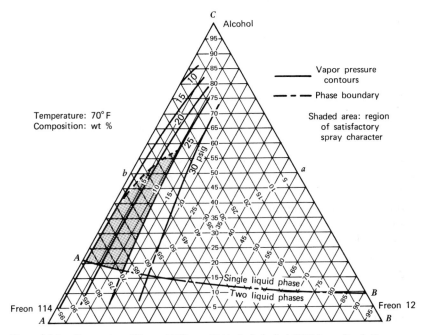

Figure 15.2. Properties of Freon 12/Freon 114/ethyl alcohol (95% by vol) solutions.

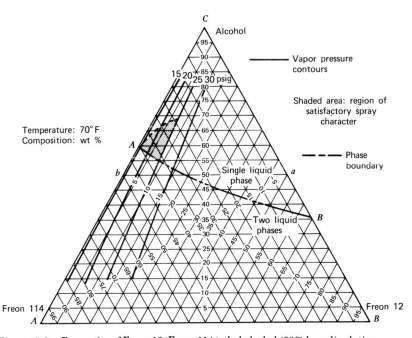

Figure 15.3. Properties of Freon 12/Freon 114/ethyl alcohol (90% by vol) solutions.

Fill   *% by Weight*
Concentrate   60.00
Propellant 12/114 (40:60)   40.00

The higher-pressure propellant indicated above is for use in a stainless steel container fitted with a metered valve. The metered valve allows the user to dispense controlled amounts of the concentrated fragrance onto the skin. On the assumption that the perfume oil is soluble and stable, modification of the system to suit a specific package or spray pattern involves the same considerations given for cologne.

Another variation of a fragrance product is the moisturizing cologne foam. An emulsion containing lanolin derivatives to moisturize as well as perfume the skin, the product is emitted in a rich, creamy foam and is rubbed into the skin. Such products can be packaged in lined aluminum containers. A typical formula is given below.

Moisturizing Cologne Foam

Concentrate	*% by Weight*
Part I	
Myristic acid (Armour)	0.25
Emersol 132 (Emery)	0.65
Polychol 40 (Croda)	1.30
Cetyl alcohol NF	0.55
Amerchol® L-101 (Amer Cholesterol)	5.00
PEG 1000 monostearate Se	3.00
Amerlate P (Amer Cholesterol)	1.00
Part II	
Distilled water	74.68
Carbopol® 934 (Goodrich)	0.37
Triethanolamine 98%	0.40
Tween® 20 (Atlas)	0.25
Tween® 80 (Atlas)	0.25
Mineral oil 65/75	0.30
Part III	
Alcohol 39C regular	10.00
Perfume oil	2.00
	100.00

Method of Preparation

Weigh out separately ingredients in Parts I, II, and III. Heat each part to 75°C. While mixing Part II at moderate speed, add I slowly in a thin, even stream. Avoiding air entrapment, continue slow mixing. At 40°C add Part III. Let stand overnight; mix slowly next day, making up lost water.

Fill	*% by Weight*
Concentrate	90.00
Propellant 12/114 (20:80)	10.00
	100.00

Bath Products

Bath oils in foam and spray form include products designed to be dispensed as floating or dispersible bath oils, either directly into the tub or applied to the body. Generally speaking, products of this type, although investigated, have not gained consumer acceptance. Bubble baths have also been introduced as aerosols but have not proved to be advantageous over existing product forms.

Some success has been achieved, however, with products designed for use as after-bath or after-shower sprays. A typical formula is given below.

After-Bath Spray

Concentrate	*% by Weight*
Perfume oil	5.0
Isopropyl myristate	6.0
Oleyl alcohol	4.0
Alcohol, SDA-40, anhydrous	84.0
Deionized water	1.0

Method of Preparation

Mix water in alcohol, add remaining ingredients, and mix thoroughly. Age 72 hours, chill, and filter.

Fill	*% by Weight*
Concentrate	60.00
Propellant 12/114 (40:60)	40.00

This formulation may be packaged in lined aluminum containers and should be fitted with a valve to deliver the desired pattern. The product is sprayed on the body after showering or bathing and rubbed in, leaving the user with a smooth, soft-feeling skin. The light, oily but nongreasy film should be evident yet not uncomfortable under clothing. Other bath-product-line extensions may also include spray powders; these are discussed more fully under "Powders."

Aerosol Shave Foams

The shave foam represents 90% of the wet shaving product forms. Basically a triethanolamine fatty acid soap, the shave foam should be formulated to achieve the following:

- Produce a reasonably coherent, cosmetically elegant foam of small-bubble-size structure.
- Spread easily and evenly on face.
- Remain wet and workable for the duration of the shave.
- Have lubricity or supply good gliding action for the blade.
- Be readily rinsed from the blade.
- Leave face smooth and comfortable.

A typical shave foam formula is as follows:

Shave Foam

Concentrate	% by Weight
Part I	
Stearic acid, T.P.	6.00
Hydrofol® acids 631 (ADM Chem.)	1.30
Lantrol®(Malmstrom)	0.40
Part II	
Glycerin, U.S.P.	3.50
Triethanolamine, 98%	4.00
Deionized water	80.80
Part III	
Mineral oil 65/75	2.50
Cetyl alcohol, N.F.	0.50
Silicone DC 200 (200 CTS) (Dow Corning)	0.50
Part IV	
Perfume oil	0.50
	100.00
Part V	
Preservative	q.s.

Method of Preparation

Weigh out separately ingredients in Parts I, II, and III. Heat each part to 75°C. While mixing Part II at moderate speed, add I slowly in a thin, even stream. Avoid air entrapment. Add Part III and continue slow mixing. At 40°C add perfume and mix until cool. Let stand overnight and mix slowly the next day, making up lost water.

Fill	% by Weight
Concentrate	96.00
Hydrocarbon propellant	4.00
	100.00

The base or concentrate is an emulsion of mixed fatty acids and triethanolamine. Other materials have been added to modify the foam properties. Cetyl alcohol helps to improve the appearance of the emitted foam, acts as an auxiliary emulsifier, and is emollient in nature; glycerin, hygroscopic in nature, acts as a humectant to retard the drying out of foam. Silicone is added to improve lubricity and slip for the razor, and other surfactants can be used to alter foam appearance or improve wetting action.

The formula can also be pressurized with mixtures of halogenated hydrocarbons such as Propellants 12 and 114. The type and concentration of propellant will have a definite effect on the properties of the emitted foam. An average of 7 to 10% of Propellants 12 and 114 is normally used in a specific blend. A particular foam can be made drier or stiffer, if desired, by increasing the amount of total propellant or the amount of Propellant 12 in the blend. Propellant effects in foam structure have been discussed in greater detail by Sanders (7).

The hydrocarbons isobutane and propane, although flammable, have been widely used in aerosol shave foams because of existing patent situations and for economy (8). Where odor is a problem, the hydrocarbon can be pruified (9).

Isobutane and propane are normally blended for use in shave foams. A common blend is A-46, which is comprised of about 84% isobutane and about 16% propane and has a vapor pressure of 46 psig at 70° F. This blend is used on the average at a 3.5 to 5.0% level in a shave foam. As with halogenated hydrocarbons, the foam characteristics of a hydrocarbon blend can be altered by changing the propellant blend ratio or concentration.

Pre- and After-Electric Shave Products

Aerosol products for application before and after shaving with an electic razor have not enjoyed commercial success because they do not offer sufficient advantages over conventional product forms. A typical formula may consist of a self-emulsifying nonionic surfactant, alcohol, emollients, water, and propellant. Several such products have been formulated as quick-breaking foams.

A pre-electric shave product is an alcoholic medium containing oily material (e.g., isopropyl myristate or silicone) which will supply lubricity and increase the speed of an electic razor head.

Powders

The aerosol system provides an easy-to-use, controlled method of applying powders to the body. Aerosol powders have been successfully marketed as body talcs, foot powders, and feminine hygiene products. The powder spray is basically a suspension of the powder in the propellant. The main ingredient is usually talc, and various additives are employed.

- *Fillers.* Other solid materials are sometimes used to alter the bulk density of the powder and to improve the suspension properties.
- *Lubricants.* Small amounts of oil-like materials (e.g., isopropyl myristate) are often added to reduce the "dusting" effect of the atomized talc. The lubricant allows more product to adhere to the target area and less to "dust" into the surrounding area. The additive also serves as a lubricant in the valve orifices.
- *Medicaments.* In deodorant talcs, foot powders, and feminine hygiene sprays, a bacteriostat or fungicide may also be added for its special qualities.
- *Surfactants.* Surface-active agents can be added to coat powder particles. This sometimes reduces an agglomeration tendency and aids suspensions.

Generally, the formulator should consider the following factors when preparing aerosol powders (10):

- *Powder Base.* The powder base containing all solid additives such as fillers should be uniformly mixed and passed through a No. 325 mesh screen. When the solid medicaments, lubricants, or perfumes are added to the powder base, it should be uniformly mixed and screened to avoid any agglomerates in the aerosol container.
- *Propellant.* The specific gravity of a propellant blend is an important factor in suspension stability. In cosmetic powders the

length of suspension required for application is relatively short. The propellant blend chosen will be dictated by economics and by the chilling effect and spray pattern acceptable for the specific product.

- *Suspension.* Aside from the normal considerations of stability, it is advisable to package test samples of the powder formula in coated glass for observation. The powder should redisperse readily and remain suspended for the time needed for application. Upon standing, the powder and the liquid phase will separate in varying degrees. The settled solid should not cake or compact lest it interfere with the redispersion of the powder during the shelf life of the product.

Shelf testing should include periodic spray testing and emptying of the container without clogging, leakage, or radical variation in spray pattern.

Hair Products

HAIR SPRAYS

A hair spray is a solution composed of a film former, additives to modify and give specific properties to the film, alcohol, fragrance, and propellant. The aerosol provides a convenient method of applying a setting mist to the hair. As the propellant and alcohol evaporate, a holding film is left uniformly distributed over the surface. The formulator not only must consider the solubility and stability of the aerosol system in a given container but also must evaluate the performance of the product both as a cast film and on hair.

The cast film may be evaluated on glass slides and should have the following characteristics (4):

- *Clarity and appearance.* The dry film should be clear and continuous and not show crazing or "orange peel" effect.
- *Flexibility.* The film should be reasonably flexible or have enough plasticity.
- *Hardness.* The degree of hardness should be such that the film is not friable or does not crack.
- *Feel.* The film should be tack-free and nonoily to the touch.
- *Sheen.* The clear film should have a brightness or sheen, and, more importantly, the sheen should be visible on hair.

Hair sprays must also be evaluated on hair swatches and, finally, in use on human heads. Generally speaking, when applied to the hair the product should exhibit the following properties:

- *Degree of hold.* The firmness and holding power should be proportional to the desired effect (i.e., soft, normal, or hard-to-hold).
- *Feel.* The hair spray should not feel too stiff, tacky, or oily.
- *Sheen.* The hair should appear lustrous but not oily or greasy.
- *Humidity.* The film should be resistant to humidity to the extent that it does not lose holding power or become tacky on humid days.
- *Removability.* The hair spray should be readily removed by shampooing.
- *Flexibility.* The film should adhere to the hair but be flexible. A film that is too hard will have a tendency to crack and fall off in pieces. These bits of dried resin will give the user a false dandruff appearance.

A number of resins have been used in aerosol hair sprays and are usually present in finished aerosols at 1 to 4% solid levels. Some of the more commonly employed materials are the following:

Polyvinylpyrrolidone (PVP). A synthetic resin (soluble in alcohol and water), PVP produces films that are clear, glossy, reasonably hard, and readily plasticized and that shampoo out easily. Since PVP is hygroscopic, it tends to be slightly tacky under humid conditions. Compatible with a large number of other resins, solvents, and plasticizers, PVP can be used in many aerosol products.

Copolymers of Polyvinylpyrrolidone/Vinyl Acetate (PVP/VA). These are also used in aerosol hair sprays and have the advantage of being less hygroscopic than PVP alone. Available in blends wherein PVP represents 30 to 70% (the remainder being vinyl acetate), these copolymers allow latitude in formulating for holding power with some resistance to humidity.

Shellac. Special aerosol grades of shellac are available and are still used in some hair sprays. Shellac is employed principally in small amounts as a resin modifier to improve the properties of a more commonly used resin. Although it is possible to formulate a hair spray based on shellac, such problems as poor solubility and difficulty of removal make the use of other resins more desirable.

Copolymer of Methyl Vinyl Ether and Maleic Anhydride. These sold under the trade name of National Starch Resyn® 28-1310, exhibits good film-forming properties. Readily soluble in alcohol, it can be neutralized to vary some of the film's properties (e.g., hardness, resistance to humidity, removability). Generally, AMPD (2-amino-2-methyl-1, 3-propanediol) is used as a neutralizer for this resin.

Copolymers of Methyl Vinyl Ether and Maleic Anhyride. These resins are produced under the name Gantrez® ES 425 (monobutyl ester) and ES 225 (monoethyl ester). The alcohol-soluble resins are readily neutralized with several available amines. A properly balanced formula will produce clear, hard, nontacky films with sufficient holding power and sheen.

Plasticizers. These materials are added to modify hardness and the degree of flexibility of the resin film. Some of the more commonly used plasticizers include dimethyl phthalate, diethyl phthalate, acetyl triethyl citrate, isopropyl myristate, polyethylene glycol, and fatty alcohols.

Other additives. A number of other additives may be employed in hair sprays to improve appearance or feel, ease of combing, rewetting properties, water sensitivity, and ease of removal. Among these are silicones, lanolin derivatives, and protein hydrolyzates.

Solvents. Ethyl alcohol (SDA-40 anhydrous) is primarily used as the solvent in hair spray concentrates. The average hair spray contains 25 to 45% alcohol in the finished aerosol. Where economics dictate or a cosolvent is required for a specific additive, chlorinated solvents such as methylene chloride or 1,1,3-trichloroethane have been used. These solvents are miscible with alcohol and halogenated propellants but have an inherent ethereal quality that should be considered in the perfuming of the finished product.

Propellants. For the most part, aerosol hair sprays employ blends of Propellants 11 and 12. As with solution systems, the final blend is dependent on the proportion of base to propellant, the pressure desired, the required spray pattern, and the container and valve used. Propellant 11, when employed in standard hair sprays in tinplate containers, requires an anhydrous system, as the presence of water can lead to hydrolysis and corrosion (11).

Hair Spray—Normal

Concentrate	% by Weight
PVP/VA E-635, 50% (GAF)	12.00
Dimethyl phthalate	0.50
Silicone 556 (Dow Corning)	0.35
Alcohol, SDA-40, anhydrous	86.50
Perfume oil	0.65
	100.00

Fill	*% by Weight*
Concentrate	30.00
Propellant 12/11 (35:65)	70.00
	100.00

Hair Spray—Regular

Concentrate	*% by Weight*
Resyn® 28-2930 (National Starch)	4.50
AMPD (Commercial Solvents)	0.33
Silicone fluid 556 (Dow Corning)	0.45
Alcohol, SDA-40, anhydrous	94.27
Perfume oil	0.45
	100.00

Fill	*% by Weight*
Concentrate	35.00
Propellant 12/11 (40:60)	65.00
	100.00

Hair Spray—Hard-to-Hold

Concentrate	*% by Weight*
Gantrez® ES 425, 50% (GAF)	12.00
AMPD (Commercial Solvents)	0.30
Silicone fluid 556 (Dow Corning)	0.30
Perfume oil	0.60
Citroflex A-2 (Pfizer)	0.30
Alcohol, SDA-40, anhydrous	86.50
	100.00

Fill	*% by Weight*
Concentrate	35.00
Propellant 12/11 (40:60)	65.00
	100.00

Hair Spray—Dry Look

Concentrate	*% by Weight*
Resyn® 28-2930 (National Starch)	2.00
AMPD (Commercial Solvents)	0.45
CSA 91 (Hodag Chemical)	0.25
Perfume oil	0.50
Alcohol, SDA-40, anhydrous	86.80
	100.00

Fill	*% by Weight*
Concentrate	40.00
Propellant 12/11 (50:50)	60.00
	100.00

Although not as commercially successful as hair sprays, a number of other hair products have been introduced. Among these are a variety of conditioners, rinses, shampoos, and depilatories.

Hair Dressings

Nonaerosol emulsion systems can be formulated to have a viscosity suitable for dispensing from nitrogen packaging or Sepro-type containers. An emulsion of this type can also be formulated to be emitted as a foam. The aerosol package provides convenience and control of application, while the hair dressing functions to make the hair manageable and lustrous.

Hair Dressing Lotion—Quick-Breaking Foam

Concentrate	*% by Weight*
Part I	
Polawax (Croda)	1.50
Ethyl alcohol, SDA-40, anhydrous	61.50
Part II	
Ucon 50 HB-660 (Union Carbide)	2.00
PVP/VA E735 resin (GAF)	2.00
Deionized water	32.75
Part III	
Perfume oil	0.25
	100.00

Method of Preparation

Heat Part I to 100 to 110°F to dissolve the Polawax. Heat Part II to 100°F, and add to I while mixing. Add Part III to mixture of Parts I and II with stirring. Fill container with concentrate while still warm (100°F).

Fill	*% by Weight*
Concentrate	92.00
Propellant 12/114 (20:80)	8.00
	100.00

Hair Sheen

The aerosol hair sheens and conditioners are used to restore manageability, gloss, and soft texture to dry hair that has been damaged by bleaching, dyeing, permanent waving, excessive teasing, or straightening.

Hair Sheen and Conditioner

Concentrate	% by Weight
Ucon LB 1145 (Union Carbide)	50.00
Isopropyl palmitate	46.00
Acetylated lanolin	3.00
Perfume oil	1.00
	100.00
Fill	
Concentrate	15.00
Propellant 12/11 (40:60)	85.00
	100.00

Hair Cleaners

In the recent past, products have been introduced as hair cleaners for between-shampoo use. These are powder sprays containing absorbent starch derivatives. The powder is sprayed on the hair and then brushed out, effecting a dry shampoo. As an alternative method, liquid spray hair cleaners can also serve as "dry" shampoos when regular shampooing is not desirable and can act as a "spruce-up" for quick removal of surface dirt or built-up hair spray; these products have added sales appeal as visible dirt removal is possible. The aerosol hair cleaner is quickly and efficiently used on either human hair or wigs.

Shampoos

Aerosol shampoos have never achieved commercial success. Basically detergent-system shampoos can be formulated in nitrogen systems or as foams and are utilized functionally in the same manner as their nonaerosol counterparts. Like any other shampoo, the product should have good lathering qualities in hard or soft water, cleanse well, and not leave the hair dull or unduly dry and unmanageable.

Depilatories

Several such products have been introduced in aerosol form. Ideally a depilatory should be nonirritating, remove hair within a short period of time, be free of offensive odor, and be readily removed from the skin.

The depilatory action of the product depends on the active agent, the pH (generally about 10 to 12.0), and the duration of action. Calcium thioglycolate is commonly employed in a product that contains some wetting agent, a filler (e.g., calcium carbonate), alkali-stable perfume, emulsifiers, and a base for pH adjustment.

For some applications, it is desirable to dispense a product without altering its physical properties. Depilatories, hand creams, hair dressings, and certain other products can be dispensed from pressurized packages in their natural forms without producing a foam or spray. One way in which this can be achieved is with the Sepro can (12), which is used in the following formulation:

Depilatory

Concentrate	% by Weight
Calcium thioglycolate	5.40
Calcium hydroxide	10.40
Cetyl alcohol	8.00
Brij 35 (ICI America)	2.00
Distilled water	63.95
Perfume oil	0.25
Sorbitol solution	5.00
Propylene glycol, U.S.P.	5.00
	100.00

Method of Preparation

Heat 60% of distilled water to 70°C. Add the cetyl alcohol, sorbitol solution, propylene glycol, calcium hydroxide, and Brij 35, with constant stirring, until completely dissolved. Discontinue stirring and allow to stand and cool to room temperature. To the remainder of the distilled water add the 5.4% of calcium thioglycolate and mix well. Add the calcium thioglycolate solution to the emulsion already made, mix well, and allow to cool to room temperature. Adjust the pH with additional calcium hydroxide to 12.0 to 12.5. Add perfume oil and mix well. With continuous mixing, fill into Sepro-type container and pressurize.

Depilatory—Spray Foam

Concentrate	% by Weight
Polawax A-21 (Croda)	4.00
Deinoized water	75.00
Calcium thioglycolate	5.00
Thioglycolic acid	5.00
Sodium hydroxide solution (adjust pH to 11.0)	
Deionized Water	q.s.
	100.00

Method of Preparation

Heat 25 parts of water and the Polawax to 60°C and stir. Add remaining water (cold) with continued agitation. Mix in remaining ingredients and adjust pH.

Fill   *% by Weight*

Concentrate	90.00
Propellant 12/114 (20:80)	10.00
	100.00

Sun-screen Products

Sun-screen products exist in a wide variety of forms. Basically, these products are designed to promote tanning of the skin while preventing burning. A number of sun-screen agents are available which act by screening out the ultraviolet light associated with erythema or sunburn, while not blocking the ultraviolet region responsible for tanning.

A sun-screen agent should have effective screening ability in the 2900 to 3200 Å range, the most active wavelength range. It should of course also be readily soluble or emulsifiable to facilitate incorporation into the various sun-screen bases. Among the agents commonly employed are the following:

Homo menthyl salicylate.
Dipropylene glycol salicylate.
Aminobenzoates.
Escalols (Van Dyk).
Giv-Tan® F (2-ethoxyethyl *p*-Methoxycinnamate) (Givaudan)

A properly formulated sun-screen product should protect against burning while permitting development of skin tan or pigmentation. The product form can vary from a spray oil to a coherent foam but in any case should leave the user with a comfortable protective coating.

The aerosol sun-screen product should serve the following purposes when applied to skin:

- Provide a continuous film of sunscreen.
- Protect for at least the equivalent of 1 hour of noonday sun in summer.
- Be resistant to dilution and removal by perspiration and water.
- Not stain skin or beachwear or be difficult to remove with soap and water.

By and large, the sun-screen aerosol is not as popular as many non-aerosol lotions and oils but has found reasonable consumer acceptance. The aerosol offers the advantages of control of application, stability to light and heat, and greater resistance to sand contamination and breakage.

Sun-tan oil sprays are solution systems containing alcohol as a solvent. When applied to the skin, the alcohol evaporates, leaving a smooth, continuous oil film that contains the sun screen.

Sun-tan lotion foams are basically triethanolamine stearate emulsions to which an ethoxylated amide (Ethomid HT/15) has been added to maintain a good foam appearance at elevated temperatures. Quick-breaking foams provide control of application by delivering the product in foam form, yet are rapidly liquefied for easy spreading by heat and by hand pressure.

Sun-Tan Lotion

Concentrate	*% by Weight*
Part I	
Giv-Tan® F	1.50
Silicone DC 200 (200 CTS) (Dow Corning)	1.00
Lanolin anhydrous (Malmstrom)	1.50
Cetyl alcohol, N.F.	0.50
Stearic acid, T.P.	4.00
Glycerin, U.S.P.	2.00
Ethomid HT/15 (Armour Chemical)	1.00
Isopropyl myristate	5.00
Part II	
Triethanolamine, 98%	1.50
Preservative	q.s.
Perfume oil	0.26
Water	81.74
	100.00

Method of Preparation

Heat Part I to 75°C, and Part II to 75°C. Add Part I to Part II on an Eppenbach-type mixer at moderate speed; mix for 10 min. Transfer to Hobart-type mixer and mix until cool. Perfume at about 45°C.

Fill	*% by Weight*
Base	96.00
Propellant A-46	4.00
	100.00

Skin Care Products

The skin offers a large surface area for which a great variety of products can be formulated to cleanse, moisturize, treat, and in general beautify. A vast collection of hand and body creams, makeup, cleaners, and treatment products exist in a number of forms (e.g., emulsions, gels, sticks). Likewise, these products have been formulated as aerosols for many of the same end uses (13–15).

The aerosol hand cream should provide the same functional and esthetic properties as the nonaerosol product. A properly formulated aerosol foam hand cream should have three properties.

1. The emitted foam should be cosmetically elegant in appearance, with optimum coherence to imply richness yet spread readily.

2. When applied, the foam should liquefy on the skin easily enough to facilitate easy coverage and without "soaping" on the skin.

3. The resultant film should be tack-free and smooth and should provide protection without a heavy or greasy feel.

Aerosol skin cleanser foam is suitable for the removal of makeup and for general cleansing. It liquefies readily on the skin and can also be used as a moisturizer lotion.

Protective Hand Cream

Concentrate	*% by Weight*
Part I	
Myristic acid (Armour)	1.00
Stearic acid, T.P.	4.00
Cetyl alcohol, N.F.	0.40
Lanolin, anhydrous (Malmstrom)	0.30
Triethanolamine, 98%	1.20
Isopropyl palmitate	1.10
Silicone fluid L-43 (Union Carbide)	2.00
Part II	
Sorbitol solution	2.10
PVP K-30 (Antara Chemical)	0.20
Deionized water	87.35
Perfume oil	0.35
	100.00

Method of Preparation

Heat Parts I and II to 70°C. Add Part II to I with stirring, after all ingredients are melted and dissolved. Stir until cool, and add perfume.

Fill	*% by Weight*
Concentrate	90.00
Propellant 12/114 (40:60)	10.00
	100.00

Aerosol Makeup

Concentrate	% by Weight
Part I	
Tegin P	0.60
Stearic acid	2.80
Amerchol® L-101	5.70
Oleic acid	0.57
Cetyl alcohol	0.48
Propyl Paraben	0.20
Methyl Paraben	0.30
Part II	
Triethanolamine, 99%	1.15
Sorbitol solution (70%)	5.60
Water	74.47
Tenox 2 (Eastman)	0.30
Part III	
Titanium dioxide	6.53
Lead oxide dye A 6205 (Whittaker, Clark & Daniels)	0.27
Suntan A 2770 (Whittaker, Clark & Daniels)	0.90
Yellow ochre 3506 (Whittaker, Clark & Daniels)	0.30
Part IV	
Perfume oil	0.10
	100.00

Method of Preparation

Mix Part I and heat to 75°C. Mix Part II and heat to 80°C. Remove one fourth of Part II. Add dyes to remaining three fourths of Part II; mix till uniform. Add Part I. Add balance of Part II to rest of base; mix for 2 hours. Add perfume oil at 45°C.

Fill	% by Weight
Concentrate	90.00
Propellant 12/114 (40:60)	10.00
	100.00

Aerosol Skin Cleanser

Concentrate	% by Weight
Stearic acid, T.P.	4.00
Myristic acid	1.00
Glyceryl monostearate	2.50
Cetyl alcohol, N.F.	1.00
Ethomid HT 15	0.50
Isopropyl myristate	6.00
Mineral oil 65/75	12.00
Lanolin AWS (Malmstrom)	2.00

Part II

Triethanolamine, 98%	1.80
PVP K-30	0.50
Deinoized water	56.40
Glycerin, U.S.P.	5.00
Polyethylene glycol 400	3.50
Propylene glycol, U.S.P.	3.50

Part III

Perfume oil	0.30
	100.00

Part IV

Preservative	q.s.

Method of Preparation

Heat Part I to 70°C, and Part II to 70°C. Add Part II to I on Eppen-bach-type mixer; switch to Hobart-type until cool. Then add Part III.

Fill	*% by Weight*
Concentrate	96.00
Propellant A-46	4.00
	100.00

THE USE OF ENCAPSULATED PERFUMES IN AEROSOL COSMETICS

A recent development in regard to cosmetic aerosol products is the use of encapsulated perfumes to achieve the release of fragrance over an extended period of time. A fragrance can be obtained as a free-flowing spray-dried powder containing 40% perfume oil on a water-soluble matrix. The encapsulated fragrance remains entrapped when suspended in an anhydrous aerosol system. When moisture is available, as on the skin's surface, the matrix dissolves, releasing the fragrance. As time passes, additional moisture in the form of perspiration dissolves other discrete particles of the encapsulated fragrance, thus creating a time-release situation.

Additional areas of potential use for these encapsulated fragrances include antiperspirants, foot products, and disinfectants. In formulating with these fragrances, one should test to determine that (a) the matrix is not soluble in the aerosol system, (b) the additional solid form required to achieve the intended perfume level does not detract from the performance or feel of the product, and (c) the fragrance is stable and adequately released over the shelf life of the product.

REFERENCES

1. J. Pickthall, Emulsions, Chapter 4 in G. Maison de Navarre (Ed.), *The Chemistry and Manufacture of Cosmetics*, Vol. 1, Van Nostrand Company, Princeton, N.J., 1962.

2. R. G. Harry, *The Principles and Practice of Modern Cosmetics*, Chemical Publishing Company, New York, 1962.

3. M. S. Balsam and Edward Sagarin, *Cosmetics: Science and Technology*, 2nd ed., Interscience Publishers, New York, 1972.

4. Technical literature, Antara Chemical Corporation.

5. *Propellants for Low Pressure Cosmetic Aerosols*, Freon® Aerosol Report FA-18, E. I. du Pont de Nemours and Company, Wilmington, Del., Dec. 15, 1954.

6. Aerosol Technical Information, E. I. du Pont de Nemours and Company, Wilmington, Del.

7. Paul Sanders, *Principles of Aerosol Technology*, Van Nostrand Reinhold Company, New York, 1970.

8. J. G. Spitzer, Irving Reich, and Norman Fine, Lather Producing Composition, U.S. Patent 2,655,480 (Oct. 13, 1953).

9. Molecular Sieve literature, Union Carbide Corporation.

10. D. C. Geary and R. D. West, New Concepts in Formulating Powders, *Aerosol Age*, **6**, 8 (August 1961), 25-27, 68-69.

11. *Freon-11 S. Propellant*, Freon® Aerosol Report A-51, E. I. du Pont de Nemours and Company, Wilmington, Del., January 1964.

12. M. A. Johnsen, W. E. Dorland, and E. K. Dorland, *The Aerosol Handbook*, Wayne E. Dorland Company, Caldwell, N.J., 1972.

13. H. R. Shepherd, *Aerosols: Science and Technology*, Interscience Publishers, New York, 1961.

14. A. Herzka and J. Picktall, *Pressurized Packaging (Aerosols)*, Butterworths Scientific Publications, London, 1961.

15. A. Herzka, *International Encyclopedia of Pressurized Packaging (Aerosols)*, Pergamon Press, London, 1966.

FRAGRANCES FOR
AEROSOLS

Perfumery, as we know it today, evolved from a rather simple procedure practiced by the priests of the ancient world, who were indeed the first perfumers. Their task was a simple one, consisting primarily of mixing together natural odoriferous oils, gums, and resins into a blend that no doubt lacked harmony and character but served the religious requirements for which it was made.

From that beginning, perfumery has grown into an important aspect of our way of life, and the task of the perfumer has become more and more difficult. Whereas, in the years gone by, perfumery was considered to be solely an art, today it is as much a science as an art and requires as many technical skills as it does artistic qualifications, if not more. This has become even more evident since the aerosol made its appearance.

The introduction of the pressure package necessitated a change in the practice of perfumery—a new concept that made the perfumer a scientist as well as an artist. An entirely new field of perfumery came into being in which the perfumer became a researcher, seeking out new knowledge about the ingredients, the tools, with which he would be working.

It has long been known that the popularity of many products depended primarily on their odors. This is equally true for the aerosol product, especially since it is difficult to test a pressurized material or even to see it in the aerosol container. Selection is often based, therefore, on the odor preference of the buyer.

The scenting of aerosol products was, from the very beginning, considered of major importance. And, in turning his attention to pressurized products, the perfumer encountered problems he had not experienced before. Moreover, since so many different aerosol products required perfuming, many diverse problems were involved. To clarify the difficulties that were present, it may be well to first outline the various aerosol products that the perfumer was called upon to perfume.

TYPES OF AEROSOL PRODUCTS REQUIRING PERFUME

Personal Fragrance Products

This category includes perfumes, colognes, toilet waters and afterbath sprays, powders, and personal sachets.

Cosmetics and Toiletries

Perfume plays an important role in many cosmetics and toiletries, including hair sprays (for both men and women), deodorants and antiperspirants, shave creams, shampoos, foot sprays, sun-tan preparations, and hair dressings.

Household Products

Fragrance is used in various household specialties for both esthetic and functional reasons. It not only serves to add a distinctive scent to such products but also masks any unpleasant inherent odors that may be present. Included in this category are room sprays, bathroom products, car deodorants, furniture and floor polishes and waxes, dusting compounds, pesticides, paints, and oven and window cleaners.

It is quite obvious, from this partial list of products in which a scent may be used, that the perfumer can encounter numerous and diverse problems during the process of developing a perfume blend. In the initial period of aerosol fragrance development, perfume blends not specifically created for pressurized products were used in such materials. This was an unfortunate error, since many perfume components that are quite suitable for products in nonaerosol containers proved to be completely incompatible in push-button packages.

Such early experiences in aerosol product perfuming revealed the new and challenging concepts that this packaging medium created for the perfumer. It was his task to determine the problems and to develop solutions before the aerosol could be successfully utilized. In 1951, this author (1) outlined the problems that faced the perfumer, including solubility, compatibility, corrosion, possible sensual reactions (irritation), fixation, and esthetic considerations. Other problems came to light as development work progressed, and, as mentioned by Sagarin (2), it also became obvious that "one cannot consider the perfuming of aerosols without more specifically discussing the nature of the finished product," as is done later in this chapter.

PROBLEMS ENCOUNTERED IN PERFUMING AEROSOLS

Solubility

It is of utmost importance that all perfume ingredients be soluble in the special environment of the aerosol container. Insolubility of any of

TABLE 16.1.   SOLUBILITY OF FRAGRANCE MATERIALS IN PROPELLANT 12/114 (CONCENTRATIONS OF 0.5%) (5)

Soluble

Hexyl cinnamic aldehyde	Heliotropine
Hydroxycitronellal	Astrotone BR synthetic musk
Eugenol	
Cinnamic alcohol	Vanillin at 0.25%
Terpineol	Aldehyde C-12 MNA at 0.05%
Linalyl acetate	
Amyl salicylate	Oil of orange, cold-pressed
Terpinyl acetate	Alpine violet
Musk ambrette	Methyl acetophenone
Oil of lavendin	Citronellol
Musk ketone	Oil of sandalwood

Insoluble

Cinnamic aldehyde	Vetivert
Anisic aldehyde	Oak moss, resin or absolute
Geraniol	Patchouly
Coumarin	Methyl ionone
Oil of lavender	Benzoin resinoid
Ylang-ylang	Propylene glycol
Benzyl acetate	Isoeugenol
Oil of lemon	Oil of orange, distilled
Oil of bergamot	Dipropylene glycol
	Triethylene glycol

the ingredients can cause odor change, as well as valve clogging and other difficulties. To achieve blends that are fully balanced and exhibit no solubility problems in the aerosol, long periods of research and experimentation were necessary to determine the solubilities of individual ingredients as well as of fragrance blends in various aerosol propellants and combinations of propellants.

During the early periods of investigation, difficulty was experienced in the adaptation of well-balanced perfumes for aerosol use. To avoid solubility and incompatibility problems, it was necessary to replace the problem-causing ingredients with materials that had proved compatible in aerosol systems. Such substitutions could cause a change in the balance of the blend, however, and necessitated complete product reformulation.

The perfumer found that such products as ambergris, civet, and castoreum could exhibit a tendency to produce deposits after standing in a mixture of alcohol and propellant for several days (3, 4). The balsams, olibanum, galbanum, and other resinoids and gum extracts could also cause solubility problems in certain aerosol systems.

Among the many testing programs conducted, Sagarin (2) mentions the unpublished work of Pantaleoni, who evaluated a number of per-

TABLE 16.2. 1% SOLUTIONS OF ESSENTIAL OILS[a] (9)

Essential Oil	Propellant 12/11 (50:50)	Propellant 114
Abies siberica	S	S
Almond, sweet, expressed	S	S
Bay	S	I
Bergamot	S	I
Cananga native	S	I
Cassia, rectified	I	S
Cedarwood	S	S
Clove, U.S.P.	C	I
Copaiba	C	I
Eucalyptus, redistilled	C	S
Fennel, U.S.P.	S	I
Lavender flowers, U.S.P. (40/42%)	C	C
Lavandin	I	I
Lemongrass, rectified	S	S
Peppermint Natural, D & O	S	S

[a] S = soluble, C = cloudy, I = insoluble.

TABLE 16.3.   1% SOLUTIONS OF AROMATIC CHEMICALS[a] (9)

Aromatic Chemical	Propellant 12/11[b] (50:50)	Propellant 114
Alcohol C-8	S	C
Alcohol C-9	S	C
Alcohol C-10	S	C
Aldehyde C-11	S	C
Aldehyde C-14	S	C
Amyl butyrate	S	S
Benzyl isoeugenol	C	I
Coumarin	I	I
Geraniol	S	S
Linalyl acetate	I	S
Methyl benzoate	S	S
Methyl coumarin	I	S
Musk ambrette	S	S
Musk ketone	S	I
Styrolyl acetate	S	S

[a] S = soluble, C = cloudy, I = insoluble.
[b] It should be taken into consideration that Propellant 11 has many more solubilizing characteristics than the other fluorocarbons.

fume materials, both natural and synthetic, in various propellants. In a later published report, Pantaleoni (5) elaborated on his earlier work, the results of which are presented in Table 16.1.

Subsequently, other researchers learned that these same materials, which had been deemed to be insoluble and were therefore rarely used in a mixture of Propellants 12 and 114, could indeed be utilized in the presence of a cosolvent, which served to render them soluble. Among the cosolvents so utilized could be included alcohol, dipropylene glycol, and isopropyl myristate. The selection of the cosolvent is completely dependent, of course, on the type of aerosol formulation involved. In addition, many aerosol products in reality are not true aerosol systems but are just pressurized cosmetic compositions (e.g., shave creams, lotions, and certain other emulsified materials), and a great many fragrance materials that may not find application in true aerosol systems can be used in these pressurized products in view of the fact that they are emulsified, suspended, in the total system.

Except for the Pantaleoni work few data have been published on the solubility of perfume materials in aerosol propellants. The present author (1) reported that problems occurred in using certain organic alcohols in an aerosol system without a cosolvent, and later (6) it was suggested that "more soluble counterparts" be used in place of "resinous constituents" and that "certain of the absolutes" such as anhydrols be substituted in aerosol formulations to overcome solubility and compatibility problems. Gilbertson (7) seemed to agree when he wrote, "If a perfume component cannot be dissolved in the aerosol composition, with or without a cosolvent, it cannot be used."

With regard to the resinous products, Pickthall (8) reported the availability from a French firm of "an excellent range of its resoins which have a guaranteed solubility and stability in various propellants." Although soluble in propellants, they were not "all completely soluble in 96% ethyl alcohol." Another series was reported to be soluble in both ethyl alcohol and propellants.

In a later article De Feo (9) reported on the solubility of various odoriferous materials in Propellants 12/11 and 114. His findings are presented in Tables 16.2 and 16.3.

Compatibility

The aerosol created a new environment never before experienced with conventional packaging media. The problem was further complicated by the fact that new compatibility problems arose as different products were pressure packaged. Physical and chemical problems that would not only affect the perfume blend itself but also cause trouble in the operation of the aerosol were observed.

It was impossible during the early stages of aerosol development for the perfumer to predict the behavior of individual perfume ingredients and blends in the aerosol. Evaluation of perfume ingedients in the presence of the various propellants became an absolute necessity to ensure complete stability of the aerosol container. It must also be noted that many more propellants are now available, from which one or a combination may be selected, than there were when the low-pressure aerosol was developed. Several of these are quite suitable for the cosmetic industry.

With regard to clogging, Kainik (10) reported that this resulted from the use of certain perfume oils, but Sagarin (2) disagreed and noted that clogging occurred only with the "metered or calibrated valve, whose delicate mechanism has sometimes been attacked by perfume oils." The development of newer and better valve systems during the last several years and the availability of more advanced aerosol technology have almost completely eliminated this problem insofar as fragrance products are concerned.

Several other problems which could occur as a result of chemical incompatibility were cited by several authors. Merle (11) mentioned two such problems.

1. Hydrolysis of propellants (i.e., their splitting in the presence of water) frequently leads to the formation of hydrochloric acid, which, in turn, tends to produce profound chemical changes in odorants or aromatic materials.

2. Under certain unfavorable conditions, aldehydes, which are frequently used in perfumery, undergo both olfactory and chemical changes in aerosols. Cyclamal and aldehyde C-12 MNA (methyl nonyl acetaldehyde) are the only ones displaying relatively satisfactory stability.

It should be stressed that the availability of propellants which are not subject to hydrolysis has almost eliminated this problem.

Since chemical incompatibility can have a definite effect on the stability of certain odoriferous materials, the perfumer must always be aware of several general factors. In this regard Merle (11) noted:

Ketones and lactones, such as ionone, methyl ionone, undecalactone (aldehyde C-14), methyl acetophenone, etc., show stability of odor and good general stability.

Esters are generally resistant, excepting formates, valerianates, and butyrates. Styrene acetate, linalyl acetate, bornyl acetate, geranyl acetate and terpenyl acetate show exceptional stability.

Linalool and citronellol are the most resistant alcohols.

Among natural products, genuine musk will change its odor, whereas macrocyclic musk substances (of the "cyclopentadecanolide" type) are highly stable.

Merle added that "musk ambrette also retains its odor without change."

It should be noted, however, that when some of the ingredients mentioned are evaluated individually, they may exhibit certain peculiar properties, such as possible instability. However, a knowledgeable and experienced perfumer can still utilize them in conjunction with other products which tend to inhibit them from undergoing certain reactions in aerosol systems.

In spite of the knowledge available in the literature, incompatibilities may continue to occur and to cause various problems for the aerosol perfumer. In the event of such chemical activity, the perfume composition is usually the first indicator, with a resulting change in odor. With proper testing, however, it is possible, by using the correct combination of aroma chemicals and natural oils, to formulate perfume blends that will undergo little or no change in the aerosol formulation.

Corrosion

Corrosion is another aspect of the problem of compatibility. Corrosion in aerosol containers, without direct reference to perfumery, was studied by researchers at du Pont (12) and General Chemical Company (13) (now Industrial Chemical Division of Allied Chemical Corporation).

The present author (1, 14) reported that there are two aspects of corrosion. The first, which can be referred to as corrosion per se, consists of a reaction between the propellant and the container or any of its components. The second may be due to the perfume, the emulsifier, the cosolvent, or another such ingredient. He went on to say:

> Because of the unpredictable behavior of various materials in an aerosol package, the addition of any one substance, although such a substance itself may not cause corrosion, requires the retesting of the entire mixture. It has been found that corrosion may be caused by one of the ingredients only in the presence of a second ingredient, the latter perhaps acting as a catalyst.

It was also noted that, when certain alcohols are used in aluminum containers, the possible formation of aluminum alcoholate should be considered (see Chapter 9).

Other authors mentioned various aspects of corrosion. Pantaleoni (5) reported corrosion induced by perfume oils in shampoos and shaving cream to be insignificant, whereas Kainik (10) noted that the problem of perfume oil causing corrosion or clogging was quite troublesome. He added that "the formula must be broken down in order to determine which element is responsible." Pickthall (15) noted that "products packed in metal, whether tin plate or aluminum, whether lacquered or not, bring forth the vital question of corrosion . . . ." Pilz (16) reported the use of corrosion inhibitors: "products which when added to the aerosol act upon the surface of the metallic interior of the container, in this way preventing corrosion by different mechanisms of reactions." He said further that "it is much safer to establish protection by a scrupulous selection of the container, avoiding metals or applying a protective coating to the interior."

Sensual Reactions (Irritation)

There is very little direct evidence that perfume can cause sensual reactions or irritation. However, the perfumer must at all times be cognizant of this potential problem, especially when an aerosol is involved. It has been reported that, because of the fine spray and particle size, materials that would normally cause only slight irritation have a decidedly different effect when aerosol dispensed. Pilz (16) commented, "Aromatic products that, according to tradition and test of compatibility in normal application (in direct tests on the skin) were known to be inoffensive, produced disagreeable effects, and have had to be excluded from the formulation of perfumes for aerosols." He cited as an example benzyl benzoate, which was also mentioned by the present author in a previous work (17). Pickthall (15) reported that some fixatives caused irritation when used in aerosols and suggested the use of newer materials, none of which he mentioned.

Roojjakkers (18) reported two types of irritation, namely, mechanical and organic. The first occurs when particles of spray impinge on sensitive body organs such as the eyes, nose, and throat tissues and cause local irritation. He reported that dry sprays, in particular those of the powder type, are most likely to produce this condition. The organic type of irritation is caused by chemical reactions between components of the product and the skin of the user.

All authorities agree that materials to be used in aerosols must be carefully screened to eliminate any found to cause sensual reaction on contact with body membranes.

Fragrance Stability

Much work has no doubt been done on fragrance stability in aerosols, but very few of the resulting data have been published. Pickthall (15) reported on such experimentation but failed to release the findings. Foresman and Pantaleoni (19), however, did present the results of laboratory work in this area, and Kainik (20) reported experimentation with several different perfume compositions. It is also interesting to note that Pickthall questioned the reliability of the data reported by Foresman and Pantaleoni.

TABLE 16.4.   STABILITY OF FRAGRANCES IN COLOGNES (19)[a]

	Room Temperature			Elevated Temperature	
RC	RF	RFF	OC	OF	OFF
1.	Not charac- teristic, not as floral		Charac- teristic		
2.					
3.					
4.					
5.	Ex.	Ex.		Ex.	Ex.
6.				Indications of very slight oxid.— vanillin note	
7.	Ex.	Ex.	Ex.	Ex.	Ex.
8.					
9.	Ex.	Ex.	Ex.	Ex.	Ex.
10.				Half evaporated	Missing
11.	Ex.	Ex.	Ex.	Ex.	Ex.
12.					
13.					
14.		Odor int. increased		Odor int. increased	Odor int. increased
15.					
16.					
17.		Odor int. increased		Odor int. increased	Odor int. increased
18.	Ex.	Ex.	Ex.	Ex.	Ex.

The work of Foresman and Pantaleoni involved 30 synthetic and natural ingredients widely used in fragrances:

1. Hydroxycitronellal
2. Phenyl ethyl alcohol coeur
3. Peach aldehyde coeur
4. Amyl cinnamic aldehyde coeur
5. Methyl diphenyl ether
6. Isoeugenol
7. Alpha ionone coeur
8. Geraniol coeur
9. Gamma methyl ionone
10. Terpineol
11. Cyclamal (α-methyl p-isopropyl phenyl propyl aldehyde)
12. Eugenol, U.S.P.
13. Linalool coeur
14. Methyl anthranilate

TABLE 16.4 CONTINUED

	Room Temperature			Elevated Temperature	
RC	RF	RFF	OC	OF	OFF
19.		Odor int. increased		Odor int. increased	Odor int. increased
20.					
21.					
22.	Ex.	Ex.	Ex.	Ex.	Ex.
23.	Ex.	Ex.	Ex.	Ex.	Ex.
24.					
25.	Ex.	Ex.	Ex.	Ex.	Ex.
26.	Slight benz-aldehyde note	Ex.	Ex.	Ex.	Ex.
27.	Ex.	Ex.	Ex.	Ex.	Ex.
28.					
29.	Decreased sweetness, turpentiney	Decreased sweetness, turpentiney		Decreased sweetness, turpentiney	Decreased sweetness, turpentiney
30.	Ex.	Ex.	Ex.	Ex.	Ex.

[a] RC = room-temperature control—unpressurized
RF =   ″   ″   ″   pressurized
RFF = duplicate room temperature—pressurized
OC = oven storage control—unpressurized
OF =   ″   ″   ″   pressurized
OFF = duplicate oven storage—pressurized.
If odor exhibited no change, no notation was made in the table.
Designation of Ex indicates excellent coverage.

15. Anisic aldehyde
16. Amyl salicylate
17. Citral C.P.
18. Phenyl acetaldehyde 100%
19. Linalyl acetate
20. Geranium Bourbon
21. Terpeneless petitgrain
22. Benzyl acetate
23. Petitgrain S.A.
24. Citronellyl acetate
25. Lavender oil 40-42
26. Methyl benzoate
27. Aldehyde C-10
28. Bergamot natural
29. Pine needle Siberian
30. 10% musk ambrette in diethyl phthalate

TABLE 16.5.   STABILITY OF FRAGRANCES IN SHAVING CREAM (19)

	Room Temperature		Elevated Temperature	
	Control: Nonpressurized Fresh Sample	Nonpressurized	Pressurized	Nonpressurized	Pressurized
1.				Generally weaker	
2.				Generally weaker	
3.					
4.				Generally weaker	
5.					
6.				Possible slight change with traces of vanillin	
7.				Discoloration Generally weaker	Discoloration
8.				Slight change	Less floral and weaker
9.				Less flowery Generally weaker	
10.				Generally weaker	
11.				Generally weaker	
12.				Slight discoloration	Slight discoloration
13.					
14.					
15.					

A simple cologne was composed of 1 g perfume oil, 14 g water, and 113 g SDA 40. The cologne was charged with Propellant 114. Another test formula was a shaving cream composed of 5.0% triethanolamine stearate and 95.0% water and pressurized with a 40:60 mixture of Propellants 12 and 114. The experimental results are indicated in Tables 16.4 and 16.5. The numbers at the left refer to the 30 ingredients listed above.

It had previously been pointed out by this author (1) that an aerosol-dispensed perfume produces a different olfactory stimulus than one

TABLE 16.5   CONTINUED

	Room Temperature		Elevated Temperature	
	Control: Nonpressurized Fresh Sample	Nonpressurized / Pressurized	Nonpresurized	Pressurized
16.			Odor change, not as nice	
17.			Great discoloration	
18.		Off-odor / Weak, flat in odor (Pressurized)	Off-odor	
			Slight off-odor	
19.				
20.				Slight off-odor
21.				
22.			Substantial degradation of odor	
23.				
24.				
25.				
26.			Noticeable degradation of odor	
27.			Slight degradation of odor	Appreciable degradation of odor—fatty
28.			Not as sweet, change in odor	
29.			Weaker due to heat—soapy	
30.			Slight change	

that is smelled from a bottle or a blotter. "The individual notes," he continued, "in the different drying stages are obtained from bottle or blotter; whereas, when the material is sprayed under pressure, complete dissemination of the entire perfume compound is accomplished in an instant." He also stated that "the same perfume compound has a different note when sprayed at different pressures and from different valve openings, and it is therefore necessary to revise a formulation when used for aerosols so that one can obtain the same note," and he suggested that a fragrance blend recommended for one type of aerosol be completely re-evaluated before it is recommended for another aerosol product. This proposal was further elaborated on in two later works (21, 22).

Additional reports on fragrance stability in aerosols confirmed the earlier contention of this author. It was reported in 1955 (23) that odoriferous products caused clogging and other problems in aerosols even after their successful use in glass-packaged products for many years. Perfumers were also cautioned on the decomposition of a perfume due to propellant hydrolysis (24) and on the possibility of a chemical reaction between the perfume and the active ingredients of the product itself.

Sagarin (2) summarized the literature on the problems involved in aerosol perfuming as follows:

1. Most materials generally used in perfumery are used in aerosol perfumery, but may give somewhat different olfactory effects.

2. This difference in the olfactory note is more discernible in the comparison of topnote, or the first fragrance, than of the latter note during the life and lingering evaporation of the perfume.

3. Resinous and solid materials may cause difficulties, but probably only rarely when used in small proportions in the perfume oil and when that oil is used in the usual small proportions in the finished can or bottle.

4. Numerous raw materials are said to be unstable, but as yet there is little agreement on these matters. Among the products of questionable value are oil of lemon, esters of aliphatic acids, primary alcohols, etc. However, the percentage used, presence or absence of other materials, and nature of finished products must be stressed.

5. Several raw materials are said to be irritating when sprayed. Again, the amounts present are important, but inasmuch as such products are often used as perfume solubilizers, they might generally be present in sufficiently large proportions to effect irritation, if they do have such irritation potential. Among such materials are diethyl phthalate, benzyl alcohol, and benzyl benzoate.

6. Solubilizers for the perfume oil may often be required; solubilizers of choice would be ethyl alcohol, dipropylene glycol, and (where the inferior odor is not a factor) isopropyl alcohol.

7. Shelf- and oven-testing of perfume in aerosol formulation is always necessary, as all data, even when confirmed, can give only an indication of the behavior of materials in a perfume compound (i.e., mixture) with numerous other ingredients, and in a given aerosol formulation. Thus, perfumery remains almost completely empirical.

In order to further illustrate the problems involved in aerosol perfumery, we now turn our attention to specific factors involved in the perfuming of individual groups of aerosol products. We should point out that the same esthetic consideration should be given to aerosol products as to materials packaged in conventional media. Although some reference to formulations will be made, our emphasis will be on perfuming, and we refer readers to other chapters for detailed data on individual products.

SPECIFIC FACTORS INVOLVED IN PERFUMING VARIOUS TYPES OF PRODUCTS

Perfumes and Colognes

The aerosol-packaged fragrance product will contain a concentration of perfume oil that differs from that used in a conventionally packaged perfume or cologne. When a fragrance is aerosol dispensed, the entire perfume is vaporized, permitting the full odor impact to be felt immediately. Pressurized perfume extracts usually contain a 3 to 5% concentration of perfume oil, and a 0.5 to 2% concentration is recommended for aerosol colognes.

Because alcohol is present in most perfume and cologne formulations and acts as a solvent, the perfumer has more flexibility in choosing his raw materials. He can utilize some that under normal circumstances, he would not have used for aerosol application for fear of insolubility. The concentration of these materials in the perfume blend depends on the percentage of finished perfume oil that is to be used and the concentration of alcohol and the type of propellant selected. In this instance the alcohol may act as a cosolvent, and for the same purpose dipropylene glycol may be added to a formulation as an aid in maintaining solubility.

Other products also offer cosolvent properties but are not recommended for use since they may cause sensual reactions in nasal passages and other body membranes. These include benzyl benzoate, benzyl laurate, and diethyl phthalate. It is also essential that only a stable propellant be used in pressurized perfumes and colognes in order to ensure good stability of the product during normal use.

Actually, the preparation of an aerosol perfume or cologne is somewhat similar to that of the conventionally packaged product. The alcoholic solutions must be allowed to stand for a period of time before chilling, filtering, and packaging. This allows waxes, resins, or other insoluble materials to settle and results in a clear solution. This procedure is of greater importance for aerosol-packaged fragrances, since the carrier is a combination of alcohol and propellants with a lower solubility factor, which may hasten the precipitation of basic components and possibly cause clogging of the valve.

In a previous work (25), the author recommended a specific procedure to be followed when preparing an aerosol perfume or cologne solution.

Phase I—Preparation of an alcoholic cologne solution

A. Incorporate desired ratio of oil in alcohol (95% SDA recommended for all propellants except those subject to hydrolization, such as Propellant 11—trichloromonofluoromethane).

B. Allow alcoholic solution to age at least 48 hours.

C. Chill solution for at least 5 hours at minus 5°C to minus 10°C.

D. While at these low temperatures, filter several times through magnesium carbonate or other suitable filtering material.

Phase II—Preparation of finished product

Recommended formula for aerosol cologne mist for use in coated glass containers:

Perfume oil	4%
Alcohol	46%
Propellant 12/114 (15:85)	50%

In order to prepare the above solution, it is recommended that an 8% concentration of perfume oil be prepared in the alcohol solution and treated as recommended above. To prepare the finished aerosol cologne mist, 50% of the 8% alcoholic solution and 50% of the propellant mixture would be used.

Metered perfume spray for use in metal containers:

60% perfume solution, 20% perfume in SDA alcohol
40% Propellant 12/114 (40:60)

The alcoholic solution given in the above formula should be treated as recommended above and charged into a metered aerosol container. Since this product contains a high concentration of perfume oil, metered valves are suggested to guard against any possible difficulties due to the particle size of the spray. It is necessary to use a higher pressure in the system to obtain proper spray characteristics. We should also mention that in view of the vast difference in specific gravities of the alcohol and the propellants used, all concentrations in these formulations are based on weight.

It has also been suggested (26) that the addition of 5 to 10% of water to fragrance formulations will enhance the finished product and simplify its use in certain aerosol containers. In glass aerosols the hydroalcoholic solution has a tendency to reduce the sharpness of the cosolvent used and also permits a better spray pattern and a lower propellant concentration. In aluminum containers the water inhibits the possible reaction between the alcohol in the formulation and the container, resulting in a longer period of stability and shelf life. In this instance, however, it is essential that the proper propellant be used to avoid the possibility of hydrolysis.

Personal Deodorants and Antiperspirants

Although the aerosol is an ideal medium for personal deodorants and antiperspirants, perfuming problems are encountered. In the first place, several of the basic ingredients that may be used in the deodorant/antiperspirant have unpleasant odors which must be masked. Second, there is always the possibility of a chemical reaction between the active ingredients, the aluminum salts, or the antibacterial agent and the perfume components.

Let us outline several of the difficulties facing the perfumer. The compatibilities of the various product components with the propellants and the container must be considered, as must the solubility of the entire formulation in the propellant mixture. He must also consider the possibility of sensitization and corrosion, especially as a result of the presence of the aluminum salts (27).

With the advent of the dry antiperspirant spray, other problems developed for the perfumer (28). The perfume components carrying a polar charge could cause jellification of the concentrate if introduced into it before loading. The end result of this would be cavitation during the actual loading process, as well as other manufacturing problems.

Among the difficulties that could be experienced because of a reaction between the perfume oils and the aluminum salts is the presence of bentonite. This can cause the release of water by catalysis, hydrolyzation of the aluminum salt, and the production of free acid. The acid can react with the perfume materials and the container, resulting in complete destruction of the product. Other problems that can occur are the following:

Direct interaction of the perfume with the aluminum salt, an odor change resulting when contact is made between the perfume oil and skin oils and prespiration.

Interactions between the perfume and the propellant.

Shelf-life instability.

Discoloration reactions and the synergistic effects that the perfume can have with the active ingredients.

With the possible utilization of other antibacterial additives, such as vitamin E or ascorbic acid, which act to prevent degradation or oxidation of oils and other components of perspiration products in the area of application, thus preventing an odor buildup, additional testing by the perfumer becomes necessary. All materials used must be restudied to make certain that they can still be utilized and will remain functional in this type of deodorant system.

Hair Sprays

When aerosol hair fixatives were first packaged, shellac was the basic ingredient in the formulation. Many problems arose, however, because of the effect that the alcohol/shellac solution had on the perfume oils as well as other ingredients and the components of the container. Current formulations are based primarily on polyvinylpyrrolidone (PVP) and its copolymers, dissolved in anhydrous alcohol with appropriate plasticizers and usually pressurized with dichlorodifluoromethane or a mixture of this compound and trichlorofluoromethane.

The resin PVP performs much better in an aerosol system and does not cause the problems encountered with a shellac hair set. Its low odor level and the fact that a minimum amount of plasticizer is necessary also permit a wider selection of fragrance types by the perfumer. There are, however, some perfuming problems that should be mentioned, not the least of which is perfume breakdown.

Because of the formulation of the hair fixative the possibility of perfume breakdown exists, and it is essential, therefore, that perfume materials be carefully selected after proper testing. One authority (29) suggested that replacing the usual grade of trichlorofluoromethane with one containing 0.3% nitromethane would prevent degradation of the product. He also mentioned that without this modification 60% of the products developed objectionable odors within a 2-month period. This claim was refuted by Eckton (30) on the basis of his own experience, although he added, "Further experience may prove that the use of a stabilized propellant will reduce the possible hazards and facilitate the development of a wider range of suitable perfumes."

It has also been found that some fragrance ingredients tend to form complexes with PVP, thus changing the balance of the perfume blend. Other perfume components have a tendency to reduce the plasticizing property of the hair fixative, reducing the effectiveness of the product. A method of avoiding this has been reported in a British patent (31).

Initially, one of the basic functions of the perfume was to mask any objectionable odors from the components of the hair-set spray. Later trends have moved toward the use of somewhat higher concentrations to achieve what might be called a "cologne–hair fixative combination"

(32). This is felt to be quite effective, since the hair serves as an ideal medium for the retention of the perfume, with each strand acting as a trap to hold the fragrance. In such a case, the highly perfumed hair set becomes an intricate part of the fragrance ensemble used by the woman. It also means that the perfumer must create a fragrance blend that will be extremely stable in the hair spray and maintain its odor characteristics for the period of use.

Foam Products

Among aerosol foam products can be included hand creams, shampoos, cologne foams, and shave creams, with the last-named accounting for the largest portion of foam aerosol production. The importance of perfume in an aerosol shave was stressed by Carsch (33), who wrote, "In aerosol shave lathers, the fragrance of a product is sometimes the sole property which distinguishes it from any number of other brands.... So important, in fact, is the fragrance of pressure packed shave products that it becomes the featured attraction of that particular brand." This, of course, is also true for other aerosol foam products.

The perfumer has a great deal of flexibility in the choice of a fragrance blend for this type of product, since the problem of solubility is not as critical as it is in other aerosol products. The nature of the foam product permits the perfume to be emulsified and suspended in the product itself, and very little valve failure or package difficulty is ever encountered. Of course, the question of coloration or discoloration must be considered, but, from his experience with nonpressurized products the perfumer can avoid any such difficulty. It is also possible that some perfume materials may affect viscosity, stability, or even foaming power, but in this case too previous experience can guide the creative perfumer.

Insecticides

The basic compositions of common and industrial pesticides, including the toxic ingredients and synergistic materials as well as hydrocarbon and aromatic solvents, present problems of both odor masking and irritation. Even with the use of toxicants, solvents, and knockdown agents of low odor level, there is a need for coverage that may tax the knowledge and experience of the perfumer.

One major problem is that, in most cases, the perfumer is confronted with the problem of scenting not just one or two basic materials, but rather a combination of components each of which has some inherent unpleasant odor. He must completely mask this combination of unpleasant odors before he can consider the possibility of reodorization.

The possible effect of the active ingredients of the insecticide on the perfume must also be considered. Khac-Man and Guihot (34) studied the effects of some 51 different perfumes in the presence of pyrethrum. They summarized this work as follows:

1. Change of odor:

For a specific perfume, there was a certain odor change in the heated and unheated insecticide samples packed in various metal containers. These changes were probably due not only to the action of the metal, but possibly also to the propellant . . . .

2. Effect of corrosion:

No corrosion was observed in the five types of metal containers utilized in our work except for minor pin-point rusting along the side seam of rolled tin-plate containers. From a corrosion standpoint, the various metal containers employed in our tests appeared to have little or no effect on the stability of the perfumes, but we did notice that the odor of certain perfumes is directly influenced by the action of the metal, particularly tin.

3. Deposit:

A slight or moderate deposit accumulated at the bottom of most aerosol insecticide samples. We believe this is some type of residue or insoluble material deriving from the pyrethrum extract.

4. Heating effect:

From the results observed in the series without propellant, we conclude that there was an effect of heating on the residue or deposit found in various containers. This effect would slightly reduce the amount of the observed residue.

5. Effect of the propellant:

According to our observations, for one specific perfume there is a possible relation between the propellant and the odor changes encountered in various metal containers.

6. Effect of certain perfumery chemicals on the pyrethrum extract:

The observations of our second test series lead us to the opinion that metal containers have an effect on certain perfumery chemicals, causing odor changes not observed in the same samples packed in glass containers.

7. Knock-down effects (i.e., KD_{50} value):

All tests to determine the insecticidal efficiency of pyrethrum in heated and unheated perfumed insecticide samples were carried out by the Pyrethrum Bureau of Kenya. From the results of these tests, we would conclude that there is no notable difference in the KD_{50} value between the heated and unheated perfumed insecticide samples. There is no significant loss of pyrethrum potency among the selected perfumery chemicals employed.

For a perfumer–chemist the aim should be not only to study the effect of various perfumes on pyrethrum extract, but also to strive for the realization of a certain number of successful insecticide perfumes which must be specially prepared for specific compatibility and odor-masking properties. In spite of most diligent and constant efforts, we realize that there were certain details in our work which both require and deserve further investigation.

The question of cost also limits the creative ability of the perfumer, since it does not permit him as wide a choice of perfume ingredients as he might wish to have. He must also be guided in his selection of odor type by the end use of the insecticide—household, industrial, or agricultural.

The limits placed on perfume ingredient selection by the cost factor make the choice more difficult insofar as solubility, compatibility, and odor coverage are concerned. The perfume blends must be formulated to permit immediate diffusion when the product is sprayed in order that the base will be covered and the coverage maintained throughout the period of use.

Space Sprays

Air deodorization can be accomplished by four methods outlined by DiGiacomo and Stoller (35):

1. *Masking.* This method involves the use of a pleasant, possibly stronger, scent to cover the unpleasant odor.

2. *Odor counteraction.* Also called odor modification or compensation, this method is based on the theory advanced by Zwaardemaker and others that if two odors are combined in certain specific proportions they will counteract or modify each other, resulting in the elimination of all odors. There are, however, differences of opinion among the chemists regarding this method and its suitability in air freshening, since it is felt that there must be chemical reaction between the deodorant and the malodor to achieve complete deodorization and no one product has as yet been developed (and some believe it never will) that could react with all types of odors.

3. *Chemical combination.* This is a method by which the air freshening product combines chemically with the malodorous molecules and reduces them.

4. *Anesthetizing.* This method affects the sense of smell, preventing it from perceiving the presence of the unpleasant odor. It can be accomplished chemically by using special ingredients, such as formaldehyde, or by utilizing a very high concentration of odor which will tend to achieve the same effect. Certain odor types also have a tendency to anesthetize.

Most commonly used space sprays are based on combinations of the above methods. A large portion of the air fresheners in use are of the odor masking type, but disinfectant sprays are becoming a significant factor in the market. There has also been a resurgence of open evaporation systems as compared to closed spray systems.

Fragrance plays an important role in air fresheners. In many cases, the perfume is the odor-combating principle. It is well known that some natural perfume ingredients act as true chemical or physiological deodorants. In other cases, the perfume adds an esthetic effect to the finished product.

In the work previously cited (35), there was a rather complete discussion of the perfuming of air fresheners. The authors wrote:

The perfuming of aerosol air fresheners presents more problems to the perfumer than other room deodorants. The fragrance composition must be compatible with and soluble in the propellant or combination of propellants utilized. Of course, the use of a cosolvent increases the variety of perfume ingredients that may be used, but it is not by far an overall panacea. The perfumer must be certain that his entire formulation is compatible to avoid any possibility of malfunction of the package.

The final odor of the air deodorant is influenced by the size of the valve orifice which determines the fineness of the spray which will be disseminated. Unless the product is specifically formulated so that a definite fragrance will be evident to the user when the product is dispensed, it is quite possible that the scent actually obtained will be different from the one desired. To illustrate: a floral fragrance prepared for use in an aerosol air freshener could go through a definite change upon dissemination. The fine particle size which is sprayed through the aerosol orifice can, if the formulation is not properly balanced, change the odor and place it in the range of nondescript types. If the true odor character of the floral fragrance is desired, the product must be formulated and balanced accordingly. Furthermore, the fragrance blend formulated for use in an aerosol must in no way affect the container or its component parts.

One other factor which the perfumer must consider is the possibility of nasal stinging when the air freshener is sprayed into the area of use. This requires evaluation of the entire product and not just individual ingredients. It is known that many ingredients which do not exhibit untoward effects when sprayed alone do have such effects when sprayed as part of a mixture. Conversely, some ingredients that do exhibit untoward effects lose this property when combined with others in a complete formulation. In any event, all suspected products should be removed from an aerosol formulation.

Sachets

Since pressurized sachets can be sprayed directly on fabrics, linens, clothing, and similar items, and since these sachets contain perfume concentrations of 0.5 to 3.0%, care must be taken in the selection of fragrance components to avoid those that can cause coloration or discoloration. Alcohol is used in the product as a carrier and cosolvent for the perfume blend. If a dry spray effect is desired, it can be achieved by selecting fragrance components that are soluble in the propellants used and by reducing or eliminating the alcohol in the basic formulation.

Powders

Among the powder products packaged in aerosols, there are many that require perfuming for both esthetic and masking reasons. The task of formulating a fragrance is more involved for an aerosol powder than for a conventionally packaged product. Although some of the basic problems are similar, the inherent peculiarities of the aerosol make the task of the perfumer somewhat more difficult.

The perfumer must study the compatibility of each individual perfume ingredient, as well as the completed blend, with the components of the active principle of the product and the propellants. This is absolutely essential when powders are involved. Gums, concretes, and certain resins must be avoided. Terpeneless oils should be used since the presence of terpenes can cause odor change and stability problems in the pressure package.

For products that require deodorization, the addition of certain odor depressants is helpful. Isopropyl myristate has been found to reduce odor levels in various aerosol products, especially insecticides and mothicides. Isopropyl myristate acts to coat the odoriferous molecules in the formulation, thus reducing the odor level and permitting better odor coverage.

There are, of course, many other products packaged in aerosols that require perfumes, including foot and shoe sprays, sun-tanning preparations, body powders, shampoos, creams, household products, and pet products. In these cases the problems involved in the addition of a perfume compound are similar to those we have already mentioned.

CONCLUSIONS

It should be emphasized that all the problems arising during the scenting of aerosol preparations, as well as those arising when the fragrance blends are added to the finished product, can be overcome. Much research and development work has already been done on the perfuming of aerosols, although very few data have been published on the subject. The importance of the factors mentioned previously—solubility, corrosion, sensual reactions, compatibility—for all aerosol products cannot be overemphasized. In addition, the importance of shelf testing to determine the stability of the perfumed aerosol must be recognized.

REFERENCES

1. Victor DiGiacomo, Problems Encountered in the Perfuming of Aerosols, *Proc. Sci. Sect. Toilet Goods Assoc.*, May 1951, No. 15, p. 9.

2. Edward Sagarin, The Odor of Aerosol Products, Chapter 9 in H. Shepherd (Ed.), *Aerosols: Science and Technology*, Interscience Publishers, New York, 1960.

3. Ernest Shiftan, Perfumers' Problems in the Manufacturing of Aerosols, *Amer. Perfum. Aromat.*, **69**, 1 (1957), 99.

4. Peter G. Dilworth, Aerosol Perfumery—the Finishing Touch, *Specialities*, **3**, (August 1967), 8.

5. R. Pantaleoni, The Perfumery of Aerosols, *Amer. Perfum. Essent. Oil Rev.*, **58**, 6 (1951), 425.

6. Problems in Adapting Cosmetics and Fragrances for Aerosols, *Givaudanian*, January 1957, p. 3 (Givaudan Corporation, Clifton, N.J.).

7. G. Gilbertson, Perfuming an Aerosol Product, *Mfg. Chem. Aerosol News*, **41**, (December 1970), 38.

8. J. Pickthall, Perfumery, Chapter 19 in A. Herzka (Ed.), *International Encyclopaedia of Pressurized Packaging (Aerosols)*, Pergamon Press, London, 1966.

9. V. De Feo, Solubility of Aromatic Substances, *Amer. Perfum.*, **75**, 10 (1960), 84.

10. Herbert Kainik, Perfuming of Aerosols, *Soap Sanit. Chem.*, **24** (October 1953), 45.

11. Antoine C. Merle, Problems Connected with the Use of Perfume in Aerosols, *Dragoco Rept.*, No. 1, 1972, p. 3.

12. H. M. Parmelee and R. C. Downing, Corrosion in Aqueous and Alcoholic Aerosol Systems (Proceedings of the Chemical Specialties Manufacturers Association Meeting, December 1950) **45**, 47–51.

13. General Chemical Corporation, Private communication cited by Victor DiGiacomo: Perfuming Aerosols, *Drug Cosmet. Ind.*, **69**, 1 (July 1951), 30.

14. Victor DiGiacomo, Perfuming Aerosols, *Drug Cosmet. Ind.*, **69**, 1 (July 1951), 30.

15. Jack Pickthall, Aerosol Perfumery Problems. *Amer. Perfum. Aromat.*, **67**, (May 1956), 40.

16. Wolfgang Pilz, Perfuming of Aerosols, *Aerosol Age*, **12** (March 1967), 24.

17. Victor DiGiacomo, The Adaptation of Fragrances for Aerosols, *Givaudanian*, April 1955, p. 3.

18. Th. W.M.I. Roojjakkers, Perfuming Aerosol Products, *Aerosol Rept.*, **9**, 9 (1970), 448.

19. R. A. Foresman and R. Pantaleoni, Odor Stability in Aerosols, *Drug Cosmet. Ind.*, **75**, (July 1954), 38.

20. Herbert G. Kainik, Fragrance Stability in Aerosols, *Soap Chem. Spec.,* **32** (August 1956), 43.

21. Fragrance and the Aerosol, *Givaudanian,* November 1953, p. 3.

22. Victor DiGiacomo, Formulating Fragrances for Aerosols, *Amer. Perfum. Essent. Oil Rev.,* **65** (April 1955), 29.

23. Perfuming Aerosols, *Soap Chem. Spec.,* **32** (January 1955), 39.

24. Aerosol Topics, *Givaudanian,* May 1955, p. 9.

25. Victor DiGiacomo, Aerosol Perfuming, *Soap Chem. Spec.,* **36** (January 1960), 61.

26. Pressurized Fragrances, *Givaudanian,* December 1957, p. 8.

27. Victor DiGiacomo, The Aerosol Deodorant, *Givaudanian,* June 1965, p. 3.

28. Victor DiGiacomo, The Technical Aspects of Perfumery, Unpublished address before joint meeting of the American Society of Perfumers and the New York Chapter Society of Cosmetic Chemists, Oct. 13, 1971.

29. Find Cause and Remedy for Aerosol Container Corrosion, *Amer. Perfum.* **76** (November 1961), 43.

30. R. E. Eckton, The Choice of Perfume for Pressurized Products, *Aerosol Age,* **9,** (April 1966), 34.

31. Brit. Patent 868,879.

32. Aerosol Hair Sets, *Givaudanian,* April 1958, p. 8.

33. Gustav Carsch, Perfuming Aerosol Shave Lathers, *Soap Chem. Spec.,* **46** (November 1970), 42.

34. Le Khac-Man and Noel Guihot, Do Perfuming Materials Affect Pyrethrum Based Aerosol Insecticides? *Aerosol Age,* **10,** 4 (April 1965), 34.

35. Victor DiGiacomo and Leonard Stoller, Deodorizing the Air, *Givaudanian,* November 1966, p. 3.

CHAPTER SEVENTEEN

LEONARD STOLLER
AND JOHN J. SCIARRA

FOOD AND PHARMACEUTICAL AEROSOLS

Aerosol packaging has been used in the food and pharmaceutical industries for a number of years. In fact, the pressurized topping, several burn-treatment products, and a local anesthetic were marketed before the major insecticidal development of World War II.

In spite of these early developments and much subsequent research into other areas of application, the use of the aerosol for foods and pharmaceuticals has not achieved what was originally expected. Many reasons could be cited, not the least of them the economic factor, particularly in the area of foods. For pharmaceuticals, various governmental regulations, as well as limitations in aerosol components, may have contributed to the disappointing market growth.

There are, however, a number of both food and pharmaceutical aerosols being sold today. This chapter offers a brief overview of both product types.

FOOD AEROSOLS

Pressure-dispensed food products represent a very small percentage of the total aerosol market. Only foam products, such as whipped cream and other toppings, have proved successful, although several other foods, including barbecue sauces, salad dressings, and syrups, have been introduced. In spite of the many failures that have been experienced and that have slowed development in this area, the future outlook is not as black as it may seem. Although no immediate increase in pressurized food development, especially during a period of economic difficulties, is expected, there is reason for optimism in regard to the future.

Historical Review

The history of "aerosol foods," which Graham (1) defined as "products packaged with a non-toxic compressed gas which serves to self-propel the food through a suitably valved closure," can be traced back to 1925, when a method for fluffing and dispensing ice cream under pressure was patented by Ashley (2). Long before that date, a patent was granted to vander Weyde and Matthews (3) covering the use of nitrous oxide in artificially aerated drinks, and in 1882 the use of compressed gases for the charging of milk products was patented (4).

In 1932 Roush (5) described a method for dispensing whipped cream by using compressed air under high pressure, and Burmeister (6) patented his use of air and oxygen under pressure for foaming cream without mechanical whipping. From 1934 to 1936 Getz and his coworkers (7–10) produced whipping cream with 200 to 650% overrun by aerating with nitrous oxide and other gases in small containers.

A patent issued to Reinecke in 1938 (11) covered his device for the charging of cream in a hermetically sealed container, using a small cartridge charged with nitrous oxide, and in 1939 Diller (12) and in 1941 Goosmann (13) patented their use of nitrous oxide and carbon dioxide to produce whipped cream. A charging device for the same purpose was also invented by Marmorek and patented in 1939 (14). Other patents were issued to Getz (15, 16) and Smith (17) covering processes for the preparation of aerated food products using nitrous oxide and other gases, and in 1949 Diamond was granted a patent (18) for an aerosol vegetable topping.

During the next decade no advances were made in the food aerosol area, although the pressurized topping was taking hold in the market place. In the mid-1950s an aerosol drink mix was introduced, and this was followed by a barbecue sauce, chocolate and table syrups, and a vitamin syrup. As the decade came to an end, a meat tenderizer, salad dressing, and several other products were added to the list. Unfortunately, there were disappointments with several of these products, as a result of which development work dwindled almost to a stop. Although other products were added during the 1960s and the sales of aerosol toppings grew to about 70 million units annually, aerosol food sales were far behind the figures for nonfood aerosols.

When aerosol milk flavorings were introduced in the early 1960s, some felt that these would provide the impetus the industry needed to move sales ahead. Together with the new pressurized cheese spreads and cake decorator products, these developments pushed total food aerosol sales to about 120 million units in 1965. Although there was reason for optimism, no major breakthrough occurred to push the food aerosol to the top. Even the most optimistic believed that food aerosols would move slowly, but they were equally certain that the marriage of the aerosol and food products would eventually be a success.

Containers

There is actually very little, if any, difference between the containers used for food and those used for nonfood products. A wide range of containers has been made available for the packaging of foods; these are discussed at some length in Chapters 7, 8, 9, and 10.

It should be mentioned at the very outset, however, that when foods are involved the container and its components, including lining materials, must be approved by the Food and Drug Administration. This, of course, places some limitations on the materials that can be utilized, particularly for lining the inner wall of the container. In addition, the packager should consider the problems of corrosion, component toxicity, and taste contamination.

Insofar as corrosion is concerned, it is important to evaluate both the product and the various types of containers. Some food products are highly reactive to tin-plate, whereas others are relatively nonreactive. For products that are relatively noncorrosive but are affected by contact with metal, the inner wall of the container should be treated with at least one layer of coating material. Highly corrosive products require not only a specially prepared, corrosion-resistant container, but also one that has been carefully lined to avoid product–metal contact.

Of major importance to the packaging of foods in aerosol form was the development of the "free piston" container, which was first used for a food product in 1962. This container prevents the mixing of the propellant with the product to be dispensed and discharges almost all of the contents of the can. The aluminum unit is rustproof as well as leakproof and has been made with phenolic and epoxy linings which further protect the product.

In addition to the piston container, the "bag-in-can" system has been considered useful for foods. In this system, the piston is replaced by a bag that acts as the medium to separate the food from the propellant.

Valves

According to Graham (19), "the valve most widely used for food aerosols today is basically the same proven design as originated for whipped cream twenty years ago." And Sciarra (20) has commented that "the present day aerosol food valve is nothing more than an adaptation of non-food aerosol valves."

There are actually several types of valves available for food aerosols; these are discussed in Chapter 11. It suffices to say here that few problems are involved in valve selection. Available valves are well suited to the products now being pressure packaged.

Propellants

Several propellants are available today for use in foods, but only the compressed gases are employed in any quantity. Among the available propellants are the following:

1. Nitrous oxide.
2. Carbon dioxide.
3. Nitrogen.
4. Octafluorocyclobutane.
5. Chloropentafluoroethane.

The fourth and fifth are the only fluorinated hydrocarbons that have been approved by the FDA. Of course, any of these propellants can be used in combination with another of the group, and this is often done.

Webster (21) listed five characteristics required of a propellant for food aerosol use.

1. Inertness.
2. No lending of adverse flavor to product.
3. Nontoxicity.
4. Nonflammability.
5. Low cost.

Both the compressed gases and the approved fluorinated hydrocarbons meet these requirements.

At present, whipped cream and other toppings utilize a mixture of carbon dioxide and nitrous oxide as the propellant system. This mixture has also been combined with a mixture of both fluorinated hydrocarbons for use in similar food aerosols. It should be kept in mind that the type of dispensing desired must be considered when the propellant system is selected. The more soluble propellants, nitrous oxide and carbon dioxide, tend to form a product that foams. Nitrogen, on the other hand, delivers the product in a nonaerated form.

There are advantages and disadvantages to both the compressed gases and the fluorinated hydrocarbons. On the negative side, the latter are far more expensive, and the former may have some effect on the taste of food. This subject will not be discussed further here, since propellants for all aerosol products are fully treated in Chapter 6.

Types

Many food aerosols have been offered in the market place, but very few have survived. Those remaining include the very-strong-positioned whipped cream and other cake toppings, cheese spreads, and milk and other flavorings. During the late 1960s and early 1970s there was a surge of interest in food aerosols which may have indicated a change in attitude by both the manufacturer and the consumer.

With the exception of the several products already mentioned, many food aerosols are just possibilities in the minds of development groups. In a paper delivered before the Chemical Specialties Manufacturers Association in 1955, Graham (22) presented a list of potential food aerosols (Table 17.1). To these could be added barbecue and steak sauces, lemon juice, meat tenderizers, popcorn and butter oil, mustard, butter sprays, drink concentrates, and others.

Of all the products packaged in aerosols, the most important thus far has been whipped cream. Graham (1, 23) reviewed the development

TABLE 17.1.   POTENTIAL FOOD AEROSOLS (22)

Cake frosting and icing	Mayonnaise
Catsup	Marshmallow topping
Cheese spreads	Meringue
Chocolate syrup	Pancake batter
Cooking oil	Peanut butter
Flavoring syrups (fruit, maple)	Salad dressing
Food pastes	Shortening
French dressing	Sour cream
Ice cream mixes	Soy sauce
Liquid butter	Toppings

TABLE 17.2.   TYPICAL AEROSOL WHIPPED CREAM FORMULA (25). (A 12-OZ CONTAINER IS FILLED WITH 7-OZ WHIPPED CREAM MIX AND 4-OZ COMPRESSED GAS).

Whipped Cream Mix	Compressed Gas
Cream (30% butter fat) Stabilizer (gelatin or vegetable gum) Vanilla Sugar (5 to 10%) Added milk solids 2 to 3% over that found in 30% cream	Nitrous oxide (100%) or Carbon dioxide (15%) and nitrous oxide (85%)

In synthetic topping, cream and protein are substituted for by hydrogenated vegetable oils and/or soya protein, other vegetable proteins, and sodium caseinate.

and manufacture of whipped cream and quoted Sommer's (24) description of the product as a dispersion of air or gas cells and clumped fat masses as contiguous phases in cream serum to form a more or less rigid foam. He went on to describe the difficulties encountered by the housewife in preparing whipped cream and mentioned the contributions of the dairy industry. A typical whipped cream formula was presented by Graham in a later work (25) and is shown in Table 17.2. Readers are referred to the works cited for more detailed information on pressurized whipped cream.

After the success of the aerosol topping, other food products were marketed, some successfully, others not. The pressurized cake decorator, introduced by Pillsbury and using the "free piston" container, proved to be successful. What contributed to the success of this unit, according to Davis (26), was the use of a multiple-mounting device that gave the housewife four different dispensing nozzles.

Milk flavorings, first marketed in 1964, reached 40 million unit sales in about a year, and sales continued to move upward, although at a somewhat slower pace. Also, one brand of cheese spreads, also introduced in 1964, continues to be marketed successfully.

Failures, of course, have outnumbered successes by a large percentage; they include such products as pressurized toothpaste, barbecue sauce, an instant coffee concentrate, concentrated beverage additives, and spray vermouth. They failed because of technical difficulties, improper formulation, poor testing procedures, and equally inadequate market studies. Anderson (27) commented that lack of a strong, well-financed, sustained merchandising program has been the downfall of most food aerosols. He also indicated several criteria that are met by the food aerosols still being successfully marketed.

- They fulfill a need; they aren't just packaging gimmicks.
- They offer convenience.
- They increase the effectiveness of the food, often by dispensing it in a new form acceptable to the consumer.
- They can be merchandised through traditional food channels.
- Their packaging costs are as low as possible in relation to total product costs.
- Their ultimate cost to the consumer must not be unreasonably higher than that of the nonaerosol food product against which they must compete.

The Coming Decades

Many of the food aerosols that failed did not utilize the newer developments in aerosol technology, and there are several products that may prove to be successful in the decades to come. Davis (26) has reported five categories for which there may be potential success.

1. Barbecue sprays, meat tenderizers, and seasonings.
2. Ice cream and pudding toppings, including heavy viscous chocolate syrups.
3. Specially flavored salad dressings, possibly spiced with wines, chestnuts, and other exotic flavors.
4. Saccharine and other sweeteners.
5. Flavored peanut butter or flavored honey.

The future of the food aerosol will depend on the development techniques of the large food companies. They must learn from the failures of the past and avoid straying into the same error-strewn path. Only with proper planning can the food aerosol come into its own.

PHARMACEUTICAL AEROSOLS

The application of the principles of aerosol technology to the dispensing and administration of therapeutic agents is of relatively recent occurrence. However, the administration of medicinals in the form of fine liquids or solid mists is not new. The efficacy of the vapor from burning leaves and herbs, and of the smoke from burning stramonium leaves, of the administration of penicillin and other antibiotics by way of the respiratory tract, and of the inhalation of epinephrine, isoproterenol, and similar antiasthmatic agents, as well as ethyl and octyl nitrites, has been definitely established (28-33).

Inhalation or aerosol therapy has been defined as "the administration via the respiratory system of medicinal agents which are in the form of fine solid particles or a liquid mist." Although this form of therapy is useful for its local effects, it also produces systemic effects, that is, effects that occur because of absorption of the drug into the general circulation. The nature of the respiratory system is such that it permits the prompt absorption of many medicinal agents, so that inhalation therapy almost approaches intravenous therapy in the rapidity of its onset of action.

With the introduction of liquefied gas aerosols, a new package was developed that made it possible for many medicinal agents to be administered by inhalation. Several advantages over the conventional-type atomizers and nebulizers were immediately noted. The medicinal agent could be mixed with the propellant, thereby forming a self-powered unit capable of producing a spray or mist of the desired particle size. Since the package was sealed, there was no danger of contamination or decomposition of the medication. If the medication was sterilized and packaged under aseptic conditions, sterility would be maintained throughout the life of the product. Through use of a metered valve, the dosage could be controlled so that there was no danger of either under- or overmedication. Finally, the package was of such design as to make it easy and convenient for the patient to use (34-36).

The packaging of many topical pharmaceutical products has been limited to a great extent. Ointments and ointment-like products have been packaged in glass, porcelain, polyethylene, or plastic wide-mouth jars or in collapsible metal tubes. Liquid preparations have been packaged in conventional screw-cap glass bottles or, more recently, in plastic or polyethylene containers. Although this type of packaging has proved to be generally satisfactory, it is not without disadvantage. Since a large area of the ointment is exposed to the deleterious effects of air and moisture, ointments containing easily oxidized medicinals or medicinal agents affected by atmospheric moisture are subject to decomposition. In this regard, the use of collapsible tubes is advantageous, but these containers do not take into account, to any large extent, such factors as ease of application, cleanliness, danger of contamination from an infected area or with foreign materials, maintenance of sterility, oxidation, waste due to the method of application, and stability of the therapeutically active ingredients.

With the introduction in 1944 of the "aerosol" insecticide, thought was given to the utilization of aerosol principles for the dispensing of pharmaceutical products. Since that time, many pharmaceutical products have been developed in aerosol form. They include burn preparations, local anesthetics, steroids, first-aid products, foot sprays, antiseptic and germicidal sprays, and vaporizer sprays. All of these products have been readily accepted by both physician and patient.

Topical pharmaceutical aerosols may be defined specifically as aerosol products that contain therapeutically active ingredients dissolved, suspended, or emulsified in a propellant or a mixture of propellant and solvent, and intended for topical administration or for administration into one of the body cavities such as the ear, rectum, or vagina. Vaporizer sprays are also included in this definition.

Pharmaceutical aerosols, used topically, have the following advantages as compared to conventionally packaged products:

1. The method of application is convenient, fast, and efficient.

2. Since the medication is applied directly from the container to the affected area of the skin, there is no waste or messiness as with the use of an applicator or cotton swab.

3. There is no danger of contamination of the product since the aerosol container is completely sealed and only the amount used can be dispensed at any one time. At no time can the unused preparation come into contact with foreign materials other than the container and valve components.

4. If the product is packaged under sterile conditions, sterility is maintained throughout the life of the product; this is especially important for ophthalmic preparations.

5. The stability of the dosage form is ensured since the aerosol container is a completely sealed unit; this also prevents "drying out" of the product, as well as keeping water-sensitive drugs (e.g., antibiotics) from absorbing moisture from the atmosphere.

6. Liquefied gas aerosol products dry rapidly because of the vaporization of the propellant, causing a cooling sensation that may be desired in certain cases. Where needed, solvents can be added to decrease this cooling effect.

7. The irritation produced by the mechanical application of an ointment over an abraded area of the skin is reduced and sometimes eliminated by a "spray-on" aerosol.

8. The medication can be applied in an extremely thin layer directly over the affected area; this results in faster absorption and a more efficient utilization of a given amount of medication.

Topical pharmaceutical aerosols have been developed as fine sprays, foams, powders, and semisolid preparations. Since they are used topically, local irritation is an important consideration. The propellants used for pharmaceutical aerosols are relatively free from toxicity and unlikely to cause local irritation.

Many different pharmaceutical ingredients have been formulated into aerosol products. Some types of products, their ingredients, and their applications are as follows.

Local Anesthetics

A variety of different local anesthetic agents have been used for this type of aerosol; benzocaine, ethyl chloride, cyclomethylcaine, and tetracaine are just a few that have proved useful. These products are convenient for use by the physician in his office as local anesthetics for minor surgery.

Antiseptic Aerosols

This type of aerosol product has been readily accepted by both physician and patient. The application of antiseptic agents by a spray has been extremely effective. Many different agents can be used in this manner, ranging from the simplest to some of the more complex antibiotics and quaternary ammonium compounds. Figure 17.1 illustrates an antibiotic containing aerosol.

First-Aid Aerosols

These aerosols combine the antiseptic values of various agents with the local anesthetic activity of compounds such as benzocaine. They are effective in preventing infection, as well as in alleviating the pain associated with minor cuts and bruises.

Burn Aerosols

A burn aerosol consists chiefly of a local anesthetic in combination with antiseptic agents such as sulfur, benzyl alcohol, or tannic acid. In

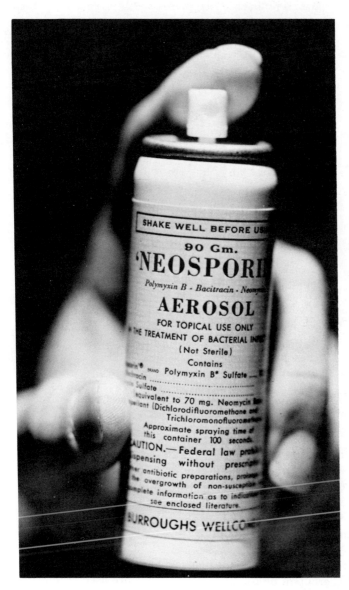

Figure 17.1. Antibiotic aerosol. (Courtesy of Burroughs Wellcome Co.)

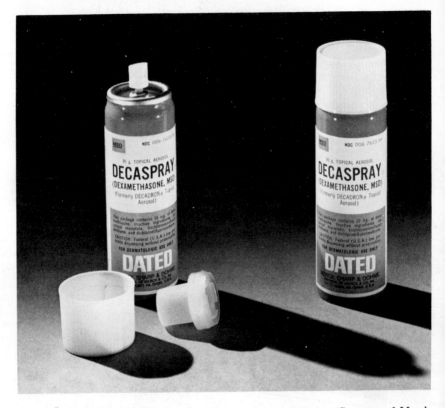

Figure 17.2. Dermatological aerosol containing dexamethasone. (Courtesy of Merck Sharp and Dohme.)

addition, cooling and soothing agents such as menthol and chlorobutanol are also present. Various antibiotic spray powders have been shown to be effective in the treatment of burns.

Dermatological Preparations

The principles of aerosol technology have found greatest application with this group of products, many of which are used as antipruritics or antiinflammatory agents or in the treatment of acne, poison ivy, related conditions, and so forth. The active ingredients used consist of steroids such as cortisone and prednisolone, antihistaminics, calamine, zinc oxide, sulfur, and many others. These products are extremely effective, since they allow for maximum penetration of the active ingredients with a minimum of waste. In most cases, a thin film of medication is applied over the affected area. With foam preparations, the medication is applied only to the affected area without waste. Figures 17.2 ⌐nd 17.3 are examples of several aerosols in this category.

Foot Preparations

The aerosol method is ideal for the application of foot preparations. In the treatment of athlete's foot, the shoes, stockings, and affected areas can easily be sprayed with medication. The usual ingredients for athlete's foot remedies are undecylenic acid and one of its salts (such as zinc undecylenate), although many other ingredients are being used. Other foot preparations utilize a talcum powder with or without an antiseptic agent.

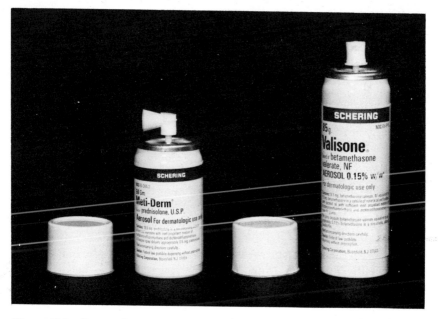

Figure 17.3. Dermatological aerosols containing prednisolone and betamethasone, respectively. (Courtesy of Schering Corp.)

Spray-on Protective Films

This type of product is a direct result of the development of the aerosol industry. The use of collodion in conventional packages to protect wounds and small cuts has been of limited value. In aerosols, a film-forming agent can be used in conjunction with antiseptic, local anesthetic, and other agents. Upon application, the solvent and propellants vaporize, leaving behind a thin transparent film, which, in addition to providing protection, permits visual inspection of the affected area. Since this is a "breathing film," the possibility of infection by anaerobic bacteria is minimized.

Vaporizers and Inhalants

When the vapor from this type of product is inhaled, instant relief is obtained from the symptoms of nasal and bronchial congestion. The vapors are generally produced through the use of therapeutically active ingredients and a vaporizer. Recently, aerosol products have become available for this purpose. In use, an enclosed area is sprayed with the medicinal ingredients. The power of the liquefied gas disperses the medicinal agent throughout the area, just as a vaporizer does. In addition, these products can be used by spraying a small amount on a handkerchief and inhaling.*

FORMULATION OF PHARMACEUTICAL AEROSOLS

Because of various FDA regulations and procedures, the development and production of any pharmaceutical, whether tablet, ointment, or aerosol, is a complex process requiring a team of highly trained personnel (37). In addition, for pharmaceutical aerosols the unique problems of the aerosol industry are combined with the exacting requirements and procedures of pharmaceutical manufacturing.

Formulation changes are nearly always required when going from conventional packaging to aerosol packaging. Before reformulating a pharmaceutical product from its original package, it is necessary to determine the economic feasibility of doing so by noting possible product improvement as to ease or speed of application, prevention of con-

---

* The reader is advised that in July 1973, the FDA ordered the recall of several vaporizer sprays containing the solvent 1,1,1, trichloroethane. Several deaths had been attributed to the use of these products.

tamination, and uniform application to the skin, and to consider potential improved marketability and the achievement of greater sales.

In the successful development of a pharmaceutical aerosol it is usually preferable to formulate for aerosol application rather than to adapt to it (38). There are many factors to consider, including the selection of a suitable aerosol system, a suitable vehicle, and the proper container, valve, and propellant system, as well as the proper pressure and the resultant spray characteristics. The usual developmental considerations, such as odor, color, physical stability, chemical stability, and probable customer acceptance, including marketability, price, and advertising, also accompany the development of pharmaceutical aerosols and usually to a greater degree than prevails for conventional packaged products.

In the development of a pharmaceutical aerosol there is a choice of five systems:

> Three-phase system.
> Two-phase system.
> Emulsion system.
> Dispersion system.
> Semisolid system.

These systems were described in detail in Chapter 4.

Several methods can be used to formulate pharmaceutical aerosols. The active ingredients can be dissolved in a suitable solvent, such as ethyl alcohol, propylene glycol, or dipropylene glycol, and mixed with a propellant consisting of a mixture of Propellants 12 and 11 or 12 and 114. This will produce a spray type of aerosol useful for topical anesthetics and antiseptics, foot sprays, spray-on bandages, vaporizers, and similar products. A modification of this system can be used for medicinals which are insoluble in most liquid solvents. In this case, the active ingredients are suspended or dispersed in the propellant and other liquids. Antibiotics, foot powders, and steroids have been formulated in this manner.

In cases where ingredients are predominately water-soluble, where the active ingredient may be irritating or harmful if inhaled, or where the application of the medicinal is restricted to an extremely limited area, a foam-type system can be developed. This is done by dissolving the ingredients in water and then emulsifying with Propellant 12/114, using a suitable surfactant. This type of system has been especially useful for dermatological aerosols. The nature of the foam can be varied to produce a quick-breaking or a stable product. In addition, nonaqueous foams can also be developed (40–44).

FORMULATION OF MEDICINAL AEROSOLS

There are two different ways whereby medicinals can be dispensed in aerosol form for inhalation, namely, the two-phase solution and the two-phase suspension or dispersion system.

In the two-phase solution system, the active ingredients are dissolved in a mixture of propellant and solvents that forms a clear solution. When the valve of the container is depressed, a mixture of active ingredients, propellant, and solvent is liberated. Because of the rapid vaporization of the propellant, the medicinal agent is finely dispersed and can be inhaled. The most popular solvent is ethyl alcohol, which is used to the extent of about 30%. The medicinal is dissolved directly in the alcohol; or, when the salt of the medicinal agent is used, it is dissolved in a small quantity of water. The alcohol/water system is miscible with the propellant.

Since most medicinal agents are not directly soluble in the propellant, the two-phase suspension system has been developed. In this sytem, the medicinal agent is finely subdivided to particles less than 10 μ in diameter. These particles are then suspended in the propellant by means of a dispersing agent. In use, the finely dispersed medicinal agent is emitted through the valve as a fine mist.

To ensure the safe use of these aerosols, specially designed valves known as metered valves have been developed. These valves are so designed that they dispense a constant amount of medication each time they are depressed. Thus they prevent overmedication as well as ensure accurate dosage control. Medicinals used in aerosol form are fitted with this type of valve.

Selection of the proper amount of propellant and the proper propellant system depends on the container to be used and the formulation in question, as well as the foam or spray characteristics desired. Either liquefied or compressed gases may be employed as propellants (the chief criterion determining which are used is their degree of toxicity and reactivity). The fluorinated hydrocarbons are widely used in both pharmaceutical and medicinal aerosol products; they are noted for their low toxicity, low reactivity, and general inertness. However, Propellants 11, 12, and 114 have been used to the greatest extent. Of these compounds, Propellant 11 is subject to hydrolysis and hence cannot be used in aqueous systems. The acids formed would be irritating to the skin and mucous membranes, as well as resulting in many incompatibilities. For the most part, Propellant 11 has been replaced by Propellant 114 for aqueous systems.

Of the nonliquefied compressed gas propellants, nitrogen has found the greatest use. Its nonsolubility and inertness have made it the propellant of choice for dispensing certain pharmaceutical products in their original forms, such as vitamins, ointments, and creams.

The pressure of the system along with the nature of the active ingredient plays an important role in the selection of a container. Although glass is preferred from an esthetic viewpoint and because of its resistance to chemical attack, its use is limited to products that have relatively low vapor pressures and where the total propellant content of the system is not more than 50 to 55%. Many suitable pharmaceutical aerosols have been formulated in this manner.

Aluminum and tin-plated steel containers have been used primarily because of their ability to resist breakage if dropped accidentally and for their relatively low cost. Incompatibility of the product with the container is a serious problem, however, and must be considered. In many instances the incompatibility can be overcome through the use of a suitable liner.

The aerosol valve, together with the propellant, is responsible for the proper delivery of the product. One must ascertain whether control of dosage is necessary and, if so, arrange for the use of a metered valve. Many specially designed valves are available for use with different types of products; for example, a valve designed to dispense powders is quite different from one intended for foams. The sizes of the various orifices and their relationships to one another are important considerations. To achieve different spraying characteristics and finer dispersion of active ingredients, a vapor tap valve may be indicated. The actuator must be carefully selected in order to achieve the desired particle size, shape of spray, and other characteristics.

The stability of the entire package must also be considered. For new drugs developed as aerosols or for older drugs being reformulated as aerosols for the first time, an Investigational New Drug (IND) application and/or a New Drug Application (NDA) may be required. The stability of the product must be indicated in the application. Aside from this requirement, it makes good sense to determine the stability of the product at various temperatures so that the limitations of storage may be known.

Several medicinal agents lend themselves to administration by inhalation. In fact, any drug given by intravenous injection can, in most cases, be reformulated into a suitable aerosol provided it is capable of being deposited in the respiratory tract and is nonirritating. Such drugs as epinephrine, isoproterenol, octyl nitrite, phenylephrine, erg-

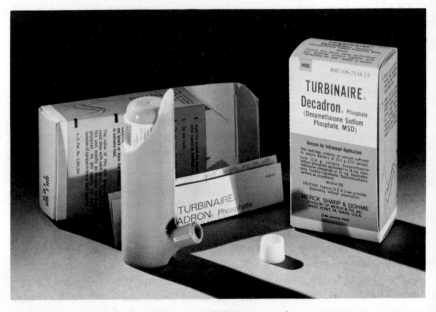

Figure 17.4.   Aerosol for nasal use. (Courtesy of Merck Sharp and Dohme.)

otamine, and dexamethasone have been successfully formulated into liquefied gas aerosols and are used by inhalation. These products are highly effective and have been readily accepted by both physician and patient. Figure 17.4 illustrates an aerosol containing a steroid intended for nasal administration.

The effectiveness of many of these medicinal agents in aerosol form has been indicated in many publications. Freedman (45), Grater and Shuey (46), Harris (47), Sedtzer (48), and Zohman and Williams (49) have all reported significant therapeutic improvements when patients suffering from chronic asthma were treated with liquefied gas aerosols containing either epinephrine or isoproterenol. The results indicate that this method of treatment is superior to ordinary aerosol therapies. It has been found that inhalation of isoproterenol aerosol immediately relieves the objective and subjective symptoms of acute asthmatic attacks and has a beneficial effect also in cases of chronic bronchial asthma. A typical aerosol package together with an oral applicator is shown in Figure 17.5.

The treatment and control of allergic diseases of the respiratory tract have depended on the use of corticosteroids (50–51). Fisch and Grater (52) evaluated an aerosol containing dexamethasone and found this preparation to be highly effective in the treatment of bronchial asthma. The symptoms of wheezing were reduced, and a corresponding increase in vital capacity was noted. Of great importance, it was found that the inhalational dexamethasone required a smaller total steroid dose than the orally administered drug. However, it should also be indicated that the FDA recently indicated that there is a lack of substantial evidence that the combination of dexamethasone sodium phosphate and isoproterenol sulfate is effective for this purpose (53).

The development and growth of pharmaceutical aerosols has been and always will be relatively slow as compared to the growth of such other aerosols as shave creams, hair sprays, colognes and perfumes, insecticides, and room deodorants. Also, the use of pharmaceuticals is limited to the segment of the population that suffers from some ailment.

In reviewing the development and growth of pharmaceutical aerosols, several facts became apparent (54).

1. There are relatively few different and distinct aerosol pharmaceuticals as compared to other aerosol product types and to the total number of nonaerosol pharmaceuticals.

2. Although several companies produce aerosol pharmaceuticals, only a few of these can be considered to be major pharmaceutical

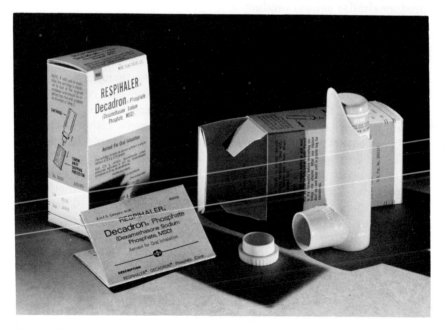

Figure 17.5. Aerosol for oral inhalation. (Courtesy of Merck Sharp and Dohme.)

companies. Of the major pharmaceutical companies offering aerosol products, the total number of such products for each company, in many cases, is only one or, at the most, two to three. Therefore the time spent on these products is limited.

3. Of the total number of pharmaceutical aerosols produced, only a few have been manufactured by pharmaceutical companies. Many of them are produced by aerosol contract fillers. This limits the facilities available for the manufacture of these products.

4. The pharmaceutical aerosols currently available are those that can be developed and formulated without much difficulty. This is not meant to imply that no problems were encountered during the development of these products. For example, epinephrine has been used for a great number of years via inhalation in the symptomatic treatment of asthma. The utilization of this drug in a pressurized package involved physicochemical techniques rather than pharmacological skills. To take a drug that has never been used via inhalation therapy and to develop it into an aerosol would present problems of much greater magnitude.

5. The number of both active and inactive ingredients used to formulate pharmaceutical aerosols has been limited. In many instances, basically the same ingredients have been used by several companies to produce similar aerosol products.

Another factor responsible for the decreased growth rate, as well as the small number of new pharmaceutical aerosol products, are the FDA rules regulating the introduction and subsequent sale of drugs. One provision of the FDA regulations covers the introduction of all-new drugs. Most aerosol products would fall into this category. The provisions of NDA result in a long, time-consuming, and costly process. The interpretation of the term "new drug" has been a rather broad one and hence covers a large number of products. For example, if one were to take an existing, well-known, fully tested drug that is given orally and reformulate it as an aerosol for inhalation therapy, it could be interpreted as a "new drug" by virtue of the change in the method of administration. A great deal of clinical testing would then be required to determine its effectiveness and toxicity, if any.

REFERENCES

1. W. Earl Graham, The History and Background of Pressurized Foods, *Food Technol.*, **12** (July 1958), 317–319.
2. Frank M. Ashley, U.S. Patent 1,548,430 (Aug. 4, 1925).
3. Peter H. vander Weyde and John Matthews, Jr., U.S. Patent 88,348 (Mar. 18, 1869).

4. Jacob W. Decastro, U.S. Patent 256,129 (Apr. 11, 1882).

5. L. Roush, U.S. Patent 1,852,267 (Apr. 5, 1932).

6. H. Burmeister, U.S. Patent 1,889,236 (Nov. 29, 1932).

7. C. A. Getz and G. F. Smith, A New Aeration Process for the Preparation of Whipped Cream, *Trans. Illinois State Acad. Sci.,* **27** (1934), 71.

8. C. A. Getz, G. F. Smith, and P. H. Tracy, Instant Whipping of Cream by Aeration, *J. Dairy Sci.,* **19** (1936), 490.

9. C. A. Getz, G. F. Smith, P. H. Tracy, and M. J. Prucha, Instant Whipping of Cream by Aeration, *Food Res.,* **2** (1937), 409.

10. G. F. Smith and C. A. Getz, Cut of Whipping Cream Acquires Scientific Finesse, *Milk Dealer,* **36** (1935).

11. M. C. Reinecke, U.S. Patent 2,120,297 (June 14, 1938).

12. Isaac M. Diller, U.S. Patent 2,155,260 (Apr. 18, 1939).

13. Justus C. Goosmann, U.S. Patent 2,250,300 (July 22, 1941).

14. L. Marmorek, U.S. Patent 2,162,096 (June 13, 1939).

15. C. A. Getz, U.S. Patent 2,294,172 (Aug. 25, 1942).

16. C. A. Getz, U.S. Patent 2,435,682 (Feb. 12, 1948).

17. A. H. Smith, U.S. Patent 2,281,604 (May 5, 1942).

18. H. W. Diamond, U.S. Patent 2,487,698 (Nov. 8, 1949).

19. Earl Graham, Food Aerosols—Past, Present and Future, Address before the Packaging Forum, New York, Oct. 4, 1966.

20. John J. Sciarra, Formulation Techniques Applicable to Food Aerosols, *Aerosol Age,* **13** (June 1968), 57, 117.

21. R. C. Webster, Compressed Gas Propellants for Food Aerosol Products, Continental Can Symposium, 1958.

22. W. E. Graham, Food Aerosols, *Chem. Spec. Mfr. Assoc. Proc. 42nd Ann. Meet.,* December 1955.

23. W. E. Graham, Packaging Whipped Cream in Pressurized Containers, *Food Ind.,* **4** (June 1950), 225.

24. H. H. Sommer, *Market Milk and Related Products,* 2nd ed., Olsen Publishing Co., Milwaukee, Wis., 1946, p. 489.

25. W. E. Graham, Food Aerosols, Chapter 13 in H. R. Shephard (Ed.), *Aerosols: Science and Technology,* Interscience Publishers, New York, 1961, p. 417.

26. Donald A. Davis, Aerosol Food Products, *Aerosol Age,* **9** (May 1964), 29.

27. Earl V. Anderson, Food Aerosols, *Chem. Eng. News,* May 2, 1966.

28. William F. Miller, Aerosol Therapy in Acute and Chronic Respiratory Disease, *Arch. Intern. Med.,* **131** (1973), 148.

29. L. Dautrebande, Physiological and Pharmacological Characteristics of Liquid Aerosols, *Physiol. Rev.,* **32** (1952), 214.

30. L. Dautrebande, *Microaerosols,* Academic Press, New York, 1962, pp. 260–285.

31. John J. Sciarra, Aerosol Therapy, *Aerosol Age,* **1** (September 1956), 14.

32. B. Idson, Inhalation Therapy, *Drug Cosmet. Ind.,* **107** (July 1970), 46.

33. S. Blaug, Oral Inhalation Therapy, *Aerosol Age,* **13** (December 1968), 48.

34. R. L. Harris and H. D. Riley, Jr., Reactions to Aerosol Medication in Infants and Children, *J. Amer. Med. Assoc.,* **201** (1968), 127.

35. J. L. Kanig, Pharmaceutical Aerosols, *J. Pharm. Sci.,* **52** (1963), 513.

36. John J. Sciarra and Vincent de Paul Lynch, Aerosol Inhalation Therapy, *Drug Cosmet. Ind.*, **86** (June 1960), 752.

37. I. Porush, Pharmaceutical Products, in H. Shepherd (Ed.), *Aerosols: Sceince and Technology*, Interscience Publishers, New York, 1961, p. 388.

38. M. J. Root, Pharmaceutical Aerosols from Idea to Product, *Drug Cosmet. Ind.*, **79** (October 1956), 473.

39. John J. Sciarra, Pharmaceutical and Medicinal Aerosols, in A. Herzka (Ed.), *International Encyclopedia of Pressurized Packaging (Aerosols)*, Pergamon Press, London, 1966, pp. 571–619.

40. John J. Sciarra, Aerosols, in J. B. Sprowls (Ed.), *Prescription Pharmacy*, 2nd ed., J. B. Lippincott Company, Philadelphia, 1970, pp. 280–328.

41. I. Porush, Pharmaceutical Products, in H. Shepherd (Ed.), *Aerosols: Science and Technology*, Interscience Publishers, New York, 1961, pp. 387–408.

42. Harvey Mintzer, Aerosols, in E. W. Martin (Ed.), *Dispensing of Medication*, 7th ed., Mack Publishing Company, Easton, Pa., 1971, pp. 928–967.

43. John J. Sciarra, Pharmaceutical Aerosols, in L. Lachman, H. A. Lieberman, and J. L. Kanig (Eds.), *The Theory and Practice of Industrial Pharmacy*, Lea and Febiger, Philadelphia, 1970, pp. 605–638.

44. C. Bloom, Pharmaceutical Aerosols, *Aerosol Age,* **8** (December 1963), 52; **9** (January 1964), 36.

45. T. Freedman, Medihaler Therapy for Bronchial Asthma, *Postgrad. Med.,* **20** (1956), 667.

46. W. C. Grater and C. B. Shuey, Medihaler in Asthma, *South. Med. J.,* **51** (1958), 1600.

47. M. C. Harris, Medihaler in Asthma, *Postgrad. Med.,* **23** (1958), 170.

48. A. Sedtzer, A Useful Device for Treating Acute Allergic Drug Reactions, *Med. Ann. Dist. Columbia,* **27** (1958), 131.

49. L. Zohman, and H. M. Williams, Comparative Effects of Aerosol Bronchodilators on Ventilatory Function in Bronchial Asthma and Chronic Pulmonary Emphysema, *J. Allergy,* **29** (1958), 72.

50. H. M. Carryer et al., Effects of Cortisone on Bronchial Asthma and Hay Fever, *Proc. Staff Meet. Mayo Clinic,* **25** (1950), 482.

51. M. L. Gelfand, Administration of Cortisone by the Aerosol Method in the Treatment of Bronchial Asthma, *New Engl. J. Med.,* **245** (1951), 293.

52. B. R. Fisch and W. C. Grater, Dexamethasone Aerosol in Respiratory Tract Disease, *J. New Drugs,* **2** (1962), 298.

53. *Federal Register,* Dec. 9, 12, 13, 1972.

54. John J. Sciarra, Pharmaceutical Aerosols, *Drug Cosmet. Ind.,* **105** (August 1969), 46.

CHAPTER EIGHTEEN

HARRY H. INCHO
AND LLOYD A. MCDONALD

PESTICIDE AEROSOLS

The beginning of the present-day aerosol industry and, particularly, the use of aerosols for dispensing insecticides and other pesticides may be traced back to the early work of L. D. Goodhue and W. N. Sullivan before and during World War II (1). Although Erik Rotheim had earlier developed and patented the process of dispensing substances by spraying solutions in flammable liquefied gases under their own pressure, his patents went unnoticed, as did the earlier contributions of Carl Iddings and others (2). It remained for Goodhue and Sullivan, working as a team on the possible use of smokes and "aerosols" for the control of various insects, to first combine pyrethrum, sesame oil, and Propellant 12 in a practical, self-contained pressurized unit. This unit demonstrated its effectiveness against houseflies, mosquitoes, and cockroaches in tests by J. H. Fales at the Beltsville, Maryland, laboratories of the U.S. Department of Agriculture, and a patent, assigned to the Secretary of Agriculture, was issued in 1943 (3). The claims specified less than 10% nonvolatile ingredients to ensure the production of an active aerosol with particles less than 50 μ in diameter.

The development of practical aerosol "bombs" by the Westinghouse company resulted in the production of over 40 million units for the armed forces before the end of World War II. Nearly all were based on the original pyrethrum/sesame oil formulation, but later units contained a very effective pyrethrum-DDT combination.

According to Goodhue (1), there were a number of reasons for the almost immediate success of the Goodhue-Sullivan aerosol, where other attempts had failed. Among these ingredients for success were:

Timing. The aerosol "bomb" became available when there was an urgent military need for such a device.

Design. It was a nontoxic, nonflammable, compact unit that met military specifications for use in combat areas.

Effectiveness. It was very effective for its primary use in the control of mosquitoes.

Recognition. The distribution of over 40 million free samples to the armed forces provided free advertising and promoted later use by civilians.

Development. The aerosol was developed in a country whose people had the means to purchase a new gadget for dispensing insecticides and an interest in doing so.

The need for a lighter, more economical insecticide aerosol for civilian use resulted in the development of commercial low-pressure units in 1946. A typical formulation at that time consisted of the following (4):

Ingredients	*% by Weight*
Pyrethrins	0.25
DDT	2.00
Piperonyl butoxide	1.00
Aromatic petroleum derivatives	5.00
Deodorized kerosine	6.75
Propellant 12/11 (50:50)	85.00
	100.00

Further developments and refinements in formulations, valves, and containers followed rapidly. Odor problems were reduced by the introduction of alternative volatile solvents in the mid-1950s. Allethrin was introduced at about the same time and, in combination with the thiocyanates, served as a partial replacement for pyrethrins in aerosols during the Korean War.

Water-based synergized pyrethrum formulations with hydrocarbon propellants such as isobutane, *n*-butane, and propane were later developed to replace the oil-based aerosol formulations. This brought a marked reduction in the cost of pressurized household sprays, and a major portion of the production of pesticide aerosols is now based on a water/emulsion toxicant system with hydrocarbon propellants (6). Resistance to DDT became apparent during this period, and most commercial aerosol insecticide formulations were altered to include the more effective combinations of pyrethrins and synergists. The recent ban on the use of DDT in household applications also served to place even greater emphasis on aerosol pesticides low in mammalian toxicity and potentially less harmful to the environment. A typical water-based, synergized pyrethrum formulation now consists of the following:

Active Ingredients	% by Weight
Pyrethrins	0.25
Piperonyl butoxide	1.00
Solvents	7.45
Inert Ingredients	
Emulsifier	0.20
Inhibitors	0.20
Deionized water	58.90
Isobutane	32.00
	100.00

Refinements in the pesticide aerosol packages took place as the newer formulations were being developed. These resulted in products such as the "house and garden" or "multipurpose" units designed for the control of insects, spiders, and mites in the home, as well as on plants in the garden or around the home. Similarly, an outdoor spray formulation containing insecticides and repellents was developed to provide protection from biting insects by the production of a heavy spray effective for a distance of up to 20 ft. "Ant and roach" pressurized sprays were designed to be dispensed from an aerosol-type container producing a long-lasting toxic residue by the deposition of a heavy spray.

The proliferation of pressurized pesticide units has continued unabated and now includes not only insecticides but also repellents, herbicides, miticides, fungicides, and bactericides, within many specialized formulations and packages. Some of the more common types are reviewed in the following sections.

From the first tentative steps with 40 million high-pressure units provided solely for military use, the production and marketing of pressurized insecticide products has had a phenomenal growth paralleling that of all aerosol products stemming from those early beginnings. By 1951, the development of the low-pressure package, with the improvements in formulations already noted, resulted in the sale of over 17 million units in the United States and Canada, according to the Chemical Specialties Manufacturers Association in its Annual Survey of Aerosol and Pressurized Products Filled. Each year this survey showed a steady growth in the production of insecticide aerosols until, 20 years later, the 1971 report lists about 109 million units filled and marketed (Figure 18.1). These survey figures cover the following types of pressurized products: (a) space insecticides, (b) residual insecticides (all

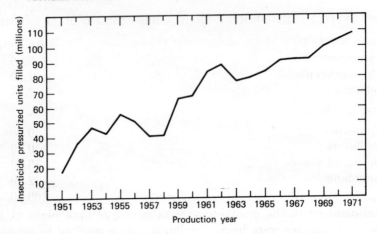

Figure 18.1. Annual production of insecticide (pesticide) aerosols, based on the CSMA Annual Survey of Aerosol and Pressurized Products Filled.

types, such as roach and ant sprays, insect repellents, and plant sprays), and (c) mothproofers (includes all fabric pests). In the later years the totals were necessarily adjusted upward on the basis of reports from valve and can manufacturers, since the actual number of insecticide aerosols marketed was not divulged by some of the largest member companies.

The extensive use of pressurized pesticides has resulted in the promulgation of many government regulations to control their formulation, labeling, shipment, and application. The following is a brief review of some of these important rules.

GOVERNMENT REGULATIONS

Aerosols and pressurized sprays that contain an economic poison as defined in:

Interpretation 3 Revision 1 of Part 362—Regulations for Enforcement of Federal Insecticide, Fungicide and Rodenticide Act, Section 362.101 Interpretation of terms included in definition of economic poison.

must be registered with the Pesticides Regulation Division of EPA [Environmental Protection Agency] for interstate distribution. The labeling of the package must conform with the requirements of the Act, and the product must bear the registration number assigned by the Division.

In addition to the federal requirements for registration, most states also require registration of pesticides that are sold within their boundaries (7).

The necessarily lengthy and costly procedures for obtaining such registrations usually are handled by the basic chemical manufacturers and involve the submission of test data to prove the pesticidal efficacy of a new chemical, a known chemical in a new type of formulation, or a known formulation with proposed label revisions. Test data (laboratory and field) are required to support claims for the control of all insects, mites, spiders, and so forth on the aerosol label and may include tests from diverse areas within the country. Toxicology data on a new chemical and on products containing the new chemical must be submitted to support its acceptability under the proposed conditions of use. Toxicity tests usually include animal studies to determine acute and chronic effects, dermal toxicity studies, inhalation studies, and long-term studies of possible carcinogenic, mutagenic, or teratogenic effects. The flammability of the product also must be determined and reported to the EPA in support of the application for registration.

When the pesticide aerosol marketer has fulfilled these efficacy and toxicology requirements for the new chemical or formulation and has determined the size of the container in which the product will be packaged, he must decide on the type of label to be used (glued paper, lithographed, silk screen, etc.). A final copy of the label is then submitted to the EPA with a "confidential statement of formula" listing all products used in preparing the finished units. Upon final acceptance of the label submitted, a registration number is issued for inclusion on the finished label. The EPA automatically reviews labels every 5 years, at which time new wording may be required. A label claim also may be challenged at any time in the light of new information on efficacy or toxicology.

Pesticide aerosol labels include a number of items required by good marketing practices and by government regulations. Some of these are as follows:

1. *Product name.* An important item for the marketer, the name is usually developed with the assistance of marketing and advertising specialists. It should convey the purpose of the product, yet be concise and appealing.

2. *"Banner" claims.* The front panels of pressurized pesticide formulations generally contain a few important features in large type to indicate the scope of control and areas of use.

3. *Ingredient statement.* All active ingredients must be listed by chemical name on the front panel of the label. Accepted chemical names have been established in various numbered interpretations of the Federal Insecticide, Fungicide and Rodenticide Act. Two options are permitted in regard to showing the quantity of each active ingredi-

ent on the label: (*a*) listing the actual percentage of each active ingredient: this is the usual procedure; (*b*) naming the active ingredients in descending order of quantity, without showing percentages. Under this option the exact percentages of active ingredients must be provided in a confidential statement to the EPA.

4. *Precautionary labeling.* All pesticide aerosol labels must have the signal word "caution," "warning," or "danger" on the front panel in 18-point type. This must be followed by "Keep out of reach of children" in 12-point type. Additional precautionary copy may appear either on the front or the back panel; if it is not on the front panel, reference must be made to additional precautions on the rear panel. In addition to toxicity and use precautions appropriate to the individual formulation, the label of every pressurized product must have a further paragraph headed by the signal word "warning" or "danger," followed by pressure and fire hazard precautions appropriate to the demonstrated degree of flammability of the product.

5. *Directions for use.* Specific locations where the product is to be applied and directions for use against each insect claimed anywhere on the label must be included.

6. *Net contents.* The aerosol net contents must be listed by weight in terms of the largest unit of measurement.

7. *Manufacturer's or marketer's name and address.* Although the actual aerosol loading may be done by a custom filler, the registrant is considered to be the manufacturer. Supplemental registrations may be granted for labels using the same copy except for the name. The term "manufactured for" or "distributed by" must be used.

8. *EPA registration number.* This number must appear on the label. Until it is issued, any sales of the product across state lines are illegal.

In addition, since pesticide aerosols are "pressurized packages," they can contain "Class B poisons," and the contents can be "corrosive" in nature, the interstate shipment of such products is controlled by Interstate Commerce Commission regulations. Item 834(a) from the Hazardous Materials Regulations of the Department of Transportation specifically states:

The ICC shall formulate regulations for the safe transportation within the United States of explosives and other dangerous articles, including radioactive materials, etiologic agents, flammable liquids, flammable solids, oxidizing materials, corrosive liquids, compressed gases and poisonous substances, which shall be binding upon all carriers engaged in interstate or foreign commerce which transport explosives or other dangerous articles by land, and upon all shippers making shipments of explosives or other dangerous articles via any carrier engaged in interstate or foreign commerce by land, air or water.

COMPONENTS OF PESTICIDE AEROSOLS

The components of a pesticide aerosol are essentially the same as those for any pressurized aerosol package, namely, product concentrate, propellant, valve and actuator, and container. Since the most important objective in developing an aerosol formulation is to obtain a product giving a satisfactory spray, the components that contribute to or affect the spray pattern should be selected with care. Insecticide sprays intended for space treatments should consist of small particles which will remain suspended in the air for as long as possible. Careful selection of solvents, propellants, valves, and actuators will determine the particle sizes and the efficiency of such an aerosol.

Product Concentrates

The product concentrate consists of the active pesticide or pesticides, in solution with other necessary agents, such as antioxidants, surfactants, emulsifiers, solvents, stabilizers, and perfumes.

When the active ingredients are soluble in the propellant and solvent, a two-phase system consisting of a vapor and a liquid is formed. Such products are generally referred to as "oil-based" aerosols. The replacement with water of all or part of the nonaqueous solvents results in products termed "water-based" aerosols or "aqueous pressurized sprays." To obtain a satisfactory spray, this formulation must consist of a dispersion of active ingredients and solvents in an emulsion system in which the propellant is in the external phase, thereby forming a three-phase aerosol consisting of propellant, water, and vapor (8).

Solvents

The primary function of a solvent is to bring the components of the product concentrate into solution to form a two-phase solution system or a three-phase dispersion system. Solvents also help to provide a spray with the particle size that is most effective for a particular application, and they act in the capacity of vapor pressure depressants when used in conjunction with a high-vapor-pressure propellant. For example, ethyl alcohol and odorless mineral spirits are pressure depressants for a high-vapor-pressure propellant such as Propellant 12 (9).

Propellants

Propellants used in pesticide aerosols are either liquefied gases (nonflammable fluorinated hydrocarbons), such as Propellant 11 or 12; flammable liquefied petroleum gases, such as propane, isobutane, or butane; or compressed gases, such as nitrogen, carbon dioxide, or nitrous oxide. Blends of flammmable, nonflammable, and compressed gases are used in pesticide aerosols to improve spray performance (9).

The concentration and type of propellant can influence the spray characterisitics of pressurized products. Generally, an increase in the concentration of propellant will decrease the particle size, and low-boiling propellants with high vapor pressures will give finer sprays than higher-boiling propellants with low vapor pressures (9).

Valve and Actuator Combinations

A wide variety of valves and actuators are now available, and it is usually possible to select a combination that will be satisfactory for delivering the pesticide in the desired form. Pesticides may be dispensed from pressurized formulations as sprays, solid streams, foams, powders, and suspensions (8).

It is difficult to select the exact valve–actuator combination that is best for a specific type of delivery, although the manufacturers generally can recommend a number of possible combinations. However, the final selection of a combination with the desired spray characteristics is generally made through a "trial and error" method of evaluation.

Containers

Aerosol containers must withstand internal pressures as high as 140 to 180 psi at 130°F.

Various types of containers have been used for packaging pesticide aerosols (8). They include the tin-plated steel container, of either two- or three-piece construction, as well as large, heavy-walled steel containers. Aluminum has not been used to any great extent for these products.

Metal containers are preferred for packaging pesticide aerosols, but the problems of corrosion must be offset by the inclusion of corrosion inhibitors in many of the formulations (8). Although glass containers are free from corrosion problems, brittleness and the dangers of breakage limit their use. Plastic containers are light in weight and extremely resistant to breakage and corrosion, but their use is limited by their permeability to gases (9).

PESTICIDES IN PRESSURIZED FORMULATIONS

The use of pesticides in pressurized formulations has expanded greatly from the early days of space insecticides designed solely for mosquito control. Many formulations now include repellents, herbicides, fungicides, and bactericides, in addition to insecticides. A number of the pesticides found in marketed pressurized products today are listed in Table 18.1. These have been grouped into (a) pyrethroids, (b) synergists and activators, (c) carbamates and chlorinated hydrocarbons, (d) organic phosphate insecticides and acaricides, (e) miscellaneous insecticides and acaricides, (f) repellents, and (g) herbicides and fungicides. Each compound has been listed under one or more accepted names. The chemical identity and structure, the acute toxicity against rats (LD_{50}), and the primary uses are given for each pesticide.

This listing is confined to chemicals of relatively low mammalian toxicities and of general use. For example, methyl parathion has not been included, since it is highly toxic and is used in pressurized products only for the treatment of greenhouses by trained personnel. Similarly, although insecticides such as DDT and Strobane found extensive applications in the past in pressurized products, they are not listed because their use is no longer permitted by government regulation.

Pyrethroid Insecticides

This group includes natural pyrethrins and the related synthetic materials. Pyrethrins, usually combined with one or more synergists, have been used as knockdown and killing agents in aerosols and other pressurized products since their inception. Pyrethrins alone produce very rapid knockdown of flying and crawling insects, but the "down" insects quickly recover unless synergists are added to promote the final killing action.

Production of synthetic pyrethroids began with allethrin about 20 years ago. This insecticide has been generally marketed in aerosol formulations with synergists and with other toxicants, but its major use was in the aerosols filled for the armed forces during the Korean con-

TABLE 18.1. SOME ACTIVE INGREDIENTS CURRENTLY FOUND IN PRESSURIZED PESTICIDE FORMULATIONS (24–26)

Name	Chemical Identity	Rat Oral Toxicity: LD_{50} (mg/kg)	Uses
Pyrethroid Insecticides			
Allethrin Allyl homolog of cinerin I Pynamin ENT 17510 FMC MGK	DL-2-Allyl-4-hydroxy-3-methyl-2-cyclopenten-1-one, esterified with a mixture of *cis*- and *trans*-DL-chrysanthemum monocarboxylic acid 	920	Flying insect control. Rapid knockdown and good kill with synergists.
D-*trans*-Allethrin Bioallethrin D-Trans® ENT 16275 MGK	DL-2-Allyl-4-hydroxy-3-methyl-2-cyclopenten-1-one, esterified with D-*trans*-chrysanthemum-monocarboxylic acid	7,071	Control of flies, mosquitoes, etc. Rapid knockdown effect.
Pyrethrins FMC MGK PENICK PRENTISS	Mixture of 6 compounds: 	584–900	Knockdown and kill of household insects when synergized. Flushing and control of crawling insects.

	R	R'
Pyrethrin I	$CH_2CH\overset{c}{=}CHCH=CH_2$	CH_3
Pyrethrin II	$CH_2CH\overset{c}{=}CHCH=CH_2$	$COOCH_3$
Cinerin I	$CH_2CH\overset{c}{=}CHCH_3$	CH_3
Cinerin II	$CH_2CH\overset{c}{=}CHCH_3$	$COOCH_3$
Jasmolin I	$CH_2CH\overset{c}{=}CHCH_2CH_3$	CH_3
Jasmolin II	$CH_2CH\overset{c}{=}CHCH_2CH_3$	$COOCH_3$

Resmethrin
SBP-1382®
Synthrin ™
NIA 17370
NRDC 104
ENT 27474
FMC
  PENICK

(5-Benzyl-3-furyl)methyl-2,2-dimethyl-3-(2-methyl-propenyl)cyclopropane carboxylate

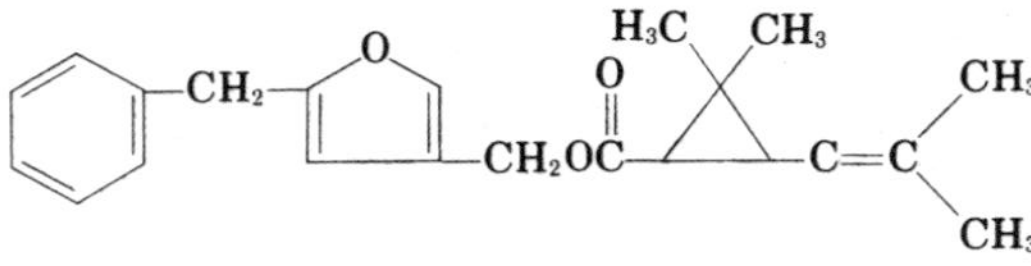

4,500

Flying insect control; kill activity superior to that of pyrethrins.

Tetramethrin
Neo-Pynamin®
Phthalthrin
NIA 9260
ENT 27339
FMC

3,4,5,6-Tetrahydrophthalimidomethanol, esterified with a mixture of cis- and trans-DL-chrysanthemum-monocarboxylic acid

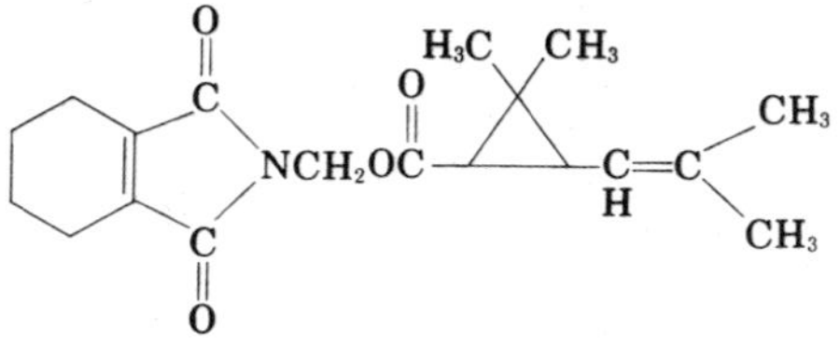

20,000

High knockdown and kill of flying insects when combined with resmethrin, synergists.

TABLE 18.1. CONTINUED

Name	Chemical Identity	Rat Oral Toxicity: LD$_{50}$ (mg/kg)	Uses
Insecticide Synergists or Activators			
MGK 264® Octacide 264 MGK	N-(2-Ethylhexyl)bicyclo[2.2.1]-5-heptene-2,3-dicarboximide	2,800	Synergist for pyrethrins, allethrin.
Piperonyl Butoxide Butacide® ENT 14250 FMC	α-[2-(2-n-Butoxyethoxy)ethoxy]-4,5-methylenedioxy-2-propyltoluene	11,500	Synergist for pyrethrins, allethrin, tetramethrin, carbamate insecticides.

Propyl isome PENICK	Di-*n*-propyl-3-methyl-6,7-methylenedioxy-1,2,3,4-tetrahydronaphthalene-1,2-dicarboxylate $COOC_3H_7$ $COOC_3H_7$ CH_3 H_2C (O, O)	15,000	Synergist for pyrethrins.
Sulfoxide Sulfox-Cide® *n*-Octyl sulfoxide of isosafrole PENICK	1,2-Methylenedioxy-4-[2-(octylsulfinyl)propyl]benzene $CH_2CHSC_8H_{17}$ CH_3 with $\overset{O}{\parallel}$; H_2C (O, O)	2,000	Synergist for pyrethrins.
Sesamin Asarinin	2,6-Bis(3,4-methylenedioxyphenyl)-3,7-dioxabicyclo-[3.3.0]octane sesame oil extractive 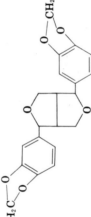		Activator for pyrethrins.
S-421 Octachlorodipropyl ether BADISCHE	Bis(2,3,3,3-tetrachloropropyl) ether $Cl_3CCHClCH_2OCH_2CHClCCl_3$	2.5 ml	Activator for pyrethrins, used in European aerosols.

TABLE 18.1. CONTINUED

Name	Chemical Identity	Rat Oral Toxicity: LD$_{50}$ (mg/kg)	Uses
Tropital® ENT 28344 MGK	Piperonal bis[2-(2'-n-butoxyethoxy)ethyl]acetal	4,000	Synergist for pyrethrins.

Carbamate and Chlorinated Hydrocarbon Insecticides

Name	Chemical Identity	Rat Oral Toxicity: LD$_{50}$ (mg/kg)	Uses
Aprocarb Baygon® Bay 39007 Propoxur ENT 25671 CHEMAGRO	o-Isopropoxyphenyl methyl carbamate	100	Residual insecticide for cockroach control. Good flushing effect.
Carbaryl Sevin® ENT 23969 CARBIDE	1-Naphthyl methyl carbamate	400	Residual insecticide in rose sprays and pet sprays.

Chlordane Octachlor® Chlordan ENT 9932 VELSICOL	1,2,3,5,6,7,8,8-Octachloro-2,3,3a,4,7,7a-hexahydro-4,7-methanoindene		457–900	Residual insecticide used for ant control.
Endosulfan Thiodan® NIA 5462 ENT 23979 FMC	6,7,8,9,10,10-Hexachloro-1,5,5a,6,9,9a-hexahydro-6,9-methano-2,4,3-benzodioxathiepin 3-oxide		100	Insect control on ornamentals.
Methoxychlor Marlate® ENT 1716 DU PONT	1,1,1-Trichloro-2,2-bis(4-methoxyphenyl)-ethane		5,000	Used in household aerosol and spray formulations to give killing effect.
Perthane® Q-137 ROHM & HAAS	1,1-Bis(4-ethylphenyl)-2,2-dichloroethane		8,200	Insecticide (mothproofer).

TABLE 18.1. CONTINUED

466

Name	Chemical Identity	Rat Oral Toxicity: LD_{50} (mg/kg)	Uses
TDE Rhothane® DDD ENT 4225 ROHM & HAAS ALLIED	2,2-Bis(4-chlorophenyl)-1,1-dichloroethane 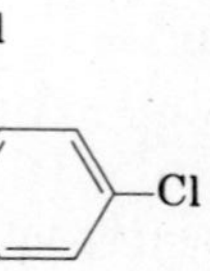	3,400	Insecticide (mothproofer).

Organic Phosphate Insecticides and Acaricides

Ciodrin® ENT 24717 SHELL	α-Methylbenzyl 3-(dimethoxyphosphinyloxy)-*cis*-crotonate 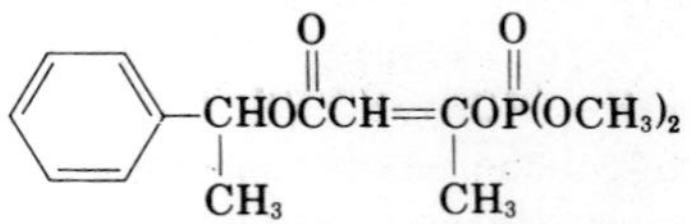	125	Controls cattle ectoparasites. Residual insecticide for fly control in dairy barns.
Coumithoate Dition® Dithion ™ ENT 24986 MONTECATINI	0,0-Diethyl 0-(7,8,9,10-tetrahydro-6-oxobenzo[c]chroman-3-yl) phosphorothioate 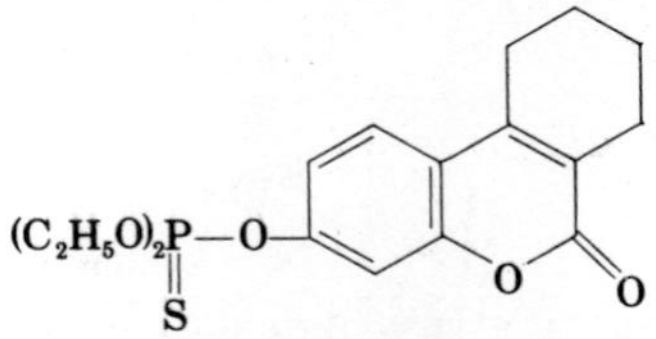	300	Insecticide, acaricide.

Diazinon Alfa-tox® Basudin® ENT 19507 GEIGY	*O,O*-Diethyl *O*-(2-isopropyl-4-methyl-6-pyrimidinyl) phosphorothioate $\text{(pyrimidinyl ring, CH}_3\text{, (CH}_3\text{)}_2\text{CH)} - \text{O} - \overset{\overset{\text{S}}{\|}}{\text{P}}(\text{OC}_2\text{H}_5)_2$	300–400	Controls cockroaches, pests of ornamentals
Dichlorvos Vapona® DDVP ENT 20738 SHELL	*O,O*-Dimethyl *O*-(2,2-dichlorovinyl) phosphate $(\text{CH}_3\text{O})_2\overset{\overset{\text{O}}{\|}}{\text{P}} - \text{O} - \text{CH}=\text{CCl}_2$	56–80	For ant and cockroach control. Short residual, but good fumigant effect.
Chlorpyrifos Dursban® ENT 27331 DOW	*O,O*-Diethyl *O*-(3,5,6-trichloro-2-pyridyl) phosphorothioate $\text{(trichloropyridyl ring, Cl, Cl, Cl)} - \text{O} - \overset{\overset{\text{S}}{\|}}{\text{P}}(\text{OC}_2\text{H}_5)_2$	135–160	New insecticide for household use against crawling insects.
Malathion Cython ENT 17034 CYANAMID	*S*-[1,2-Bis(ethoxycarbonyl)ethyl]*O,O*-dimethyl phosphorodithioate $(\text{CH}_3\text{O})_2\overset{\overset{\text{S}}{\|}}{\text{P}} - \text{S} - \overset{\underset{\displaystyle\text{CH}_2\text{COOC}_2\text{H}_5}{|}}{\text{CH}}\text{COOC}_2\text{H}_5$	4,000	Effective in the control of flies, mosquitoes, agricultural and ornamental pests.

TABLE 18.1. CONTINUED

Name	Chemical Identity	Rat Oral Toxicity: LD_{50} (mg/kg)	Uses
Naled Dibrom® ENT 24988 CHEVRON	O-(1,2-Dibromo-2,2-dichloroethyl)O,O-dimethyl phosphate 	430	Residual insecticide with some fumigant effect.
Ronnel Korlan® Trolene® fenchlorophos ENT 23284 DOW	O,O-Dimethyl O-(2,4,5-trichlorophenyl) phosphorothionate 	1,700	Residual insecticide for the control of flies, cockroaches, etc.
Trichlorfon Dipterex® CHEMAGRO	O,O-Dimethyl (1-hydroxy-2,2,2-trichloroethyl) phosphonate 	450–500	Fly and mosquito control; veterinary use for animal ectoparasites.

Miscellaneous Insecticides and Acaricides

Ammonium fluosilicate	Ammonium silicofluoride $(NH_4)_2SiF_6$		150 (guinea pigs)	Insecticide coating on silica aerogel.
Aramite UNIROYAL	2-(4-*tert*-Butylphenoxy)-1-methyl ethyl 2-chloroethyl sulfite	$(CH_3)_3C$ —— O—CH$_2$CH$_2$—O—S—OCH$_2$CH$_2$Cl (with CH$_3$ and O on the sulfite group)	3,900	Adulticide for use on ornamentals; toxic to phytophagous mites.
Chlorbenzilate GEIGY	Ethyl 4,4′-dichlorobenzilate	Cl———C(OH)(COOC$_2$H$_5$)———Cl	3,100	Control of mites on ornamentals.
Lethane 384® ROHM & HAAS	2-Butoxy-2′-thiocyanodiethyl ether C$_4$H$_9$OCH$_2$CH$_2$OCH$_2$CH$_2$SCN		90	Knockdown agent in household aerosols.
Ovex Ovotran® Ovochlor® chlorfenson DOW	*p*-Chlorophenyl *p*-chlorobenzene sulfonate	Cl———S(=O)(=O)—O———Cl	2,000–2,050	Mite ovicide.

TABLE 18.1. CONTINUED

Name	Chemical Identity	Rat Oral Toxicity: LD_{50} (mg/kg)	Uses
Thanite® HERCULES	Isobornyl thiocyanoacetate (82%) and related compounds	1,600	Knockdown agent and insecticide.
Silica aerogel Dri-Die SG-67 SG-77 GRACE	Very finely divided particles of silicon dioxide SiO_2		Residual insecticide for long-term control of cockroaches, dry-wood termites.
Rotenone Derris FMC PENICK PRENTISS MGK	Active insecticidal principle of several species of tropical and subtropical plants, notably the genera *Derris*, *Lonchocarpus*, *Tephrosia*, and *Mellittia*	132–1,500	Control of garden insects, animal ectoparasites.

Chemical structure for Thanite®:

$$CH_3-\!\!\!\underset{H_3CCCH_3}{\overset{|}{\bigcirc}}\!\!\!-O-\overset{O}{\overset{\|}{C}}-CH_2S-C\equiv N$$

Repellents

Butoxypolypropylene glycol Stabilene® Crag® Fly Repellent FMC CARBIDE	$C_4H_9[OCH(CH_3)CH_2]_nOC_4H_9$	11,200	Biting insect repellent for use in dairies, on livestock, in patio and outdoor sprays.
Deet Delphene® meta-Delphene® Detamide ENT 20218 OLIN HERCULES UNIROYAL	*N,N*-Diethyl-*m*-toluamide 	2,000	Personal repellent for gnats, ticks, fleas, flies, mosquitoes, and chiggers
Dimethylphthalate DMP NTM ALLIED	Dimethylphthalate 	8,200	Personal repellent for mosquitoes and chiggers.
Indolone® FMC	Butyl 3,4-dihydro-2,2-dimethyl-4-oxo-1,2*H*-pyran 6-carboxylate 	7,400	Personal repellent for mosquitoes, black flies, chiggers.

TABLE 18.1. CONTINUED

Name	Chemical Identity	Rat Oral Toxicity: LD$_{50}$ (mg/kg)	Uses
MGK repellent 11® Phillips R-11 ENT 17596 MGK	1,5a,6,9,9a,9b-Hexahydro-4a(4H)-dibenzofuran carboxaldehyde	2,500	Repellent for houseflies, cockroaches, mosquitoes.
MGK repellent 326® Dipropyl isocinchomeronate Phillips R-326 MGK	Di-n-propyl 2,5-pyridinecarboxylate	6,230	Housefly repellent.
MGK repellent 874® 2-Hydroxyethyloctyl sulfide Phillips R-874 MGK	2-(Octylthio)ethanol $CH_3(CH_2)_7-S-CH_2CH_2OH$	8,500	Residual cockroach, ant repellent.

Oil of citronella RITTER	Volatile oil from *Cymbopogon nardus*, which grows in Ceylon and Java		Mosquito repellent.
Rutgers 6-12 6-12 Insect Repellent® Ethyl hexanediol CARBIDE	2-Ethyl-1,3-hexanediol $CH_3CH_2CH_2CHCHCH_2OH$ $\quad\quad\quad\quad\quad OH \quad\; C_2H_5$	1,400	Personal repellent for mosquitoes, biting flies, chiggers.
Tabatrex® GLENN	Di-*n*-butyl succinate $CH_2COOC_4H_9$ $CH_2COOC_4H_9$	8,000	Fly repellent.

Herbicides and Fungicides

Actamer® Vancide® VANDERBILT	2,2'-Thiobis(4,6-dichlorophenol)	6,627	Fungicide.
Amitrole CYANAMID AMCHEM	3-Amino-1,2,4-triazole	14,000	Controls annual and perennial grasses and broadleaf weeds, poison ivy.

TABLE 18.1. CONTINUED

Name	Chemical Identity	Rat Oral Toxicity: LD$_{50}$ (mg/kg)	Uses
Captan Orthocide® ENT 26538 STAUFFER CHEVRON	N-(Trichloromethylthio)-4-cyclohexene-1,2-dicarboximide	10,000	Fungicide; controls scabs, blotches, rots, mildew, etc., on flowers.
2,4-D amide Emid ™ AMCHEM	2,4-Dichlorophenoxyacetamide	300–1,000	Herbicide for broadleaf plants.
Dichlone Phygon® FMC UNIROYAL	2,3-Dichloro-1,4-naphthoquinone	1,500	Fungicide for foliar treatment of fruits and vegetables.

Dichlorophen Dichlorophene Preventol GD® G-4® GIVAUDAN	2,2′-Dihydroxy-5,5′-dichlorophenylmethane 	2,000 (dogs)	Fungicide, bactericide in pet sprays.
Dinocap Karathane® Arathane® ENT 24727 ROHM & HAAS	2-(1-Methyl-*n*-heptyl)-4,6-dinitrophenyl crotonate 	1,000	Acaricide, controls powdery mildew on fruits, vegetables.
Folpet Phaltan® STAUFFER CHEVRON	*N*-(Trichloromethylthio)phthalimide 	10,000	Controls scab, black spot, rose mildew, and other plant diseases on flowers, ornamentals.
Mecoprop MORTON	2-(4-Chloro-2-methylphenoxy)propionic 	930	Herbicide

TABLE 18.1. CONTINUED

Name	Chemical Identity	Rat Oral Toxicity: LD$_{50}$ (mg/kg)	Uses
PCP, Sodium salt Dowcide® DOW MONSANTO	Pentachlorophenol, sodium salt		Contact herbicide and wood preservative.
Silvex DOW HERCULES	2-(2,4,5-Trichlorophenoxy)propionic acid	650	Control of young broad-leaved weeds such as chickweed, hen-bit, lamb's quarters.

NAA AMCHEM

α-Naphthaleneacetic acid

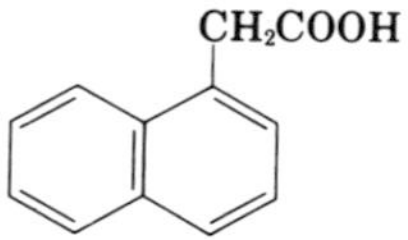

1000   Plant growth regulator

Chlorflurenol MERCK

Methyl 2-chloro-9-hydroxyfluorene (9)-carboxylate

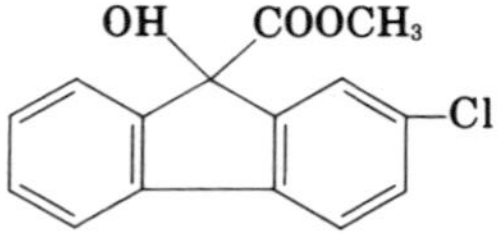

5500   Herbicide, plant growth
        regulator

TABLE 18.1a. MANUFACTURERS OF PESTICIDES

Abbreviation	Company	Address
Allied	Allied Chemical Corp., Agricultural Div.	P.O. Box 2061R Morristown, N.J. 07960
AmChem	AmChem Products, Inc.	Ambler, Pa. 19002
Badische	BASF Wyandotte Corp.	100 Cherry Hill Road P.O. Box 181 Parsipanny, N.J. 07054
Carbide	Union Carbide Corp., Consumer Products Div.	270 Park Avenue New York, N.Y. 10017
Chemagro	Chemagro Corp.	P.O. Box 4913 Hawthorn Road Kansas City, Mo. 64120
Chevron	Chevron Chemical Co., Ortho Div.	411 E. Marlton Pike Cherry Hill, N.J. 08034
Cyanamid	American Cyanamid Co., Agricultural Chemicals Div.	P.O. Box 400 Princeton, N.J. 08540
Dow	Dow Chemical Co.	P.O. Box 1706 Midland, Mich. 48640
Du Pont	E. I. Du Pont de Nemours & Co., Inc., Industrial & Biochemicals Dept.	1007 Market Street Wilmington, Del. 19804
FMC	FMC Corp, Agricultural Chemical Div.	100 Niagara Street Middleport, N.Y. 14105
Geigy	Geigy Agricultural Chemical, Div. of Ciba-Geigy Corp.	Saw Mill River Road Ardsley, N.Y. 10502
Givaudan	Givaudan Corp.	100 Delawanna Avenue Clifton, N.J. 07014
Glenn	Glenn Chemical Co.	2735 Ashland Avenue Chicago, Ill. 60620
Grace	W. R. Grace & Co., Agricultural Chemicals	100 N. Main Street P.O. Box 277 Memphis, Tenn. 38101
Hercules	Hercules, Inc., Synthetics Dept.	900 Market Street Wilmington, Del. 19899

TABLE 18.1a. CONTINUED

Abbreviation	Company	Address
Merck	Merck Chemical Div., Merck & Co., Inc.	Lincoln Avenue Rahway, N.J. 07065
MGK	McLaughlin Gormley King Co.	1715 S.E. Fifth Street Minneapolis, Minn. 53414
Monsanto	Monsanto Chemical Co., Agricultural Div.	800 N. Lindbergh Boulevard St. Louis, Mo. 63122
Montecatini	Montecatini, Div. Prodotti, Agricoltura	Largo Donegani 1 Milano, Italy
Morton	Morton Chemical Co.	110 North Wacker Drive Chicago, Ill. 60606
Olin	Olin Chemical Corp., Chemicals Div.	P.O. Box 991 Little Rock, Ark. 72203
Penick	S. B. Penick & Co., Div. of CPC International, Inc.	100 Church Street New York, N.Y. 10007
Prentiss	Prentiss Drug & Chemical Co., Inc.	101 W. 31st Street New York, N.Y. 10001
Ritter	F. Ritter and Co.	4001 Goodwin Avenue Los Angeles, Calif. 90039
Rohm & Haas	Rohm & Haas Co.	Independence Mall West Philadelphia, Pa. 19105
Shell	Shell Chemical Co., Agricultural Chemicals Sales Div.	110 W. 51st Street New York, N.Y. 10020
Stauffer	Stauffer Chemical Co.	229 Park Avenue New York, N.Y. 10017
Uniroyal	Uniroyal Chemical, Div. of Uniroyal, Inc., Agricultural Chemicals, Research & Development	Bethany, Conn. 06525
Vanderbilt	R. T. Vanderbilt Co., Inc.	230 Park Avenue New York, N.Y. 10017
Velsicol	Velsicol Chemical Corp.	341 E. Ohio Street Chicago, Ill. 60611

flict. It produces acceptable knockdown effect on flying insects but gives poor kill activity and inadequate cockroach control. The _D-trans_ isomer of allethrin (_D-trans-allethrin_) has recently become a marketable insecticidal component of household aerosols. It has a more rapid knockdown effect and is about 2 times as active as allethrin for housefly kill (10).

Recent work by Elliott in England (11, 12) and by workers at the Sumitomo Chemical Company in Japan has made available resmethrin (NRDC 104, SBP-1382®, Synthrin^{TM}) and tetramethrin (the active ingredient of Neo-Pynamin®), two new, highly active, synthetic pyrethroid-type insecticides for use in pressurized formulations. These pyrethrin-like materials have exhibited differing but complementary effects when applied to insects such as the housefly (13). As noted in the original Japanese studies with tetramethrin (14) and in later USDA studies (15), this compound has knockdown characteristics superior to those of pyrethrins or allethrin, but only about equal kill activity. Conversely, although Elliott (11, 12), Brooks (16), and Incho (13) found resmethrin to be a highly potent, broad-spectrum insecticide many times more active than pyrethrins, its rate of knockdown was considerably lower than that of the latter compounds.

The two complementary types of activity are now being utilized in combinations of resmethrin and tetramethrin marketed as Tetralate^{TM} (13). Resmethrin alone at higher concentrations also appears in a number of pressurized products, and tetramethrin serves as a knockdown agent in products of the "patio and outdoor" type.

Insecticide Synergists or Activators

The use of synergists and activators has been the major factor in the marketing of aerosols and other pressurized products containing prethrins and allethrin. Without these cheaper materials, capable of markedly increasing the kill effectiveness of high-cost pyrethrins, these natural products with their many advantages could not have been properly utilized.

Piperonyl butoxide and combinations of piperonyl butoxide with MGK-264 have been the major materials used as synergists for pyrethrins and allethrin. Other methylenedioxyphenyl-type compounds with synergistic activity, such as propyl isome, sulfoxide, and sesamin, also have been marketed in combinations with pyrethrins, but to a more limited extent. Tropital®, a recently developed methylenedioxyphenyl derivative, has appeared as a new synergist for pyrethrins, reportedly

with good cockroach control activity and low nasal irritation effects
(17).

Carbamate and Chlorinated Hydrocarbon Insecticides

Carbamate and chlorinated hydrocarbon insecticides provide control
of a large variety of insects when incorporated in pressurized formulations. For example, aprocarb has good flushing and residual action
against cockroaches, carbaryl is effective against ectoparasites on pets,
chlordane controls ants, methoxychlor enhances the knockdown and
kill of flying insects, and endosulfan is active against pests of ornamental shrubs. In addition, two chlorinated materials, Perthane® and
TDE, have found extensive use as mothproofers in pressurized formulations.

Organic Phosphate Insecticides and Acaricides

A number of the less toxic organic phosphate insecticides are components of pressurized insecticides. Materials such as Ciodrin®, coumi-
thoate, malathion, naled, trichlorfon, and ronnel provide long residual
activity for the control of flies and mosquitoes. Diazinon®, dichlorvos,
Dursban®, and ronnel (fenchlorphos) are effective against cockroaches, when used alone or in combinations (such as aprocarb plus
dichlorvos) in pressurized formulations designed to put down a long-
lasting residue.

Miscellaneous Pesticides

Three materials for the control of phytophagous mites and a number of
insecticides have been included among the miscellaneous pesticides in
Table 18.1. The miticides (Aramite®, chlorbenzilate, and ovex) are
usually included in house- and garden-type pressurized formulations to
give control of mites attacking plants in and around the home.
Lethan 384® and Thanite® were substituted for pyrethrins as knockdown agents and toxicants in early aerosol formulations, but their relatively low levels of effectiveness have limited their application in more
recent insecticide aerosols.

Silica aerogel (Dri-Die), usually with a coating of ammonium fluosil-
icate on the individual particles, acts as a residual insecticide for the

control of cockroaches. The silica aerogel has a special affinity for oils and waxes, absorbing up to 300% of its own weight. It functions as an insecticide by absorbing the insect's waxy cuticular coating, causing the loss of internal moisture and literally drying the insect to death. The long-lasting residual activity of Dri-Die has been combined with the immediate flushing and knockdown effect of pyrethrins synergized with piperonyl butoxide in Drione®, a product being marketed in a number of powder aerosol formulations used for cockroach control.

Rotenone is a complex insecticidal chemical derived from tropical plants grown primarily in South America. It is effective against fleas, lice, ticks, mites, houseflies, cockroaches, bedbugs, ants, and other pests and is widely used as an agricultural and veterinary insecticide (18). It is often included in house and garden aerosol formulations to provide added control of plant-infesting insects and in pet aerosols for the control of dog and cat ectoparasites. Rotenone is extremely toxic to cold-blooded animals and has also found extensive use as a fish killer, but it is relatively harmless to man and domestic animals.

Repellents

A number of effective personal repellents, alone and in various combinations, are found in pressurized formulations. Mallis (18) notes the following observations by Wilkes on the desirable characteristics of a repellent, approached, but not as yet achieved, by these materials:

> A substance to be generally acceptable as a repellent should possess . . . effectiveness over a relatively long period of time against one or more of such biting or blood sucking pests as mosquitoes, flies, fleas, chiggers, and the like. To this end, it should be relatively stable chemically, and not readily dissipated by evaporation or vaporization, although it is believed that at least some degree of volatility is essential. For dermal application, a repellent should be non-irritating and easy to apply . . . . For use under wet-skin conditions . . . the repellent preferably should be relatively insoluble in water.
>
> Repellents which may affect clothing by staining, bleaching or weakening of the fiber, or which leave an objectionable "oily" appearance or feel on the skin, are limited in their usefulness. Preferably, the repellent should be free of odor, especially such odors as may be regarded as unpleasant or disagreeable, and difficult to mask. Preferably, also, the repellent substance should have little or no solvent action on various finishes, paints, varnishes, lacquers and the like.

Some of the more common repellents are listed in Table 18.1. Citronella, an old standby for many years, has been largely replaced by more effective new materials but may be found in some aerosols and in repellent candles. Butoxypolypropylene glycol (Stabilene®) is particularly effective against biting flies on livestock and serves as a repellent in "patio and outdoor spray" formulations. Tabatrex® is an effective fly repellent generally used on livestock.

Among the personal repellents, Deet is probably the most popular chemical. It has a broad range of effectiveness and is one of the best repellents for use against mosquitoes, fleas, biting flies, chiggers, and ticks. This repellent does not need to be mixed with others, and it is readily diluted with alcohols to the desired concentration. Deet is resistant to rubbing and to removal by perspiration and water. It may be applied as an aerosol or pressurized spray to both the skin and the clothing, with only the usual precautions not to get it in the eyes or on the lips.

Dimethyl phthalate is a colorless and practically odorless liquid with a specific gravity of 1.19. It is only 0.5% soluble in water. Dimethyl phthalate is the best repellent for the common malaria mosquito (*Anopheles quadrimaculatus* Say) and for sand flies and is a preferred repellent for chiggers.

Indalone is a light yellow to reddish brown liquid with a specific gravity of 1.06. It is especially effective against stable flies and ticks.

Rutgers 6-12 is a slightly viscous, colorless liquid with a mild odor. It has some solvent effect on spar varnish and will soften shellac, but it is one of the most desirable repellents for *Aedes* and other pest mosquitoes.

MGK repellent 11 and MGK repellent 326 are found in combinations with deet in pressurized personal repellents. These combinations provide superior repellent effect against mosquitoes and black flies (17). The synergist MGK 264 is often included in the formulations to aid in insect repellent activity. Repellent 874 also is a component of the popular "yard-guard" type of formulation and is applied as a residual to repel cockroaches and ants.

Herbicides and Fungicides

Pressurized herbicide formulations for the control of broad-leaved plants in the lawn now include materials such as 2,4-D, mecoprop, and silvex, usually formulated as combinations to provide control of all weeds. Where complete control of grasses and weeds is desired along sidewalks or in driveways, amitrole and PCP are applied. Plant growth regulators, such as NAA and chlorflurenol, are applied in pressurized products to control tree sprouting.

A number of fungicides are included in pressurized formulations designed for application to indoor and outdoor plants. Dichlone, folpet, dinocap, actamer, and captan are active against scab, black spot, and rose mildew and are included in aerosol sprays for ornamentals and flowers, such as roses. Dichlorophen is a fungicide and bactericide added to insecticide sprays for pets.

Many aerosols and other pressurized products now being marketed include the materials listed in Table 18.1. Some examples of the various product types are discussed in the following section.

REPRESENTATIVE PESTICIDE AEROSOL FORMULATIONS

Pressurized Flying Insect Killers

The typical pressurized insecticide formulation designed for the control of flying insects around the home is an oil-based aerosol applied as a space treatment. The directions for use generally state that the aerosol should be sprayed into the air for a specified length of time, based on the estimated room volume; doors and windows should be closed; and the room should remain closed and unoccupied for 15 min.

Although effectiveness against other insects is usually claimed on the labels of these aerosols, they are primarily designed and intended for use against flies, mosquitoes, moths, and other flying insects. The use of pyrethrins plus synergist provides an effective household-type spray characterized by quick knockdown and good kill of most flying insects.

A representative oil-based formulation containing these materials is as follows:

Active Ingredients	*% by Weight*
Pyrethrins	0.50
Piperonyl butoxide	2.00
Petroleum distillate	17.50
Inert Ingredients	
Propellant 12/11 (50:50)	80.00
EPA Reg. No. 5316-27*	100.00

A similar water-based formulation contains the following:

Active Ingredients	*% by Weight*
Pyrethrins	0.20
Piperonyl butoxide	1.00
MGK 264	1.00
Petroleum distillate	0.80
Inert Ingredients	97.00
	100.00

EPA Reg. No. 4822-34

* Owners of EPA registrations are given in Table 18.2.

TABLE 18.2. IDENTIFICATION OF EPA REGISTRATIONS

Formulation Name	EPA Reg. No.	Registration Issued to:
Macleans Aerosol Insect Spray	5316–27	Neil A. Maclean Co., Inc.
Raid Flying Insect Killer	4822–34	S. C. Johnson & Son, Inc.
NIA 17370 Aerosol Insecticide 0.35	4816–374	FMC Corp.
NIA 17370 Aqueous Pressurized Spray 0.35	4816–383	FMC Corp.
General Purpose Aerosol Insect Killer	279–2912	FMC Corp.
Multi-Purpose Insect Killer	279–2790	FMC Corp.
Wasp and Hornet Killer No. 1	4816–337	FMC Corp.
Pyrenone Baygon Roach and Ant Spray	4816–371	FMC Corp.
Drione Insecticide Spray	4816–367	FMC Corp.
Pyrenone House and Garden Insect Spray	4816–332	FMC Corp.
House and Garden Insect Spray	4816–360	FMC Corp.
0.25% Tetramethrin 0.25% NIA 17370 Pressurized Spray	279–2907	FMC Corp.
Your Brand Automatic Sequential Aerosol Insecticide	4816–393	FMC Corp.
D-Con Indoor Fogger	3282–23	The d-Con Company, Inc.
6-2-2 Triple-Mix Insect Repellent	4816–339	FMC Corp.
Niagara Patio and Outdoor Spray	279–2835	FMC Corp.
Raid Mothproofer	4822–6	S. C. Johnson & Son, Inc.
SLA	1603–15	Reefer-Galler Co.
SLA Mothproofer	1603–14	Reefer-Galler Co.
Flea and Tick Killer for Cats and Dogs	4816–386	FMC Corp.
Pyrenone Dairy Aerosol	4816–365	FMC Corp.
Ortho-Weed B-Gone	239–2334	Chevron Chemical Co.
Ortho-Weedone	264–189	Chevron Chemical Co.
Rose and Floral Spray Residual	4816–348	FMC Corp.
Asplundh Sprout Gard	10114–1	Asplundh Tree Expert Co.
Maintain A	1624–101	U.S. Borax & Chemical Co.

The newer synthetic pyrethroid insecticides also are formulated alone and in various combinations as pressurized oil- and water-based products to control flying insects around the home. A representative oil based formulation utilizing the highly effective resmethrin (NIA 17370, SBP-1382®) alone is as follows:

Active Ingredients	% by Weight
Resmethrin	0.35
Related compounds	0.05
Aromatic petroleum derivatives	0.46
Petroleum distillate	19.12
Inert Ingredients	
Propellant 12/11 (50:50)	80.00
Others	0.02
	100.00

EPA Reg. No. 4816-374

A similar water-based formulation contains the following:

Active Ingredients	% by Weight
Resmethrin	0.35
Related compounds	0.05
Aromatic petroleum derivatives	0.46
Inert Ingredients	
Inhibitor	0.10
Deionized water	67.56
Propellant	30.00
Others	1.48
	100.00

EPA Reg. No. 4816-383

Combinations of tetramethrin and resmethrin have been formulated in aerosols to utilize the excellent knockdown effectiveness of the former and the high kill activity of the latter material in the control of flying insects. The following is an example of an oil-based formulation:

Active Ingredients	% by Weight
Tetramethrin	0.25
Related compounds	0.03
Resmethrin	0.11
Related compounds	0.01
Petroleum distillate	18.50

Inert Ingredients

Propellant 12/11 (50:50)	80.00
Others	1.10
	100.00

EPA Reg. No. 279-2912

The following formulation is water based:

Active Ingredients	*% by Weight*
Tetramethrin	0.25
Related compounds	0.03
Resmethrin	0.11
Related compounds	0.10
Petroleum distillate	9.00

Inert Ingredients

Inhibitor	0.10
Deionized water	59.90
Propellant	30.00
Others	0.51
	100.00

EPA Reg. No. 279-2790

Wasps, hornets, and yellow jackets are controlled around the home by the application of aerosols and pressurized sprays containing pyrethroids for fast knockdown and a carbamate- or phosphate-type insecicide for good kill. An example of this type of product (oil-based) is as follows:

Active Ingredients	*% by Weight*
Pyrethrins	0.07
Piperonyl butoxide	0.19
Carbaryl	0.50
Petroleum distillate	24.24

Inert Ingredients

Methylene chloride	25.00
Propellant 12/11 (50:50)	50.00
	100.00

EPA Reg. No. 4816-337

The heavy spray from this formulation is directed at the insects, if possible, as well as under eaves and into nests, cracks, holes, or crevices where the insects are noticed. When applied until the surface is moist, the insecticide will leave a toxic residue to control returning insects that could not be sprayed.

Crawling-Insect Sprays

Pressurized insecticide formulations designed to produce coarse, wet sprays, which upon drying, leave a residue of active materials, are used to control crawling insects such as cockroaches, ants, silverfish, fleas, and bedbugs (19). The carbamate and organic phosphate insecticides combined with synergized pyrethrins are particularly effective in this type of formulation.

Active Ingredients	*% by Weight*
Aprocarb	0.50
Pyrethrins	0.05
Piperonyl butoxide	0.26
Petroleum distillate	70.21

Inert Ingredients	
Propellant 12	20.00
Others	8.98
	100.00

EPA Reg. No. 4816-371

Similarly, a Drione® formulation utilizes the long-lasting dessicant action of silica aerogel and the more immediate effect of pyrethrins plus synergist.

Active Ingredients	*% by Weight*
Pyrethrins	0.10
Piperonyl butoxide	1.00
Dri-Die®	4.00
Petroleum distillate	4.90

Inert Ingredients	
Methylene chloride	30.00
Chlorothene NU	10.00
Propellant 11	30.00
Propellant 12	20.00
	100.00

EPA Reg. No. 4816-367

This formulation provides a wet spray that soon dries to a white film toxic to cockroaches for a long period of time. The valve is selected for a narrow spray pattern or a jet spray. An injection tube actuator may be added to the valve to enable the user to spray cracks, crevices, and other openings where crawling insects hide and where the residue will be undisturbed.

House and Garden Pressurized Sprays

Double-duty pressurized insecticides for use in the home and in the garden can be formulated by proper selection of specific solvents to prevent plant phytotoxicity (20). Dual-purpose units employing pyrethrins and synthetic pyrethroids, both as pressurized solvent and aqueous spray formulations, are effective in the control of flying insects and as garden sprays against beetles, thrips, aphids, and mites. Synergists, such as piperonyl butoxide, MGK 264®, and Tropital®, are used in combination with these insecticides to provide adequate knockdown and kill.

A typical example of a solvent-based house and garden type of aerosol formulation employing synergized pyrethrins is as follows:

Active Ingredients	*% by Weights*
Pyrethrins	0.20
Piperonyl butoxide	1.00
Petroleum distillate	0.80
Inert Ingredients	
Perfume	0.10
Isopropanol	17.19
Propellant 12/11 (50:50)	80.00
Others	0.71
	100.00

EPA Reg. No. 4816-332

An example of a similar water-based synergized pyrethrins house and garden type of insect spray is as follows:

Active Ingredients	*% by Weight*
Pyrethrins	0.25
Piperonyl butoxide	1.00
Petroleum distillate	1.00
Inert Ingredients	
Chlorothene NU	6.35
Emulsifier	0.20
Inhibitor	0.30
Deionized water	58.90
Propellant	32.00
	100.00

EPA Reg. No. 4816-360

An example of the use of synthetic pyrethroids in a water-based house and garden type of aerosol is the versatile multipurpose spray containing the product Tetralate. This aerosol incorporates two synthetic pyrethroid insecticides in the following formulation:

Active Ingredients	*% by Weight*
Tetramethrin	0.25
Related compounds	0.03
Resmethrin	0.25
Related compounds	0.03
Petroleum distillate	8.80
Inert Ingredients	
Inhibitor	0.10
Deionized water	59.90
Propellant	30.00
Others	0.64
	100.00

EPA Reg. No 279-2907

Intermittent Dispensers

Intermittent aerosol dispensing is a popular method for controlling infestations in industrial and food-processing plants. The intermittent dispensing devices are usually equipped with electrical or electronic timing mechanisms to periodically depress the actuator buttons and with metering valves to deliver a predetermined weight (45–100 mg) of formulation with each depression of the button. Since the aerosol formula is usually a highly concentrated combination of active insecticide and synergist, the periodic spraying soon produces an insecticidal atmosphere to control invading flying insects (21).

A typical automatic sequential aerosol formulation is as follows:

Active Ingredients	*% by Weight*
Pyrethrins	0.90
Piperonyl butoxide	9.00
Petroleum distillate	10.10
Inert Ingredient	
Propellant 12/11 (50:50)	80.00
	100.00

EPA Reg. No. 4816-393

Total-Release Insect Foggers

As the title implies, an aerosol of this type is designed so that the entire contents of the unit are released with a single depression of the valve actuator. This is accomplished by means of an actuator designed to lock into place when depressed, resulting in complete release of the pressurized contents.

The total-release aerosol propels a high concentration of insecticide in the form of a fine mist (fog) that quickly fills the entire area and dissipates into voids, cracks, and crevices where insects hide. The closed area remains exposed to the pesticidal effects of the total-release aerosol for several hours, thus eliminating any flying or crawling insects that emerge from their hiding places.

A typical total-release formulation consists of the following:

Active Ingredients	% by Weight
Pyrethrins	0.50
Piperonyl butoxide	1.00
MGK 264	1.67
Petroleum distillates	11.83
Inert Ingredients	85.00
	100.00

EPA Reg. No. 3282-23

Repellent Aerosols

Insect repellents are chemicals that, when added to an aerosol unit, do not kill, but serve to keep insects away from treated surfaces. When applied to clothing or the skin, the volatile vapors from most repellants tend to prevent insects from alighting, resulting in protection from insect bites.

Oils such as those of citronella, cedar wood, eucalyptus, and turpentine have been used extensively for many years, but their effectiveness is limited. The two modern chemicals most commonly used as repellents are deet and Rutgers 6–12. Other repellents for use alone or in combinations are listed in Table 18.1.

These insect repellent materials can be formulated in a number of different ways for consumer use, for example, as pressurized sprays or foams.

PRESSURIZED SPRAYS

Pressurized sprays incorporate the active repellent chemical in alcohol. The drying effect of the alcohol can be reduced somewhat by using water as a partial replacement. Propellant 12 is generally used as the propellant.

A typical pressurized spray type of repellent formulation is as follows:

Active Ingredients	% by Weight
Dimethyl phthalate	12.77
Indalone	3.77
Rutgers 6-12	3.46
Inert Ingredients	
Isopropyl alcohol	10.00
Propellant 12/11 (50:50)	70.00
	100.00

EPA Reg. No. 4816-339

FOAM TYPES

Foam-type repellent aerosols are patterned after the typical aqueous emulsion products, formulated as oil-in-water emulsions. The oil-in-water emulsions generally produce rigid foams, such as shaving lathers, and utilize foam actuators and valves for dispensing the product either as a foam or as a liquid that subsequently expands into a foam.

The use of foams permits low concentrations of propellant, thus providing room for high concentrations of the active ingredients in the formulation. Foams also are easy to use, can be applied directly to the desired area, and are attractive in appearance.

Pressurized foam repellents are generally prepared by dissolving the active repellent chemical and an emulsifier in ethyl alcohol. This "oil phase" is then added to the "water phase," consisting of deionized water and a corrosion inhibitor, with agitation to form a stable oil-in-water emulsion. Propellants such as isobutane, propane, and Propellant 12 are used in pressurized foams.

A typical example of a pressurized foam type of repellent is as follows:

Active Ingredient	% by Weight
Deet	20.00
Inert Ingredients	
Ethyl alcohol	37.25
Emulsifier	3.00
Deionized water	32.25
Propellant 12	7.50
	100.00

Patio and Outdoor Sprays

Patio and outdoor sprays are specialty insecticide aerosol products, first introduced in 1969 by S. C. Johnson & Son as the Raid® "Yard Guard" Outdoor Fogger. Intended for the treatment of backyards, patios, and picnic areas with a heavy spray projected as much as 20 ft, patio and outdoor fogger formulations contain insecticides and repellents that are effective against most flying and crawling insects. A typical formulation incorporates the synthetic pyrethroid tetramethrin in combination with piperonyl butoxide, providing fast knockdown and kill activity. The formulation also contains methoxychlor, which contributes to kill and residual activity, and an insect repellent such as MGK repellent 874 or butoxypolypropylene glycol.

A typical water-based aerosol formulation representative of patio and outdoor sprays is as follows:

Active Ingredients	*% by Weight*
Tetramethrin	0.20
Related compounds	0.03
Piperonyl butoxide	1.00
Methoxychlor	1.00
Butoxypolypropylene glycol	1.00
Petroleum distillate	15.25
Inert Ingredients	
Inhibitor	0.10
Deionized water	44.90
Propellant	30.00
Others	6.52
	100.00

EPA Reg. No. 279–2835

The proper method of handling this type of pressurized unit is printed on the label as follows:

Shake before using; point valve opening away from face. For best results spray when air is calm. Allow a few minutes for the product to take effect. Spray from center of area to be treated when not breezy. Spray with wind if breeze is blowing.

Mothproofers

Since aerosols have been generally successful in controlling flying and crawling insects, it was only natural for insecticide manufacturers to apply this principle to the application of insecticides for mothproofing. Many of the pressurized mothproofers now on the market contain from 50 to 65% petroleum distillates, 30 to 40% propellant gases, and varying amounts of toxicants, such as pyrethrins or pyrethroids plus synergist, perthane, or TDE as the active ingredients. Because the pressures are usually below those of conventional insecticide aerosols, and the valves are designed to produce a wet spray, mothproofer aerosols are surface rather than space sprays. Three examples of pressurized mothproofers are as follows:

Active Ingredients	*% by Weight*
Perthane	4.40
Related compounds	0.60
Petroleum distillate	61.83
Inert Ingredients	33.17
	100.00

EPA Reg. No. 4822–6

Active Ingredients	*% by Weight*
Pyrethrins	0.15
Allethrin	0.15
Piperonyl butoxide	0.50
MGK 264	0.75
Cedar oils	0.50
Petroleum distillate	52.95
Inert Ingredients	45.00
	100.00

EPA Reg. No. 1603–15

Active Ingredients	*% by Weight*
TDE	4.75
Related compounds	0.25
Petroleum distillate	56.33
Inert Ingredients	38.67
	100.00

EPA Reg. No. 1603-14

Pet and Livestock Aerosols

Pressurized products have been found to possess many advantages as partial replacements for powder or tank-dip applications in controlling pet and livestock pests. Ease of application, improved liquid adherence to the animal's coat, and savings to the user in time and cost of application are some of the advantages associated with the adoption of such pressurized spray formulations.

Fleas, lice, and ticks on cats and dogs can be controlled by the following formulation, which incorporates dichlorophen as a bactericide and animal deodorant, and butoxypolypropylene glycol as an insect repellent.

Active Ingredients	*% by Weight*
Allethrin	0.10
Piperonyl butoxide	0.60
Carbaryl	0.50
Butoxypolypropylene glycol	5.00
Dichlorophen	0.10
Inert Ingredients	
Isopropanol	14.56
Methylene chloride	9.00
Propellant 12/11 (50:50)	70.00
Others	0.14
	100.00

EPA Reg. No. 4816–386

For the rapid control of flies and mosquitoes on cattle and horses, the following formulation is a typical example:

Active Ingredients	*% by Weight*
Pyrethrins	0.50
Piperonyl butoxide	5.00
Petroleum distillate	2.00
Inert Ingredients	
Isopropyl alcohol	12.50
Propellant 12/11 (50:50)	80.00
	100.00

EPA Reg. No. 4816–365

The inclusion of butoxypolypropylene glycol or MGK repellents 11 and 326 at 2 to 5% will add repellency to the above formulation to aid in the overall protection of livestock from flies and mosquitoes.

Herbicide and Fungicide Sprays

Several pressurized herbicide, fungicide, and plant regulators are marketed as aerosols for the convenience of home owners.

HERBICIDES

Two general types of herbicides are usually recognized: selective and nonselective. Selective herbicides will kill certain types of plants (weeds) without serious injury to other types growing in the same area.

The discovery of 2,4-D made available an excellent selective herbicide, and this compound is now widely used in various forms and formulations, including pressurized sprays. The several forms of 2,4-D have the property of killing most broadleaf plants without injury to grasses, cereals, and other monocotyledonous plants. An aerosol formulation of 2,4-D in combination with mecoprop utilizes two selective herbicides to produce a spray effective against a large range of weeds.

Active Ingredients	*% by Weight*
2,4-D (diethanolamine salt)	0.75
Mecoprop (diethanolamine salt)	0.75
Inert Ingredients	98.50
	100.00

EPA Reg. No. 239-2334

Nonselective herbicides are those that destroy nearly all forms of plant life. They are useful in eradicating completely all herbage from crop lands overgrown with undesirable species. An example of this type of herbicide in an aerosol formulation is as follows:

Active Ingredient	*% by Weight*
Amitrole	1.00
Inert Ingredients	99.00
	100.00

EPA Reg. No. 264-189

FUNGICIDES

Most fungicides are applied before the diseases appear and serve to kill the fungi or inhibit their growth when they arrive at the material to be protected. For example, PCP (pentachlorophenol) is used for the preservation of wood, and folpet serves as a protective fungicide on foliage. The following pressurized plant spray is an example of a combination insecticide and protective fungicide:

Active Ingredients	*% by Weight*
Pyrethrins	0.02
Piperonyl butoxide	0.26
Rotenone	0.13

Other cube extractives	0.24
Carbaryl	1.00
Folpet	0.70
Petroleum distillate	0.05
Inert Ingredients	97.60
	100.00

EPA Reg. No. 4816-348

Plant Regulators

Plant regulators (22), such as auxins, gibberellins, and cytokinins, are substances that through physiological action accelerate or retard the rate of growth or rate of maturation of plants or otherwise alter their behavior. The plant regulator activity associated with various classes of naturally occurring substances has, in most cases, been duplicated in synthetic compounds. Two examples of pressurized formulations of synthetic plant regulators are as follows:

Active Ingredient	*% by Weight*
NAA	1.00
Inert Ingredients	99.00
	100.00

EPA Reg. No. 10114-1

Active Ingredients	*% by Weight*
Chloroflurenol	0.18
Closely related compounds	0.07
Inert Ingredients	99.75
	100.00

EPA Reg. No. 1624-101

These formulations are applied as paints on the wound or pruning cut of a tree or shrub, and the synthetic plant regulators control the regrowth of sprouts.

TESTING

Pesticide aerosols are evaluated by a number of physical, chemical, and biological tests. These tests determine level of quality, compliance with Department of Transportation regulations as well as possible additional local regulations, and biological efficacy of the aerosols.

The following listing of tests compiled by the Aerosol Division of the CSMA is a recognized aid to vendors and fillers in the development and safe handling of aerosols and pressurized packages. These tests, as well as other procedures that have been developed for assessing the quality, performance, and safety of pressurized pesticide products, are complied in the CSMA *Aerosol Guide*. Copies of this guide are available at a nominal cost from the Chemical Specialties Manufacturers Association, Inc., 50 East 41st Street, New York, New York 10017.

A. Flammability and combustibility.
   1. Flash point test.
   2. Flame extension test.
   3. Closed drum test.
B. Physiochemical evaluations.
   1. Moisture content determination.
   2. Identification of propellants.
   3. Determination of concentrate/propellant ratio.
   4. Density determination.
   5. Internal-pressure determination.
   6. Method for storage tests of aerosols.
   7. Test for aerosols containing pesticides.
   8. Premarketing product check.
   9. Filler safety manual.
C. Performance.
   1. Method for spray patterns.
   2. Method for determining delivery rate of aerosol products.
   3. Particle-size determination.
   4. Method for determining pickup efficiency of residual aerosol insecticides.
D. Biological evaluations.
   1. Test method for flying insects.
   2. Cockroach aerosol method.

Flammability Test

To indicate the effect of a self-pressurized container on the extension of an open flame and to determine the flammability of the contents, a standard flame projection test is followed as outlined in the *Aerosol Guide* (7). The test methods are included in Chapter 13.

The contents of self-pressurized containers are classed as "extremely flammable" if the flash back (a flame extending back to the dispenser) is obtained at any degree of valve opening and the flash point is less than 20°F.

The contents of self-pressurized containers are classed as "flammable" if a flame projection exceeding 18 in. is obtained at full valve opening or a flash back is obtained at any degree of valve opening.

Bioassay Procedures

Standard procedures for the biological evaluation of pesticide products to be used in the home and on plants in or around the home have been established by the Chemical Specialties Manufacturers Association. These methods are periodically updated and printed in the *Soap and Chemical Specialties* "Blue Book" (23). The following standard biological tests apply to pressurized products.

AEROSOL AND PRESSURIZED-SPACE-SPRAY INSECTICIDE TEST METHOD FOR FLYING INSECTS

This procedure establishes the following definitions for aerosols and pressurized sprays:

1. *Aerosols*. The spray from aerosol dispensers should be in finely divided form in which 80 percent or more of the individual spray particles have a mean diameter of 30 microns or less, and none of the spray particles has a diameter of more than 50 microns. Aerosols must be no less effective in biological performance ... than the Official Test Aerosol (OTA) when tested against house flies at the same dosage or less.

2. *Pressurized Sprays*. These products deliver mist sprays intermediate between aerosol-type sprays and those which are intended to deposit an insecticidal residue of a chemical. They must be no less effective than the Official Test Aerosol (OTA) when tested at no more than twice the dosage specified for the OTA.

As noted in the procedure, this technique for testing aerosols should be regarded as a practical method designed for the comparison of formulations in the dispensers in which they will be employed by the consumer.

Apparatus

Test insect. The housefly (*Musca domestica* L.), reared from strains mixed under the supervision of the CSMA, is the test insect for this method. In addition to the specified CSMA nonresistant strain, a standard, DDT-resistant strain (F58W) is available and is now being used in most laboratories. Detailed rearing and handling procedures for this insect are outlined in the complete method.

Reference insecticide. The reference insecticide is the current Official Test Aerosol, which may be purchased from the CSMA.

Aerosol test chamber. The test chamber is specified as a Peet-Grady chamber, or a larger chamber meeting the specifications of a Peet-Grady chamber. This is basically a box with 6′ × 6′ × ′6 interior dimensions and a smooth inner surface; it is fitted with a door, exhaust ports, and viewing ports. An exhaust fan is attached to the chamber to remove air at a rate of not less than 1000 ft³-min.

Procedure

Before the test is started, the test chamber is cleaned, clean paper is placed on the floor, all ports and openings are closed, and the temperature is adjusted to 82 ± 2°F. Only flies capable of flying are then released into the chamber. In the small-group method a sample of about 100 flies is used in each test; in the large-group method, a sample of approximately 500 flies, usually all those in a single cage.

After liberating the flies in the chamber, a total of 3.0 ± 0.5 g of aerosol mixture per 1000 ft³ for a Grade A aerosol or pressurized spray evaluation is applied in a continuous flow. For a Grade B pressurized spray evaluation, a total of 6.0 ± 0.5 g of mixture per 1000 ft³ is applied in a continuous flow.

Counts are made of the number of flies "down" (paralyzed) at 5, 10, and 15 min after insecticide application. At the end of 15 min, the ports are opened and the chamber is ventilated with the exhaust fan while the "down" flies are collected and counted.

The "down" flies are transferred immediately to clean holding cages, provided with food (5% sugar solution), and placed in the rearing room for subsequent 24-hour mortality counts. The "up" (unparalyzed) flies in the chamber at the end of the 15-min exposure period are counted and discarded.

The number of "up" flies counted and recorded at the end of the 15-min exposure period plus the "down" flies noted at that time (or when the 24-hour mortality counts are taken) yields the total flies in the test. The "aerosol test knockdown mortality" is the percentage dead of the total flies, since the "up" flies at the end of the 15-min exposure period are considered alive after 24 hours. The "aerosol test knockdowns" are the percentages "down" (or paralyzed) of the total flies at 5, 10, and 15 min.

Conditions for Official Evaluation

The experimental aerosol or pressurized space spray tested at 3.0 ± 0.5 g dosage can be reported as meeting the standard if its knockdown at 5, 10, and 15 min and its 24-hour aerosol test knockdown mortality are equal to, greater than, or not less than 5 percentage points below the corresponding values of the OTA included in the same series of tests at a dosage of 3.0 ± 0.5 g.

The percentage knockdown and the mortality of the test formulation, provided it "meets the standard," can be designated by letter in accordance with the following grading system:

Grade	*Formulation*	*"Equal to" OTA at 3.0 ± 0.5 g/1000 ft³ at Dosage Indicated*
A	Aerosol or pressurized space spray	3.0 ± 0.5 g/1000 ft³
B	Pressurized space spray	6.0 ± 0.5 g/1000 ft³

COCKROACH AEROSOL TEST METHOD

This official CSMA method is designed for evaluating the relative insecticidal effectiveness of aerosols as direct sprays against cockroaches. The term "aerosol" applies to pressurized formulations containing 20% by weight (or less) volatile ingredients and 80% or more propellant. This method has not been officially adopted for evaluating formulations containing more than 20% low-volatile ingredients, such as the water-based aerosols.

Apparatus

Test insect. The test insects are last-nymphal instars of the German cockroach (*Blatella germanica* L.). Detailed rearing and handling procedures are outlined in the complete method.

Reference insecticide. The reference insecticide is the current Official Test Aerosol (OTA) prepared by the CSMA.

Spray chamber. The spray chamber consists of an open boxlike structure of solid material measuring 18 in. square and 25 to 30 in. high. The open floor of the chamber is covered with 1/2-in.-mesh wire hardware cloth, and suitable guides are attached to permit centering the treatment container in relation to the aerosol nozzle.

An adjustable hinged shelf for holding an aerosol unit is attached to the outside of the center of the back upper edge of the spray chamber. Any suitable shelf that will permit the OTA and test aerosol dispensers to be held in a standard fixed position can be used.

Treatment container. The treatment container is a metal container 3 1/2 in. in diameter with 3-in. side walls. Sixteen-mesh screen is attached to the bottom of the container in such a manner that the entire bottom is completely open.

Recovery dishes. Glass crystallizing dishes measuring 125 mm in diameter and 65 mm in height are used as recovery dishes. The bottoms of the recovery dishes are not covered with paper. Sixteen-mesh wire screen covers may be used as recovery dish covers during the 48-hour holding period.

Test Procedure

Before use, the OTA and test aerosol dispensers are calibrated to determine their spray rates in grams per second. The dispensers are aligned on the adjustable shelf so that the aerosol mist is directed into the open top of the treatment container, which rests on the floor of the spray chamber 24 in. from the dispenser spray nozzle. The treatment container rests on a 5″ × 5″ square of unsized, nonglazed, absorbent paper such as brown kraft or gray bogus paper. Except for the 5″ × 5″ square of paper, the chamber floor is open.

Immediately before spray application, test groups of 20 last-instar German cockroaches are transferred to the treatment containers, which are suitably treated on the interior surfaces to confine the insects to the container floors. The treatment container is removed from the aerosol spray 30 sec after the start of spray application, and the cockroaches are transferred to recovery dishes, where they are held without food or water during the 48-hour observation period.

In evaluating a test aerosol, a minimum of 10 test groups is used in conjunction with 10 test groups treated with the OTA. The dose is established as that which will result in an average of 50 to 75% dead and moribund at 48 hours with the OTA. Test dispensers are weighed before and after the treatment of each test group, and the weight of material used is recorded.

Test aerosols are evaluated on the basis of the percentages of insects normal, moribund, and dead 48 hours after spray application. Any insect showing signs of life but incapable of normal locomotion is considered moribund. The 48-hour average percentages dead and moribund of the test sample are reported as "meeting the standard" if they are equal to, greater than, or within 10 percentage points less than the corresponding values for the OTA.

DETERMINATION OF PLANT SAFETY OF PRESSURIZED FORMULATIONS FOR USE ON PLANTS IN THE HOUSE OR GARDEN

An official method for evaluating the plant safety of aerosol and pressurized sprays used on common plants in and around the home has been adopted by the CSMA. The introduction to the method states:

The following procedure is designed to consider the test package as an entire unit so the test (will) be a valid rating of the product. It is realized that the phytotoxicity of many chemicals varies from one plant species to another, but the three species designated as "indicator plants" will detect most pressurized plant sprays that are too phytotoxic to be marketed. In some cases, practical field evaluation on a large variety of plants (will) also be necessary before a product (can) be registered.

Apparatus

Indicator plants. The test indicator plants are designated as Bountiful snap beans (7 to 10 days old), Early White Spine cucumbers (14 to 18 days old), and Rutgers tomatoes (21 to 26 days old). Detailed descriptions of the plants and their culture are given in the complete procedure.

Testing room. The testing room should be maintained at 72 to 78°F and around 50% relative humidity. A hood is provided to confine the spray during treatment. The plants and the pressurized plant spray are allowed to come to testing-room temperature before the spray is applied. After treatment, the plants are held in the testing room out of direct sunlight until the spray deposit is dry.

Holding greenhouse. After treatment, the plants are kept in a holding greenhouse or suitable plant culturing room for observation.

Procedure

The indicator plant is placed in the hood and tilted slightly forward at an angle of 15°. The pressurized plant spray is held in a fixed position with the nozzle 12 in. from the terminal bud of the indicator plant and is positioned to deliver spray directly at the terminal bud. Three plants of each species are sprayed for 1-, 2-, 3-, 4-, 6-, and 8- sec periods. The entire process is repeated at 18 in., and tests may also be conducted at a 6-in. distance in some cases of known low phytotoxicity. Phytotoxic symptoms will appear in 24 hours or less with many plant sprays; however, some chemicals may cause more subtle effects, such as dwarfing or chlorosis, which are slower to appear. All tests are replicated 3 times, either on the same day or on different ones.

Test results are recorded by a letter or numerical system ranging from O: No plant injury (0) through M: Moderate plant injury (5–7) to D: Death of plant (10). A rating of the injury produced on each plant is recorded in tabular form, and a short description of the type of injury produced by the plant spray is included in the report.

Conditions for Official Evaluation

A product that produces no plant injury to any of the three indicator plants when sprayed for 4 sec at 18 in. and no injury at 24 in. at higher dosage is considered to have "passed the test." The product label should give directions to ensure safe, effective control of the listed insect species on the plant species named.

Any product that fails this test, but does not produce more than traces of plant injury on two of the three indicator plants sprayed for 3 sec at 18 in., is considered as "tentatively passed," provided that the label bears a designated cautionary statement.

A product not passing or not tentatively passing the tests may be accepted for use on a specific plant if the plant is not injured by a 3-sec spray application from 18 in. However, the label must bear designated precautions limiting the use of the product to the uninjured plant species.

REFERENCES

1. L. D. Goodhue, How It All Began, *Aerosol Age,* **10**, 5 (1965), 30–33, 118–119.

2. A., Edelstein, Erik Rotheim, Aerosol Pioneer Extraordinary, *Aerosol Age,* **2**, 3 (1957), 18.

3. L. D. Goodhue and W. N., Sullivan, Method of Applying Pesticides, U.S. Patent 2,321,023 (1943).

4. J. E. Lee, Aerosol Insecticide Formulations Up to Date, *Soap Chem. Spec.,* **43**, 4 (1967), 62, 64, 66, 70, 72, 193.

5. A. E., Schober, The Use of Chlorothene for Aerosol Insecticides, *Aerosol Age,* **4**, 8 (1959), 32, 35, 36, 72.

6. G. D. Glynne-Jones and A. J. S., Weaning, The Formulation and Activity of Water-Based Aerosols Containing Pyrethrum, *Aerosol Age,* **11**, 3 (1966), 21.

7. *Aerosol Guide,* 6th ed., Chemical Specialties Manufacturers Association, New York, 1971.

8. J. J. Sciarra, Pharmaceutical Aerosols, in L., Lachman, H. Lieberman, and J. Kanig (Eds.), *The Theory and Practice of Industrial Pharmacy,* Lea & Febiger, Philadelphia, 1970, pp. 605–637.

9. P. A., Sanders, *Principles of Aerosol Technology,* Van Nostrand Reinhold Company, New York, 1970, p. 418.

10. J. H., Fales, et al., Relative Effectiveness of Pyrethroid Insecticides, *Soap Cosmet. Chem. Spec.,* **48**, 5 (1972), 60, 62, 66, 68, 90, 92.

11. M. A., Elliott, et al., New Pyrethrin-Like Esters with High Insecticidal Activity, *Nature,* **207** (1965), 938–940.

12. M. A., Elliott, et al., 5-Benzyl-3-furylmethyl Chrysanthemate, a New Potent Insecticide, *Nature,* **213** (1967), 493–495.

13. H. H., Incho, New Insecticide for Aerosols, *Soap Chem. Spec.,* **46**, 2 (1970), 37–40, 68, 69, 72.

14. K., Buei et al., Evaluation of Pyrethroid in Kerosene and Deobase Against Adults of the Common House Fly, *Musca domestica vicina* Macq., by Settling Mist Method. II: Studies on the Biological Assay of Pyrethroids, *Botyu Kagaku,* **30** (1965), 37–44.

15. J. H. Fales and O. F., Bodenstein, Evaluation of "Neo-Pynamin" as an Insecticide, *Soap Chem. Spec.,* **42**, 6 (1966), 80–88; **42**, 7 (1966), 66–67, 104–106.

16. I. C., Brooks et al., SBP-1382—a New Synthetic Pyrethroid, *Soap Chem. Spec.,* **45**, 3 (1969), 62–74; **45**, 4 (1969), 49, 50, 79.

17. J. E., Lee, Aerosol Insecticide Formulations—Review of Recent Developments, *Soap Chem. Spec.,* **44**, 4 (1968), 58, 60, 62, 171.

18. A., Mallis, *Handbook of Pest Control,* 5th ed., MacNair-Dorland Company, New York, 1969, p. 1158.

19. A. Herzka, and J. Pickthall, *Pressurized Packaging (Aerosols),* Academic Press, New York, 1958, p. 509.

20. H. R., Shepherd, *Aerosols: Science and Technology,* Interscience Publishers, New York, 1961, p. 537.

21. J. F., Odeneal, A New Method of Control for Flying Insects, *Aerosol Age,* **4**, 3 (1959), 20, 21, 61, 62.

22. Plant Growth Regulators, Monograph 31, *Soc. Chem. Ind. London, S.W.I.,* 1968, p. 251.

23. Blue Book Reference and Buyers' Guide, *Soap Chem. Spec.,* **47**, 4A (1971), 161, 164, 173.

24. D. E. H., Frear, *Pesticide Index,* 4th ed., College Science Publishers, State College, Pa., 1969, p. 399.

25. D. E. H., Frear, *Pesticide Handbook,* 24th ed., College Science Publishers, State College, Pa., 1972, p. 279.

26. J., Neumeyer, D., Gibbons, and H., Trask, Pesticides, *Chem. Week,* Apr. 12 and Apr. 26, 1969.

HOUSEHOLD, AUTOMOTIVE, AND INDUSTRIAL PRODUCTS

The aerosols currently marketed in greatest numbers include cosmetics and toiletries, personal hygiene products, pesticides, and space sprays. There is, however, a group of pressurized products that do not individually contribute significant sales but are important as a whole and should be considered. In this group are such products as antistatic sprays, artificial snow, cleaners and polishes, and deicers. Some were made possible by the availability of the aerosol, while others represent novel applications. In this chapter, we discuss this group of diverse products and their formulations.

ADHESIVES

The aerosol adhesive market has not grown at the same rate as other household markets. One major reason is the fact that the overspray of an aerosol adhesive product is difficult to contain. Therefore a disposable background must be used when spraying an adhesive, and this eliminates part of the convenience.

The adhesives marketed are generally composed of polymeric resins that are tenacious in nature, dissolved in suitable solvents. Evaporation of the solvent results in a resin film. Some resins can be modified to produce flexible films as shown below.

Ingredients	% by Weight
Resyn 76-4680 (National Starch)	3.7
Sucrose acetate isobutyrate	1.6
Acrylic resin B-66 (Rohm & Haas)	2.0
Methylene chloride	3.7
Acetone	9.0
Propellant 12	80.0

A single resin/solvent formula is:

Ingredients	% by Weight
Resyn 30-1211 (National Starch)	5.6
Methylene chloride	19.4
Propellant 12/11 (70:30)	75.0

A pressure-sensitive type, generally used on paper, utilizes tacky Resyn 76-4347 (National Starch and Chemical Corporation) at a ratio of 5 parts in a 1:1 Propellant 12/11 solution. For ease in filling, the resin is dispersed in Propellant 11 and Propellant 12 is added under pressure.

Spray adhesives have found use in the artistic field, where fast drying is necessary. Artistic creations of papier-mâché can be made rapidly and conveniently by the use of aerosol glues and adhesives.*

ANIMAL MARKING FLUIDS

When culling herds and flocks, ranchers have had a difficult time identifying the animals selected for the market place. It is essential that animal markings can be readily removed in processing, but are not easily removed by rain. The spray container has done much to provide the rancher with an easy method of marking. A product that meets these requirements contains a water-soluble resin similar to the ones used in hair sprays. In order to provide some water repellency, lanolin, easily emulsified during scouring, is added.

A second key to a successful product of this type is to use a water-solvent dye, as in the following formulation.

Ingredients	% by Weight
Polyvinylpyrrolidone/vinyl acetate	2.5
Isopropyl alcohol	32.5
Anhydrous lanolin	14.7
Dye	0.3
Propellant 12/11 (30:70)	50.0

The dye should be dissolved in the alcohol, and this solution filtered before adding the other ingredients.

---

* On August 17, 1973 the Consumer Products Safety Commission indicated that some spray adhesives might be linked to birth defects. At this time, it is not known whether any of the above ingredients are implicated. Several marketers have voluntarily recalled these products from the market place.

ANTIFOGGING AGENTS

Anti-fogging agents are designed to prevent condensation of moisture, which forms droplets on glass, rendering it translucent. Therefore materials that lower the surface tension of the droplets of water, causing them to coalesce, will eliminate fogging. Soaps, surfactants, glycerin, and glycols will generally prevent fogging of windows. However, the formulator is confronted with the problem of spraying these materials in a thin film without causing a foam to develop. One product that prevents fogging for about 2 hours is made with a surface-active agent and a siloxane-type product.

Ingredients	% by Weight
Aerosol OT (American Cyanamid)	2.5
XF-1-1037 fluid (Dow Corning)	2.5
Isopropyl alcohol	35.0
Propellant 12/11 (50:50)	60.0

Hygroscopic film formers also reduce fogging of windows. These materials absorb moisture, thus reducing droplet formation. However, problems occur with the use of these moisture-absorbing films. They become tacky when wet and if rubbed will smear, blocking vision.

ANTIGLARE SPRAYS FOR GLASS AND PLASTIC

Reduction of glare from smooth shiny surfaces is often difficult to accomplish without distorting vision or causing color changes on the surface. Colloidal clays such as Attagel 40, manufactured by Engelhard Minerals and Chemical Corporation, can be used with a film former to reduce glare. The sheen of the film can be controlled by varying the amount of clay added. Suitable solvents, such as alcohol, are used to dissolve the film former. Since the film, or surface, will generally collect dirt and dust, it has been found that the use of a water-soluble resin, properly plasticized to produce a clear, hard film works best. The propellant combination can be any fluorocarbon that produces the proper pressure for spraying and atomizing the product.

ANTIOXIDANTS

Antioxidants have long been used in the paint and varnish industry to prevent "skinning" of paints after packing. Antioxidants also prevent the drying or skinning of inks on printing presses. Several types of compounds can be used for this purpose. Amine, oxime, and thio compounds readily absorb oxygen, preventing the oxidation of resins in coatings and inks. These compounds, in pressurized products, have been marketed for spraying over the top layer of liquid in a paint can after use and before storage. Also, printing plants, using resins to bind pigments to paper, employ a spray product on the presses to prevent skinning of the ink fountains while the press is down for short periods of time. The formulator may find the following a good starting point:

Ingredients	% by Weight
Cyclohexanone oxime	5.0
Mineral spirits	13.0
Light mineral oil	2.0
Propellant 12/11 (30:70)	80.0

In place of cyclohexanone oxime, another compound, methyl ethyl ketoxime, can be used. Either of the two compounds will act as a dispersant for highly polymerized products to prevent coalescing of the particles and will absorb oxygen without altering its volatility.

ANTISTATIC SPRAYS

The polyamines are particularly effective for dissipating static charges that develop on highly polished surfaces and smooth resins. Polyoxyethylene fatty amine, such as Atlas Chemical Company's Compound G-3780A or Onyx Chemical's Aston RC and AP, used in a concentrations of about 5% in an aerosol product, reduces the static electrical charge buildup on solids.

ANTITARNISH PRODUCTS

Different surfaces become dull for different reasons and therefore require specific treatments. For instance, the application of a barrier coating such as an acrylic or nitrocellulose will prevent the oxidation of metals. Dulling by oxidation can be reduced by applying an oxygen-absorbing medium to the surface.

When developing a product to use on anything that may come in contact with foods, it is best to use film formers with a very low order of toxicity. Polyvinylpyrrolidone or copolymers with the pyrrolidone have proved to be satisfactory for such applications.

ARTIFICIAL SNOW

The early artificial snows were combinations of film formers or resins so formulated in solvents as to produce an almost dry product when sprayed. Formulating in this manner produced a flock or expanded lump of resin, which settled on the article sprayed. Since artificial snow was among the first consumer products, its development generated many U.S. patents. One of the first patents issued on a composition of this type was U.S. Patent 2,659,704, dated 1953; however, there is evidence that particles of polymers were dispensed to represent snow as early as the turn of the century.

Technology has advanced to the point where the ingredients in artificial snow formulations today cost less than 6 cents. Many of those produced are based on water emulsions and are propelled with hydrocarbon propellants. The following represents current advances in technology:

Ingredients	% by Weight
Acrylic resin	6.0
Methylene chloride	10.0
Emcol 14 (polyglycerol ester) (Witco Chemical)	1.4
Water	74.6
Isobutane	8.0

OR

Ingredients	
Polyvinyl Chemical Resin B-722	3.50
Toluene	7.70
Emcol 14 (Witco Chemical)	0.48
Epoxal 9-5 (Swift)	0.40
Stearic acid	4.00
Water	77.42
Propellant A-46	6.50

The Epoxal has been added as a corrosion inhibitor to allow the packaging of the product in a plain tin-plate container. Emcol 14 is a good emulsifier for water-in-oil emulsions with a low order of corrosion.

ARTIST'S FIXATIVE

This product, designed to protect works of art ranging from charcoal renderings to oil paintings, has a limited market. It is difficult to develop products that will not cause bleeding of pigments and dyes in charcoal or softening and sagging of oils. Solvents must be carefully selected, and the evaporation rate must be rapid. Resins such as acrylics and cellulose butyrates produce a film adequate to protect the art work; however, many tests should be made before applying protective coatings to expensive renderings. A product using these resins could be formulated as follows:

Ingredients	% by Weight
Tenth-second cellulose acetate butyrate (Eastman Chemical)	1.8
Acetone	3.0
n-Butyl alcohol	1.8
Methylene chloride	19.8
Isobutyl acetate	25.2
n-Propyl alcohol	8.4
Propellant 12	40.0

OR

Acryloid B-72 (Rohm & Haas)	8.5
Toluene	51.5
Propellant 12/VC (80:20)	40.0

The distance of the spray from the object is of critical importance. If sprayed too near the art work, the solvent system will cause damage. Therefore a minimum distance of 12 in. is recommended when using such products.

ATHLETE'S GRIP SPRAY

Because sports are generally very active, physical endeavors, perspiring hands can impede an athlete's ability to perform. Hence athletes use rosin to achieve a better grip with the hands. This can be incorporated in a pressure product that will dispense an even coating. Pine for rosin is used at a 3% level with an appropriate solvent such as isopropyl alcohol and methylene chloride. The solution thus formed can be propelled with either a fluorocarbon blend or a blend of fluorocarbons and hydrocarbons.

AUTOMOTIVE PRODUCTS

Some automotive products in pressurized packages are similar to industrial products; however, others are primarily designed for use specifically in the automotive area. The latter type, such as polishes, soundproofing products, traction sprays, and starting fluids, are covered in this section. However, products that find general application in the household or industrial area, even though they may be used also in the automotive field, are covered at the proper place in the alphabetical listing of the chapter. Therefore this section does not include all products that may be used in, on, or around the automobile.

Cleaners and Polishes

The consumer expects his car to look better after he has used a cleaner and polish. Also, he strives to select the product that produces lasting beauty for the longest period of time. Therefore, when formulating a cleaner and polish, the chemist should include chemicals and compounds that remove the oxidized paint film and road tars, are easily applied and buffed, and produce a durable gloss. Recently, detergent resistance has become a necessity with the increase of automated car wash establishments.

The chemist should develop a product with the fewest ingredients possible because such formulations are the easiest to manufacture. A simple car polish can be made of silicone oils, wax, mineral spirits, and propellants. However, this product, even though it produces a high gloss, cannot be applied unless the car finish has been cleaned before use. The following is an example of a simple car polish:

Ingredients	% by Weight
Polyethylene AC-629 (Allied Chemical)	3.0
Silicone fluid, 500 cs (General Electric)	1.8
VM&P naphtha	50.2
Propellant 12	45.0

The addition of mild abrasives coupled with solvents helps to remove the oxidized paint film. Correct balance of ingredients, such as the solvent abrasive ratio, is necessary to eliminate smearing and removal of nonoxidized paints. Ratios between 1.5 and 2 to 1 seem to work best. As a starting pressurized formula, the following is offered:

Ingredients	% by Weight
Silicone oil, 1000 cs	4.5
Silicone oil, 30,000 cs	1.0
Carnauba wax	1.8
Beeswax	1.8
Stearic acid	2.3

Mineral spirits	10.8
Kerosine	7.2
Triethanolamine	2.3
Kaopolite SFO (Harold M. Johnson)	12.0
CMC-7H (Hercules Powder)	0.3
Water	46.0
Isobutane	10.0

The most recent advance in automobile polishes has been the addition of detergent resistance. Dow Corning Corporation reported a new advance in silicones at the Chemical Specialties Manufacturers Association meeting in May 1968. It has been reported that amine-modified polydimethylsiloxanes are superior bonding agents and form a protective layer over painted surfaces. Further, continued polymerization takes place when these amine-modified compounds are exposed to moisture, producing gelling of the polymer. A suggested formula is as follows:

Ingredients	% by Weight
531 fluid (Dow Corning)	2.70
530 fluid (Dow Corning)	0.45
Mineral spirits	9.00
VM&P naphtha	47.85
Propellant 12	40.00

Wax can be added to this formula; however, adequate dispersion and possible solution of the wax should be obtained. As mentioned above, moisture in the formula should be avoided to prevent gelation of the siloxanes. Therefore dry solvents should be added.

Deicers

Auto windshield deicers had phenomenal growth for many years and then leveled off. Deicer products are designed to melt ice or frost on the windshield and contain ingredients such as alcohols and glycols, which lower the freezing point of water. For economy and to avoid federal regulations, isopropyl alcohol is used in these formulations. Ethylene glycol, the basic ingredient in many automotive cooling system antifreeze solutions, is also a primary constituent of windshield deicers. Water is added to the formulation to reduce the viscosity as well as to adjust the freezing point depression as desired.

Since deicers are likely to be used in extremely cold weather, a lique-fied propellant does not exert sufficient pressure to discharge the product. Therefore a compressed gas such as nitrous oxide is used as the propellant. Nitrous oxide is suitable also because it is quite soluble in alcohols and glycols; its reported solubility in isopropyl alcohol is 0.65 mole % at 75°F and 1 atm pressure. By comparison, nitrous oxide is soluble at a rate of only 0.04 mole % in water under the same conditions.

Starting Fluids

Engine-starting fluids have a limited market, estimated at $10 to $11 million annually. Starting fluids are composed of extremely flammable liquids with relatively high volatility because they are intended for use in engines using low-volatility fuels such as diesel oil. Starting fluids are used primarily when the ambient temperature is too low to volatize the diesel oil or other low-volatility fuel. The first U.S. patent of record for an engine-starting fluid in a pressurized container is No. 2,928,435, dated March 15, 1960, even though these fluids were in use before this date.

These products, like deicers, are intended for use in cold weather, and again the propellant is nitrous oxide. The base concentrate is generally diethyl ether or petroleum ether.

Tire Inflators and Sealants

It is a harrowing experience to have a flat tire on a portion of highway where changing the tire is neither safe nor practical. For this reason, tire inflators and sealants in pressurized containers have become standard items in many automobiles. Many types of liquefied propellant have been used in these products. The chemist must keep in mind the fact that the container must hold a sufficient volume of liquid to expand into a gas. This gas must be capable of producing adequate pressure to allow operation of the vehicle with reasonable safety.

The propellants in use are Propellants 12, 12/propane (91:9), 22/CO_2, and 501 (a blend of 22 and 12), and a blend of propane and isobutane. The last-named is not recommended, however, because of its extremely flammable nature.

Extensive tests were performed on the following formula. A butyl latex emulsion made with Enjay Chemical Company butyl latex 80-21 is used as the sealant. The composition of this portion is as follows:

Ingredients	% by Weight
Butyl latex 80-21 (55% solids)	30
Cellosize QP (Union Carbide Chemicals)	5
Tamol 731 (Rohm & Haas)	1
Ethylene glycol	20
Asbestos floats (Lake Asbestos)	15
Water	29

The combination of 85% by weight of Propellant 12 and 15% sealant is filled in a pressure container to a weight of 12 oz. A valve and adaptor appropriate to transfer the contents of the container through the tire valve stem is used. One supplier of a suitable adaptor is Schrader Valve Company.

In the tests performed, a tire of 8.00 × 14 size, normally inflated to 24 lb air pressure, was used. A regular tenpenny nail 3 in. long was driven into the tread area of a mounted, inflated tire. The nail was removed, and the tire allowed to go flat until no measurable pressure was evident. One 12-oz can of the formula given above was added to each tire tested. The instantaneous reinflation pressure was 15 lb at the ambient temperature of 60°F. After several rotations of the tire the pressure rose to 20 lbs. The vehicle was driven less than 1/2 mile and allowed to stand for 3 hours, after which it was driven over a stop-and-go course for a total of 35 miles.

The pressure in the tire after the 35-mile test drive was 23.5 lb. The tire and rim were removed from the car and allowed to stand for 72 hours, after which the tire pressure was 17.5 lb.

Traction Sprays

A recent product is an aerosol used to spray automobile tires to produce more positive traction on ice and snow. Since the product is relatively new to the market, little is known of its potential. Although many attempts were made in the past to produce a workable product of this type, few succeeded. Previously, formulators tried to incorporate grit or solid particles that would produce traction. This approach failed because of valve clogging.

The most recent approach is to spray a solution of wood rosin, similar to that used by athletes, on the tire to give better grip. This is a reasonable approach since loss of traction on ice or snow results from a film of water between the tire and the ice or snow. A suggested formula includes a pale-colored polymerized rosin manufactured by Hercules Powder Company and sold under the trade name Poly-Pale Resin. This resin can be dissolved in either higher alcohols such as isopropyl, aromatic solvents, or hydrocarbon solvents without crystallization occurring.

Ingredients	% by Weight
Poly-Pale® Resin (Hercules)	20.0
Isopropyl alcohol, anhydrous	50.0
Propellant 12	30.0

Undercoating

The formulating of aerosol automotive undercoating is relatively simple since this product is made in the same manner as the conventional brushed-on type. Both contain an asphalt compound, solvents, and a filler. One major difference is that the filler in the brush-on type need not be uniform in size. Also, a greater variety of fillers can be used, such as asbestos, wood fibers, or clay. These fillers are necessary for the ultimate performance of the product, since it is designed to protect the metal as well as to reduce the noise level when small stones or other objects strike the underpart of the vehicle. The filler absorbs the shock, or impact velocity, of an object striking the undercarriage. A spray product suggested to accomplish the desired results is as follows:

Ingredients	% by Weight
Crystal asphalt (Celotex)	42.0
HB Snowflake (Archer, Daniels, Midland)	3.0
Propellant 12/11 (37:63)	55.0

A higher ratio of propellant to concentrate than is normally found in automotive products is used to produce some foaming of the asphalt when applied. The "hills," or foam bubble irregularities, serve as a cushion to deaden the sound when objects strike the metal.

BELT DRESSINGS

In the industrial field, drive belt slippage on pulleys means lost power and higher costs. Therefore it is desirable to apply materials that will prevent slippage. Although such materials are available in sticks and liquid form, the easiest and safest method of application is to use a pressurized product. A suitable antislip product contains the following:

Ingredients	% by Weight
Wood rosin	2.5
Tallow	0.5
Isopropyl alcohol	47.0
Propellant 12/11 (50:50)	50.0

Some recent reports claim wood rosin accelerates the deterioration of rubber belts. Resins of the ·chlorinated rubber type may be used in place of wood rosin to lessen deterioration.

BIRD REPELLENTS

Instead of driving starlings away by playing phonograph records of their own frightened shrieks, a gelatinous material in ribbon form can be applied to likely landing places. The compound used in this pressurized product is Polybutene No. 24, manufactured by Chevron Chemical Company. The compound is harmless to birds, but it sticks to their feet and is more disagreeable than a piece of chewing gum on the sole of a shoe. The following is a suitable formula:

Ingredients	% by Weight
Polybutene No. 24	60.0
Castor oil	3.0
Hexane	7.0
Propellant 12/11 (50:50)	30.0

The unique feature of this product is that it is effective against all birds, not just starlings.

CLEANERS (ALL TYPES)

Among the most heavily consumed products in the household line are cleaners of all types in pressurized containers. This general category includes bathroom, fabric, mirror, oven, rug, tile, toilet bowl, and window cleaners.

All household cleaners contain either an emulsifying agent or detergent and with few exceptions are water-based products. As such, they can be propelled with the lighter-than-water, liquefied petroleum gases such as isobutane and propane.

Emulsifying and detergent agents should be noncorrosive to the container. In some cases, it is difficult to select a noncorrosive agent without sacrificing cleaning power. For this reason, most cleaners of the water-based type are packaged in lacquer-lined containers. For instance, a typical bathroom cleaner includes a quaternary ammonium chloride salt for a broad spectrum of sanitization and disinfecting, coupled with surface-active agents for the removal of household dirt. An example follows:

Bathroom Cleaner

Ingredients	% by Weight
BTC 2125 (Onyx Chemical)	0.296
Sodium silicate, meta	0.231
Tergitol nonionic, NPX (Union Carbide)	1.000
Triton X-100 (Rohm & Haas)	0.500
Diethylene glycol monoethyl ether	3.600
Antifoam AF-60 (General Electric)	10.000
Water	74.173
Perfume	0.200
Propellant A - 46	10.000

This spray product foams on contact with a hard surface.

Similarly, a blackboard cleaner contains a surface-active agent, water, and a solvent for grease removal.

Blackboard (Chalk) Cleaner

Ingredients	% by Weight
Ultrawet K (Atlantic Refining)	0.2
Isopropyl alcohol	15.8
Water	79.0
Propellant 12/114 (40:60)	5.0

Oven Cleaners

These products are slightly different from most cleaners in that they generally contain strong solvents to cut through cooking oils and fats, yet contain surface-active agents for cleaning. A typical formula is as follows:

Ingredients	% by Weight
Neutronyx 834 (Onyx Chemical)	1.5
Methylene chloride	37.5
Crude scale wax	2.0
Ethyl cellulose	2.0
Methyl alcohol	9.1
Cellosolve acetate	9.6
Trichloroethylene	13.3
Propellant 12	25.0

In order to prevent the solvent portion from evaporating too rapidly, thickners such as waxes and ethyl cellulose are added. The addition of these retarders keeps the solvent and surface-active agent in contact with vertical surfaces for longer periods of time. Certainly, competitive products such as brush-ons are less expensive. However, the aerosol product affords greater safety, is faster acting, and is more easily applied.

During the last several years, alkaline-type oven cleaners have been introduced. These products contain hydroxides, either sodium or potassium, to saponify the cooking oils and fats. A typical product contains the following:

Ingredients	% by Weight
Water	64.5
Propylene glycol	20.0
Alkali hydroxide	3.0
Renex® 648 (Atlas Chemical)	2.5
Propellant A-46	10.0

The Renex® is a heavy-duty detergent generally used in metal-cleaning operations. Some degree of hazard exists when spraying caustic or in using a hydrocarbon propellant in a confined space, such as an oven. Thus it is preferable to use a fluorocarbon propellant in oven cleaners.

Rug Shampoos and Cleaners

These products have been on the market for about 15 years. Recent promotion has caused an increase in consumer acceptance.

Ingredients	% by Weight
Sipon LT/5 (American Alcolac)	72.0
Pine oil	0.9
Water	17.1
Propellant 12/114 (40:60)*	10.0

This product has a pH near 7, which is noncorrosive; however, because of imperfections in tin plate, iron oxide can appear and lined containers are recommended. The Sipon LT/5 is a 45% active liquid of triethanolamine lauryl sulfate. Pine oil is added to produce a clean, fresh fragrance.

The above formula can be changed as follows to include solvent-soluble surface-active agents that may absorb oil stains:

Ingredients	% by Weight
Sarkosyl NL-30 (Ciba-Geigy)	28.7
Triethanolamine lauryl sulfate	4.5
Alrosol C (Geigy Chemical)	2.7
Perfume	0.3
Water	53.8
Propellant 12/114 (40:60)*	10.0

The Sarkosyl NL is lauroyl sarcosine that is sparingly soluble in water and has been emulsified into a water-in-oil emulsion by use of a nonionic fatty amide with a pH of 9.2.

The most recent rug cleaners introduced to the market are, in the most practical sense, the same as the early products. The similarity can be seen by the following:

Ingredients	% by Weight
Sodium lauryl sulfate 30% active	35.0
Sarkosyl NL-30 (Ciba-Geigy)	23.0
Water	17.0
Propellant 12/114 (20:80)*	25.0

Window or Glass Cleaners

Although sales of these cleaners increased by 130% during the past 10 years, for the past 5 years the growth rate has been only 10%, showing

---

* Although a fluoronated hydrocarbon propellant is illustrated, a hydrocarbon propellant can also be used.

a leveling off and market saturation. The latest advertising approach is to attempt to convince the consumer that the chemicals (ammonia in this case) used by housewives of an earlier generation are the best cleaning agents. The most recent window cleaners marketed contain ammonia. Without a doubt, the newer products reduce streaking, and ammonia unquestionably adds something to the product besides odor. However, today's product contains wetting agents, solvents, and water, as did window cleaners marketed in 1956. The earlier products were formulated thus:

Ingredients	% by Weight
Triton X-155 (Rohm & Haas)	0.10
Ultrawet K (Atlantic Refining)	0.10
Silicone 200 (10 cps) (Dow Corning)	0.02
Isopropyl alcohol	9.78
Water	86.00
Isobutane	4.00

By changing the solvent system and using a more modern wetting agent, as well as adding ammonium hydroxide, today's product is formulated as follows:

Ingredients	% by Weight
Tergitol 15-5-9 (Union Carbide)	0.045
Butyl cellosolve	2.700
Isopropyl alcohol	4.500
Ammonium hydroxide (25% in water)	0.270
Perfume	0.180
Water	82.305
Propellant 12/n-butane (50:50)	10.00

Note the mixture of fluorocarbon and hydrocarbon. Allied Chemical Corporation has reported propellant mixtures that produce a propellant density approaching that of the concentrate. As a result, the propellant remains emulsified for a greater period of time than either light-density hydrocarbons, heavy-density hydrocarbons, or heavy-density fluorocarbons. Consequently, the consumer does not have to shake the container constantly while using the product. This principle has been applied with good success to many emulsified products.

CUTTING OILS (INDUSTRIAL)

Metal working, which requires the use of cutting tools such as lathes, knives, and drill bits, is based on two principles: first, the formation of the metal chip from the metal by the cutting tool; second, the movement of the metal cut up the face of the cutting tool. Heat is evolved during this process. As a result, machinists use fluids capable of removing heat to cool the cutting tool and to prevent loss of hardness and wear. The fluid used for cooling also prevents distortion of the metal being worked. The same cooling fluid is expected to lubricate both to reduce heat formation and to improve the finish of the surface being formed. Excessive heat and lack of lubrication can cause cut metal to weld to the tool points and to the cut previously made.

Oil-soluble petroleum sulfonates have been used to produce cutting oil emulsions. These oil-soluble sulfonates serve as emulsifying agents as well as rust preventatives. From the above description, the following formulation was developed:

Ingredients	% by Weight
Sodium lauryl sulfate, 30% active	35.0
Sarkosyl N-30 (Ciba-Geigy)	23.0
Water	17.0
Propellant 12/114 (20:80)	25.0

When extreme pressure is applied to a cutting tool in use, a molecular barrier is required to prevent galling and heat deformation of the cutting tool and work piece. Additives containing active sulfur, chlorine, and phosphorus are usually incorporated into the cutting oil formula. Fluorocarbons have demonstrated excellent lubricity at high friction points because of thermal decomposition into active fluorine and chlorine compounds. Therefore the use of a fluorocarbon propellant is recommended.

A second formula containing a "mahogany soap" composed of sulfonates, mineral oil, and water uses Petronate® HL as the main ingredient.

Ingredients	% by Weight
Petronate HL (Witco Chemical)	13.5
Water	87.5
Isobutane (mercaptan added)	10.0

DESIGN LAYOUT INK (INDUSTRIAL)

Generally, it is difficult to mark metal with a sharp writing instrument. Even though it is possible to use a scribe or sharp metal tool for this purpose, such marks cannot be easily removed if an error is made while drawing a pattern on metal. Machinists have found that a nitrocellulose lacquer containing a dye, when applied to metal, can be marked easily with a sharp hard pencil and can be removed or recoated if an error occurs in the layout of the design. The application of the lacquer in an aerosol spray is much more convenient than brushing.

The base lacquer is made as follows:

Ingredients	% by Weight
RS Nitrocellulose cotton, 27% solids, 1/4-sec cut in methyl isobutyl ketone	8.80
Sucrose acetate isobutyrate	2.60
Dioctyl phthalate	0.75
Victoria blue B.O. base (Allied Chemical)	0.67
Acetone	50.18
Toluene	37.00

Once the base has been made and well mixed, this concentrate is used in the aerosol formulation as follows:

Ingredients	% by Weight
Concentrate	60.0
Propellant 12	40.0

The film produced is easily scratched and can be removed with acetone or other ketones.

DETERGENT PRELAUNDRY SPOT REMOVER

Increased emphasis is being placed on presoaking stains in washable clothing, sheets, or table linens before laundering. It is not necessary to soak the entire item when a stain or spot is concentrated in one part of it. Application of an oil solvent as well as a concentrated detergent or emulsifier usually yields good results. An aerosol product offers controlled application and ease of storage. These features, coupled with good cleaning power, may result in an expanded market in the future.

A product demonstrating the foregoing attributes is as follows:

Ingredients	% by Weight
Ninol 1281 (Stepan Chemical)	21.0
Perchloroethylene	49.0
Propellant 12	30.0

The Ninol is a nonionic detergent and chemically is coconut diethanolamide. The perchloroethylene is a good grease and oil solvent.

DRY FLY SPRAY (FOR FISHING)

Fishermen use lures that are intended to float; however, those made with feathers or animal hair become wet after repeated use. The water-repellent properties can be restored by spraying a material such as silicone onto the lure. Formulas containing silicone of a low viscosity, dissolved in alcohol and propelled with a fluorocarbon, are quite satisfactory.

DUST MOP SPRAYS

The ancient practice of collecting dust on a cloth continues today. Oils applied to cloth or rag mops aid in removing dust from floors. Formulations for such products do not require much investigation. Small quantities of mineral oil, usually 2 to 3%, blended with a silicone (1%) in naphtha or kerosene and propelled with a 60% 12/11 fluorocarbon blend, produce a good wet spray that has all the desired properties for removing dust from hardwood and metal.

ENGINE CLEANERS

Cleaning engines requires products that are good at cutting and removing grease. Solvents may be used effectively; however, they must be applied with cloths and rubbed. Incorporating a detergent or surfactant with the ability to emulsify the grease and dirt aids in removing soil. Nacconol® 90F, a solid linear alkylate sulfonate sodium salt coupled with a strong degreaser, such as methylene chloride, can be applied to the dirty engine and flushed off with water after 1 hour. The aerosol product is formulated as follows:

Ingredients	% by Weight
NACCONOL® 90F (Allied Chemical)	5.0
Orthodichlorobenzene	20.0
Methylene chloride	25.0
Propellant 12	50.0

This formula is enhanced if Ozene® is used in place of the orthodichlorobenzene. Ozene is also manufactured by Allied Chemical Corporation and is an emulsified mixture of chlorobenzenes that functions as a solvent for removing heavy grease, oils, and tars.

FILTER (AIR CONDITIONING AND HEATING) CLEANERS

Many heating and air conditioning units employ reusable filters. Cleaning these filters is a problem, and they seldom function adequately. Spraying a dust collector and an emulsifying agent onto the filter improves its efficiency, as well as making it easier to clean. A suggested starting point is the following:

Ingredients	% by Weight
Polyethylene glycol	20.0
Igepal® CO- 520 (General Aniline & Film)	2.0
Hyamine® 1622 (Rohm & Haas)	2.0
Isopropyl alcohol, anhydrous	26.0
Propellant A (12/11/isobutane, 45:45:10)	50.0

A quarternary ammonium type of bactericide is added to impart some sanitizing properties to the filter; however, over long periods of time it loses its effectiveness.

FIRE EXTINGUISHERS

Aerosol fire extinguishers have been on the consumer market for over 10 years. However, none has the approval of the Fire Chiefs Association. Since many are small in size, generally 1 lb, the association is concerned that people may become totally dependent on an inadequate extinguisher instead of calling the fire department at the outset of a fire. Furthermore, none of the extinguishers on the market has a device to indicate that the unit is in operating condition. Devices to show liquid level, or pressure, are desirable features. Several years ago, straight fluorocarbons were used. However, during the last 5 years fire departments have banned the sale of fluorocarbon and halocarbon fire extinguishers because toxic decomposition products were produced under high temperatures.

The type of extinguishers referred to above may be formulated as follows:

Ingredients	% by Weight
Propellant 11	75.0
Propellant 12	25.0

OR

Propellant 11 pressurized with carbon dioxide to 100 psig.

During the last several years, a powder type of extinguisher containing potassium bicarbonate and a water/detergent mixture has been marketed successfully. Each is propelled with dichlorodifluoromethane. Since the propellant is not the main extinguishing agent and is present in small concentrations, this type of extinguisher is gaining acceptance in the market place.

FLOOR WAXES

Obviously, it is uneconomical to wax a complete floor with an aerosol product, but a touch-up aerosol wax used to rewax areas that receive abnormal wear is quite economical and convenient.

Ingredients	% by Weight
Neocryl® B-705 (Polyvinyl Chemical)	3.50
Shanco 300 (Shanco Plastics & Chemical)	1.00
Hardwax E-DM (Dura Commodities)	0.25
Myristic acid	0.25
Methylene chloride	10.00
VM&P naphtha	35.00
Propellant 12/11 (50:50)	50.00

The Neocryl is an acrylic-based polymer that imparts hardness to the waxes. The above product is easily removed with commercial floor cleaners.

FRYING PAN SPRAY

Nonstick cooking with minimal fat is accomplished by the use of polytetrafluoroethylene (PTFE)-coated pots and pans. To really enjoy this no-stick feature, all cookware should be of this new treated type. Rather than replacing present cookware, the housewife can use a formulated product that can be sprayed onto steel, aluminum, or glass. It has been tested for food release after cooking and has proved to be quite successful. The product contains Halon® PTFE powder and a silicone resin. The key to proper formulation is the particle size of the Halon (polytetrafluoroethylane); the polymer must have a particle size of less than 15 μ but larger than 10 μ. The formula is as follows:

Ingredients	% by Weight
Resin SR-417 (General Electric)	7.0
Halon® XG-800, 15 μ PTFE (Allied Chemical)	1.4
1,1,1-Trichloroethane	46.6
Propellant 12	45.0

The maximum release and slip properties of the product can be achieved by curing in an oven at 500°F for a period of 1.5 to 2 hours. When cured, the home-applied coating has a kinetic coefficient of friction of 0.202, compared to a value of 0.203 for a commercially PTFE-coated pan. The surface hardness is the same as that of a commercially PTFE-coated pan.

The product, when applied and cured, releases cooked-on residues of such foods as cheese, eggs, and meat. The product can also be used to coat ice trays and, as mentioned above, glass Pyrex dishes.

FUNITURE POLISHES AND WAXES

A good furniture finish requires continuous care if it is to retain its beauty. Polishes should be formulated to remove dirt and grease and to restore the luster. The polish should be almost completely removable except for a residual film of waxes to protect the finish. However, the residual film should be composed of compounds that do not attract dust or reveal fingerprints. Since furniture finishes are generally made up of resins applied in a solvent base, polishes should not contain harmful solvents. For instance, nitrocellulose lacquers are attacked by ketones and some chlorinated solvents; therefore their use in polishes should be avoided.

The earlier furniture polishes contained waxes dissolved in low-solvency solvents such as varnish makers and paint naphtha. Although these polishes performed satisfactorily, they required considerable rubbing to remove or to spread the wax to a thin film. One of the first formulations to be used during the latter part of the 1950s was the following:

Ingredients	% by Weight
A-C® polyethylene 629 (Allied Chemical)	3.0
Silicone 200 (500 cs) (Dow Corning)	1.8
VM&P naphtha	25.2
Propellant 12/11 (50:50)	70.0

The A-C polyethylene produces good gloss, is resistant to water and ultraviolet light, and at the same time is not as slippery as other waxes.

The naphtha in the above formulation serves only as a carrier or solvent for the wax. The formulator recognized that, if the wax could be applied via another less expensive vehicle, the same end result could be accomplished more economically. The result was the introduction of emulsified furniture polishes containing water and a smaller percentage of naphtha as the vehicle, for example:

Ingredients	% by Weight
Carnauba wax	1.60
Ceresine wax	1.20
Silicone 200 (500 cs) (Dow Corning)	2.10
Span® 80 (Sorbitan Monooleate) (Atlas Chemical)	0.85
VM&P naphtha	13.00
Water	66.25
Propellant 12	15.00

A hard wax, carnauba, is used to produce a high gloss and lasting beauty. Coupled with the carnauba is ceresine, a soft wax to aid in spreading the hard wax and to facilitate removal. The silicone produces water resistance.

Recently, marketers have revived the furniture polishes of yesteryear, citing the advantages of lemon oil products. The majority of polishes sold with a lemon odor years ago were composed of mineral oil to which oil of citronella was added as an odorant. Sometimes a small quantity of yellow dye was added to produce the color of lemon. Today, this same technique is used in formulating to give a lemon odor to the popular "lemon oil" pressurized furniture polishes. The following is an example:

Ingredients	% by Weight
Part A	
Carnauba wax	0.45
Flexowax C (Glyco Chemical)	1.50
VM&P naphtha	15.45
Silicone SF-96 (1000 cs) (General Electric)	1.00
Arlacel® 186 (Atlas Chemical)	0.50
Part B	
Water	80.35
Part C	
Lemon oil perfume	0.75

This formulation has been shown in several parts because it is essential that the waxes be melted and dissolved in the naphtha with the emulsifier. Once this solution is made, the emulsified product is completed by adding the water and perfume while under constant agitation. The emulsion is then added to the aerosol container, and 7 to 10% isobutane is added as propellant. The product contains a hard and a soft wax, as in the preceding formula, and performs in an excellent manner.

GLASS FROSTING SPRAY

Clear glass is often installed in a window or door where privacy later becomes desirable. Formerly, in such cases the glass had to be removed and replaced with a translucent pane. Today the same result can be obtained by spraying the clear glass with a combination of a film former containing nitrocellulose and the key ingredient, 1 to 2% wet ground mica. The mica plates out into a continuous sheet, allowing light to be transmitted but rendering the surface frosted.

IGNITION SYSTEM DRIERS

Electrical systems rely on insulation to prevent conduction of current along paths that will cause a short circuit. Water is a good conductor and can cause shorting. The electrical circuit of an ignition system can be coated with resins to prevent shorting, or the water can be displaced with solutions that dissolve and flush away the moisture. Alcohol is a good desiccant for water; however, it is highly flammable and may ignite in the presence of a spark. The flammability of alcohol can be reduced by use of a nonflammable solvent such as trichlorotrifluoroethane. The following simple formula will dry ignition systems:

Ingredients	% by Weight
Isopropyl alcohol, anhydrous	23.0
Propellant 113 (trichlorotrifluoroethane)	42.0
Propellant 12	35.0

The alcohol/fluorocarbon blend takes up water and, since it is heavier, displaces the water, leaving a dry surface.

LEATHER DRESSING AND PRESERVATIVES

Leather has been used by mankind for various purposes from the beginning of time, and its preservation for extended life has long been practiced. Leather contains a controlled amount of moisture and oil, and removal of either causes leather to become stiff and crack. Therefore leather cleaners should remove embedded dirt and restore oils and moisture. Saddle soap, a blend of tallow and rosin saponified with a mixture of potassium and sodium hydroxide, has been used with small quantities of water to clean and restore the flexibility of leather over the past century. Since it is desirable to control the quantity of water and solvent applied, the aerosol foam soap product is ideal for cleaning leather. A typical product formulation is shown below:

Ingredients	% by Weight
Emulsion SM-2032 (General Electric)	2.85
Stepanol® WAT (Stepan Chemical)	4.75
Isopropyl alcohol	9.50
Glycerine	1.90
Roccal® NC-114 (Hilton-Davis Chemical)	0.45
Propellant 12	5.00
Water	75.55

The SM-2032 is an emulsion of dimethyl polysiloxane fluid, which renders the leather water-repellent after drying. Glycerin has been added as a humectant to prevent moisture from evaporating from the leather. The additive Roccal® is both an antistatic agent and a sanitizer that prevents dust adhesion as well as fungus growth.

A simple formula to treat leather can be made with neatsfoot oil, silicone, and any weak solvent such as naphtha propelled with fluorocarbons.

LUBRICANTS

Lubricants in pressurized packages have the same broad applications as conventional products. They can be composed of simple mineral oil silicones or greases; each type is dispensed with ease. Applications such as lubricating wooden drawer runners, sewing machine needles for high-speed sewing, and television tuners have broadened the market beyond the traditional use on door locks, rubber moving parts, and moving gears.

The simplest and most commonly used lubricant in a pressurized package is a low-viscosity silicone oil, usually present in a ratio of 3 parts to 15 parts of isopropyl alcohol. The remainder is a fluorocarbon propellant.

Molybdenum disulfide can easily be used in a pressurized product as shown:

Ingredients	% by Weight
Molybdenum disulfide	5.0
Naphtha	25.0
Methylene chloride	20.0
Propellant 12/11 (30:70)	50.0

This formula can be used on rubber parts, metal door locks, hinges, and bearing surfaces.

Dry lubricants incorporating polytetrafluoroethylene have been gaining favor in the market place because of their lack of wetness and their good lubricity. The formulations usually use a carrier which is a high-boiling fluorocarbon or chlorinated hydrocarbon.

Ingredients	% by Weight
Polymist® F-7 (Allied Chemical)	2.0
1,1,1-Trichloroethane	53.0
Propellant 12	45.0

This formula can be used on any surface requiring lubrication.

Open-gear lubricants require heavier-bodied oils and must remain on the moving parts at all times. For this application, an asphalt compound has been dissolved in a heavy oil as shown below:

Ingredients	% by Weight
Crystal asphalt coating (Celotex)	42.0
HB Snowflake oil (Archer, Daniels, Midland)	3.0
1,1,1-Trichloroethane	25.0
Propellant 12	30.0

In each of the above formulations, if rust inhibition is desired, a highly purified oil-soluble petroleum sulfonate can be used.

METAL POLISHES AND CLEANERS

Metal products, whether useful or ornamental, form oxides and require cleaning from time to time. After chemical reaction has taken place to produce oxides, the original luster can be restored only by removing this film. All metal polishes and cleaners contain a mild abrasive to cut through the oxide film and expose the base metal underneath. Once the base metal has been exposed, however, oxidation will again take place rapidly unless protection is provided. As a result, metal polishes generally contain waxes or silicones to prevent oxidation.

A typical formulation that can be used to clean aluminum, copper, brass, silver, and stainless steel is shown below:

Ingredients	% by Weight
Beeswax	3.0
VM&P naphtha	20.0
Stearic acid	2.5
Water	58.3
CMC-70, high-viscosity (Hercules Chemical)	0.2
Super Floss® (Johns Manville)	14.0
Ammonium hydroxide (28%)	2.0

This base formula is manufactured in steps. After the first three components are heated and dissolved together, they are added to the water. The thickening agent, CMC-70, and the abrasive are then dispersed into this solution. Finally the ammonium hydroxide is added. Since this is a product based on water, a hydrocarbon propellant such as isobutane can be used.

The abrasive should be a diatomaceous silica with particles less than $3\,\mu$ in size and of the nonscratching variety, that is, particles that crush when pressure is applied.

OIL PAINTING PROTECTIVE COATING

Several types of coatings were covered under the heading "Artist's Fixative." However, because of the special nature of the paints used on new and old oil paintings, an additional product formulation will be given:

Ingredients	% by Weight
Dammar varnish resin (dewaxed)	35.0
Wood turpentine	5.0
Toluene	5.0
Propellant 12/11 (70:30)	55.0

The dammer resin is a hard natural resin compatible with the linseed oil paints used for oil paintings.

PHOTOGRAPHER'S AIR SPRAY

Dust and lint that collect on camera lenses, photographs, and negatives are difficult to remove by wiping. Compressed air is generally used in the photographic shop, but air, high in moisture content, results in condensation on lenses that can cause damage. Vapors from low-moisture halocarbons can also be used; however, if liquid comes into contact with the photographic components, rapid cooling can occur, again causing moisture condensation. It has been learned that fluorocarbons will gel with certain soaps and be released slowly, as a vapor, when the pressure is reduced. A small market for such products, even though specialized, has developed. A suggested product is as follows:

Ingredients	% by Weight
Aluminum octoate	4.0
Methylene chloride	47.0
Propellant 12	49.0

Once this product is filled into an aerosol container and shaken, the entire mass becomes gelatinous.

RELEASE AGENTS

The largest market for release agents is in the molding industry, whether the molds are plastic or rubber. Siloxanes, which have excellent heat resistance, are used almost exclusively in mold-release products. These products are formulated similarly to lubricants and contain a very small concentration of the silicones with solvents and propellants. A complete line of siloxane products is available, from one that produces a glossy surface to one that can be painted after it is removed from the mold. Since considerable literature is available on the subject of mold releases, they will not be covered in depth.

SKI WAXES

The earliest known pressurized product was a ski wax produced by Erik Rotheim in Norway over 50 years ago. The avid skier, like Dr. Rotheim, continues to seek a better method of applying wax to skis than by rubbing with a hard block of the substance. Unfortunately, of the formulations developed, the slip produced from a pressurized package is not as good as that obtained with a hard wax block. Hard waxes such as carnauba are dissolved in naphtha and propelled by fluorocarbons. Unless a soft wax is used in the formula, settling of the hard wax occurs, causing valve clogging.

A suggested formula not using carnauba is as follows:

Ingredients	% by Weight
Kryax® A (Air Reduction Chemical)	5.0
Silicone 200 (350 cs) (Dow Corning)	0.5
Naphtha	14.5
Propellant 12/11 (30:70)	80.0

STARCHES

Spray-on laundry starches were introduced to the market in 1958 and 1959 and enjoyed a steady growth until permanent press clothing began to take over a large segment of the clothing market. Nevertheless, in 1972, over 175 million units of spray starch were sold.

The spray product is a modification of the conventional dip product and affords the convenience of a ready preparation when needed. Spray starch products must be evaluated for stability in the container, scorch resistance, ease of ironing, and feel of the fabric after spraying and ironing.

The containers are generally tin-plated with a lacquer lining of several coatings. Some container manufacturers recommend combinations of two types of coatings to prevent corrosion within the container. In addition to using coated containers, corrosion inhibitors such as ethyl ammonium phosphate and sodium nitrate are added to the formula.

Scorch resistance is a function of the type of starch used. The highly oxidized starches have a greater tendency to scorch than the thin-boiling products. Thin-boiling starches are acid-treated raw starches and have less tendency to gel on the fabric, so that they give better penetration.

Ease of ironing is related to the slip properties of the iron over a starched textile. It is a function of the starch used in combination with an ironing aid. Ironing aids can be silicone emulsions or sulfonated castor oils; both work well to reduce iron drag.

The feel of the textile is of extreme importance to the wearer and involves a subjective test in the laboratory. The addition of borax increases the stiffness of the fabric. A proper balance of starch and borax is essential in producing a product acceptable to the consumer.

A generalized formula should contain the following:

Ingredients	% by Weight
Starch	6.0
Water	92.6
Ironing aids	0.5
Preservatives	0.2
Corrosion inhibitors	0.5
Additives (optical brighteners, perfume, antifoam agents)	0.2

Preservatives are added to prevent bacterial growth. They include formaldehyde or parabens which belong to the bacteriostat class.

The usual method of manufacture is to use one half the water to make a superconcentrate. Once the batch has been heated to 125°F and all ingredients except the optical brighteners, perfume, and other additives have been dissolved, the batch is cooked at a temperature of 210°F for more than 10 but less than 15 min. The rest of the water is then added; and when the mixture has cooled below 110°F, the additives are mixed with the other ingredients.

The diluted or finished mix is added to the aerosol container in a ratio of 95 parts to 5 parts of a hydrocarbon propellant.

TREE WOUND DRESSINGS

Removing bark or cutting a limb from a tree exposes the fibers that carry the fluids throughout the growing structure. Consequently, fluids can be lost by evaporation or weeping. Also, exposed fibers are more susceptible to disease and attack by insects. A recommended formula for a tree dressing is as follows:

Ingredients	% by Weight
Crystal asphalt compound	40.0
VM&P naphtha	20.0
Propellant 12	40.0

The crystal asphalt compound is similar to a creosote-based material, having been produced from coal tars. The asphalt coats the exposed fibers, sealing the ends and preventing weeping, as well as forming a barrier that is not penetrated by insects.

URETHANE FOAM AEROSOL

Urethane foam is produced by reacting toluene diisocyanate (TDI) with a polyester or polyol containing reactive hydrogens in the presence of a blowing agent and catalyst. When these materials are mixed, the reaction usually begins in a matter of seconds and is complete within 20 min. Foam density is controlled by the quantity of blowing agent added. These products are used in the packaging of small, as well as large, delicate parts. In addition, they can also be used as insulation material.

A two-can system, one containing the TDI and a portion of the blowing agent and the other a resin, catalyst, and the remaining blowing agent, has been developed. The components shown in the following formula were selected because the reaction time is greater than that of other systems, allowing time to discharge the contents from the container before complete reaction:

Ingredients	% by Weight
TDI Container	
H-102-N-T (Nopco Chemical)	77.28
Genetron® 11 SBA* (Allied Chemical)	22.72

* Gentron SBA is Genetron to which has been added α-methylstyrene.

Resin Container

H-102-N-R (Nopco Chemical)	64.12
Tetramethyl 1,3-butanediamine (Catalyst)	0.27
Genetron® 11 SBA	35.61

Pressurize to 100 psig with nitrogen.

The two products are held in separate containers until used. A transfer, mother-daughter type of valve is used on one container and its mate on the other. The resin formula is transferred to the TDI container, hence the nitrogen pressure in the resin container. The weight ratio of TDI to resin must be 1:1; that is, if 115g of H–102–N–T is used, the same weight (115 g) of H-102-N-R is put into the resin can. The other ingredients are increased proportionately. Consider the following example.

Assume that 115 g N-T and 35 g Genetron 11 SBA are in the TDI container, and 115 g N-R, 0.49 g catalyst, and 64.4 g Genetron 11 SBA are in the resin container. The two parts are combined in a 12-oz container holding the TDI. The resultant mixture, after shaking until warm, will produce approximately 500 in.³ of rigid urethane foam with a density of 1.5 lb/ft³. Containers meeting the 2Q specification are recommended.

WATER REPELLENTS

Textiles used to make wearing apparel to be worn in rain and snow are treated to repel water. Dry cleaning these fabrics tends to remove the coating that imparts water resistance, and further treatment is required. Silicones, designed to cure or cross-link in the presence of atmospheric moisture, have been found useful in this application. The air-cured silicones are soluble in aliphatic, chlorinated, and aromatic solvents and retain their activity when packaged in an anhydrous system. Two manufacturers of fabric, water-repellent silicones are Dow Corning Corporation and General Electric Company; the specific registered trade names are Dow Corning C–2–0563 and GE SS–4098, respectively. The manufacturers recommend the use of 5 lb of water repellent compound dissolved in 20 lb of solvent. This concentrate is packaged in a pressurized container in a ratio of 25% to 75% Propellant 12/11 (50:50). Upon drying overnight, the fabric will achieve maximum water repellency.

WEED KILLERS

Weed killers, in pressurized containers, are designed to treat weed plants that grow in isolated parts of the lawn. Costwise, the aerosol package cannot compete with the granular product in the treatment of large weedy areas. Several common weed killers soluble in alcohols and oils can be used in the pressurized package. The butyl ester of 2,4-D and the isopropyl ester of 2,4-D can be dissolved in base oils and packaged with 25% Propellant 12. Very low concentrations of the 2,4-D ester are required. Generally, 0.2 to 0.3% is adequate to kill the common broadleaf weeds. Water-in-oil emulsions can be made with the commercial weed killers to produce foams that are easily seen when sprayed. These also reduce cost.

ZIPPER LUBRICANT

Metallic zippers become corroded when subjected to the alkalinity of laundry soaps and detergents. When corrosion products build up on the metal parts, the slide becomes difficult to operate. The application of a silicone lubricant in a pressurized spray reduces the drag of the slide on the teeth of the zipper. Polytetrafluoroethylene (PTFE) suspended in 1,1,2-trichloro-1,2,2-trifluoroethane propelled with Propellant 12 provides lubricity without affecting the cloth. Small quantities of PTFE, usually 3 to 5%, are adequate for a good lubricant spray.

DEDICATION

The author would like to dedicate this chapter to his son Dean, who assisted him by doing a major share of the library research.

BIBLIOGRAPHY

Product Information Aerosol Files, Allied Chemical Corporation, Specialty Chemicals Division, Morristown, N. J.

Commercial Organic Chemical Names, SOCMA Handbook, American Chemical Society, Washington, D.C., 1965.

Raymond E. Kirk and Donald F. Othmer (Eds.), *Encyclopedia of Chemical Technology*, Vols. 1–15, Interscience Publishers, New York, 1947.

Milton A. Lesser, *Modern Chemical Specialties*, McNair-Dorland Company, New York, 1950.

O. T. Zimmerman and Irvin Lavine (Eds.), *Handbook of Material Trade Names*, Industrial Research Service, Dover, N.H., 1953; *Supplements*, 1956, 1957, 1960, 1965.

A. Herzka (Ed.), *International Encyclopedia of Pressurized Packaging (Aerosols)*, Pergamon Press, London, 1966.

AIR FRESHENERS AND
SPACE BACTERICIDES

Offensive odors have plagued man since Neanderthal days, and his efforts to cope with the problem have been many and varied. His caves and other rude dwelling places were always selected with a view to ventilation. Cooking was done near a good draft, and a downwind, remote area was reserved for spoiled foods and other waste products. As civilization progressed, people became more enclosed and more sensitive to the increasing need for odor control. Since ventilation was not always practical, many other methods were developed. Fragrant oils were made by heating flower petals in animal fats so that they could be evaporated from wicks or burned. Bags of pine and cedar bark were often hung near ceilings, where the warm air could hasten the vaporization of terpenes and other scented ingredients. The modern sachet is a holdover from these bygone days. Somewhat more flexibility could be achieved by burning incense, but such unpredictable soporific and hallucinogenic side effects due to inhalation of certain pyrolysis products often resulted that these powders became surrounded with a rather surrealistic mystique. Sometimes they were banned. More often, however, they played an important part in religious ceremonies, medicinal practices, and even in the so-called black arts. All these early methods were either so slow, so cumbersome, or so ineffective that it was generally conceded that odor problems were rarely so serious as to justify the cure.

When disposable aerosols arrived in 1947, what could seem more natural than to formulate a perfumed spray and market the product as an air freshener? The idea became a reality in 1948. Within a few short years the aerosol spray became the principal means of combatting household odors. Aerosols worked quickly and could be sprayed about with a minimum of effort and discomfit. They were even rather fashionable. A truly magnificent future was predicted.

The earliest air fresheners were simple blends of perfume, solvent, and a preponderance of propellant. Actually, they were "reodorants" more than anything else. From about 1950, chemical deodorizing ingredients were included in the formulas of some products in order to enhance their capabilities. In fact, during the next 10 years or so a number of very elaborate compositions and testing methods were developed as several major marketers scrambled to attain leadership in this rapidly growing field. Most of the products were represented as air deodorants. In the case of the straight perfume types this might be considered technically insupportable, except on the premise that the introduction of a strong odor tends to physiologically reduce the impact of other ones.

By 1960 competition had driven prices to very low levels, and vigorous programs were initiated with the aim of reducing the cost of formulation. Water-based formulas of various types had been under study for several years but were fraught with operational problems—plus the need to use a flammable propellant consisting either largely or entirely of isobutane. When an emulsion type of air freshener was introduced nationally in late 1960, economic comparisons were so shattering to the standard products that other marketers had no recourse but to follow suit. The formula sophistications of the 1950s were quickly discarded, so that these earlier products became practically nonexistent except for door-to-door and other premium-priced outlets.

The water-based air fresheners were characterized by rather loud sprays, with a somewhat coarse particle size. Because of their slower delivery rates, cans lasted longer than before. These factors largely offset the low prices and slowed the market growth to only a few per cent each year. In 1964 a new deodorant appeared: the spray disinfectant. Assisted by heavy brand name advertising, these products brought about a profound renewal of consumer interest in air deodorants and, incidentally, a unit sales increase of nearly 50% in 1965.

Since this spurt of activity, the market has once again settled down to a growth rate of about 3.5% per year. Low selling prices preclude any large advertising or promotional thrusts. Also, no dramatic innovations appear to be at hand. Hence most marketers predict a simple continuance from the estimated 1970 base of about 180,000,000 units per year in the United States and Canada. This amounts to about 7% of the total aerosol production in these two countries.

PSYCHOPHYSICS OF ODOR RECEPTION

The psychology of olfaction is a subscience that has been extensively studied. It remains a very elusive and difficult area, however, and one in which research can be singularly unrewarding. In the case of air fresheners it is generally agreed that the ingredients may act to physically or sometimes chemically change the properties of malodorous molecules. In addition, they usually perform a physiological function on the olfactory, trigeminal, and taste receptor cells, so that the overall sensory impulses that they develop and send to the brain will have a more pleasing interpretation.

The psychological interrelationships of sensory phenomena are well defined. By using sensory overload techniques, it is possible to dull or even eliminate pain by loud shouting, dipping the feet in ice water, or listening to "white sound" pulsating through earphones. In the same fashion, it is possible to reduce or eliminate the effects of many offensive odors by simply flooding the brain with sensory images resulting from the reception of more acceptable ones. These effects will diminish in time as the brain accommodates to the new situation, but by then the odor problem may have abated. This is probably the principal *modus operandi* of air fresheners.

By electron micrographs and other techniques it has been shown that, in order to be detected, the invading odorous molecules must pass through a film of surface moisture and a layer of fatty tissue and into the extracellular solution of the olfactory cell. They must then bring about a sharp decrease in the electrical resting potential in order to generate a response signal. Ascorbic acid has been shown to function as a cofactor required for olfaction. In its presence an odorant triggers a conformational change in the epithelial protein, rendering the ascorbic acid available for oxidation. The acetylcholine/cholinesterase system is also thought to participate in odor perception mechanisms, but the methodology is still unclear.

These revelations have established, in a general way, the conditions that molecules must satisfy in order to have a detectable odor. They must either be reasonably volatile or else get dispersed into the air in very finely divided form. This is necessary so that the odorant can reach the olfactory area. They must then possess both hydrophilic and lipophilic properties, in order to assure permeation of the aqueous and fatty layers surrounding the cell. Actually, this is not a critical situation, for the layers are quite thin and somewhat hybridized. Finally, the odorant molecules must have an active site—generally intraatomically or intramolecularly unsaturated. This has been termed the osmophoric area and is necessary for cell excitation. If the osmophore is buried, complexed, absorbed, or otherwise shielded, the physiological response is reduced. The detection threshold may increase thousands, even millions, of times. The odor will become less pungent—perhaps even pleasant.

Despite these apparent restrictions, almost every reasonably volatile substance has a definable odor. This is largely due to the extreme sensitivity of the sense of smell. Mercury metal has a dry alliaceous odor detectable by many people, although its vapor pressure is only about 1.0×10^{-8} atm at room temperatures. Perhaps the minimum identifiable odor threshold for man is reached with methyl selenol, $CH_3 \cdot SeH$; persons chronically sensitized to this substance because of a history of selenium poisoning can detect as little as 1.0×10^{-14} g. The ultimate in perception sensitivity may have been reached in the insect world. Responses to certain aliphatic sex attractants appear to extend down to 10^{-18} or even 10^{-20} g levels, for example, a few hundred molecules. Minimum taste thresholds are less attenuated. The bitter taste of Bitrex and related quaternary compounds can be picked up at about 2.0×10^{-8} g, and this appears to be about the sensitivity limit.

The mechanical detection of air-borne odorants depends almost exclusively on gas chromatographic separations, with columns operating under cool conditions and reduced pressures. Special techniques are used to isolate trace components from huge volumes of air. After a sufficient amount has been collected, the operating conditions are adjusted so that individual odorants can be perked through and identified, using infrared spectrophotometry or other analytical techniques.

SOURCES OF ODOR

Odorous materials find their way into the environment as a result of cooking (steam distillation), pyrolytic degradation, and bacterial putrefactions. Most household odors are generated in the kitchen. They may be divided into three classes: those from fresh foods, from cooked foods, and from stale or putrified food products. Fresh fruit fragrances are said to be the most pleasant of all odors but are usually rather weak. Except for their limited solubility in fluorocarbons, these odors would actually be preferred for use in breath freshener aerosols, instead of the more common mixtures of peppermint and spearmint oils with additions of menthol. Other fresh food odors are generally well received, although again they are rather weak. Certain raw meat, liliaceous, and spicy smells sometimes have an unsettling effect on susceptible people, with the magnitude depending on their psychological outlook, the odorous pungency, and many other factors.

Cooking changes and greatly intensifies nearly all food odors. The more volatile ingredients are steamed off preferentially, so that natural balances are distressed. Also, chemical reactions take place during cooking. They include oxidations, hydrolyses, transesterifications, semiacetal formations, condensations, and enzyme degradations. Any of these can promptly produce a more or less profound change in the odor complex. The pre-eminence of garlic, onion, cabbage, squash, and liver odors is due partly to the formation of minute amounts of various simple sulfur compounds during cooking.

Without a doubt the most offensive kitchen odors are those brought forth by the decay of foods. Spoilage processes are generally caused by the splitting actions of microorganisms: bacteria, molds, yeasts, viruses, and so forth. Less importantly, decay may be caused by simple chemical instability, oxidation, or enyzme-catalysed decompositions. Meat products spoil, forming amino acids, amines, putrescine, and other compounds, many of them highly offensive. Fish residues spoil and give off ammonia, simple amines, skatole, and various other indoles. Eggs form ammonia, hydrogen sulfide, and a collection of thiols. Dairy products spoil by processes that include ester hydrolysis, forming butyric, valeric, and caproic acids and other compounds. Finally, vegetables decay to produce a vile effluvium rich in thioethers, thioamides, and organic sulfides.

After kitchen odors, those that originate in bathrooms are next in importance. Here the odor mix is due to urine, feces, and flatus, plus a much smaller contribution from axillary perspiration. Hydrogen sulfide, ammonia, and the alkyl analogs of these two gases are the principal offenders. In particular, the urine content in bathroom or nursery diaper pails can be rendered almost unbearably odoriferous by the action of *Brevibacterium ammoniagenes*, *Proteus vulgaris*, and similar bacteria that split urea and amino acids. The action of these organisms can generate large quantities of ammonia and amines within just a few hours.

Another important household odor is that of tobacco smoke. It is not localized to any special areas. In addition to the tars and nicotine so often talked about, the smoke contains aldehydes, methylethylpyridine, furfuryl thiol, and hundreds of other ingredients. Many of them are further oxidized in air, so that the composition of stale smoke is quite different from that of fresh smoke.

The most frequently overlooked household odors are those that originate in the breath and perspiration. The average person transpires

TABLE 20.1. CHEMICAL CONSTITUTION OF HUMAN PERSPIRATION (% BY WEIGHT)

Sodium ion	0.017–0.400	Chloride ion	0.036–0.995
Potassium ion	0.007–0.145	Sulfate ion	0.004–0.0170
Calcium ion	0.0003–0.0118	Phosphorus ion	0.00000–0.00048
Magnesium ion	0.00002–0.00332	Sulfur ion	0.00008–0.00737
Copper ion	0.000006	Bromide ion	Present
Manganese ion	0.000006	Fluoride ion	0.00002
Iron ion	0.000064–0.000200	Iodide ion	0.00000072–0.0000095
Formic acid	0.0001	Asparaginic acid	0.00005–0.00006
Acetic acid	0.0005	Glutaminic acid	0.00006–0.000065
Propionic acid	0.0002–0.020	Serine	0.00001
Butyric acid	0.0001–0.0010	Glycine	0.00007
Valeric acid	0.00018–0.0047	Threonine	0.00108–0.00011
Isovaleric acid	0.0002–0.0100	Alanine	0.000076–0.00006
Capric acid	0.0006–0.002	Valine	0.000055–0.00005
Caprylic acid	0.0008–0.030	Isoleucine	0.00004
Caproic acid	0.0005–0.002	Leucine	0.00007
Lauric acid	<0.0001–0.0005	Citrulline	0.00019–0.00172
Myristic acid	Present	Histidine	0.00002
Palmitic acid	Present	Arginine	0.00003–0.00032
Stearic acid	Present	Lysine	0.00003
		Taurine	<0.00001

Urea	0.0235–0.4000		Cystiene	0.00001
Uric acid	0.0725–0.2520		Tyrosine	0.00002
Creatine	0.0047–0.0085		Phenylalanine	0.00002
Creatinine	0.0007–0.00190		Proline	<0.00001
Choline	0.0003–0.0015		Methionine	0.00001
			Tryptophane	0.000014
Glucose	0.0010–0.0256		Hydroxyproline	<0.00001
Lactic acid	0.033–0.306		Ornithine	0.00009–0.00226
Pyruvic acid	0.00090–0.00695		α-Aminobutyric acid	0.00080–0.002
Folic acid	0.00006–0.00013			
Inositol	0.001		Urocanic acid	0.0031
Pantothenic acid	0.0002–0.00045		17-Ketosteroids	0.0001
Thiamin (B$_1$)	0.00001–0.00048		Cortisone	<0.0000001–0.0000570
Riboflavin (B$_2$)	0.0005			
Nicotinic acid (B$_5$)	0.0004–0.00105		Carbon dioxide	Present
Pyridoxin (B$_6$)	0.000008		Hydrogen sulfide	Present
Biotin (B$_7$)	<0.00001			
Cobalumin (B$_{12}$)	0.00001		Water	ca. 96.50
Ascorbic acid (C)	0.0015–0.0066		Methyl alcohol	0.0001
Dehydroascorbic acid	0.0041		Ethyl alcohol	0.0002 min.
p-Aminobenzoic acid	0.0005		Formaldehyde	0.00005
			Acetaldehyde	0.00024

about 3 lb of water daily, plus smaller amounts of salt and other products. The organic residues are easily acted on by body bacteria to produce a diversity of new products, many of them distinctly malodorous.

A person with a clean, healthy body, who exercises adaquately, usually has a minimum odor level in his breath and perspiration. The faint odors that do exist are sweet and animalistic. They even possess moderate sexual overtones, possibly as a result of trace amounts of higher aromatic aldehydes and macrocyclic ketones. Bacteria, resident on the skin and also present in the mouth and lungs, contribute specific compounds to the odor mix. These substances often assist in the development of an interesting overall body odor. Thioethers, for instance, add intriguing by-notes to the breath odors of animals and people but are absent in the breath of germ-free animals. Table 20.1 lists the chemical composition of human perspiration.

A truly large collection of chemical substances is present in the breath, and an even greater number are found in perspiration. The compositions are quite variable, changing rapidly to reflect the health, activities, and cleanliness of the host. The presence of unique odorants and other substances can be related to the effect of certain diseases with surprisingly accurate results. Thus acetone can be a factor in diagnosing diabetic acidosis, and *trans*-2-methylhexanoic acid can be detected in the sweat of schizophrenics. Pyridines are produced as a result of nervous tension or strenuous exertion. The breath of a person with uremia has a fecal odor. Nitrites signify abscesses or, sometimes, intestinal obstructions. The particular body chemistry of Mongoloids leads to the production of uric acid, so that the content in their perspiration is at least 15 times the usual level. The use of liquor and tobacco can alter the composition of the breath so predictably that product brand names can be determined by the use of chromatographic techniques.

A goodly number of products are used for the reduction of bad breath and body odors. Many of them are aerosols. The list includes breath fresheners, personal deodorants, antiperspirants, feminine hygiene deodorants, the newer male hygienic deodorizers, and even perfumed spray talcum powder products. Almost invariably, these formulations include a bacteriostatic agent. Chlorinated phenols are most commonly used, followed by ammonium quaternaries and a group of polyhalogenated organics. The aluminum chlorohydroxide complexes found in most antiperspirants are moderately antibacterial because of their acidity—pH of 2 or 3, as a rule. In a few instances, breath fresheners depend on the chemical deodorancy of spearmint, amyl esters (Juicy Fruit), chlorophyllins, and other special compounds, rather than true bacteriostats. A few rather unusual substances, such as copper(II) gluconate and endodextranases, have been tested as oral bacteriostats with good results. All these products, when used on a regular schedule, will help to reduce the contribution of breath and perspiration odors to the total environmental odor level.

Unfortunately, household odors are often of a fairly permanent nature. In some instances, active malodors can permeate into furniture and porous house surfaces, after which they release at slow rates that depend on temperature and humidity. Many off-odors originate from molds in damp basements, shower stalls, and woodwork exposed to condensations from water pipes. Mice, rats, and untrained household pets create rather permanent and undesirable odors. Termites and other insects produce acidic odors. Combustion products from fireplaces, furnaces, water heaters, and so forth generate odors that can be smoky, pyroligneous, and sulfurous in nature. Remote combustions are also contributory. The average large city dwelling contains a permanent level of about 1 g of metallic lead dust in the air space, as a result of gasoline combustion.

Except under unusual conditions the amount of foreign gases and vapors will amount to only 0.5 to 125 ppm by volume of the total household air. This is far below the asphyxiation level. In fact, these concentrations are so low that only a few very poisonous gases could ever produce any toxic effects. Nevertheless, many people complain of headaches, nausea, and other enervating or generally distressing symptoms. In most cases they are thought to be the physiological result of emotional reflexes, which cause vasoconstriction and subsequent interference with proper blood circulation. The total effect then becomes one of mild oxygen starvation.

CHEMICAL PRINICIPLES OF ODOR CONTROL

Since nearly all odorants contain unsaturated or low-valence osmophoric sites, the simplest method for destroying them would appear to be through direct oxidation. Obviously, this can be done with fire. In this case, not only is the osmophore destroyed, but usually the rest of the organic molecule as well. Fire has less effect on inorganic odorants. Certainly such procedures as cremations, flame sterilization, and combustion-type waste disposal prove the importance of fire as a large-scale agency for the destruction of odors. In the home, of course, the use of open fires is restricted to the fireplace and is not a very effective method for odor control.

The most common natural oxidant is ozone gas, a highly reactive, dangerous, and irritating product formed by the ultraviolet irradiation of oxygen. It can also be produced electrically, both in nature and in the laboratory. Unfortunately, the properties of ozone are such that it cannot be used as a household deodorant.

Ozone can be reacted with unsaturated organic compounds to form ozonides. With many ketones it gives trimeric peroxides. Thiols are oxidized to disulfides and then to sulfones. Thioethers react to form sulfones and sometimes sulfoxides. The ozonides and peroxides are very strong oxidants, often reacting instantly with odorant molecules to cause the release of oxygen and the destruction of the osmophore. Both types of compounds are fairly dangerous and unstable, especially the ozonides. Olive oil ozonide has been prepared as a 5% solution in water. Although it did not appear to acutely affect the skin, the solution eliminated many offensive odors and also killed numerous bacteria and viruses.

Substances such as peracetic acid (also known as acetyl hydroperoxide), peroxidized geraniol, and peroxidized perfume oils have been tested in aerosol form as chemosterilants and deodorants. Such products have never been marketed, however, because of instability problems.

Aliphatic Aldehydes as Air Deodorants

Most aldehydes are very active compounds capable of reacting rather easily with a great diversity of other organic molecules, with oxygen to form the corresponding carboxylic acid, with selected inorganic compounds to give addition products, and even with themselves under certain conditions to produce complex polymeric mixtures. The oxygen atom is relatively ionic in nature. This allows the simpler aldehydes to exhibit good solubility in polar solvents such as water.

Many aldehydes can be detected by odor, down to about 0.5 ppm in moist air. Above about 20 to 100 ppm the simpler aliphatic aldehydes exert both an irritating and a sensitizing action on the nose and throat mucosa. These unpleasant effects begin to lessen at about C_{10} and are greatly suppressed upon reaching C_{15}. The primary aldehydes are more irritating than their secondary counterparts.

Aldehydes from about C_6 upward have been used in the preparation of perfume bases designed for industrial odor control. Certain ones are more effective than others. As a rule, aldehydes in the C_9 to C_{11} range form the best deodorants, particularly if the aliphatic chain is modified by the inclusion of methyl groups, a hydroxyl group, a double bond, or some similar feature. These aldehydes have rather pungent green notes, plus a characterizing secondary note that may be grassy, citrus, fatty, leafy, pine, or lily-like in type. The overall sensation is one of almost overpowering freshness and cleanliness.

Formaldehyde, the simplest member of the aldehyde family, is a reactive gas, soluble in water to over 40% at room temperature and pressure. It has been tested as an aerosol deodorant in both water-based and anhydrous systems. Although there is no doubt that it can chemically destroy certain odors, such as those of the amine and sulfide varieties, the sharp, penetrating odor of the gas itself far outweighs any benefits. Added problems have been posed by corrosion, trimerization, and elastomer embrittlement reactions. The gas is not used in air fresheners at this time.

The next member of the series, acetaldehyde, is similar in behavior, but only about one third as reactive. It has a pungent, choking odor with a somewhat fruitlike overtone. Aldehydes such as acetaldehyde are often formed as a result of corrosion reactions in alcoholic aerosol products, but they are not used as deodorants.

Aldehydes from C_3 to C_8 sometimes find marginal applications as minor ingredients in aerosol perfumes, but here the trend is more toward the creation of a final odor than anything else. At C_9 there is one of the more famous true deodorants: 3,5,5-trimethylhexanal. This compound is more or less representative of many polymethylated intermediate aliphatic aldehydes in the C_8 to C_{12} range. It achieved early prominence largely as a result of the relative ease of preparation, being readily formed by oxo processes starting with inexpensive petroleum intermediates. It was used by the military as early as 1953 to reduce hospital odors, particularly that of blood. During late 1954 the crude material was released to industry. It was purified to a minimum assay of at least 95% total aldehydes by several perfume houses and then sold as a chemical deodorant for various uses, including the preparation of deodorizing aerosols. The material available in 1973 runs at least 99% total aldehyde, with over 93% of this amount consisting of actual 3,5,5-trimethylhexanal. Nonaldehydic impurities are related higher alcohols, water, and a substituted phenol oxidation inhibitor. The odor is very sharp, long-lasting, and pungent, rather like that of a handful of freshly cut grass.

3,5,5-Trimethylhexanal is a colorless liquid that yellows upon exposure to air. Eventually, it becomes rather cloudy. It has a specific gravity of 0.831 at 70°F and boils at 273°F. The pH varies from 3.8 to 6.2, depending on the degree of oxidation to the corresponding 3,5,5-trimethylhexanoic acid and the amount of water present in the sample. Water solubility is 0.59% at 77°F, and the pure compound generally contains from 0.03 to 0.25% water in solution.

The deodorant is applied at levels of about 0.01 to 0.035% when formulating aerosols. The actual limit is determined by tolerance to the sharp odor more than by any other factor. Formulas with high levels of perfume (especially pine types), as well as those that deliver as coarse sprays or metered sprays, can tolerate more deodorant. Many perfumes contain ingredients capable of chemically reacting with 3,5,5-trimethylhexanal, so they should be tested on a basis of proven long-term compatability. Ideally, tests should be conducted for at least a month, using both room-temperature and 100°F storage. In the past, aerosols were sold that employed 3,5,5-trimethylhexanal (and perhaps also n-lauryl methacrylate) as the deodorant, but contained no perfume to provide a final, pleasing fragrance. These products were uniformly unsuccessful and demonstrated the need to include perfumes in all air fresheners.

3,5,5-Trimethylhexanal has demonstrated good compatibility with alcohols, water, petroleum distillates, various propellants, and other chemical deodorants. However, it is incompatible with reducing or oxidizing agents, amino acids, amphoteric detergents, triethanolamine and triethanolamine fatty acid soaps, many spice-type perfume components, and certain metals. Aluminum apparently reduces it slowly to the alcohol. In the presence of oxygen, tin and, to a lesser extent, lead drive the aldehyde to the acid and then react to give insoluble salts, such as tin(II) 3,5,5-trimethyl hexanoate. These crystalline deposits are capable of plugging aerosol valves. Also, the formation of by-products detracts from the deodorant efficacy of the aerosol. Generally, they can be avoided by using double-lined aerosol cans with a side-seam stripe. Ideally, for these formulas cans should be vacuum-crimped at 20 in. Hg° minimum.

A great number of other aliphatic aldehydes have been tested as chemical deodorants; among them are p-ethoxpropionaldehyde, glutaraldehyde, 2.6-nonadien-1-al (violet leaf aldehyde), 2,4,4-trimethylheptanal, n-octanal, and the higher, citrus-type aldehydes, which begin at C_{11}. These last compounds require perfumes that complement the clean, citrus notes displayed by the deodorant. In general, it is found that the more highly branched chain compounds are more effective as deodorants. This carries over into the aromatic aldehydes. Vanillins, for instance, are a principal component of the well-known Alamask blends. The branched aldehydes are also more water soluble, indicating increased lability of the carbonyl oxygen atom. By making the alkyl moiety more electrophilic, the deodorant ability is also enhanced.

The methods by which the higher aldehydes control odors are only partly understood. The odorous level of these aldehydes acts as a limit to the amount that can be introduced into a given air space. This is considered to be about 2.0 ppm as a localized, temporary, maximum concentration, right after spraying. The concentration then levels out to about 0.1 ppm before eventual dissipation. Since such low concentrations of aldehydes are present, it is contended that these deodorants must have a selectivity pattern which permits them to seek out and destroy more obnoxious odors. Usually, these particular odorous molecules would be the most chemically reactive and thus more susceptible to attack. Mechanisms such as oxidation, addition, hemiacetal formation, and physiological odor compensation have all been suggested, but in each instance the supporting evidence has been fragmentary and controvertable.

Double-Bond Compounds as Deodorants

Unsaturated compounds have been of interest as chemical deodorants on the basis that many reactive odor-causing molecules could be eliminated by addition across the double bond. Unfortunately, many of these proposed unsaturates are themselves possessed of rather vile odors. In contrast to ethane and propane, which are nearly odorless in the pure form, ethylene and propylene have pronounced phosphoretic smells that are quite revolting. This relationship extends in unmitigated form all the way up the respective series to about C_{20}. Acetylene has a foul odor. Allene (1,2-propyldiene) has a characteristic odor, not too unpleasant, especially when masking-type perfumes are present. It is claimed to function as a potent deodorant at 0.3% levels in aerosol formulas. The liquid boils at $-28°F$. The gas is unstable to water and alcohols because of slow solvolysis. Compositions based on the use of petroleum distillate solvents and fluorocarbon propellants appear to be the most promising, but the high cost of the allene and several other factors have led to a virtual neglect of its apparent abilities. Butadiene gas has a very bad odor, and the conjugated position of the double bonds leads to a stability factor that would probably compromise the ability of this compound to destroy odors in any event.

Perhaps the most interesting groups of double-bond deodorants are the esters of fumaric acid and the esters of substituted fumaric acids, such as citraconic acid, mesaconic acid, and the two aconitic acids. The double bonds in these acids and esters are bracketed between carbonyl groupings, which gives them a considerable electron deficiency and hence a high level of reactivity.

These unsaturated esters readily undergo simple addition reactions with thousands of thiols, halogen compounds, aldehydes, and so forth, producing substituted succinates in the case of the fumaric esters and similar compounds for the related esters. They also undergo the Diels-Alder reaction with conjugated dienes.

Organoleptic examinations of several fumarate esters have shown that these substances are essentially nonirritating and nontoxic and have quite pleasant, mild odors. This mildness appears to extend to the mesaconic esters and related compounds.

The alpha-unsaturated dicarboxylic esters have been sold to some extent for aerosol deodorant applications. Dihexyl fumarate is available for this purpose, usually in combination with a complementary air deodorant, geranyl crotonate, under the trade name Neutrolair D-7 (Fritzsche-Dodge & Olcott, Inc.).

Possibly the most important class of aerosol deodorants consists of the alpha-unsaturated monocarboxylic esters. They have many properties in common with the esters just discussed, but are generally considered to be more versatile and reactive. Typical examples are the esters of the following acids: crotonic, acrylic, methacrylic, tiglic, and senecionic. In each of these compounds, the conjugated system $-C{=}C-C{=}O$ is present, with the carbonyl group acting to stabilize the molecule generally, but meanwhile augmenting the singular reactivity of the double bond by functioning as an electron sink. The first of these esters to be used for aerosol deodorant applications was n-lauryl methacrylate. Since 1955 this chemical has been sold under the trade name Metazene (Motomco Inc.). Table 20.2 illustrates the use of these compounds.

Metazene is a very light amber liquid of low volatility. It has a very mild, clean odor, somewhat like that of extremely dilute laurylaldehyde. Older samples may pick up a fatty odor through splitting at the ester linkage. The commercial material is essentially 100% n-alkyl methacrylate. The alkyl portion is derived from commerical lauryl alcohol, which contains mostly C_{12} and C_{14} alcohols, with lesser percentages of the C_{10} and C_8 alcohols as well. When alcohols of less than about C_8 complexity are present to any extent, the methacrylate ester composite becomes more reactive and also more odoriferous. Partitioning generally takes place in the final distillation step, where the forerun is carefully removed and discarded.

The selection of commercial lauryl alcohol for the alkyl moiety was based on organoleptic testing. It is now recognized that some water solubility is a prerequisite for all deodorants. In the case of the methacrylates, water miscibility decreases drastically as the alcohol group passes beyond C_8 becoming almost zero upon reaching C_{14} to C_{16}. Because the higher esters cannot penetrate odors locked within airborne condensation nuclei, and since they are generally too chemically inactive, they do not function as effective air deodorants.

Methacrylic acid and its esters homopolymerize and copolymerize readily, and usually quite rapidly, under the influence of peroxides, oxygen, and other catalytic influences. This property is strongest in the methyl ester and decreases slowly as the chain length of the alkyl substituent increases. In contrast, crotonic acid and its esters do not homopolymerize at all beyond a dimer and copolymerize only with some difficulty. Both of these isomeric series are inhibited against polymerization by the addition of compounds such as hydroquinone. The ease of addition reactions across the double bond is about the same for the methacrylates and crotonates, but occurs less readily for the acrylates, tiglates, and (presumably) senecioates. In selecting

TABLE 20.2.   FORMULATIONS FOR DEODORIZING AIR FRESHENERS

	Formulation (%)		
Ingredients	I	II	III
3,5,5-Trimethylhexanal	0.03		0.025
Metazene, 100%		2.00	1.600
Perfume oil	0.47	0.30	0.375
Odorless mineral spirits	24.60	27.70	22.500
Methylene chloride, stabilized		10.00	
Propellant 11	30.00	10.00	65.000
Propellant 12	45.00	50.00	
Propane A-108			10.500
Pressure (20 in. Hg° crimp) (psig at 70°F)	48	51	49
Delivery rate (g/sec at 70°F)	1.1	1.1	1.0
Flame propagation at 70°F (in.)	16	17	17
Closed drum test at 70°F (sec)	68	62	61
Flash point (°F) (TOC-Mod.)	137	130	< -75
Container	Single lined	Plain	Single lined

compounds as air deodorants it is important to pick those that do not homopolymerize readily but will copolymerize with odorants or else add them easily across the activated double bond. These, then, are the considerations that have been used to explain the exceptional abilities of n-lauryl methacrylate, geranyl crotonate, 2-ethyl hexyl crotonate, and similar specific compounds as aerosol deodorants.

The alpha-unsaturated monocarboxylic esters undergo well-defined, rather rapid reactions with specific odorants. With ammonia, for instance, the crotonates react to give esters of β-aminobutyric acid. Amines and aniline add on by means of similar nucleophilic reactions. With hydrogen sulfide, n-butyl cinnamate produces di-n-butyl-β,β'-diphenyl-β,β'-thiopropionate:

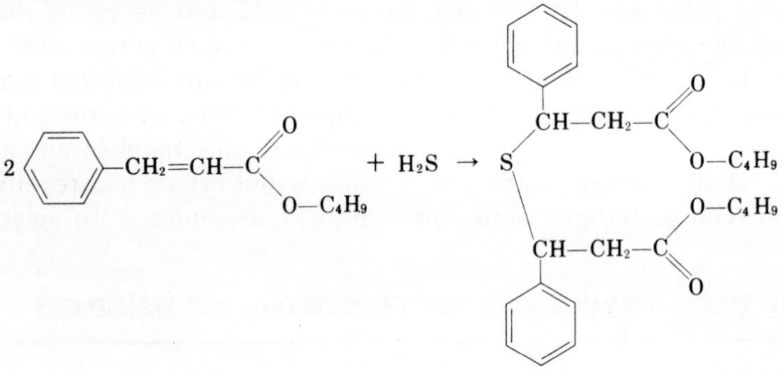

Hypochlorous acid adds to n-lauryl methacrylate by an electrophilic mechanism, producing n-lauryl α-chloro-β-hydroxyisobutyrate and a much smaller amount of n-lauryl α-hydroxy-β-chloroisobutyrate:

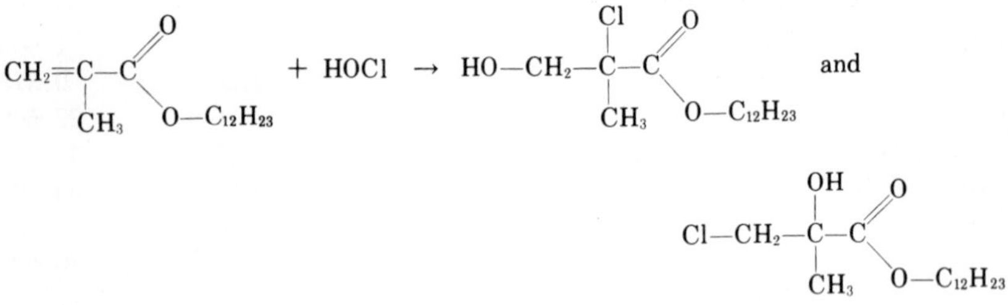

Iodine vapors react with *geranyl* crotonate by a similar electrophilic reaction to give geranyl α,β-diiodobutyrate:

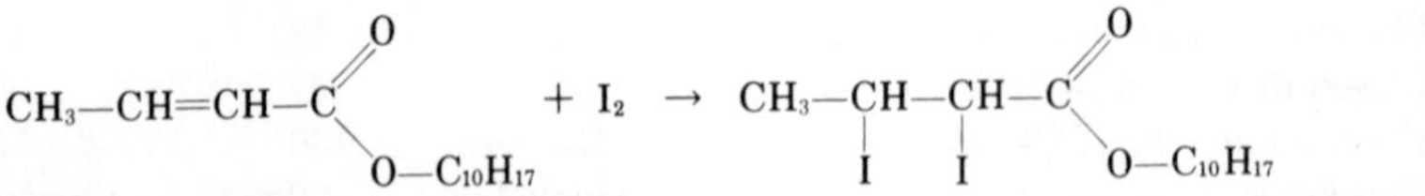

Butadiene reacts with crotonates by a Diels-Alder route to yield the corresponding tetrahydro-o-toluic acid esters:

$$
\begin{array}{c}
\text{HC} \overset{\text{CH}_2}{\underset{\text{CH}_2}{\diagup}} \\
\end{array}
\quad + \quad
\begin{array}{c}
\text{H} \quad \text{CH}_3 \\
\text{C} \\
\| \\
\text{C} \quad \text{O} \\
\text{H} \quad \text{C} \\
\qquad \text{O—R}
\end{array}
\quad \rightarrow \quad
\begin{array}{c}
\text{H} \quad \text{H} \\
\text{C} \quad \text{CH}_3 \\
\text{HC} \quad \text{C—H} \\
\text{HC} \quad \text{C—H} \quad \text{O} \\
\text{C} \quad \text{C} \\
\text{H} \quad \text{H} \quad \text{O—R}
\end{array}
$$

It has been found that the alpha-unsaturated esters control the more reactive odors most effectively. As a rule, these are the most intolerable odors. For instance, ammonia, with its strong electronegativity, reacts more rapidly than methylamine. Similarly, methylamine reacts faster than acetamide. Like comparisons may be made with various osmophoric series of sulfur, selenium, and phosphorus compounds.

Metazane is generally used in aerosols at a level of 2.0%. The Neutrolair D-7 blend is suggested for use at either 1.0% or 2.0%. If an aerosol containing 2.0% deodorant is sprayed into a 1000-ft^3 room so that 10 g of the formula is eventually spread evenly throughout the air space, calculations show that the deodorant content will only be about 6 ppm by weight. Perhaps only about 20% of this will actually exist in "viable" form, that is, not polymerized, bound within air-borne droplets, attached to air-borne dust motes, or already descended out of the scene of action. Some minor fraction of gaseous deodorant remaining in the air will probably react with active, nonodorous molecules and also be lost. It necessarily follows that the concentration of deodorant that is finally left for odor control is in the magnitude of only about 1 ppm by weight. The fact that so little is present leads to the conclusion, already advanced for the aldehydic deodorants, that the alpha-unsaturated esters react preferentially with the more villainous odor-causing molecules, removing them from a greater concentration of total odorants and causing a greater decrease in odor than would normally be anticipated on the basis of deodorant concentrations alone.

Any assessment of the deodorizing ability of a particular compound or mixture must be made organoleptically. For preliminary screening and rough evaluation work, Fritzsche-Dodge & Olcott, Inc., have devised a glass apparatus consisting of a 12-liter flask with top, side, and bottom necks. An air stirrer is inserted though the top, and a standard malodor is introduced through the bottom plug, using a solid or liquid source, which is then removed. The aerosol is then sprayed in through the bottom of the flask and the stirrer started up. At appropriate intervals, the stirrer is stopped and the side plug is removed so that experts can organoleptically assess the progress of deodorization. Several refinements have recently been made in controlling the amount (very small) of aerosol spray introduced and in placing a slight positive pressure in the flask so that the treated air actually wafts out of the side opening during evaluations.

A considerably more complex deodorant evaluation procedure is given in Interim Federal Specification P-D-00200(GSA-FSS), dated September 27, 1956. Since that time the government has become less interested in air deodorizers, and no updating has been carried out. In the method, two 500-ft^3 test rooms are constructed, preferably lined with polished stainless steel throughout. A 3-sec spray of one of five standard malodor aerosols is introduced into both enclosures. After this, a 3-sec spray of deodorant is dispensed into one of the rooms. At 1 min and 5 min after spraying, a panel of at least five trained people must walk through the test room and the control room and certify that odor reduction is at least 80% against at least four of the five standard malodorants tested. The malodor aerosols are prepared according to specific formulations such as the following:

It is somewhat doubtful that the General Services Administration has looked into this area as carefully as the testing method might indicate, because the program for any given deodorant composition is extremely time consuming and the results are not nearly as reproducible as might be anticipated. For example, certain sulfide odors actually seem to lock substantively onto the stainless steel surfaces of the test chamber and are almost impossible to remove completely. In addition, several of the malodor ingredients are no longer available and would have to be synthesized. All things considered, this procedure is almost never used any more.

PHYSICOCHEMICAL PRINICIPLES OF ODOR CONTROL

There can be no sharp distinction between deodorant chemicals that diminish odors by chemical means and those that do so by physical means. Nearly all deodorants operate by both routes. The physicochemical forces that can be considered here include solvation, electrostatic charge reduction, dipolar attraction, van der Waals effect, hydrogen bridging activity, and Heitler-London interactions. Several deodorants operate predominantly by means of these phenomena.

Most odor particles are dispersed into the air as very dilute aqueous solutions, such as those cast off from the body, from cooking, or by evaporations. As the water evaporates, the shrinking droplet becomes charged. Size stabilization occurs within the 0.1 to 10.0 μ diameter range. Negatively charged particles, mainly oligophilic entities such as bacteria, body cells, fats, amides, and pituitous gel structures, predominate.

Tobacco Smoke Odor

Ingredient	Per Cent
Propylene glycol	0.0024540
Furfural	0.0002454
Furfural mercaptan	0.0000061
Methylethylpyridine	0.0006130
Thiovanic acid	0.0000247
Creosote	0.0001842
Pyroligneous acid	0.0014726
Dipropylene glycol	4.9950000
Propellant 11	47.5000000
Propellant 12	47.5000000

Body Odor No. 1

Ingredient	Per Cent
Caproic acid	0.0037222
Isovaleraldehyde	0.0003515
Phenylacetic acid	0.0002068
Skatole	0.0000041
Indole	0.0000020
p-Cresyl phenyl acetate	0.0000931
p-Cresyl isovalerate	0.0002068
Thiovanic acid	0.0004136
Dipropylene glycol	4.9950000
Propellant 11	47.5000000
Propellant 12	47.5000000

Kitchen Odor

Ingredient	Per Cent
Diacetyl	0.0000893
Amylguaiacol	0.0000447
Butylguaiacol	0.0002232
Furfurylmercaptan	0.0000893
Diallyl sulfide	0.0002232
Dimethyl sulfide	0.0022329
n-Heptaldehyde	0.0000893

Oil of anise	0.0002232
Propionic acid	0.0008929
Oil dill seed	0.0004465
Caraway	0.0004465
Dipropylene glycol	4.9950000
Propellant 11	47.5000000
Propellant 12	47.5000000

When these air-borne nuclei are impacted by other electrically unbalanced particles, a number of transitions take place, several of which lead to a reduction of odor. In some cases head-to-tail effects cause thin films to develop, so that the odorant molecules are more effectively locked in than before. Electrical polarities are reduced, and size is often increased. These changes in physical and electrical characteristics can often be studied by the scattering of a laser beam in a photometer. Diffraction of the beam can also be used if the particle is sufficiently transparent.

Certain substances, notably oxo-process hexadecyl alcohol, have been used in order to reduce odors by these means. In some cases a prominent odor can be greatly reduced, but in others there may be little overall effect. New odors are sometimes generated. In fact, the picture is complicated by interphase partitioning of the odor molecules, formation of condensation nuclei, humidity effects, and so forth.

A deodorant substance known as Meelium, sold exclusively by the Prentiss Drug and Chemical Company since 1956, is the most important chemical in this category. During the period from 1951 to about 1963 it enjoyed fairly wide use, but application has now dwindled to a low level. Pure anhydrous Meelium is a dark brown, glassy solid, very difficult to prepare. It is said to consist of various mono- and disulfonates of polyhydrophenanthrene. The commercial 60% product is made by simply diluting the raw materials with Stoddard solvent (a deodorized kerosine) as they discharge from the sulfonation reactor. The composition is approximately as follows:

50% Polyhydrophenanthrene sulfonates and related sulfonates
2% Polyhydrophenanthrene and related compounds
3% Sodium sulfate
5% Water
40% Stoddard solvent (sometimes with a trace of oleic acid as a coupler)

TABLE 20.3. FORMULATIONS FOR AIR FRESHENERS, SHOWING DIFFERENT PROPELLANT SYSTEMS

Ingredients	Formulation (%)			
	I	II	III	IV
Metazene, 100%				1.60
Neutrolair D-7	0.50			
Meelium, 100%		1.50		
Perfume oil	0.25	0.50	0.50	0.40
Odorless mineral spirits		18.00	19.50	6.00
Ethanol, anhydrous	28.00		2.50	
Water, preferably deionized				45.30
Sodium nitrite				0.10
Sodium benzoate				0.10
Emulsifier (e.g., Emcol 14)				1.50
Methylene chloride, stabilized		15.00	15.00	10.00
Triethylene glycol, ATG	4.25			
Dipropylene glycol	2.00			
Propellant 11	20.00	29.25	55.00	
Propellant 12	45.00	29.25		
Isobutane		6.50		35.00
Carbon dioxide			7.50	
Pressure (20 in. Hg° crimp) (psig at 70°F)	50	36	112	37
Pressure (20 in. Hg° crimp) (psig at 130°F)	133	108	159	104
Delivery rate (g/sec at 70°F)	1.1	0.9	1.4	0.7 max.
Flame propagation at 70°F (in.)	16	11	8	0
Closed drum test at 70°F (sec)	65	>70	60	63
Flash point (°F) (TOC-Mod.)	69	132	136	(Inapplicable)
Container	Single lined	Plain	Plain	Plain

The deodorizing ability is considered to arise from two reactive centers. One is a "frozen styrene" sector within the ring structure; the other is located at the base of each sulfonate group. Conjugate double bonds are present at each site. The chemical contribution to odor control probably involves both addition and chain-transfer polymerization, but the physicochemical effects of van der Waals forces and hydrogen bridging mechanisms also participate in deodorization. Meelium molecules are thought to form films over the surface of charged malodor particles. Because of the variable nature of the Meelium materials and the complexity of the deodorizing process, there is no way to define the relative importance of structures or mechanisms in terms of total deodorizing activity. Constructions can be based only on generally accepted theories of chemical interactions.

The deodorizing ability of Meelium is well documented. Tests run at independent laboratories have shown that as little as 0.1 ppm by volume of the deodorant will completely eliminate the odors of cigar smoke, fried onions, boiled cabbage, decaying fish, and Limburger cheese.

The toxicity of Meelium is very low indeed. Unperfumed Meelium aerosols have a slight, sweetish chemical odor, and for this reason from about 0.05 to 0.25% perfume is always included. The moisture content of Meelium must be considered. At the common use level of 1.60% deodorant, moisture will be 0.13%. This is enough to separate and to cause can corrosion in certain systems, so that it is always advisable to market Meelium products in inside-lacquered cans, unless adaquate storage tests have shown cans with plain interiors to be satisfactory.

Most of the work on Meelium deodorants has been done in systems where either odorless base oils or odorless mineral spirits make up most of the concentrate. When such distillates are replaced with ethanol, isopropanol, or water-in-oil emulsion systems, the deodorizing efficacy of Meelium is said to be reduced, although the extent is unknown.

Formulations for air fresheners, showing different propellant systems, are given in Table 20.3.

PHYSIOLOGICAL PRINCIPLES OF ODOR CONTROL

The chemical deodorants used in air fresheners are invariably assisted by the addition of at least a small amount of perfume. In fact, a very substantial part of the entire market is now held by products that employ perfumes as the sole means of ameliorating odors. In some of these formulas the perfume may play a double role, creating a pleasing fragrance, on the one hand, while functioning as a mild to moderate deodorizer, on the other.

Perfumes as Chemical Deodorants

The principal function of any perfume is to mask unpleasant odors and leave an appealing fragrance of its own. Chemical deodorancy may occur in the masking process, particularly if the perfume has ingredients with aldehydic or ester groupings. Vanillin, an aromatic aldehyde, has been used for hundreds of years to reduce kitchen odors. It is a common ingredient of Alamask and other deodorizing perfumes. Geranyl butyrate, alkyl anthranilates, terpene aldehydes, and methyl nonylaldehyde (lime fragrance) all have pronounced fragrances and deodorizing abilities. Isocoriandrol and certain ionones reduce isopropanol and acetone odors, while citrus oils and pine derivitives serve to subdue the odors of urine and certain offensive chemicals.

Many deodorizing compounds are effective at levels as low as about 10 ppb in air. This equates to only about 0.02% of such materials in an air freshener formula. Quite often, if the level of deodorizing fragrance ingredients amounts to 10% of the perfume, the straight perfume formula will do a very adaquate job of deodorization. In fact, in actual practice true deodorizers are no longer used to any great extent, with most modern products deriving their deodorancy solely from the perfumes included. Such terms as "reodorizer," "masking agent," "neutralizer," and "deodorant" can all be applied to the perfume, as well as to the pure deodorizing chemicals. This consideration and others have led marketers to largely eliminate true deodorants in favor of more perfume or of higher-quality perfumes.

Perfumes as Physiological Deodorants (or Reodorants)

Even air fresheners formulated with completely "nondeodorizing" perfumes are effective to some extent in reducing the apparent level of malodors. This suppressive effect increases as the percentage of perfume increases. It is the result of a preponderance of fragrance which impinges upon the olfactory surfaces so that other odor responses are drowned out. As mentioned previously, the phenomenon is a common one, demonstrated when one shouts to lessen pain, when "white sound" is used in dentistry, and when syrups are utilized to mask the bitterness of many medicines. Although perfumes can mask other odors,

they cannot cover up associated responses, such as pungency or stinging, acrid, and bitter sensations, nearly as well.

The aerosol device is unique in that all the odorous ingredients of the
perfume top notes, middle notes, and fonds are caused to become air-
borne and to strike the olfactory center at the same time. The perfumer must create his fragrance with the method of application clearly in
mind, so that clashes and dominations will be minimized.

Every air freshener has a floral background. The florals give a distinct impression of quality and elegance, even to basic pines and
spices. Clean, crisp lavenders and citrus perfumes still retain a background of carnation, rose, and other florals. On the other hand,
straight, nondescript florals are often avoided, especially if they are
excessively sweet or cloying.

Specialty air fresheners have capitalized on unusual fragrance types.
Fresh aldehydic and cashmere bouquet perfumes are used for nursery
sprays. Institutional sprays use Siberian pine florals and sometimes
those with Oriental citrus overtones. Lavenders are still the favorite for
sachet sprays. There are even sprays to make a man's den seem more
masculine by giving it the odor of fine cigars. In the final analysis,
many a perfume is selected simply on the basis that its odor is pleasing
and of the type desired, and that the proper strength can be obtained
within a certain cost per thousand cans.

BACTERICIDAL PRINCIPLES OF ODOR CONTROL

The deodorant aspects of air-sterilizing chemicals have been recognized
for over a century. In 1941 air treatment programs were used to protect
the health of people crowded together in London air raid shelters, and
during this time a superior spray was devised based on a concentrate of
10% hexylresorcinol and 90% propylene glycol. Investigations proved
that the persistent mists of propylene glycol were useful in keeping the
hexylresorcinol air-borne. Later it was found that only a slight reduction in biocidal efficacy resulted when the hexylresorcinol was removed
from the formula. The propylene glycol acted by vapor condensation
onto the droplets carrying bacteria, so that an aqueous antiseptic solution that killed the pathogens was formed.

Further tests disclosed that triethylene glycol was the most effective
of all the common glycols and polyglycols, with as little as 0.20 g being
sufficient to create nearly sterile conditions in a room of 1000-ft³ volume. By comparison, 2.0 g of dipropylene glycol or about 18.0 g of
propylene glycol were needed for the same results. Under ideal humid-

ity conditions of 15 to 40% the triethylene glycol vapors destroyed 85% of the air-borne bacteria within 1 min and essentially 100% within 3 minutes.

The use of glycols in aerosol air fresheners was described as early as 1948. Acceptance was mediocre at best, partly because many marketers decided to use isopropanol as the solvent, along with a minimum of perfume, and to advertise the products as if they were for sickroom or bathroom areas alone. During the late 1950s sprays were formulated in which the glycol air sanitizers were backed up with quaternary ammonium salts or sometimes with substituted phenols. They did not become important, however, until about 1964, when a spray based on *o*-phenylphenol and ethanol was introduced by a major marketer. The product had three advantages: it was protected by U.S. and other patents, it was registered by the USDA (now the EPA), and, most importantly, it was supported by very vigorous national advertising. Sales have been so dramatic that they were estimated at about 92,-000,000 units per year in 1971, for this type of product.

Triethylene Glycol Deodorants

Since triethylene glycol is most effective in the vapor form, formulation attempts have focused on discharging the proper level of this key ingredient in a very finely divided spray pattern. The material has a vapor pressure of only 0.001 mm Hg at 68°F; this means that a 1000-ft³ room, saturated with vapor, would hold only 0.2 g. Above this level the droplets cannot vaporize and therefore are of little benefit except perhaps as a reserve supply.

Calculations show that a typical aerosol containing 2.7% of triethylene glycol can bring a 1000-ft³ room to vapor saturation when sprayed for 10 sec at a rate of 0.75 g/sec. Of course, this assumes perfect conditions all around. In practice we must consider such things as uneven distribution, fallout of some particles and slow evaporation of others, air flow out of the room, and spray periods of less than 10 sec. A contingency figure of 50% has been applied to the 2.7% concentration, to give a working minimum of 4.0% triethylene glycol. The maximum level is generally set at about 7.0%. Undesirable fogging effects are noticed when severe overuse takes place, but this is rare.

In order to achieve the very-fine-particle-size sprays conducive to total evaporation of the triethylene glycol, the propellant should contain a relatively high amount of dispersive ingredients. There should be at least 40% Propellant 12 or a minimum of 11% Propane A-108, for example.

The manner in which glycols sanitize air is still under intense study. It is recognized that the destruction of bacteria is not in itself a deodorizing operation, but only a means of eliminating future odors that may arise as the result of putrefactive processes. Glycols have an affinity for sulfur-containing compounds and are able to form complexes with them, but the importance of this characteristic to deodorancy has never been determined. High humidities strongly reduce both the antibacterial and air-deodorizing actions of triethylene glycol. Airborne dust also interferes. Relatively low control is obtained unless the glycol is at least 50% of air saturation.

Although the di-, tri-, and polypropylene glycols are soluble in aerosol propellants, the more useful glycols are sufficiently polar that cosolvents must be employed. As a rule as little as about 7% ethanol or isopropanol may be used. Methylene chloride is less effective, sometimes requiring up to 30%. The exact minimum of cosolvent required depends on the type and percentage of the glycols to be solubilized, as well as the temperature of filling. Moisture pickup can be a problem during filling, because of the hygroscopic nature of the alcoholic glycol concentrates.

Therapeutic Air Fresheners

Although the simple glycol-based sprays have some therapeutic value, the term "therapeutic" has now become associated with those specifically designed as decongestants. They always contain vasodilating ingredients such as camphor, menthol, thymol, and oil of eucalyptol, superimposed on an alcoholic glycol formulation. The special ingredients constitute only 1 or 2% of the total formula; otherwise the odor level would be overpowering. They are also relatively expensive.

The advantages of these products include quick relief from congestion due to colds, hay fever, coughs, and pollen allergies, caused by enlargement of irritated breathing passages. The final particle sizes are such that penetration is gained into the farthest recesses of the lungs. Isopropanol is commonly used as the primary solvent. Ethanol should be evaluated with more caution, from a can compatibility standpoint. Water-based formulations have been developed that contain little or no alcohols, modest amounts of glycols, and ample quantities of the decongestant terpenes. So far the sale of such formulations has been extremely limited, despite the significant economic advantages they offer. Marketers have explained their reluctance to change formulas in terms of the expense required to confirm the medicinal properties and safety of the new compositions. This is particularly true when one considers the small size of the therapeutic room-vaporizer market as a whole: less than 1,000,000 cans per year.

Table 20.4 lists typical decongestant formulas.

TABLE 20.4. DECONGESTANT FORMULAS

Ingredients	Formulation (%)		
	Anhydrous	Water-Base	Water-Base
Triethylene glycol, ATG	3.5	0.5	0.5
Dipropylene glycol	3.0	. . .	. . .
Camphor, U.S.P.	0.6	a	0.7
Thymol, N.F.	0.1	a	0.1
Eucalyptol, U.S.P.	0.4	a	0.4
Menthol, racemic, U.S.P.	1.3	a	1.3
Perfume—Lavender	. . .	a	. . .
Cold Relief Blend, H&R-20428	. . .	1.5	. . .
Atlox 4600	. . .	0.1	0.1
Emulsifier PF	. . .	0.5	0.5
Span 20	. . .	0.3	0.3
Methylene chloride (Aerothene MM)	. . .	4.0	4.0
Odorless mineral spirits	. . .	9.5	9.5
Alcohol SD 40-2, anhydrous	17.0	. . .	. . .
Deionized water	. . .	49.4	48.5
Isobutane A-31	. . .	34.0	. . .
Hydrocarbon blend A-46	. . .	. . .	34.0
Propellant 11	37.3	. . .	. . .
Propellant 12	37.3	. . .	. . .
Pressure (20 in. Hg° crimp) (psig at 70°)	42	37	52
Delivery rate (g/sec at 70°)	1.2	0.5	0.6
Flame projection at 70° (in.)	13	0	0
Closed drum test at 70° (sec)	60	72	64
pH	. . .	8.3	8.4
Container, 0.50/0.25 lb ETP plain with	2/98	Tin	Tin

a These items are contained in Cold Relief Blend H & R-20428.

Phenolic and Quaternary Disinfectant Deodorants

These products easily represent the largest single category in the field of air fresheners and therefore deserve special attention. Sales during 1971 were estimated as somewhat less than 100,000,000 units, with phenolics occupying about 90% of the market, and with one product alone (Lysol) thought to enjoy sales of about 75,000,000 units.

The first significant disinfectant deodorant was Duel, introduced by National Laboratories Division about 1960. The formula contained 2-phenylphenol, which has remained a favorite to this day, as the principal bactericidal ingredient. The product had a heavy-duty label connotation and was oriented toward institutional uses. It also utilized a custom-made spray-dome which, in those days, was something of a novelty and helped to focus attention on the package.

The "sanitizer-deodorant-disinfectant" products were designed to function in a variety of ways. They were regarded as air sanitizers, capable of drastically reducing the numbers of air-borne bacteria; as air deodorizers, so that noxious odors were minimized; and finally as surface disinfectants, capable of destroying various bacteria, fungi, and other microorganisms. Since the regulation of disinfectants, sanitizers, and similar products recommended for use in the inanimate environment comes within the purview of the Federal Insecticide, Fungicide and Rodenticide Act, these aerosols must be registered with the Environmental Protection Agency. The Pesticides Regulation Division of the EPA renders judgments as to compliance with the efficacy requirements of the act, largely on the basis of laboratory data. Before being offered for sale, every economic poison must be approved on the basis of efficacy, proper labeling, safety, and other criteria. The words "EPA Reg. No.———" must appear on the label. Complete formula disclosure must be made in confidence, and the active ingredients must be listed on the label.

The assessment of air-sanitizing ability is very complex. A standard method has yet to be determined, although several procedures have been described. The EPA currently evaluates air-sanitizing activity on the basis of simulated use studies. A test method developed by R. J. McGray of S. C. Johnson and Son, Inc., during 1969 and 1970 is thought to be the most reliable, in that a complete, self-controlled evaluation can be performed simply and accurately, with a high degree of reproducibility. The McGray method gives results which show that triethylene glycol formulas give good air sanitization, but that 2-phenylphenol, the polybromosalicyanilides, and the usual quaternary ammonium salts do not. In these tests commercial products were used;

therefore it should be mentioned that the glycol concentrations were from 50 to 100 times the levels of the 2-phenylphenol and other biocides.

The McGray procedure involves the use of a 1000-ft^3 sanitary enclosure, fitted with temperature and humidity controls, a filtering facility to provide laminar-flow air changes to simulate in-home use conditions, an atomizer with a Binks No. 50-100 nozzle for the introduction of bacteria, and a Reyniers Air Slit Sampler, designed to rotate an agar Petri dish at 1 revolution per hour. A standard 24-hour culture of *S. lysodeikticus, S. aureus, E. coli,* or other microorganism is prepared, and 1 ml is atomized into the chamber. After an equilibration period of about 15 min, during which the sampler moves through the first rotational quadrant, a 10-sec spray of the aerosol test product is introduced. The decrease in population density is then observed through the next 45 min and compared with the projection of the first 15-min control curve to determine the percentage reduction of bacteria with time. Impingement of glycol-and phenolic-type sprays onto the agar plate does not cause bacteriostasis, but with quaternary-based sprays this does occur to an advanced degree. This gives the appearance of an air-sanitizing effect, which is not actually present. By adding a mixture of lecithin, Tween 80, Tamol SN, and the ferric sodium salt of 1-nitroso-2-naphthol-6-sulfonic acid to the recovered agar, the bacteriostatic activity of impinging quaternary salts is neutralized almost immediately upon contact, so that true results, showing that no air sanitization occurs, are obtained.

The implications of the McGray method are of considerable importance, since it is realized that scores of phenolic- and quaternary-based products, now designated as "air sanitizers," "space bactericides," and so forth (in addition to other claims), do not actually qualify for such labeling. Although there are no hard rules as yet, it is generally agreed that an "air sanitizer" should be able to reduce the air-borne bacterial population by 90% or more when used in an average-size room in accordance with the directions printed on the label. In contrast, an "air bactericide" is expected to diminish the vegetative bacterial population of the air by at least 99%, under in-use conditions. This 99% minimum figure, however, is virtually unobtainable. New bacteria can be expected to enter the room rapidly enough to balance out the killing rate of the aerosol long before the 99% figure can be attained. Many aerosols designated as "air sanitizers" have such directions for use as "Direct spray upward, toward center of room, for 10-15 seconds." In such cases the 10-sec spray time must be the one used in any qualification tests. At this time the EPA Microbiological Laboratory is denying "air sanitizer" registration to aerosol products that do not contain at least 5% triethylene glycol. This agency is also suggesting that the chamber tests run to authenticate "air sanitizer" claims be conducted to a degree that gives a 95% confidence level in the results. It is not, to our knowledge, approving any aerosol products as "air bactericides" or "air sterilants." However, the distinction between these terms and the lesser designation "air sanitizer" is lost on the final purchaser of the product, so that there is no real incentive to strive for the higher rating.

The air-deodorizing capabilities of the phenolic- and quaternary-based sprays have been strenuously advertised. Many experts feel that they are overrated. Since these bactericides are used at levels of only about 0.14%, the maximum concentration discharged into the air under normal use conditions will be about 35 ppb by volume, or roughly 480 pg/ml. It is difficult to imagine that these products could exert much of an effect at such a low level.

The deodorizing properties of bactericidal quaternary salts have been described in a meager and unsatisfactory fashion. Although these compounds are reported to be effective against a number of household odors, no concentrations are mentioned. In other investigations, aerosols containing up to 0.10% quaternaries were said to cause only slight reductions in fish, onion, and cabbage odors. In all cases a somewhat bitter chemical odor of the disinfectant was left in the air.

No air-deodorizing abilities have been described for 2-phenylphenol and related phenolic disinfectants. However, nearly all the phenolic formulas contain about 0.04% N-alkyl (C_{18} 92%, C_{16} 8%)-N-ethylmorpholinium ethyl sulfates, since this mysterious ingredient appeared on the label of Lysol, Duel, Amphyl, and the other disinfectant deodorants sold by the Lehn & Fink Products Corporation, causing many other firms to follow suit. The quaternary was first described in 1950 as an aerosol disinfectant and deodorant, but the EPA has now disallowed the disinfectant claim, so that it may no longer be listed on the label as an active ingredient. There is also some question regarding the extent of its deodorizing properties. Its primary function appears to be an amelioration of the harsh, biting odor of the air-borne phenolics.

In the light of this rather clouded and contradictory situation, how is it that these products have been so heavily advertised and well accepted as effective air deodorants? The answer may be quite simple and essentially physical in nature. Assisted by their antistatic properties, the rather large hydroalcoholic particles of deodorant, each with an autophobic monolayer surface of positive charge, slowly sift down through predominantly negative-charged odorous particles, causing opposite charge collisions, elimination of charge on the odor particle, diffusional film transfers, and often a complete integration. This

TABLE 20.5. DISINFECTANT/DEODORANT AIR FRESHENERS

Ingredients	Type		
	Phenolic and Fluorocarbon (%)	Phenolic and Hydrocarbon (%)	Cationic and Fluorocarbon (%)
Dowicide 1 (chlorodioxin free)	0.1458	0.1800	
Dowicide 4	0.0325		
p-tert-Amylphenol, 98% min.		0.0450	
BTC-2125 (50% in water) onyx			0.2880
QAI-50 (Hollingshead)	0.1000	0.0400	
Atlas G-271 (35% in water)	0.1142		
Morpholine			0.1600
Sodium nitrite			0.2000
Perfume oil	0.1000	0.3000	0.5000
Alcohol, SD 40-2, anhydrous	67.6460	53.4600	59.8270
Deionized water	11.8615	25.9750	19.0250
Propellant 12	20.0000		20.0000
Propellant A-40 blend		20.0000	
EPA Label Declaration			
2-Phenylphenol	0.143	0.176	
2-Chloro-4-phenylphenol	0.029		
p-tert-Amylphenol		0.044	
n-Alkyl (60% C_{14}, 30% C_{16}, 5% C_{12}, 5% C_{18}) dimethyl-benzylammonium chloride			0.072
n-Alkyl (50% C_{12}, 30% C_{14}, 17% C_{16}, 3% C_{18}) dimethyl-ethylbenzylammonium chloride			0.072
Essential oils		0.300	0.500
Alcohol	67.646	53.460	59.827
Inert ingredients	32.182	46.010	39.529
Pressure (20 in. Hg° crimp) (psig at 70°F)	39	42	49
Delivery rate (g/sec at 70°F)	1.0	0.5	0.8
Flame projection at 70°F (in.)	17	12	15
Closed drum test at 70°F (sec)	67	>70	>70
Flashpoint (°F) (TOC-Mod.)	83°	−80°	102°
Container	Double lined	Double lined	Plain, plus side stream stripe

sequency of events is thought to rather effectively flush most of the odorous material down to floor level. Some chemical fixation may also take place. The phenolics and quaternaries are often sufficiently acidic that loose, saltlike compounds may form with amine-type odorants. Data published on the N-alkyl (C_{18} 92%, C_{16} 8%)-N-ethylmorpholinium ethyl sulfates material indicate that the substance is quite ineffective as a deodorant in the absence of humidity, and that maximum deodorizing effects are obtained when the air contains over 300 parts of moisture to 1 of the quaternary compound. Typical disinfectant/deodorant air freshener formulas are given in Table 20.5.

Perfumes are used in all the disinfectant deodorants, generally in the 0.12 to 0.20% range. Some of these perfumes may possess deodorizing properties, as discussed earlier, especially the lavenders, which are a popular choice. They also help to mask the chemical odors of the ethanol and disinfectant ingredients.

The final attribute of the disinfectant/deodorant spray products is their ability to disinfect environmental surfaces when applied according to directions. This area is of extreme interest to the EPA, so that it has become necessary to adjust all formulas, conduct all tests, and develop all labels and advertising copy with the central aim of obtaining product registration from this agency. Unregistered products cannot be sold in interstate commerce—or even intrastate, when state agencies are also considered.

The process of obtaining registration is very painstaking, time consuming, and often frustrating. Many hundreds of "official AOAC germicidal spray tests" must be conducted in order to convince the EPA that the formula will kill specified or implied bacteria at a confidence level of 95% or better. The label must also be printed in exact accordance with EPA stipulations, many of which have never been published. A request for registration is then submitted to the Environmental Protection Agency, Washington, D.C., with the required supporting data.

A first action from the EPA may be expected in 2 to 6 months; the time will vary according to the agency's workload. When new federal laws have made label changes necessary, the EPA is automatically called upon to approve revisions and possible updatings for all 65,000 economic poison labels it now controls. Applications for products making claims beyond the scope of those for simple economic poisons may be submitted to other governmental agencies by the EPA before final clearance is given, and, this is very time consuming indeed. During much of 1971 the EPA was running on a 4-month response basis. Recently the agency stipulated that all labels must be resubmitted for review every 5 years, and the immediate result was thousands of requests for continuing approval of registration. This work has been added to the regular activities of the EPA.

The result of the EPA first action is usually a letter listing from 5 to 40 specific points of objection. The objections take three forms: additional test data are needed, further descriptive data are required, or label revisions are stipulated. The applicant has 6 months during which to respond; otherwise his file is discarded.

If a satisfactory response is provided to cover the EPA objections, the agency will issue a provisional registration number. The applicant may then arrange for the preparation of final labels or lithographic plates, so that samples of the finished label copy may be submitted for final approval.

It is not at all uncommon for the registration process to require 1 to 2 years for completion. In some instances, the EPA requirements will change during this period, so that material found acceptable during the first-action phase will be considered objectionable during the second. The applicant will then be asked to conduct further tests, submit new data, or make label revisions that were not anticipated during the initial phase of the program. It is obvious that such long or unanticipated delays have frequently held up complex time tables for new product introductions, usually at great expense to the applicant. In other instances the applicant has simply been forced to abandon the registration program or else to revise the product so that it can no longer be classed as an economic poison. In some cases the original formula may be placed on the market, using a label that carefully cloaks the fact that the product is truly a disinfectant. When registration is finally secured, the label can then be changed accordingly. The EPA objects to any listing of active ingredients on the label of an unregistered product, feeling that such declaration implies that the product is registered as an economic poison.

In its review of economic poison labels the EPA takes special notice of words and phrases that describe biological potency. It disallows any claims that state or imply efficacy beyond the proven capabilities of the product. Some typical descriptive words are defined as follows:

a. Sterilant. Destroys all forms of life, including microorganisms, bacterial and fungal spores, and viruses. This is too strong a claim for all but a few highly specialized aerosol products, such as those containing peracetic acid or ethylene oxide. These are far too drastic for household uses.

b. Disinfectant.

Will kill 100% of most vegetative bacteria, although not capable of destroying bacterial endorspores. It is the strongest claim that can be made for a routinely used germicidal product.

c. Sanitizer.

Will reduce the number of bacteria to relatively safe levels, normally to 5% or less of the original. The term is commonly used with reference to food, dairy, and laundry applications. It is an awkward concept, carrying with it the inference of cleaning, as well as the removal of infection. The word may no longer be applied to aerosol room sprays to indicate minimal antimicrobial activity.

d. Antiseptic.

Uniquely limited to products used to inhibit or kill bacteria found on living tissue, such as the skin. It may not be applied to economic poisons, but only to products regulated under the FDA.

e. Fungicidal.

Kills 100% of most fungi.

f. Germicidal.

Kills all microorganisms other than spores, by definition. But in practice is generally considered as capable of killing pathogenic bacteria.

g. Mildewcidal.

Kills 100% of most fungi. Auxiliary claim.

h. Pseudomonacidal.

Kills 100% of most *Pseudomonas* bacteria, a genus of gram-negative rods often resistant to broad-spectrum bactericides. Auxiliary claim.

i. Sporicidal.

Destroys bacterial endospores. The mechanism usually involves inducement of germination, followed by toxicity to the germinated spore. All sporicidal actions are slow at room temperatures. *Bacillus subtilis* spores can survive a 4-hour exposure to either 8% formaldehyde or 8% phenol solutions, provided temperatures are below 100°F. Iodine complexes are sporicidal, but quaternaries are only sporostatic at best. Routine economic poisons are never sporicidal.

j. Staphylocidal.

Kills 100% of most species of *Staphylococcus*, a genus of gram-positive bacteria. Auxiliary claim.

k. Streptocidal.

Kills 100% of most species of *Streptococcus*, a genus of gram-positive bacteria. Auxiliary claim.

l. Tuberculocidal.

Kills 100% of *Mycobacterium tuberculosis*, a species of gram-positive bacteria responsible for over 90% of tuberculosis infections in man. Auxiliary claim.

m. Virucidal.

Kills or inactivates viruses. This terms must be modified for labeling purposes, as "Virucidal against many viruses, such as Type 2 adenovirus, influenza A$_2$, vaccinia, Type 1 polio, and herpes simplex." The product must have been proved effective against all viruses listed. Auxiliary claim.

When minimal claims, such as "germicide," "disinfectant," or "bactericide," are all that are needed, and areas of recommended use are limited principally to the control of gram-negative bacteria, the EPA requires testing against only a single organism: *Salmonella cholerasius*. In this case, minimum requirements for registration involve the AOAC germicidal spray test, using 60 replicates on three preparations, with 30 replicates on duplicate samples of one of these preparations after a 60-day waiting period. This adds up to 240 carriers. The testing program ensures, with 95% confidence, that the product will kill 100% of gram-negative bacteria over a reasonable shelf-life period. These compositions are known as "Class A disinfectants."

For more general disinfection, where claims of "broad spectrum" and so forth are made, further testing is required. Data are also needed on tests using the same procedure, again with three preparations, this time employing *Staphylococcus aureus* and one other organism of the applicant's choice. In this case 30 replicates for each preparation will suffice. The total program is 420 carriers. It ensures, with 95% confidence, that the product will kill 100% of gram-negative bacteria over a reasonable shelf-life period, and that it also has good efficacy against gram-positive bacteria. Such compositions are known as "Class B disinfectants."

When germicides are recommended for disinfection in hospitals, clinics, dental offices, and veterinarian establishments, the primary test organism becomes *S. aureus*. Testing must include 60 replicates on three preparations, and also 30 replicates each on duplicate samples of one of these preparations after a 60-day storage period. In addition, tests using both *Salmonella cholerasuis* and *Pseudomonas aeruginosa* must be submitted on each of the three preparations, using 30 replicates each. The total program is 420 carriers. Such compositions are known as "Class C disinfectants."

A "Class C" product may be accepted for registration even though it is ineffective or only slightly effective against *P. aeruginosa*, provided that the label displays a prominent disclaimer to this effect. If "Class A" or "Class B" compositions are found to be effective against *P. aeruginosa* on the basis of tests involving 30 replicates on at least one preparation, an auxiliary claim of "pseudomonacidal" (or the practical equivalent) may be placed on the label. Other specific bacteria may be listed in auxiliary label claims of any germicide by submitting equivalent test data. If a claim for effectiveness against the organisms causing tuberculosis is made, results on one preparation using the AOAC tuberculocidal test must also be submitted.

If a claim for efficacy against pathogenic fungi is made, data on duplicate samples of one preparation must be submitted. The test is a modification of the AOAC fungicide method, except that the test slides are sprayed, using the aerosol. *Trichophyton mentagrophytes* is generally used as the test organism, replacing *Trichophyton interdigitalis,* which has become hard to obtain as a standard culture.

In some instances the word "fungicidal" or "mildewcidal" may appear on the label of products where the list of environmental surfaces for application includes wood, leather, composition board, or other semiporous materials. The EPA then requires ancillary labeling in accordance with the results of special tests, such as the "wood block mildew fungistatic test," issued March 1969. In this test, ten small pine blocks are sprayed with the test aerosol. They are then resprayed with an inoculum of *Aspergillus niger* and suspended from hooks through lids of 16-oz jars, over an inch or so of water. The test extends for up to 1 month at 80°F, with weekly evaluations for possible mold growth. If the blocks remain free of mold for only 7 days (as is very common), the product label must carry directions for use that specify retreatment of the wood every 7 days. A "leather mildew fungistatic test" is available for leather goods claims, and a "fabric mildew fungistatic test" is used for products making fabric treatment claims. These last two procedures employ *Penicillium glaucum* in addition to *A. niger.*

Claims for effectiveness against viruses must be supported by data obtained according to accepted virological techniques, using at least one preparation for each specific virus listed on the label. The most difficult problem in developing a standard test for virucidal activity is the toxicity of the germicides when applied to the carrier cell cultures. For instance, a germicide may completely inactivate a virus at 50 ppm but may become toxic to the cell culture at 100 ppm. This makes it necessary to use special assay methods. In one method, 10 min after treatment with the spray the virus material is recovered in a carrier culture medium. Cellular debris is removed by 0.22-μ filtration. Virus potency is then determined by making tenfold dilutions to extinction and testing each dilution for virus. The potency is expressed as the log $TCID_{50}$ (Tissue culture infective dose—50% level). This approach is needed to construct accurate differentiations between the cytotoxic effects of the aerosol spray and the cytopathogenic effects of the virus. The EPA is presently trying to standardize techniques for virucidal assays and is accepting data from established virological studies performed in duplicate against a single virus strain. If the TCID shows a three to four log reduction (or better), a claim may then be made on the label for effectiveness against the particular virus in question.

The EPA uses a number of guidelines in its approach to the labeling of economic poisons. Claims are held rigorously within the scope of test data and reasonable extensions thereof. Statements required by other federal laws are fully considered. Claims or comments of an impractical nature are not allowed, and ambiguous statements are rejected. Specific formats are required for the presentation of certain data, such as the ingredients declaration. Type sizes, separation, contrast, and printing style are considered. Specific definitions have been developed to handle certain situations, and these must be followed on the label.

A number of specific examples will serve to illustrate these various points.

a. Claims such as "...to aid in home protection against most disease germs..." are disallowed. Claims in the context of the protection of health or mitigation of unhealthful conditions are unacceptable and must either be revised to place the claim in another context or be deleted.

b. Claims such as "...perfect for any household disinfecting job..." are disallowed. The EPA does not consider any product as perfect or ideal. (Almost any other language is acceptable as descriptive.)

c. The words "Keep out of the reach of children" must appear on the principal label panel.

d. An unqualified name and address, such as "XYZ Company," is taken to mean that of the manufacturer.

e. The words "ACTIVE INGREDIENT(S)" and "INERT INGREDIENT(S)" must be printed in type of the same size, given equal prominence, and aligned to the same margin.

f. Perfumes or essential oils when present to the extent of 0.2% or more are considered as "active ingredients" and must be so listed in the ingredient statement.

g. The claim "Disinfects all washable surfaces on contact" is too broad. It should be revised to "Disinfects on contact" or, better, "Disinfects the recommended surfaces on contact."

h. Claims for disinfecting awnings, whitewall tires, and lawn furniture are disallowed on the basis of impracticality.

i. The claim "...kills such germs as staph. on contact" is revised to "...kills many germs, such as staph., on contact."

j. Claims such as "No additives" are disallowed unless clarified and qualified as to intent.

k. Claims as "...kills most bacteria" must be changed to "...kills most bacteria on environmental surfaces."

l. Wording such as "...protects against formation of new bacteria" must be revised to "...inhibits the growth of bacteria."

m. Results of the flame extension test must be submitted, so that they can be used to confirm the nature of the fire hazard warning.

n. Directions for use must include spray times and schedules, as well as the direction of the spray.

o. Specific directions must be given if the disinfection of surgical instruments is claimed. They must specify pretreatment to remove all blood and serous exudates.

p. Directions for use to control mildew (mold or fungi) must include such things as:

1. Naming the surfaces involved.
2. Directions to clean surfaces prior to application.
3. An outline of the method of application.
4. A figure of the coverage needed (for example: "Apply until surface is completely wetted").
5. Directions to repeat application in a certain number of days, as necessary to maintain control.

q. In the precautionary section, the toxic hazards must be separated from the physical hazards, container precautions, etc. Always precede the toxic hazards with "Caution:."

r. If nursery uses are included, the caution "Rinse with water anything that will come in contact with baby's mouth" must be shown.

s. Statements of a precautionary nature which pertain to human safety must always be separated and given precedence.

t. Use of different systems of nomenclature, such as "ortho-benzyl-para-chlorophenol" and "2-chloro-4-phenylphenol," is disallowed on the same label.

u. "Air sanitizer" must be replaced with "air refresher" unless specific air control data are submitted, except for specific formulas.

v. Statements such as "Do not wet varnish, painted or plastic surfaces with this spray" should be kept under directions for use and not under precautionary statements.

w. Statements such as "Avoid contact with eyes; on foods" must be revised to the more complete form, such as "Avoid contact with eyes, skin and foods."

x. The contents declaration must appear in the bottom 30% of the principal panel, in words such as "NET WT. 17 OZS. (1 lb. 1 oz.)." The statement in parentheses may be deleted if the weight is less than one pound.

y. Unless the product is manufactured by the marketer, the words "Sold by," "Distributed by," "Manufactured for," or a similar phrase must appear above marketer's name.

z. No disclaimer statements may be used, such as "The information furnished hereon is provided gratuitously by the manufacturer, who assumes no responsibility whatever for the effectiveness or safety of the product regardless of whether or not it is used as directed," unless they are true representations designed to protect the seller against damages from careless or improper handling. An acceptable statement would be:

"Buyer assumes all risks of use, storage or handling of this material not in strict accordance with directions given herewith."

z1. Ingredient-listing statements may go on either the front or back panels. All items listed must be aligned to the margin and should be given in descending order of percentage present.

A manufacturer, such as a contract filler, may apply for registration of any number of economic poisons, provided that they are well differentiated by code on the label, to prevent confusion. He may then list distributors by submitting the name and address of each one and the product name to be used on that distributor's label. The distributor's product name and label must not differ in substance from the basic label accepted for the manufacturer, and the product must remain in the manufacturer's original, unbroken container as long as it is in the channels of trade.

Two basic approaches are used for the formulation of these disinfectant/deodorant products. The so-called Lysol type is by far the more common, accounting for perhaps as much as 98% of all sales. The formula consists of less than 1% of 2-phenylphenol and other special ingredients, to which about 67% ethanol, 12% water, and 20% Propellant 12 are added. The alternative formula is much less expensive but requires a somewhat lower delivery rate in order to maintain nonflammability when tested by the standard flame propagation and closed drum methods. The same 1% or so of special items are used, but the rest consists of about 52% ethanol, 27% water, and 20% Propellant A-46 (88 wt % isobutane plus 12 wt % propane). Both formulas are equally effective if the level of 2-phenylphenol and similar items is adjusted to compensate for the difference in delivery rate.

Water is not classed as an active ingredient in these products. Yet, if it were absent or significantly reduced, bactericidal efficacy would decrease and also become quite erratic. The presence of water brings with it the chance of corrosion; indeed, container perforations have been noted with many of these formulations. By the use of 1.00/0.25 lb ETP cans and 20 in. Hg° (minimum) vacuum crimping, this problem can be substantially reduced. Added protection is obtained by the introduction of trace amounts of specific corrosion inhibitors, such as hydroxypropylammonium nitrite, $CH_3 \cdot CH(OH) \cdot CH_2 \cdot NH_3^+NO_2^-$, sold as a 50% water solution by the R. M. Hollingshead Corporation in New Jersey (QAI-50) and the R. H. Marston, Ltd., concern in Canada.

Standard package sizes are 7 oz in a 202 × 509 can, 14 oz in a 211 × 604 can, and 17.5 oz in a 211 × 713 container. More recently, fills of 21-oz have been packed in 300 × 709 cans. In each of these examples the can is filled to slightly less than the maximum safe volume—roughly 90 vol % at 70°F. For the hydrocarbon type of formula, with its much lower density, these filling weights would have to be decreased about 20% or so.

THE FUTURE OF AIR FRESHENERS

Most marketing analysts feel that air fresheners will continue to grow by about 4% per year from now until at least 1976. Most of the activity will involve the disinfectant/deodorant types, which will grow at about 8 to 10% per year, whereas the simple perfume products, which comprise nearly all the rest of the market, will experience either zero growth or, more probably, a slight decline.

One inevitable result of competition is reduction in the cost of goods. Since cans, valves, and covers are all fixed in price, the only way left is to decrease the cost of the formulation itself. For the straight perfume products such costs have already been reduced to the limit, with the formulas consisting almost entirely of water and hydrocarbon propellant. But nearly all the disinfectant/deodorant products are based on the use of at least 20% Propellant 12. As mentioned earlier, ways are available to replace this fluorocarbon with a hydrocarbon blend, provided that a portion of the alcohol is also replaced with water. Much activity can logically be expected in this area during the next few years.

2-Phenylphenol, 2-chloro-4-phenylphenol, and similar bactericides will probably continue to be used in most disinfectant/deodorants. The glycol products can be used only for air refreshing and air sanitization, whereas surface disinfection is now considered an attribute of equal importance. The quaternaries have been subject to problems of dwindling efficacy with shelf life and are often rather sticky or film-forming when applied to surfaces as disinfectants. No "superbactericides" appear to be in the offing as a replacement for the phenols.

Container sizes will probably increase to some extent. The 207.5 diameter can will be emphasized for the simple perfume types, and the 211 × 713 and 300 × 709 cans for the disinfectant/deodorants.

Other than these modest changes the air freshener industry should remain fairly static during the next few years.

REFERENCES

1. D. Tucker, Physiology of Olfaction, *Amer. Perfum.*, 76 (June 1961), 48.
2. C. P. McCord and W. N. Witheridge, *Odors: Physiology and Control*, 1st ed., McGraw-Hill Book Company, New York, 1949, p 29.

3. D. A. M. Mackay, D. A. Lang, and M. Berdick, The Objective Measurement of Odor. III: Breath Odor and Deodorization, *Amer. Perfum. Aromat.*, 74 (December 1959), 44.

4. K. Gabelein, The Air We Breathe for Food, Health and Stimulus, *Dragoco Rept.*, No. 8 (1959), p. 135.

5. E. R. Weaver, Control of Odors, *Natl. Bur. Stand. Circ.*, 491, 1950, p. 8.

6. H. Bohme and H. Fischer, Uber die Einwirkung von Ozon and Thioäther, *Ber.*, 75 (1942), 1310.

7. H. Taube and W. C. Bray, Chain Reactions in Aqueous Solutions Containing Ozone, Hydrogen Peroxide and Acid, *J. Amer. Chem. Soc.*, 62 (1940), 3357.

8. H. Taube, Chain Reactions of Ozone in Aqueous Solutions: The Interaction of Ozone and Formic Acid in Aqueous Solutions, *J. Amer. Chem. Soc.*, 63 (1941), 2453.

9. M. Stoll and A. Rouve, Etude des Reactions Secondaires de l'Ozonolyse d'une Liaison Ethylenique, *Helv. Chim. Acta*, 27 (1944), 950.

10. G. Cronheim, Organic Ozonides as Chemotherapeutic Agents. 1: Chemical Studies, *J. Amer. Pharm. Assoc. Sci. Ed.*, 36 (1947), 274.

11. F. P. Greenspan, M. A. Johnsen, and P. C. Trexler, Peracetic Acid Aerosols, *Chem. Spec. Mfr. Assoc. Proc. Ann. Meet.*, December 1955, p. 59.

12. A. H. Gee, Space Deodorants . . . Opportunities Ahead, *Amer. Perfum. Aromat.*, 75 (June 1960), 61.

13. E. Guenther, *The Essential Oils*, Vol. 2, D. Van Nostrand Company, Princeton, N. J., 1949, p. 313.

14. Lloyd E. Weeks, U.S. Patent 2,715,611 (Aug. 16, 1955).

15. A. H. Gee, Odor Control for Air Evacuation Aircraft, *Air Force Tech. Rept.*, 6565, Part 4, September 1952 (U.S. Department of Commerce, National Technical Information Service, Springfield, Va.).

16. Harry A. Stansbury, Jr., and Raymond W. McNamee, U.S. Patent 2,648,703 (August 11, 1953).

17. R. G. Woolford, A Grignard Condensation of Glutaraldehyde, *J. Org. Chem.*, 23 (1958), 2042.

18. P. Bachman, Chemical Products Having A Musk Odor, *Givaudanian*, May 1959, p. 3 (Givaudan Corporation, Clifton, N. J.).

19. Saul Kaye, U.S. Patent 2,891,838 (June 23, 1959).

20. Lowell B. Kilgore, U.S. Patent 2,544,093 (Mar. 6, 1951).

21. *Schimmel Briefs*, No. 252 (1956), p. 1.

22. Lyle D. Goodhue and Wm. N. Sullivan, U.S. Patent 2,321,023 (June 8, 1943).

23. H. R. Snyder, J. M. . Soc., 68 (1946), 1422.

24. Victor DiGiacomo, Aerosol Formulations, *Amer. Perfum. Aromat.*, 69 (January 1957), 72.

25. E. Shiftan, Perfumer's Problems in the Manufacturing of Aerosols, *Amer. Perfum. Aromat.*, 69 (January 1957), 99.

26. A. Ph. Krenowsky, Riechstoffe in Aerosolen, *Parfum. Kosmet.*, 37 (1956), 236.

27. P. Checkovich, Development of an Aerosol Room Deodorant, *Chem. Spec. Mfrs. Assoc. Proc. Mid-Year Meet.*, May 1955, p. 86.

28. A. Demeilliers and Y. Gutsatz, Use of Aerosols in Perfumery, *Amer. Perfum. Aromat.*, 68 (August 1956), 38.

29. J. D. Besse, F. D. Haase, and M. A. Johnsen, Safe Use of Hydrocarbon Propellants: Comprehensive Analysis of Aerosol Systems Composed of Propane, Propellant 11, and Odorless Mineral Spirits, *Chem. Spec. Mfr. Assoc. Proc. Mid-Year Meet.*, May 1955, p. 56.

30. A. E. Schober, The Role of Chlorothene in Personal Product Aerosols, *Chem. Spec. Mfr. Asso. Proc. Mid-Year Meet.*, May 1958, p. 51.

31. P. A. Sanders, The Reaction of Trichloromonofluoromethane with Ethyl Alcohol, *Chem. Spec. Mfr. Assoc. Proc. Mid-Year Meet.*, May 1960, p. 66.

32. D. C. Geary, Low-Cost Water-Based Aerosols, *Chem. Spec. Mfr. Assoc. Proc. Ann. Meet.*, December 1959, p. 84.

33. Winston H. Reed, Methylene Chloride in Aerosol Production, *Aerosol Age*, 1 (June 1956), 27.

34. Montfort A. Johnsen, Butane and Propane as Aerosol Propellants.II: A Symposium of Opinions on Relative Safety, Danger to the Consumer, etc., *Aerosol Age*, 4 (October 1959), 67.

35. J. D. Fulton, M. E. Nichols, J. Woehler, C. L. Shrewsbury, L. D. Goodhue, and P. Wilkins, Germicidal Aerosols, *Soap Chem. Spec.*, 24 (1948), 125.

36. Montfort A. Johnsen, The Technical Aspects of Aerosol Air Fresheners, *Aerosol Age*, 5 (October 1960), 66; 5 (November 1960), 20; 5 (December 1960), 52.

37. H. Wise, Dispersal of Triethylene Glycol Vapor with Aerosol Bombs, *Ind. Eng. Chem.*, 41 (1949), 633.

38. L. S. Stuart and J. L. Friedl, Testing Aerosol Products for Germicidal and Sanitizing Activity, *Chem. Spec. Mfr. Assoc. Proc. Mid-Year Meet.*, May 1955, p. 93.

39. L. D. Polderman, Triethylene Glycol for Air Sanitation, *Soap Sanit. Chem.*, 25 (December 1949), 129.

40. K. Bergwein, Phyto-balneology and Cosmetics, *Dragoco Rept.*, No. 8 (1961), p. 194.

41. J. H. Mills, Deodorant Aerosols, *Soap Sanit. Chem..*, 25 (August 1949), 111.

42. W. Eibel, Disinfectants and Deodorants in Cosmetics, *Dragoco Rept.*, No. 6 (1961), p. 147.

43. Sophie L. Plechner, Antiperspirants and Deodorants, Chapter 32 in E. Sagarin (Ed.), *Cosmetics: Science and Technology*, Interscience Publishers, New York, 1957, p. 732.

44. A. R. Whitehall, Evaluation of Some Liquid Antiseptics, *J. Amer. Pharm. Assoc. Sci. Ed.*, 34 (1945), 219.

45. B. V. Alfredson, J. R. Stiefel, Frank Thorp, Jr. W. D. Baten, and M. L. Gray, Toxicity Studies on Alkydimethylbenzylammonium Chloride in Rats and in Dogs, *J. Amer. Pharm. Assoc. Sci. Ed.*, 40 (1951), 263.

46. Earl L. Richardson, U.S. Patent 2,719,129 (September 27, 1955).

47. V. DiGiacomo, Aerosol Preparations, *Givaudanian*, September 1961.

48. H. B. Stecker and Richard E. Faust, A New Antiseptic Brominated Salicylanilide Composition for Soaps and Cosmetics, *J. Soc. Cosmet. Chem.*, 11 (1960), 347.

PAINTS AND
ALLIED PRODUCTS

During the recent past, the technology of aerosol coatings has developed to the point that nearly any air-dry coating can be packed in an aerosol container. The exceptions are emulsions or water-soluble coatings, as well as some coatings made from very highly polymerized resins that are incompatible with all practical, existing propellants. In addition, catalyzed systems are available wherein a catalyst is injected into the pressurized coating mixture.

Although the most common coatings are probably those made from vinyltoluene-modified alkyd resins, a wide variety of other resins are also used, for example, styrenated alkyds, pure alkyds of various oil lengths, epoxy esters, silicones, cellulose esters, vinyls, acrylics, styrenes, polyurethanes, and phenolics. They can be formulated into enamels, lacquers, or varnishes suitable for use on a wide variety of objects ranging from boats to fresh-cut flowers. The coatings appear in the form of gloss and flat enamels and lacquers, gloss or satin varnishes, fluorescents, metallics, wrinkles, hammers, primers, and stains. The parameters that must be dealt with in the formulation of satisfactory coatings are presented in the following sections of this chapter.

BASICS OF COATING FORMULATION AND PRODUCTION

A brief discussion of the formulation of coatings in general will be helpful to the understanding of aerosol coatings. Readers who are interested in pursuing this subject in depth are referred to books by Payne (1),

Meyers and Long (2), Turner (3), Nylen and Sunderland (4), Taylor and Marks (5), and others on this subject.

A coating can be divided broadly into three components: pigment, resin, and solvent. A clear coating has only two components: resin and solvent. A fourth component, additives, is important in the formulation, but for purposes of this presentation will be largely ignored. Additives are pigment dispersants, antisettling agents, drier-catalysts, antiskinning agents, plasticizers, mar-resistant agents, mildewcides, and antiflooding agents, plus others used in emulsion coatings, which are outside the scope of aerosol technology at the present time.

The function of the pigment is to provide color, opacity, sheen control, and film strength. Certain pigments, such as zinc oxide and carbon black, also protect against ultraviolet degradation of the resin, in exterior exposure.

The function of the resin component is to provide physical and chemical protection for the substrate, to bind the pigment securely to the substrate, and, in the case of gloss and semigloss finishes, to enhance the appearance of the coated object.

The function of the solvent is to reduce the viscosity of the resin and pigment mixture to a level suitable for application. Certain specialized resins developed for spray-gun application have viscosities so low that solvents are unnecessary to bring the finished product to spraying viscosity.

In combination, the resin and solvent are called the vehicle, and the resin solids are often referred to as binder solids or as the nonvolatile vehicle portion of the formula.

It has been well established that the ratio of pigment solids, by volume, to total solids content, by volume, controls all manner of physical properties of the dried film. This ratio is called the pigment volume concentration. As the ratio increases, gloss decreases and film porosity and hiding power increase. Gloss enamels and lacquers characteristically have pigment volume concentrations ranging from less than 1% for black to 15 to 20% for white. Primers are formulated at a pigment volume concentration of 35 to 40% for optimum performance/cost characteristic .

The point at which a formula has just sufficient vehicle solids to fill the voids between the pigment particles, in their closest packing pattern, is referred to as the critical pigment volume concentration. Any increase in pigment volume beyond this point results in a serious decrease in coating quality. Porosity increases, resulting in an increase in rusting, film brittleness, and, in the case of primers, poor enamel holdout.

The first order of business in the manufacture of a pigmented, nonmetallic coating is the preparation of a stable pigment dispersion. Ideally, this would be a suspension of discrete, primary pigment particles homogeneously and uniformly distributed in a liquid medium—in this case, a portion of the total vehicle formula. As received from the manufacturer, pigments are in the form of agglomerates rather than discrete, primary pigment particles. Aggregates are tightly compacted groups of primary pigment particles which are formed during the manufacturing process. They are held tightly together by natural, physical attractive forces. Agglomerates are clusters of aggregates and primary pigment particles. They are, therefore, larger and more loosely bound than aggregates. Perhaps a better understanding of these relationships can be achieved by thinking in terms of bunches of grapes.

The primary pigment particle is represented by the individual grape. Aggregates are the grapes on a bunch, and agglomerates could be considered as several bunches of grapes loosely packed together. The purpose of dispersion is not to make smaller grapes but to first separate one bunch of grapes from the others, then to remove the grapes from the bunch, and finally to keep individual grapes separate and suspended in the liquid medium.

Pigments vary widely in their ease of dispersion, that is, it is harder to separate some types of grapes from the bunch than others. Pigment manufacturers have succeeded in applying surface treatments to certain classes of pigments, thus making them relatively easy to disperse. Titanium dioxide and the leaded pigments, chrome yellow, chrome green, and molybdate orange, are examples of such pigments. On the other hand, organic pigments generally are much more difficult to disperse because they are usually more resistant to wetting by the liquid medium than are the inorganics.

Wetting is the first stage in the dispersion process. It involves replacement of the air and moisture that surround every pigment particle with the liquid medium. Wetting can be accomplished by simple mixing, which permits intimate physical contact between the exposed pigment surface and the liquid medium. In the second stage of the dispersion process, agglomerates and aggregates are reduced in size and primary particles are separated from one another. The pigment manufacturer produces his product, as nearly as possible, in the particle range that optimizes the performance of the pigment. Therefore, even if it were possible to reduce particle size, no value would be gained from doing so. The primary pigment particle size, as manufactured, is either equal to or below the particle size which is required to be achieved during dispersion. During the particle separation stage of the dispersion operation, reduction of the sizes of agglomerates and aggregates is the principal goal. This is followed by separation of the individual primary pigment particles. After agglomerate and aggregate sizes have been reduced, it is necessary that the pigment particles be stabilized in the system, that is, it is desirable that the particles not reform into loose clumps referred to as flocs. This can occur as a result of an imbalance of electrical forces surrounding the pigment particles. Flocs, when formed, are reversible, and upon the application of a moderate shearing force the primary particles are readily separated. When the shearing forces are removed, the flocs will reform. By proper formulation of the mill base and appropriate attention to the physical aspects of dispersion, it is possible to produce a stable disperation which then forms the base of a stable enamel or pigmented lacquer.

A number of machines are available to manufacturers for the production of pigment dispersions. However, the most commonly used are the high-speed disperser and the sand mill. The high-speed disperser consists of a single vertical shaft to which a circular blade is attached. The shaft rotates at a speed sufficiently high to give the blade a peripheral speed of 4500 to 6000 fpm. At this speed, maximum dispersion takes place in approximately 15 min, provided that a pigment manufactured for this type of dispersion has been selected and the mill base formulation has been properly balanced. Dispersers are available with both fixed and variable speed, a wide range of horsepower, and a wide range of blade diameters. Blade design varies among manufacturers; some blades have simple sawtooth edges, whereas others are formed from a stack of "thin doughnuts" that are separated to allow for the passage of the mill base through the annular slots. The dry pigment, a portion of the vehicle solids and solvent, and chemical dispersion aids are charged directly to a large tank into which the shaft and disk are inserted. Dispersion takes place by the shear and impact that the spinning blade imparts to the liquid dispersion base.

A sand mill can be described as a single vertical shaft with more than one circular blade operating in a cylinder. The blades are separated by a distance of about 7 in. Usually three to five blades are found on the shaft. The shaft turns at about 1600 rpm. The cylinder is usually filled with 20-mesh sand or glass beads. The mill base is pumped into the cylinder from the bottom; and, as the shaft rotates, the disks impart a swirling motion to the sand and mill base. Dispersion is accomplished primarily by shear. The mill base flows upward through the sand and finally overflows from the top through a screen that keeps the sand inside the cylinder. By controlling the speed of movement through the sand by a variable-speed pump, the degree of dispersion is regulated. The mill is water jacketed to maintain control of internal heat. Daniel (6, 7) and Patton (8) have discussed mill base formulation for sand mill and high-speed disperser operation, and readers are referred to their writings for a detailed discussion.

The dispersion of metallic pigments, either aluminum or gold bronze, is a much simpler process than that described for nonmetallics. Since each small metallic particle is given a stearate coating during manufacture, the surface takes on a lipophilic nature. This allows dispersion by a simple, low-speed mixing operation. When coatings are made from metallic powder rather than paste, the procedure is to dissolve any hard resin component or to dilute the liquid resin in the solvent blend. The powder is then added and mixed only long enough to uniformly disperse the pigment. Excessive mixing will strip the stearate coating from the powder and convert it from a leafing to a nonleafing type. This results in loss of the high metallic sheen characteristic of leafed metallic coatings.

If the coating is made from metallic paste, the procedure is to add a small amount of the liquid resin to the paste, and, with mixing, slowly lower its viscosity with additional quantities of vehicle until a complete viscosity reduction has taken place. Failure to handle paste in this manner will usually result in an incomplete dispersion of metal particles and will often cause small seeds of paste to remain in the finished paint. These can be removed by filtration.

Whether the coating will be an enamel or a pigmented lacquer, pigment dispersion must first take place. In nitrocellulose lacquer formulations, pigments are dispersed in the nondrying alkyd portion of the formula or in the plasticizer—dioctyl phthalate, for example. Manufacturers may choose to buy the pigment dispersions from dispersion manufacturers rather than to disperse the pigments in their own facilities. However, in either way, pigment dispersion must take place.

COMMON TYPES OF AEROSOL COATINGS

Table 21.1 lists the most common aerosol coatings. Since the standard types of enamel and lacquer constitute at least 75% of all aerosol coatings sold, the distinguishing physical and chemical characteristics of these two classes of coatings are now discussed.

TABLE 21.1.   CLASSIFICATION OF COMMON TYPES OF
AEROSOL COATINGS

   I. Enamel
      A. Standard
         1. Gloss
         2. Flat
      B. Specialty
         1. Hammer
         2. Wrinkle
         3. Colored metallics
         4. Antirust
         5. Heat resistant
  II. Lacquer
      A. Standard
         1. Gloss
         2. Flat
         3. Clear—water-white or transparent colored
      B. Specialty
         1. Metallic
         2. Fluorescent
 III. Varnish
      A. Gloss
      B. Satin
 IV. Stain
      A. Penetrating
      B. Varnish
  V. Primer
      A. Lacquer type
      B. Enamel type

Enamel

By definition, an enamel is a pigmented varnish. Since the late 1920s, when alkyd resins became commercially available, enamels have been made from these "varnishes" rather than from varnishes prepared from naturally occurring resins and oils. The alkyd resins used in air-dry coatings are prepared by combining a polyhydric alcohol with a polybasic acid and modifying with the fatty acids of a drying vegetable or fish oil. A typical example is the resin derived from the reaction of phthalic anhydride and glycerol modified with such oils as linseed or soybean. For the modification of nitrocellulose lacquers, the nondrying type of alkyd resin is normally used. This would be a resin prepared from coconut or other nonoxidizing oil. Alkyd resin "varnishes" are not the only type used in aerosol enamels; epoxy ester resins and oil-modified polyurethanes also find application.

Enamels are characterized by a period of lingering tackiness during their drying cycle. This tacky period may vary from 15 min to 6 or 8 hours, depending on the composition of the resin. Enamels with a very short "tacky" time can be made by modifying an alkyd resin with styrene, vinyltoluene, or acrylic monomers. The drying time will be governed by the quantity and type of modifier, as well as by the method of incorporation into the base resin. These quick-drying alkyd resins constitute the most popular class used in aerosol enamel. They can be pigmented with a wide variety of types and colors of pigment and, with the addition of extender pigment, which is used for adjusting gloss, can also be formulated into flat (low-gloss) enamels.

The types of enamels listed under the classification "speciality" fall into the enamel category by virtue of the fact that they are pigmented "varnishes." The varnish typically used for each type of coating in this group is as follows:

1. Hammer: styrene-modified alkyd resin.
2. Wrinkle: China wood oil phenolic varnish.
3. Colored metallic: nonmodified alkyd resin or epoxy ester resin.
4. Antirust: nonmodified alkyd resin.
5. Heat resistant: silicone/alkyd resin or silicone/acrylic combination.

Lacquer

As used today, the term "lacquer" refers to any coating that dries by evaporation of organic solvents. Traditionally, "lacquers" designated coatings that had cellulosic binders. However, in recent years, acrylics, vinyls, and styrene resins have become common raw materials for the formulation of lacquers. Therefore the broad definition based on the type of drying is preferred. Since lacquers dry by evaporation, there is no "tacky" period such as is characteristic of enamels. Metallic lacquers, usually formulated from acrylic resins, dry tack free in 3 to 4 min. A nonmetallic pigmented lacquer will be tack free in 7 to 15 min, depending on whether it has a gloss or flat sheen. Cellulosics, by themselves, produce films or coatings with inferior properties; therefore they are usually used in conjunction with modifying agents, either plasticizers or resins. The main purpose of adding plasticizers to cellulosics is to increase the flexibility of the coatings. At the same time, they soften the coating and decrease its tensile strength. The resins may be natural or synthetic types and are used to increase adhesion of the film, improve light and weather resistance, enhance gloss, increase moisture and water resistance, and reduce costs.

In the case of acrylic resins for lacquers, the thermoplastic variety is customarily used. Acrylic resins are available in several grades of hardness, depending on the monomer that is the basis of the resin. These resins can be blended to obtain a satisfactory hardness, or they may be plasticized with chemical plasticizers. The acrylics have higher performance characteristics than do the cellulosics. As a rule, they have better resistance to mild chemicals, heat, sunlight, and weather.

The pigmentation of lacquers is somewhat more complex than that of enamels. All cellulosic base lacquers and many acrylic lacquers have a certain percentage of very strong organic solvents. These tend to dissolve certain types of organic pigments. In addition, when dealing with nitrocellulose, it is important to avoid pigments of an alkaline nature since they react and produce gas containing nitrogen oxides. Organic dyes are often used to color clear lacquers. Most dyes, however, are not light fast, and therefore their use is ordinarily confined to lacquers intended for interior purposes. Acrylic lacquers are the preferred base for gold, bronze, and aluminum finishes because of the very low acid numbers available in acrylic resins. Florescent lacquers are also prepared from acrylic bases because of their water-white color and outstanding color retention.

CHOICE AND FUNCTION OF COATING RAW MATERIALS

Resin

The principal factor governing the selection of a resin is the service that the coating will be expected to perform. Coating performance is very largely dependent on resin selection. Table 21.2 outlines the choices that the formulator has in his selection of a resin for a coating required to perform a specific function. There are other types of coatings or services in addition to the ones shown; Table 21.2 is only meant to be suggestive of typical examples. When more than one resin is suitable, the selection will be based on cost and the relative availability of raw materials. It would obviously be preferable to use a raw material already on hand than to formulate a product calling for one that would have to be purchased. Cost is an ever-present problem and influences the choice of raw materials to a great degree.

TABLE 21.2. FACTORS GOVERNING RESIN SELECTION: TYPE OF COATING OR SERVICE EXPECTED

---

1. Nonmetallic—quick dry—general service—moderate exterior durability
   Nitrocellulose/alkyd/plasticizer
   Vinyltoluene-, styrene-, or acrylic-modified alkyd or oil
2. Maximum interior durability—resistance to marring, abrasion, and handling
   Epoxy ester
3. Maximum exterior durability
   Silicone/alkyd
   Isophthalic medium oil alkyd
4. Metallic—Fully Leafed
   Acrylic (gold bronze, copper, and aluminum)
   Petroleum resin/oil (aluminum)
   Styrene (all colors)
5. Fluorescent—quick dry
   Acrylic
   Vinyltoluene-modified alkyd
6. Stain
   Pure alkyd/oil
   Acrylic
   Urethane
7. Hammer
   Styrenated alkyd
   Pure alkyd/silicone addition
8. Wrinkle
   Phenolic/tung oil
9. Heat resistant
   Silicone
   Silicone/acrylic
   Silicone-modified alkyd
10. Primer
    Nonmodified alkyd
    Nitrocellulose
    Styrene-, acrylic-, or vinyltoluene-modified alkyd

Subclassifications may also govern selection of the resin. Suppose, for instance, that it is necessary for a coating not only to be quick drying but also to be suitable for styrene plastic and styrofoam. This would eliminate most cellulosics because of the strong solvents required for that type of coating. It would also eliminate certain types of styrene and acrylic-modified alkyds for the same reason. As a further example, consider the selection of a resin for a coating that must have excellent exterior durability. There are several choices from a technical point of view: isophthalic alkyd resin, a silicone-modified alkyd resin, and a chain-stopped alkyd resin. If cost is a major consideration, the silicone alkyd will probably be eliminated. However, if the requirements include, in addition to exterior durability, good chemical and/or heat resistance, the silicone alkyd will be the resin of choice.

After the type of resin has been selected, it is desirable to pick, from the particular class, the specific resin that has the highest tolerance for low-solvency thinners. This is necessary in order to minimize the problem of compatibility with any of the commercially available propellants. All of these propellants are poor solvents. Finally, the solids/viscosity relationship must be taken into account. Within a given class of resin, the quality of an aerosol coating is governed by the total solids in the can. Since the amount of solids is governed by the limitations of spraying viscosity, the resin that provides the highest solids at the lowest viscosity should be used.

Pigment

In coatings that can use petroleum distillate solvents, the selection of a pigment is not complicated. It is based simply on the color requirement, the service expected of the coating (interior or exterior), the amount of hiding power needed, and the cost. In other cases, such as those involving epoxy ester resins or nitrocellulose lacquers, the formulator must take into account the solubility characteristics of certain organic pigments. For example, some types of hansa yellow are dissolved by a mixture of toluene, acetone, and methylene chloride. Many organic pigments are soluble in ketones and esters. Under these conditions, pigment selection becomes more difficult.

The selection of leaded pigments is usually governed by marketing considerations. Most leaded pigments will deliver more hiding power per dollar than their nonleaded equivalents. However, in view of the severity of current lead regulations and the labeling that is required, many companies have decided against the use of this class of pigment. Technically, leaded pigments offer few problems and can be used effectively in aerosol coatings.

The total quantity of pigment which can be used is controlled by the viscosity increase that added pigment causes and by considerations of cost. The use of a resin with a very good solids/viscosity relationship, as previously mentioned, will make possible a higher percentage of pigment before viscosity and gloss are adversely affected than would otherwise be the case. Gloss, as mentioned earlier, is controlled by the ratio of pigment to vehicle solids, and viscosity by the total solids content. The selection of a pigment for an aerosol coating is based on the same considerations as would apply if a nonaerosol coating were being formulated. For example, if maximum durability is expected, a red enamel would not be made from a calcium lithol toner, since this pigment has poor fade resistance even though it offers high brightness and good hiding power. On the other hand, it may not be possible to use a toluidene toner, although this pigment has excellent exterior durability, good brightness, and acceptable hiding power, because to achieve a haze-free gloss, particularly if the resin involved is a vinyltoluene-modified alkyd, may prove impossible.

Solvent

The choice of solvent is determined primarily by the class of resin that is selected, and secondly by the quantity of propellant that is used in the product. If the resin has infinite dilution characteristics with low-solvency solvents, the choice of solvent is based nearly exclusively on evaporation rate. However, if the tolerance of the resin for low-solvency solvent is limited, then, in order to accommodate the large volume of propellant that is associated with aerosol coatings, it will be necessary to choose strong solvents in order that the solvent power of the propellant/solvent mixture will be high enough to keep the resin in solution.

Figures 21.1 and 21.2 show graphically the dilution effect of propellant addition to a coating formula. Figure 21.1 shows the volume relationship of the principal components before the propellant is added. Figure 21.2 combines propellant and coating components. Since the propellant is an integral part of the formula, being present as a liquid except for the portion that vaporizes to fill the can space above the liquid level, it becomes, in effect, a low-solvency thinner. For purposes of discussion, the propellant choice will be limited to aliphatic hydrocarbons because of their predominance in the field.

A discussion of solvent selection must take into account the effect of the propellant on the entire system, and therefore the solvency aspects of the propellant will be considered concurrently with those of the solvent. "Solvent power" is not a precise term but is generally understood to mean the ability of a solvent to form a stable, homogeneous solution with another substance. However, the mere ability of a volatile liquid to dissolve the film former is only the first consideration. Other aspects are the maximum amount of film former that the solvent will dissolve

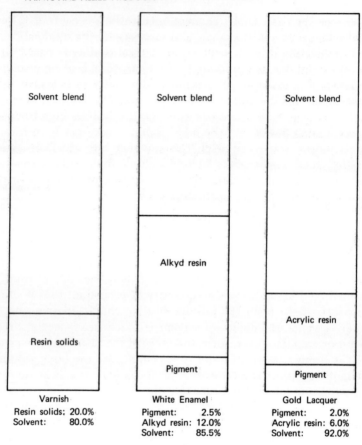

Figure 21.1.   Volume relationships of pigment, resin and solvent for typical aerosol varnish, enamel and gold lacquer before addition of propellant. (All percentages by volume).

and the concentration of the solution. Of even greater importance in coating formulation is the viscosity of the solution at a given concentration; this is the most common basis for rating solvents as to solvency for a particular resin. Viscosity increases with resin concentration. At a given concentration, viscosity is governed largely by the degree to which the dispersion approaches molecular size; the greater the size, the higher is the viscosity. Even with strong aromatic solvents and easily soluble alkyd resins, it has been shown (9) that the molecular aggregates may be 40 to 800Å in diameter and composed of 40 primary alkyd molecules.

When a solvent is added to a resin in increasing amounts, there is usually a point at which the resin precipitates or "kicks out." This limit is useful primarily as an indication of the degree of compatibility of the resin and the solvent. Long-oil, medium-oil, and vinyltoluene-modified alkyd resins are not critical to solvency and are soluble in all proportions in weak solvents. However, short-oil, styrenated alkyds and acrylic resins, for example, have critical solvency and require highly aromatic solvents for solubility in all proportions.

A great number of attempts to develop measures of solvent power have been made, most of them based on laboratory tests of the solvent.

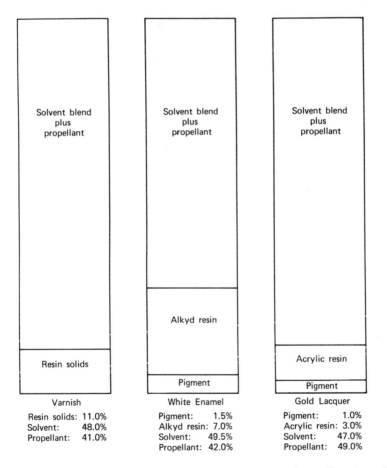

Figure 21.2.   The products shown in Figure 21.1 plus addition of propellant, in typical quantity. (All percentage by volume).

The kauri-butanol (KB) value, for example, has been used for many years in the coating industry as a measure of solvent power. This method of test is confined to hydrocarbon solvents and determines their relative solvent powers. The kauri-butanol value is the number of millileters of the solvent at 77°F required to produce a defined degree of turbidity, when added to 20 g of standard solution of kauri resin in normal butyl alcohol. For KB values of 60 and over, the standard is toluene, which has an assigned value of 105. For KB values under 60, the standard is a blend of 75% n-heptane and 25% toluene, with an assigned value of 40. Higher values indicate greater solvent power. The KB values of products that are classified as regular mineral spirits normally vary between 34 and 40. Xylol is 95, and the aromatic naphtha solvents range all the way from 55 to 108.

The KB factor of a solvent mixture is the sum of the products of the volume percentages of all constituents and their KB factors. This relationship makes possible the calculation of the KB of a mixture of hydrocarbon propellant and hydrocarbon solvents.

Consider the example of a resin described by a manufacturer as being 50% solids in VM&P naphtha* and having "infinite solubility in mineral spirits." Since mineral spirits is a low-KB solvent having a factor of 34 to 40, there should be no problem in reducing the resin to a spraying viscosity with VM&P naphtha or hexane, both of which also have low KB values, and packing the enamel prepared from such a resin with hydrocarbon propellant. The following calculation shows how to handle this particular problem. What will be the KB of a mixture of 80% hexane (KB 28) and 20% mineral spirits (KB 35)?

Example 1. Let X = volume per cent hexane, and Y = volume per cent mineral spirits.

$$\text{KB of mixture} = (X)(28) + (Y)(35)$$
$$= (0.80)(28) + (0.20)(35)$$
$$= 29.4$$

This computation shows the KB of the mixture to be 29.4, which may not be high enough to keep the resin in solution at the solids required for spraying. This computation, of course, must take into account the solvent that accompanies the resin as received from the supplier. A laboratory test of this particular solvent mixture would be required to ascertain whether or not it would keep the resin in solution. The formulator might choose to increase the KB of the solvent mixture by the

* A special grade of naphtha (varnish makers and painters naphtha).

addition of toluene (KB 105). He may wish to end with a KB of 40 in order to provide a safety factor against the precipitation of the resin.

Example 2. As before, X = volume per cent hexane and Y = volume per cent mineral spirits; now Z = volume per cent toluene. Three equations representing the various relationships can be written:

$$40 = (X)(28) + (Y)(35) + (Z)(105)$$
$$1.0 = X + Y + Z$$
$$X = 4Y \text{ (From Example 1)}$$

Solving the equations results in the solution:

$$X = 68.8\%, \qquad Y = 17.2\%, \qquad Z = 14.0\%$$

This procedure can be taken one step further with the calculation of the KB value of the solvent and propellant blend. Since the propellant is, in effect, a weak solvent for the paint system with a KB factor of about 16, it can be included in the calculation.

Aerosol enamels are commonly packed with 25 to 30% propellant by weight (the effect of the percentage of propellant on sprayability is discussed in a later section). For purposes of this example we assume that it is desired to calculate the KB factor of the combined solvent mix specified in Example 2 and the propellant, which is in the package at 27% by weight of total contents, a typical value. This mixture contains 46% propellant, 37% hexane, 9% mineral spirits, and 8% toluene, all by volume. Referring to the equation for calculating the KB factor shown in Example 1, but using four components in place of two, we develop the relationships shown in Example 3.

Example 3. As before, W = volume per cent propellant (KB 16.5), X = volume per cent hexane (KB 28), Y = volume per cent mineral spirits (KB 35), Z = volume per cent toluene (KB 105).

$$\begin{aligned}
\text{KB of mixture} &= (W)(16.5) + (X)(28) + (Y)(35) + (Z)(105) \\
&= (0.46)(16.5) + (0.37)(28) + (0.09)(35) + (0.08)(105) \\
&= 29.4
\end{aligned}$$

The KB value of 29.4 obtained from this calculation may not be high enough to keep the resin in solution and should be confirmed by experimentation. If cloudiness develops when the resin is diluted and pressure packed in a test glass bottle, it will be necessary to increase the KB of the solution by increasing the percentage of toluene. It is important to note the effect of the propellant, which will always appear in large volume, on the solvency of the mixture. A high volume of low-solvency "solvent" always has a very decided and often harmful effect on the solvency of the mixture. Formulators must continually take this into account.

While the above calculations demonstrate the method of using the KB factor to determine the solvency of mixtures of solvents and propellants typical of enamel-type aerosol coatings, computation of the solvency of solvent blends used in nitrocellulose, acetate butyrate, or vinyl lacquers is more complicated. The KB value does not apply to esters, ketones, and alcohols, all of which are used in lacquer solvent blends. Therefore no easy calculation can be made to establish the proper ratio of solvents to be used in connection with aerosol lacquers. Solvent mixtures for nitrocellulose lacquers contain three classes of volatile liquids: active solvents (ketones and esters), latent solvents (alcohols), and diluents. A typical mixture contains 30% ketones and esters, 15% alcohols, and 55% diluents. The diluents are normally aliphatic or aromatic hydrocarbons. In formulating nitrocellulose aerosol lacquers, the diluent portion is replaced by the propellant and adjustments are made in the percentage of active solvents. This requires a "cut and try" series of experiments to determine the proper percentage of each class of solvent and diluent.

A more precise method of handling solvency was developed by Burrell (10) and others, who applied the solubility parameter concept of Hildebrand and Scott (11) to resin/solvent systems. The solubility parameters of solvents are numerical constants that can be accurately calculated from measurable physical properties, such as molecular weight, density, and heat of vaporization. Most resins and polymers, however, are amorphous mixtures of a number of different molecular species with inexact thermodynamic values. Their solubility parameters cannot be calculated but must be determined experimentally and a value assigned on the basis of the midpoint of the solubility parameter range for the solvents in which the resin or polymer is soluble. Solubility parameters for solvents, resins, and polymers are commonly designated by delta (δ). The basic principle involved in the application of solubility parameters is a simple one. When the value for a solvent falls within the parameter range of the resin or polymer, the resin or polymer will be soluble in that solvent.

In applying the solubility parameter concept the hydrogen bonding strength of the solvent must be considered. Burrell (10) and the CDIC Paint & Varnish Production Club (12) classify solvents and resins according to their hydrogen bonding tendencies as follows:

Class 1: poorly hydrogen bonded—aliphatic, aromatic, and chlorinated hydrocarbons and nitrohydrocarbons.
Class 2: Moderately hydrogen bonded—esters, ethers, and ketones.
Class 3: Strongly hydrogen bonded—alcohols, amines, and acids.

Hydrogen bonding is generally expressed as a numerical parameter, represented by gamma (γ), in one of two ways. In the first, solvents are assigned numbers from an arbitrary scale in proportion to their Burrell classifications. Refinements are later made in the laboratory. In a method developed by du Pont (13), solvents are initially assigned numbers from 1 to 10: 2.5 for weakly bonded, 5.5 for moderately bonded, and 8.5 for strongly bonded solvents. The value of hydrogen bonding so assigned to a solvent can be adjusted upward or downward, depending on its performance in the laboratory as compared to the other solvents in its class.

To determine hydrogen bonding values for polymers, the same approach as outlined for solubility parameters holds; the polymer has a hydrogen bonding index range corresponding to the hydrogen bonding values of the solvents in which it is soluble. However, since this range also varies with solubility parameter, it is most convenient to use

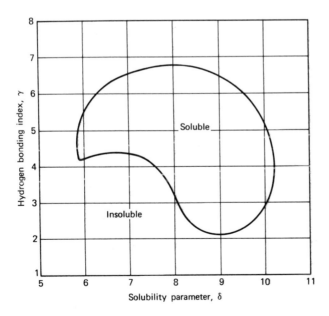

Figure 21.3.   Solubility map of a resin.

a graphical means of expressing solubility parameter and hydrogen bonding for polymers. For example, by using a number of solvents and blends, it can be experimentally determined in what range the solubility parameter and hydrogen bonding index of a solvent must be in order to dissolve a given polymer. When this is done, a map like the one shown in Figure 21.3 can be drawn. Such a map gives the information that, if the solubility parameter and hydrogen bonding index of the solvent fall within the perimeter outlined by the solid line, the solvent should dissolve the resin that the map describes.

In the search for a more exact method of expressing the solubility relationships of solvents and solutes, Eastman Chemical Products (14) has added a third parameter: dipole moment, designated by mu (μ). In addition, hydrogen bonding index is determined by a method devised by Gordy (15), who found hydrogen bonding strength to be proportional to wavelength measurements in spectroscopic analysis. This

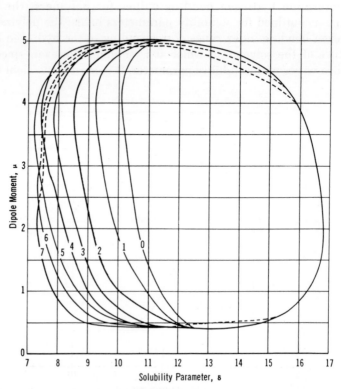

Figure 21.4.   2-D solubility map of RS 1/2-sec. nitrocellulose (γ from 0 to 7).

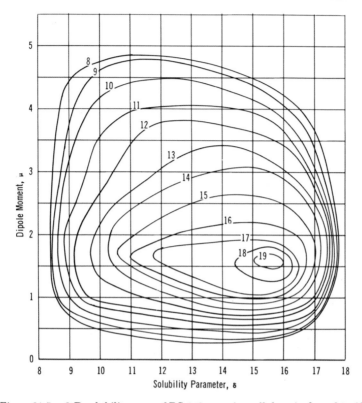

Figure 21.5.   2-D solubility map of RS 1/2-sec. nitrocellulose (γ from 8 to 19).

method provides hydrogen bonding indices ranging from 0 to about 18, although some solvents run higher. The dipole moment of a particular solvent is readily available in the literature, and values range from 0 to approximately 5. Dipole moments for polymers are assigned in the same manner as solubility parameters and hydrogen bonding indices.

The inclusion of this third variable results in a three-dimensional graphical presentation, rather than the two-dimensional map developed by du Pont. A solvent whose coordinates (δ, γ, and μ) fall within the solid model should form a solution with the polymer represented by the solid. A simpler method of handling the three-demensional approach is to use a solubility plot in two dimensions and to employ topographical contour lines to represent the shape of the model at different values of the third dimension. Figures 21.4 and 21.5 illustrate such graphical representations of the three-dimensional solid.

TABLE 21.3. SOLUBILITY CHARACTERISTICS FOR PROPELLANT/SOLVENT MIXTURE (Example 4)

	Weight (g)	Weight (%)	Volume (cc)	Volume (%)	δ	γ	μ
Propellant	100	0.36	188	0.47	6.0	0	0
MEK	35	0.13	43	0.11	9.3	7.7	2.7
MIBK	35	0.13	43	0.11	8.4	7.7	2.7
Butyl acetate	35	0.13	40	0.10	8.5	8.8	1.9
Butyl cellosolve	10	0.04	11	0.03	8.9	13.0	1.6
Isopropanol	35	0.13	44	0.11	11.9	18.7	1.7
Butanol	24	0.08	27	0.07	11.4	18.7	1.7
	274	1.00	396	1.00			

Example 4 shows how this approach to solubility determinations can be used.

Example 4. The problem is to determine whether the solvent/propellant composition shown in Table 21.3 can dissolve RS 1/2 sec. nitrocellulose component of an aerosol lacquer. The parameters δ, γ, and μ of the mixture are determined by adding the products of the parameter and volume per cent of each component.

$$\delta_{mixture} = (\delta_1 \times V_1) + (\delta_2 \times V_2) + (\delta_3 \times V_3) + \text{etc.}$$
$$\gamma_{mixture} = (\gamma_1 \times V_1) + (\gamma_2 \times V_2) + (\gamma_3 \times V_3) + \text{etc.}$$
$$\mu_{mixture} = (\mu_1 \times V_1) + (\mu_2 \times V_2) + (\mu_3 \times V_3) + \text{etc.}$$

The subscripts 1, 2, and 3 refer to components 1, 2, and 3 in the mixture. Solving these equations with the data from Table 21.3 yields:

$$\delta_{mixture} = 7.99, \qquad \gamma_{mixture} = 6.34, \qquad \mu_{mixture} = 1.14$$

From Figure 21.4, knowing the solubility parameter and dipole moment, the hydrogen bonding value is found to be 6.5. Since this value is higher than the calculated value, 6.35, this solvent/propellant mixture will not dissolve RS 1/2-sec nitrocellulose. If the calculated hydrogen bonding is within ±0.5 of the hydrogen bonding value read from the map, there will be some question regarding the validity of the solubility prediction. It is usually advantageous for the formulator to consider such values as representing borderline solubility.

Example 4A. An examination of the data in Table 21.3 and the equations shows that by lowering the quantity of propellant an increase in all three coordinates will result. By trial and error, it will be found that, if the propellant is decreased approximately 35%, the mixture will have coordinates that will result in solubility of the nitrocellulose. An example of such a mixture is shown in Table 21.4. Using the same mathmatical procedure as outlined in Example 4, we obtain the following results:

$$\delta_{mixture} = 8.60, \qquad \gamma_{mixture} = 7.22, \qquad \mu_{mixture} = 1.47$$

The solubility parameter and dipole moment values for this second blend are plotted, and a hydrogen bonding index of 3.8 is read from Figure 21.4. Since this value is less than the calculated hydrogen bonding value, 7.22, this blend will dissolve RS 1/2-sec nitrocellulose. It will also dissolve any resin in the formula that is compatible with RS 1/2-sec nitrocellulose, since compatible resins have similar solubility parameters.

As can be seen from the above examples, the propellant plays a major role in determining the compatibility of the volatile portion of the formula with the resin of the coating system, because of its high volume and low solubility characteristics. Unfortunately, such solubility maps as those appearing in Figures 21.4 and 21.5 are not always available from resin suppliers. However, some manufacturers have published solubility maps as part of their methods for applying the

TABLE 21.4. SOLUBILITY CHARACTERISTICS FOR PROPELLANT/SOLVENT MIXTURE (Example 4a)

	Weight (g)	Weight (%)	Volume (cc)	Volume (%)	δ	γ	μ
Propellant	64	0.220	123	0.31	6.0	0	0
MEK	45	0.155	57	0.14	9.3	7.7	2.7
MIBK	45	0.155	57	0.14	8.4	7.7	2.7
Butyl acetate	45	0.155	52	0.13	8.5	8.8	1.9
Butyl cellosolve	13	0.045	14	0.04	8.9	13.0	1.6
Isopropanol	45	0.155	58	0.15	11.9	18.7	1.7
Butanol	32	0.115	35	0.09	11.4	18.7	1.7
	289	1.000	396	1.00			

concept of parameters of solubility.* Also, it is possible for formulators to develop their own maps, if necessary, from laboratory data. Once a given method of application is selected, the resin characteristics obtained from laboratory determinations can be plotted. A solvent and nonsolvent can be found for a given resin if the parameters are known. Various blends of these solvents will produce varying degrees of solubility. The parameter of the solubility map is determined by the critical blend in which a change in volume percentage of either solvent will result in borderline solubility or insolubility. The coordinates thus obtained define the parameter of the solubility map. By following this procedure for a number of combinations of different solvents and nonsolvents, a solubility map for the resin can be calculated.

Propellant

The propellant for an aerosol coating, being an integral part of the formula, must have mutual solubility with the coating solvents. The combination of solvents and propellant must serve as a solvent for the resin in the formula. Therefore an aerosol coating is a two-phase system. The first phase is the liquid/solid coating mixture; the second is the gaseous propellant which exists in the head space above the mixture. Compressed gases do not make good propellants for aerosol coatings because of their limited solubilities in the solvents. Although nitrous oxide and carbon dioxide are employed in other types of aerosol products, they have not been successfully used with coatings. The propellants most commonly used in coatings are aliphatic hydrocarbons such as propane, isobutane, or mixtures of the two, and halogenated hydrocarbons such as Propellant 12 and Propellant 11. Economics strongly favor the use of the aliphatic hydrocarbons, and as a result the large majority of coatings are packed with this type of propellant.

As seen from Table 21.5, the solubility characteristics of isobutane, propane, and Propellant 12 are quite similar. Propellant 11 has a substantially higher KB value than the others. It should be possible, therefore, to replace, on a volume basis, one propellant for the other, if only

---

*These maps are available from the following manufacturers:

1. Pennsylvania Industrial Chemical Corporation, Clairton, Pa., "Solubility Contours for Picco Resins."

2. Rel-Rez Division, Reliance Universal, Inc., Louisville, Ky., "Resin Solubility Studies."

3. Ashland Chemical Co., Division Ashland Oil, Inc., Columbus, Ohio, "Predicting Resin Solubilities."

solubility characteristics are considered. In normal practice, a blend of isobutane and propane having a pressure of 70 psig at 70°F has replaced Propellant 12, which has the same vapor pressure characteristic.

From a formulation point of view, perhaps the most significant difference between the aliphatic hydrocarbon and the halogenated hydrocarbon propellants is the vast difference in density of the two classes. Halogenated hydrocarbons have a density more than twice that of the aliphatic hydrocarbons. Therefore a volume replacement of halogenated by aliphatic hydrocarbons results in a significant decrease in the density of the coating/propellant mixture. To offset this density difference, it is common practice to select methylene chloride as one of the solvents in an aerosol coating when aliphatic hydrocarbons are intended for use. There may be resin compatibility limitations, of course, that dictate the omission of methylene chloride; but wherever compatibility will allow, the high density of this compound, approximately equal to that of the halogenated hydrocarbon propellants, is helpful in minimizing the density difference between two types of propellants. The rapid evaporation rate of methylene chloride is also useful.

Although the flammability difference between the two classes of propellants is not a factor that the formulator usually disregards, it may make a difference in the labeling which is required by the Hazardous Products Labeling Act. The type of hazard statement is governed by the results of certain specified flammability tests, and these results are controlled by propellant type, solvent selection, and sprayhead choice. In some products, a higher flammability rating may be obtained with the use of aliphatic hydrocarbon propellants than is given by halogenated hydrocarbon propellants. From a practical point

TABLE 21.5. SOLUBILITY CHARACTERISTICS FOR SOME PROPELLANTS

Propellant	δ	γ	μ	Kauri-Butanol Value
Isobutane	6.17	0[a]	0	17.5
Propane	5.80	0[a]	0	15.2
Propellant 12	6.1	2.5	0.51	18.0
Propellant 11	7.8	2.5	0.45	60.0

[a] Method of Eastman Chemical Products, Inc.

of view, however, since all aerosol coatings contain a large volume of volatile solvents, the majority of which are flammable (methylene chloride being the major exception), the use of a flammable propellant does not materially increase the hazard to the consumer. However, there is a substantial increase in danger to the aerosol packer, who is confronted with the engineering task of safely handling many thousands of gallons of a flammable propellant, if he chooses to replace halogenated hydrocarbons with aliphatic hydrocarbons.

Although, as mentioned above, it is common practice to use a blend of propane and isobutane as the propellant, the formulator has the option, when the finished product requires it, of using isobutane as the only propellant. This will lower the vapor pressure in the package considerably since the vapor pressure of isobutane is approximately 32 psig at 70°F. Likewise, if it is desirable to have the highest pressure allowed by the U.S. Department of Transportation, or if, because of solubility characteristics, as outlined earlier, it is necessary to reduce the quantity of propellant, the formulator may choose propane as the only propellant. Since the vapor pressure of propane at 70°F is approximately 110 psig, it is necessary to limit the quantity so that the container pressure resulting from the blend of propane and solvents does not exceed the maximum value allowed under current regulations. The quantity of propane customarily used under these conditions is roughly 15% by weight of net contents or about 25% by volume.

Valves and Sprayheads

A satisfactory valve for aerosol coatings is one that can be cleaned of any pigment which may accumulate and cause blockage of the paint stream. This is most readily accomplished with a female valve since the metering orifice is in the side of the stem of the sprayhead, which can be removed from the valve at any time for cleaning. The two most commonly used valves of this type for aerosol coatings are the B-14, made by Newman-Green, Inc., and the AR-74, made by Aerosol Research Corporation. A third valve of different design, the Danvern valve, is made by Sprayon Products, Inc. This is a proprietary valve available only to Sprayon and not sold on the general market.

A variety of sprayheads are available for use with both the Aerosol Research and the Newman-Green valves. A mechanical breakup sprayhead is seldom used with aerosol coatings because the propellant alone does an exceptionally good job of breaking the paint stream into tiny droplets, and because there is a greater likelihood of plugging when used with pigmented coatings. The most commonly used spray-heads are those that conduct the stream straight up and out, and those that have an internal orifice which aids in additional atomization of the liquid stream. By choosing the style of sprayhead, slot size, and external orifice diameter, it is possible to obtain a wide variety of spray patterns from a given spray pack. Thus the sprayhead becomes a part of the formulation in a very real sense, since wide variations in performance can be achieved by selection of one or the other of the many sprayheads available.

IMPORTANT CHARACTERISTICS OF AEROSOL COATINGS

Sprayability

The essential element that distinguishes aerosol coatings from those applied with a brush is sprayability. This characteristic may be defined as the capability of being applied to a surface from a spray of small droplets with minimum finger fatigue, and of producing an attractive appearance when dried. Thus sprayability has two parts: application and appearance. The variables that affect sprayability are viscosity, coating/propellant ratio, solvent blend, sprayhead selection, and freedom from valve clogging.

VISCOSITY

If the viscosity of the coating formula is too high, the propellant will not be able to break the stream into satisfactorily small droplets. This may result in a coarse or "orange peel" appearance. In addition, propellant evaporation will be impaired since it will be more difficult for the propellant to break free from the large, high-viscosity droplets. This often results in bubbling as the propellant evaporates after the coating has reached the surface. On the other hand, a coating that has an unnecessarily low viscosity will have good atomization but excessively low solids. In addition, sagging may be encouraged, although this can often be corrected by increasing the percentage of fast-evaporating solvents. Since coating quality for any given group of raw materials is a function of the percentage of solids, it is desirable to keep the solids at the highest possible level commensurate with good sprayability. This means that viscosity should not be below the point just required to give satisfactory breakup. A viscosity in the range of 25 to 35 sec, measured on a No. 1 Zahn viscometer,* will usually give satisfactory performance for normal enamel or lacquer. However, for certain specialty coatings, such as hammer or wrinkle finishes, viscosity may be as high as 40 to 50 sec.

The No. 1 Zahn cup is a desirable viscometer because of its ability to differentiate viscosities in the very low ranges required for aerosol coatings. The No. 4 Ford cup, which is sometimes mentioned in the literature, is not satisfactory for measuring viscosities below 20 sec. The ASTM Standard D 1200-58 limits the use of any orifice in the Ford cup series to a minimum of 20 sec and recommends the use of the next lower number orifice for viscosities below 20 sec. Twenty seconds on a No. 4 Ford cup represents too high a viscosity for aerosol coatings to provide optimum breakup and freedom from bubbling.

COATING/PROPELLANT RATIO

For any given viscosity, the higher the propellant/paint ratio, the smaller will be the droplet size, and the dryer will be the coating when it reaches the surface. This is the result of rapid solvent evaporation from the smaller droplets. Also, as the propellant ratio increases, the quantity of paint in the can must necessarily decrease, assuming the can to be nominally full. At some point, the quality of the product will be so impaired, as a result of having such a small quantity of paint in the can, that it will provide an inadequate area of coverage per can. In practice, the quantity of propellant will vary from 15 to 30% by weight (22 to 45% by volume) of net contents. The low figure is typical for lacquer formulations, which can use only a relatively small amount of propellant because of its limited compatibility with nitrocellulose. At the lower level, a sprayhead capable of excellent breakup is required to give a smooth finished appearance. At the higher level, sprayhead choice is not as critical since the quantity of propellant is high enough to effect good atomization.

SOLVENT BLEND

Of critical importance to the appearance of a sprayed coating is the solvent blend. With the exception of metallic finishes, which can suc-

---

* The Zahn viscometer consists of a bullet-shaped, stainless steel cup, with a precision-drilled hole in the bottom. It has a 12-in. looped handle, fastened parallel to the long axis of the cup. This allows immersion of the cup into the liquid paint. At the top of this handle is a small ring which aligns the cup in a vertical position when withdrawing it from the liquid paint. It provides readings in terms of a unit called the Zahn-second, which is the time required for 44 cc of paint, the total volume of the cup, to flow through the hole in the bottom. Zahn viscometers are available with holes of five different diameters and are numbered from 1 (diameter 0.078 in.) to 5 (diameter 0.208 in.).

cessfully use a single solvent, such as toluene, clear coatings and nonmetallic enamels and lacquers require a combination of solvents, varying in evaporation rate, in order to ensure a smooth, sag-resistant finish. The rapidly evaporating solvents leave the coating during the time when the droplets are moving from the sprayhead to the surface. This increases their viscosity and dryness, thereby minimizing the likelihood of sagging. The dryer the film, the less will be the propensity to sag. Conversely, the wetter the film, the more likely it is that sagging will take place. Wetness and dryness, of course, can be influenced by the distance at which the can is held from the surface. The farther away, the greater will be the evaporation of solvents and hence the dryer the film. However, the farther away the can is held, the greater will be the overspray, and hence the area of coverage per can will be lower. Solvent blends usually consist of approximately 60 to 70% fast-evaporating, 25 to 35% medium-evaporating, and 7 to 10% slow-evaporating solvents. For any coating, it is important that the slowest-evaporating component be a true solvent for the resin. Otherwise the resin will precipitate when only the last component remains. These solvents are included in rather small amounts so as to promote sufficient flow to allow the individual droplets to flow together into a smooth film, but not to provide so much flow as to encourage sagging.

SPRAYHEAD SELECTION

The sprayhead must be considered as an element in the formulation, just like any chemical raw material. Sprayhead selection determines delivery rate, which in turn controls the distance at which the can must be held from the surface, and droplet size, which influences the appearance of the dried film. An excellent formulation can be made unsatisfactory by selection of the wrong sprayhead.

Freedom from Valve Clogging

Any product that clogs the sprayhead cannot be considered to have good sprayability. To a certain degree, freedom from clogging is controlled by the formulation. Good pigment dispersion and incorporation of proper antisettling agents will minimize pigment settling and hence reduce the likelihood of valve clogging as a result of undispersed pigment being pulled into the mechanism. However, to a large degree, clogging depends on the user. If he does not shake the can adequately before use, or, when finished, does not invert the can and spray until only propellant is emitted, he may find that the sprayhead or valve is plugged and is no longer operable. At this point, it becomes desirable for the user to be able to clear the sprayhead of accumulated pigment. In a female valve, the metering orifice is in the stem of the removable sprayhead. It is then a relatively simple matter for the user to clear the slot with a thumbnail and open the exterior orifice with a small wire or pin, thus reactivating the container. If, however, the blockage is in the valve mechanism itself, there is no way for the can to be rejuvenated. It is desirable to build into the formulation sufficiently good antsettling characteristics so that only 10 to 20 sec of shaking will reincorporate the pigment into the vehicle/propellant solution. It is unlikely that a customer will shake the can for the 1 to 2 min specified on most labels.

Hiding Power

ORGANIC PIGMENTS

The hiding power of any coating is determined by the quantity of pigment in the formulation. The limiting factors, in addition to cost, are the gloss requirement of the dried coating and the viscosity, which increases as pigmentation rises. As a rule, organic pigments do not have exceptionally good hiding power, that is, they do not have the same ability to hide a substrate that inorganic pigments of the same color have. For example, hansa and benzidine yellows do not equal the hiding power of chrome yellow of the same shade; dinitranaline orange does not have the same hiding power as molybdate orange. Organic pigments do have high brightness and in many cases excellent fade and chemical resistance. Since organic pigments are more expensive than inorganic pigments of the same shade, formula cost seriously limits the hiding power available with a coating having this type of pigment.

INORGANIC PIGMENTS

In general, pigments of this class have excellent hiding power. However, many of the leaded pigments are considered to be unsuitable for aerosol coatings because of the regulatory restrictions against the use of lead in paint. Therefore, such colors as chrome yellow, chrome green, and molybdate orange, all excellent hiding and rather inexpensive pigments, are not likely to be found in aerosol coatings. Titanium dioxide is almost the exclusive choice for white or tint bases to which organic colors are added for pastel shades. Iron oxides, which occur in various shades from muddy yellow through a range of browns to black, are also used to the extent that the final color will allow. They have excellent hiding power and are relatively inexpensive. Metallic pigments, also inorganics, also have excellent hiding quality but are limited, of course, to uses where a metallic effect is desired. Nonleafing varieties of metallic pigments can be combined with organic pigments to produce colored metallescent coatings. Taken by themselves, the fully leafed class of gold bronze, copper, or aluminum pigments provide high-hiding, attractive metallic finishes.

Drying Time

The drying time of a coating is determined by the choice of film former. Nitrocellulose and other lacquers that dry by evaporation provide coatings with tack-free times ranging from 3 to 15 min. Vinyltoluene, styrene, or acrylic-modified alkyd resins nearly duplicate the dry time of lacquers; typically, such modified alkyd resins have drying times of 15 to 20 min. Consumers have learned to associate aerosol coatings with quick drying time, and as a result the majority of coatings sold for general household or industrial touchup use exhibit dry times of less than 30 min. However, certain classes of enamels such as those made from nonmodified alkyd resins, characteristic of "rust-inhibitive" products, and enamels made from epoxy esters have dry times ranging from 2 to 8 hours. Generally speaking, the slower-drying coatings are more durable than the very-fast-drying ones.

General Performance

The performance to be obtained from any given formulation is directly related to the percentage of solids in the formula. Higher solids mean greater thickness of dry film over a given area, which, in turn, defines the amount of protection that the coating will provide the substrate. Higher solids in a pigmented coating, all other variables being constant, mean higher hiding power and hence more coverage per can. Good performance requires that the coating be properly designed for the job for which it was intended. A quick-dry touchup product may have decidedly lower rust inhibition characteristics than a product designed specifically for rust prevention. On the other hand, a rust inhibitive product is probably not usable on styrene plastic or styrofoam, for which the interior touchup enamel may be admirably suited. A product which is not sold for heavy-duty service should not be compared to types that are. Impact, mar resistance, and adhesion are important for finishes intended for touching up wagons, bicycles, and porch furniture but probably are not important for use on decorative craft pieces. Nevertheless, in any category, the higher the solids, the better will be the performance and the higher the quality.

RESEARCH AND DEVELOPMENT TEST METHODS

A number of techniques for developing new formulations have proved to be very helpful in evaluating raw materials and predicting the shelf lives of products. Pressure-resistant glass bottles varying in size from 6 to 12 oz, which are equipped with screw-type fittings capable of attaching standard valves, are very helpful in evaluating the compatibilities of various propellant/solvent resin blends. By discharging the propellant from a pressure burette into a valved and product-filled bottle, the point of incompatibility, if any, can be readily determined. In addition, long-term observation in a pressurized glass bottle will pinpoint the development of serious settling or, if evaluation is in the clear portion only, the development of precipitated resin solids.

If pressurized bottles are not available, a modified technique using an equivalent volume of isoparaffinic solvents, which have parameters of solubility approximately equal to those of hydrocarbon propellant solvents, in a nonpressurized container, can be substituted. In an unpigmented system, resin incompatibility will be easily evaluated.

It is well known (13) that there is a pigment/solvent interaction in paint formulation such that the selection of certain solvents will substantially minimize pigment settling. In addition, an effect between solvent and pigment similar to that between solvent and polymer delineated by the solubility paramater is thought to exist. By using the technique outlined by the Toronto Society (16), it is often possible to eliminate from a formulation a solvent that will cause serious pigment settling. If this cannot be done, because of solubility or evaporation characteristics, at least the formulator is made aware of the effect that such a solvent will have and can take steps in his formulation to offset the settling that will develop. This simple technique involves shaking together in a test tube a certain portion of dry pigment with the solvent to be evaluated and putting the stoppered tube aside for a period of days to observe the degree of settling that takes place.

Accelerated aging of finished packages or pressurized glass bottles is a technique which has developed over the years. Formulators vary in their choice of temperature, which can range from about 100 to 130°F.

All agree, however, that accelerated aging tends to promote reactions at a faster rate than would be expected to prevail in normal storage. The reactions referred to are pigment settling and deleafing of metallic pigments. Deleafing is particularly accelerated by high-temperature storage. An improperly formulated product will deleaf in 2 to 4 weeks at accelerated aging temperatures, as compared to 12 months under standard shelf conditions.

FORMULATION TROUBLE SHOOTING

Problem	*Corrective Action*
Sagging	1. Increase the percentage of fast-evaporating solvents by reducing medium- and slow-evaporating ones. 2. Increase the percentage of propellant. 3. Select a sprayhead that does a better job of breaking up the stream of paint.
Bubbling	1. Reduce viscosity. 2. Add a small quantity of water-miscible solvent, such as isopropyl alcohol or glycol ethers of the proper evaporation rate. 3. Increase the percentage of slow-evaporating solvent, such as mineral spirits, by about 5%.
Blushing	1. Decrease the percentage of fast-evaporating solvent and replace with medium-evaporating solvent. 2. Add approximately 3% by volume of lacquer-type blush retarder; butyl cellosolve.
Low Gloss	1. Check the compatibility of all solid ingredients in the unpigmented state. Spray or pour out on glass. 2. Dispersion may not be fine enough. Check on grind gauge.*

* A fineness-of-grind gauge is a hardened and chrome-plated steel block containing one or more tapered grooves varying in depth from a maximum at one end to zero at the other. Calibration may be according to depth (mils or microns) or in arbitrary units.
A test is made by placing a small volume of liquid paint in the deep end of the groove. The sample is drawn toward the shallow end of the groove with a straight-edge scraper. Ratings are made in terms of the scale where particles first appear in substantial concentration. ASTM Method D 1210-54, Fineness of Dispersion of Pigment-Vehicle Systems, describes this test.

3. Low gloss may be caused by blushing, particularly in hot, moist conditions. Add 3% butyl cellosolve or similar blush-retarding solvent.

4. Examine under microscope to determine whether loss of gloss is really a fine wrinkle. If so, review drier content of formula. If panels show slight wrinkle when dried vertically, improve sag resistance by methods under "Sagging" above, and add flow control and antisag agent to basic formula.

"Orange peel" appearance of dried film

1. Increase the amount of slow-evaporating solvent to promote better flow-out of wet paint.

2. Change sprayhead to one that will atomize to smaller droplet size.

3. Replace part of the fast-evaporating solvent with medium-evaporating solvent. This will also help the flow.

TYPICAL FORMULATIONS

The formulations that follow are typical of those commercially available. They have been prepared to take into account the various formulation pitfalls and requirements that have been referred to in this chapter.

White Enamel

Ingredients	% by Weight
Titanium dioxide	10.0
Vinyltoluene alkyd resin	12.0
VM&P naphtha	13.0
Toluene	20.0
Mineral spirits	6.0
Methylene chloride	12.0
Aliphatic hydrocarbon propellant	27.0
	100.0

Aluminum Acrylic Lacquer

Ingredients	% by Weight
Acrylic resin	6.0
Aluminum paste	2.6
Toluene	46.4
Methylene Chloride	20.0
Aliphatic hydrocarbon propellant	25.0
	100.0

Polyurethane Varnish

Ingredients	% by Weight
Linseed-oil-modified urethane resin	15.0
VM&P naphtha	12.0
Mineral spirits	15.0
Methylene chloride	31.0
Aliphatic hydrocarbon propellant	27.0
	100.0

Clear Acrylic Lacquer

Ingredients	% by Weight
Acrylic Resin	7.0
Toluene	38.0
Methylene chloride	28.0
Aliphatic hydrocarbon propellant	27.0
	100.0

White Nitrocellulose Lacquer

Ingredients	% by Weight
Titanium dioxide	7.0
Nitrocellulose	2.5
Coconut alkyd resin	5.0
Ester gum	1.5
Dioctyl phthalate	1.0
Methyl ethyl ketone	20.0
Ethyl amyl ketone	5.0
Butyl acetate	12.0
Isopropanol	5.0
Butanol	6.0
Toluene	10.0
Methylene chloride	10.0
Aliphatic hydrocarbon propellant	15.0
	100.0

REFERENCES

1. H. F. Payne, *Organic Coating Technology,* Vols. I and II, John Wiley and Sons, New York, 1954.
2. R. R. Meyers, and J. S. Long, *Film Forming Compositions,* Marcel Dekker, New York, 1967.
3. G. P. A. Turner, *Introduction to Paint Chemistry,* Chapman & Hall, London, 1967.
4. P. Nylen, and E. Sunderland, *Modern Surface Coatings,* John Wiley and Sons, New York, 1965.

5. Oil and Colour Chemists Association, *Paint Technology Manuals,* C. J. A. Taylor and S. Marks (Eds.), Vols. I–VI, Reinhold Publishing Corporation, New York, 1966.

6. F. K. Daniel, *J. Paint Technol.,* **38** (1966), 534.

7. F. K. Daniel, *Paint Varnish Prod.,* **60** (May 1970), 29–34.

8. T. C. Patton, *Paint Flow and Pigment Dispersion,* John Wiley and Sons, New York, 1964.

9. W. W. Reynolds, and E. C. Larson, *Offic. Dig. Fed. Soc. Paint Technol.,* **34** (1962), 311.

10. H. Burrell, *Offic. Dig. Fed. Paint Varnish Prod. Clubs,* **27** (1955), 726.

11. J. H. Hildebrand, and R. L. Scott, *The Solubility of Nonelectrolytes,* 3rd ed., Reinhold Publishing Corporation, New York, 1950.

12. CDIC Paint and Varnish Production Club, *Offic. Dig. Paint Varnish Prod. Clubs,* **29** (1957), 1069.

13. *A New Dimension in Solvent Formulation,* E. I. du Pont de Nemours & Company, Wilmington, Del.

14. J. D. Crowley, G. S. Teague, Jr., and J. W. Lowe, Jr., *Three-Dimensional Approach to Solubility,* Eastman Chemical Products (copyright by Federation of Societies for Paint Technology).

15. W. Gordy, *J. Chem. Phys.,* **7** (1939), 93–99; **8** (1940), 170–177; **9** (1941), 204–214.

16. Toronto Society for Paint Technology, *Offic. Dig. Paint Varnish Prod. Clubs,* **35** (1963), 1211.

PHARMACOLOGY AND TOXICOLOGY OF INHALED AEROSOLS

Pharmacology is the science that studies the effects of chemical substances, usually pharmaceuticals for human or veterinary use, on various physiologic and biochemical functions in mammalian and other species. Usually these chemicals are administered in doses that approximate the amounts considered necessary to produce a therapeutic response. Toxicology, on the other hand, is concerned with the effects of these same chemical substances in very high doses, often approaching the lethal level.

Because of the widespread use of medicinal and household chemicals in aerosol form, their properties are of interest to the pharmacologist-toxicologist. This assumption holds true whether or not they are intended for human or animal ingestion. In fact, many of the more toxic aerosolized chemicals have never been proposed for human ingestion or application; their toxicologic manifestations are, however, often seen in this most knowledgeable of species because of either their inherently hazardous nature or their indiscriminate use. Their toxicologic and pharmacologic manifestations, therefore, become of interest to the medical and scientific community, as well as to their manufacturers. This interest is directed toward many avenues: the treatment and prevention of toxic hazards; the expanded therapeutic use of this often-underestimated dosage form; and the elimination of many of the toxicologic hazards associated with the use of aerosols through remedial changes in formulation and particle size. This chapter attempts to put these items into their proper perspective, after an overview of the physiologic processes that are operable whenever aerosols are administered by the oral route for medicinal purposes. In addition, the laboratory techniques utilized in the pharmacologic and toxicologic evaluation of aerosols are also discussed.

ANATOMY AND PHYSIOLOGY OF THE RESPIRATORY SYSTEM

Respiration is a ubiquitous physiologic function, and its complexity increases as one ascends the phylogenetic scale. From the simple diffusion of gases through the cellular membrane of the protozoa to the anatomically complex lungs of the mammalia, however, certain similarities are notable. Universally in the phyla, respiration serves the following general purposes: assimilation and dispersion of the necessary gases for metabolic functions; removal of wastes, chiefly gases; and, in higher phyla, heat dispersion and maintenance of the acid-base balance (1).

The deposition of aerosol particles in the respiratory tract depends on the particle size and velocity and on the transit times for air and other gases in the respiratory tract. Thus the anatomical relationships of the various structures in the respiratory system must be considered in an analysis of aerosol deposition.

Aerosolized particles enter the mammalian respiratory system either through the month or at the nares or nostrils; from the nares, particles then flow to the vestibules, which are slightly expanded chambers lined with sebaceous glands and coarse hairs (Figure 22.1). The latter curve downward to guard against obstruction of the lower respiratory system by large particles. The nasal cavity itself is divided by a septum into the right and left passages, which are covered with a sticky mucous membrane that serves to warm and moisten the inspired air; particles that come into contact with the mucous lining are removed at this point and do not penetrate further into the respiratory system (2).

The pharynx, the major part of the larynx, and the vocal cords within are all covered with a similar mucous membrane; the lower portion of the laryngeal cavity is covered with a mucus-secreting ciliated membrane (2). If aerosolized particles get past the above structures, they enter the trachea. From this point, the remaining portion of the respiratory system is composed of tubelike structures of ever decreasing internal diameter. This results in a progressively increasing surface area which reaches its maximum in the alveoli. The velocity of air flow, however, steadily decreases (3). The trachea itself, which consists of rings of smooth muscle and cartilage, extends from the larynx to the bifurcation of the main bronchi and is lined with a ciliated, mucus-secreting epithelial membrane (2, 4).

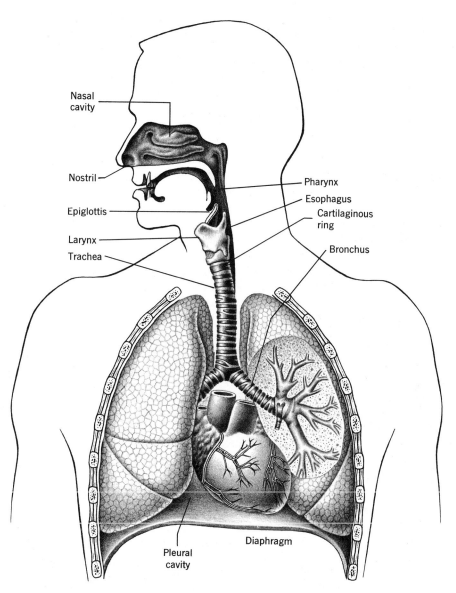

Figure 22.1.   The human respiratory structures.

The trachea connects with the main bronchi, which bifurcate at their junction. The right bronchus divides into three branches to supply air to the three lobes of the right lung, while the left bronchus has two divisions. Each of these main bronchi continues to subdivide until nearly a thousand smaller branches are formed (5). These smaller branches, called bronchioles, eventually lose all cartilage and are composed primarily of smooth muscle fibers, arranged circularly in the walls. They are also lined with mucus-secreting, ciliated epithelium (4). As these structures subdivide further, most of the smooth muscle walls disappear; and the teriinal bronchioles are the last structures to have this regular smooth muscle wall. The terminal bronchioles then divide to form the respiratory bronchioles, which empty into the alveolar ducts and alveoli. The latter structures are composed almost exclusively of a nonciliated epithelial membrane. The alveolar walls contain a dense network of capillaries and connective tissue fibers. There are approximately 300 million alveoli in the human lungs with a total surface area of about 80 m^2 (3). The alveoli are the predominant respiratory structures, and it is here that gas exchange and absorption of aerosols into the systemic circulation largely take place.

The lungs have two separate circulations. The bronchial circulation, which involves small systemic arteries from the aorta, supplies oxygen for the relatively high metabolic needs of the lungs. The pulmonary circulation, which serves respiratory function, begins in the pulmonary artery, bringing venous blood from the right atrium. The pulmonary arteries arborize extensively and finally terminate in a dense capillary network around the alveoli. Venous blood returns to the left atrium via veins which coalesce and eventually form the pulmonary venous system. The venous blood from the bronchial circulation returns to the systemic circulation via the azygous and pulmonary veins (6).

Particulate matter that enters these lower regions of the respiratory "tree" may exert local effects on surrounding respiratory structures and/or may be absorbed and exert systemic effects because of the extreme vascularity of the alveoli. The latter phenomenon occurs with great frequency, particularly in the case of water-soluble substances.

Respiration occurs when, in response to inspiratory nervous impulses from the medulla oblongata (located at the upper end of the spinal cord at the base of the brain), the external intercostal muscles and the diaphragm contract, enlarging the thoracic cavity. The lungs enlarge because of the negative intrathoracic pressure thus created, and air rushes into the alveoli. Expiration occurs when the inspiratory center ceases to discharge and the lungs recoil because of their elasticity. This type of passive expiration is most common and occurs because of the

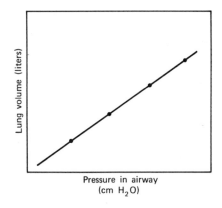

Figure 22.2. Typical Compliance Plot. Lung volumes are plotted against changes in airway pressure and the slope of the line is calculated.

"compliance" of the pleural membranes and the surface tension of the alveoli. When the airways become obstructed or when the patient is hyperventilating, the internal intercostals and the abdominal muscles may contract, resulting in "active" expiration (3).

Compliance is of interest because of the ability of certain drugs to affect it, as will be discussed below. It is a measure of the distensibility of the chest and lungs and is expressed as the volume change per unit change in intra-alveolar pressure (the values are expressed as liters of lung volume per centimeter water pressure) (Figure 22.2). Its reciprocal, "elastance," measures the increase of airway pressure for a given increase in lung volume (1).

Experimentally, compliance can be measured in large animals after tracheal cannulation and plethysmographic determination of changes in lung volume. Increasingly larger volumes of a known gas are forced into the tracheal cannula, while the other arm of the cannula is connected to a pressure transducer for recording on a polygraph. In this manner the static pressure of lung recoil is recorded, along with the change in volume of the lungs. The slope of the line thus obtained is the compliance of the lungs.

INHALATION AND CLEARANCE OF AEROSOLIZED PARTICLES

The respiratory system serves as a portal of entry into the body for a great number of both gaseous and particulate substances. Depending on the adequacy of the respiratory structures and the chemical and physical natures of the aerosolized particles, they may exert profound or insignificant effects on the respiratory system and the body as a whole. Usually the rate at which particles penetrate the respiratory tree is inversely proportional to the particle size and the solubility in water. In other words, smaller particles of water-insoluble substances tend to penetrate deeper into the respiratory system than do large, water-soluble compounds. These observations, it must be remembered, hold true only for aerosolized particles; gaseous substances usually penetrate into the alveoli (7, 8).

Many processes are involved in either aiding or preventing aerosolized substances from penetrating to the alveoli. Coarse particles tend to be trapped in the nasal chambers after collision with the nasal hairs; the mucous secretions of the vestibules trap still other, larger particles (2).

Particles that travel further into the lungs are deposited at varying rates, depending on particle size, aqueous solubility, rate of inhalation, and the concentration of the substance in the inhaled mist. It follows that smaller particles tend to penetrate deeper and that the alveoli will receive a higher concentration of the drug if a dense mist is inhaled. Systemic and local pulmonary effects after aerosol inhalation are also dependent on similar factors. Therapeutic responses more often follow the inhalation of water-soluble materials; lipid-soluble or water-insoluble particles are usually restricted to local pharmacologic effects in the bronchial tree (7). Isoproterenol aerosols, for example, exert direct bronchodilatory effects on the smooth muscle of the lungs while simultaneously stimulating cardiac action after systemic absorption from the highly vascular alveoli. Insoluble substances, such as quartz (silica) dust, on the other hand, exert only local toxic effects, the primary manifestations being those of tissue sequestration of the noxious material; the subsequent tissue reaction is responsible for the toxic effect (2). Other properties of substances that render them more susceptible to removal and sequestration from the inspired air in the upper portions of the respiratory tract include hygroscopic nature (particles enlarge and are deposited because of the moist air in the respiratory passages), uniformity of particle size of inhaled aerosol (maximum deposition of particles in man occurs if most particles have a size of about 1 μ), and good penetrating characteristics of the vehicle employed (2). The last is obvious; if the vehicle has properties that prevent penetration, the aerosolized medication will be unable to reach the alveoli. Gaseous substances, such as the fluorocarbon propellants, penetrate readily to all parts of the pulmonary system and can thus exert both local and systemic effects, depending on the propellant in question.

Not only do the amount of material that reaches the alveoli and its chemical and physical properties affect the pharmacologic response to aerosolized substances, but the length of time that the compound is allowed to remain in the lung also influences the eventual host response. Thus the efficiency of the mechanisms of pulmonary clearance discussed in the next section directly influences the onset, magnitude, and duration of the pharmacologic–toxicologic response to the aerosolized compounds.

PULMONARY CLEARANCE MECHANISMS

The lung has three primary mechanisms for ridding itself of unwanted or foreign substnaces: solubility in the moving mucus blanket that lines the entire respiratory system, followed by either expectoration or systemic ingestion into the gastrointestinal system; sequestration after phagocytosis by specialized alveolar cells and the subsequent tissue reactions which may occur; and, finally, transport of the inhaled substance to the lymph nodes that are frequently found in the connective tissue of the respiratory tract. These processes are operative only for insoluble particulate matter; soluble compounds may either be absorbed directly into the pulmonary capillaries or enter the moving blanket of mucus and thus enjoy a fate similar to that of insoluble substances (2). In either case, the rate at which these clearance mechanisms function will alter significantly the biological response to the administered aerosol.

PHARMACOLOGICAL METHODS FOR EVALUATION OF AEROSOLIZED MEDICINALS

When the majority of particles in a medicinal aerosol are between 0.5 and 5.0 μ in size, the optimal therapeutic response will be produced (7, 10). The host response to inhalation of an aerosol must be considered from both a local (lung) and a systemic viewpoint. A certain amount of almost any aerosolized medication will be absorbed, the amount varying with the water solubility, the particle size, and the other factors discussed above.

Local responses to aerosols usually fall within several categories; bronchoconstriction and bronchodilatation are probably the most common ones, with the latter effect especially desirable in diseases

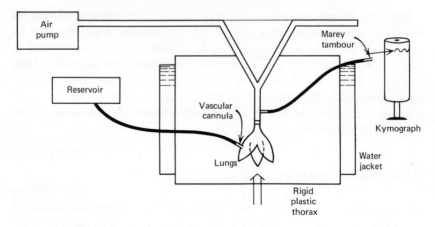

Figure 22.3.   Schematic diagram of the isolated guinea pig lung apparatus of Delaunois, Dautrebande, and Heymans. [From *Arch Int. Pharmacodyn*, **108** (1956), 238.]

such as asthma. Removal and liquefication of mucus secretions, and other local actions such as bactericidal and antiinflammatory effects, are less commonly desired. On occasion, wetting or lubrication of the respiratory mucous membranes with an aqueous spray may be wanted. Other types of chemotherapy, although once promising, have since proved to be disappointing in the clinic.

The local effects of aerosolized drugs in the lungs have been estimated by various laboratory techniques. Some of the methods that have been used are gross observation of respiratory movements in unanesthetized, immobilized animals (11); recording of changes in respiratory patterns (rate and depth); photographic observation of movements of the subdivisions of the lungs; measurement of variations in gas exchange, hematologically, before and after exposure to the aerosol; and study of the mechanics of breathing in normal and treated animals (including compliance, airway resistance, and pulmonary mixing of gases). Finally, much work has been done, especially by Lucien Dautrebande of the University of Liege, Belgium, using isolated mammalian lungs maintained in an artificial environment. This technique has been so widely studied and imitated that a brief discussion of its relative merits is warranted.

The method, using isolated guinea pig lungs, was developed by Delaunois, Dautrebande, and Heymans in 1956 (12). After a guinea pig is sacrificed by cervical dislocation, the intact lungs are removed and placed in a rigid plastic chamber kept at 37C (Figure 22.3). The lungs are then connected to a pump, via a tracheal cannula, which allows for exposure to various aerosols. Another valve in the tracheal cannula is used to record changes in the lung volume during aerosol exposure, using a tambour and kymograph. The rigid plastic chamber is connected to an air pump which keeps the negative pressure in the chamber constant, so that the lungs remain inflated. In addition, the lungs are perfused, via the vascular system, with Tyrodes solution containing 3% dextran and are aerated with a mixture of oxygen and carbon dioxide.

With this preparation, measurements of intrapulmonary pressure, vascular resistance, and relative lung volumes can be accurately recorded after administration of a medicinal aerosol. By this method, acetic acid, serotonin, histamine, acetylcholine, carbachol, and various metallic dusts were found to be bronchoconstrictors, while isoproterenol, ephedrine, and epinephrine produced bronchodilatation. Atropine has been shown to partially antagonize the constrictor effects of many of the cholinergic drugs, but it is more effective when perfused through the vascular system of the isolated guinea pig lung than when administered as an aerosol.

The effects of a large number of antibiotics and other compounds on this preparation were studied (10), and it was found that most of the substances tested in aerosol form, including penicillin, tetracyclines, chloromycetin, and PAS, produced a severe bronchoconstriction which was readily antagonized by isoproterenol. It was also shown that isoproterenol, administered with these antibiotics, reduced the degree of bronchoconstriction and facilitated deeper penetration of the aerosol into the lung.

In a further modification of the artificial thorax, Dautrebande and Heymans (13) connected the thorax to an electrical relay and then to the shutter of a camera, so that photographs could be taken of the isolated guinea pig lung when the negative pressure produced by the respirator was at a maximum and when the atmospheric pressure was completely restored in the plastic thorax, that is, at the peak of the inspiratory expansion and also at maximal expiratory depletion. Thus photographs of the lung surface in response to normal and drug-modified conditions could be taken, and volumetric measurements of the enlargements used to demonstrate the areas of the lung involved in response to drug effects. In general, cholinergic bronchoconstrictors, such as carbachol, produced a decrease in the ratio between the inspiratory and expiratory (I/E) surfaces, regardless of the absolute surface as measured at the peak of the maximal expiratory constriction. Bronchodilating agents, on the other hand, produced increases in the I/E ratio. These observations held true whether the drugs were administered as aerosols or were injected into the vascular perfusion fluid.

Clinically, much information regarding patient response to broncho-constricting and bronchodilating drugs can be gained from either X-ray examination of the chest during both inspiration and expiration or from spirometry. The latter method for clinical evaluation of pulmonary function utilizes a large, inverted metallic drum that can record a variety of respiratory volumes. It is an especially important tool in large-scale screening of patients with suspected respiratory impairment, but can also be used to measure changes in lung volumes in response to drug therapy. Of the many respiratory volumes measured spirometrically, vital capacity is one of the more important. Vital capacity is the volume, in liters, of air that enters and leaves in a maximal respiratory cycle. Bronchodilating aerosols, such as isoproterenol, increase the vital capacity, whereas the cholinergic bronchocon-strictors tend to reduce it.

In evaluation of the clinical response to aerosolized drugs, the technique of administration of the medicinal substance is of great importance. Stalport (14) has shown that the slower the rate of respiration, the greater is the quantity of aerosolized particles deposited and fixed in the lungs. He also showed that, after maximal inspiration with the breath held, the percentage of inhaled particles retained in the lungs is increased even further. (This forms the basis for the directions that accompany many commercial nembulizers today.) Using tidal volume as a reliable clinical index, isoproterenol aerosols have been shown to double the tidal volume, while a few breaths of a 0.5% carbachol aerosol reduced normal values (amount of air entering and leaving the lungs in a normal respiratory cycle) by 50% within 48 min after exposure (10).

In all cases in which the lungs are constricted by a parasympathomimetic aerosol such as carbachol, the bronchoconstriction is easily reversed by several breaths of either isoproterenol or atropine (less effective). In one study (10) the subjects were required to inhale 10 deep, slow breaths of a 0.5% carbachol aerosol (mean particle size, 0.7 μ). Within several minutes after the last breath, breathing became rapid and superficial and was soon characterized by short periods of acute reactions, such as a series of extremely short inspirations with inefficient expiration. Within an hour, the utilization coefficient of the inspired air dropped from 69 to 39%. When all of the symptoms were well defined, the subject was required to take 10 slow, deep inspirations of a 5% amphetamine aerosol. Within 17 min after the last inhalation of the amphetamine aerosol, the respiratory frequency had returned to normal and the utilization coefficient of the inspired air had risen to 85%. At that moment, however, all of the general systemic effects of the carbachol (salivation, perspiration, reddening of the skin, and borborygmi) persisted. Only the bronchoconstriction had been relieved. Further inhalation of amphetamine caused tachypnea (increased respiratory rate) as a result of an action on the respiratory center.

TOXICOLOGY OF AEROSOLS

The large number of aerosolized products available to the consumer for both medicinal and household use dictates that any aerosol preparation that may come in contact with either human skin or eyes, or that may be inhaled, either accidentally or purposefully, be tested for toxicologic potentiality. Methods for predicting human toxicity based on animal data are not usually 100% reliable. In the case of aerosols intended for systemic use, this generality also holds true. However, it has been demonstrated that a good degree of predictive reliability has been established for relating the data obtained from topical application of medicinal and household substances in animals to the toxic potentialities of the same compounds in human beings (15, 16). Although the more elaborate methods for studying pharmacological responses to medicinal aerosols previously outlined are also applicable here, Draize (11) developed a relatively simple test for predicting the acute toxic potential of aerosols either when applied topically or when inhaled. Rabbits, rats, or guinea pigs are immobilized and placed in a suitable exposure chamber. With the eyes covered, the animals are subjected to 30 sec of continuous spray release and then allowed to remain in the sprayed atmosphere for 15 min. The procedure is repeated at 30-min intervals for a total of 10 times, and the animals are then cleansed of extraneous material and placed in metabolism cages. They are then observed for a period of 4 days; observations include symptomatology, food intake, body weight, and hematology. After sacrifice, the tissues are examined histopathologically. Subacute or chronic toxicity can be determined by administering fewer exposures per day for periods of 90 days or longer. By using a similar technique, toxic effects on the eye and skin can be evaluated.

A more elaborate version of the above test is recommended by the regulations of the Federal Hazardous Substances Labeling Act, passed in 1961, for the evaluation of the toxic potential of household substances (17). For this test a more complex exposure chamber in which rodents are held at three levels is recommended. The chamber is maintained at constant temperature and humidity, and the aerosolized substance is introduced into the chamber via an inlet for constantly moving and filtered air. The amount of aerosol entering the chamber, as well as the concentration of the substance in the exposure chamber atmosphere, must be determined experimentally. If a substance is to be considered "safe," rats must not be killed after exposure to 20,000 ppm for 1 hour. In addition, the dosage lethal to 50% of the animals tested must also be determined.

In all of these tests, the observational skills of the experimental pharmacologist are of utmost importance in evaluating the toxic potential of the product in question. Not only are the local effects of the aerosol on the lungs critical, but the systemic response to the portion of the aerosolized product that is absorbed into the blood stream must also be evaluated. Although precise determinations of systemic effects of absorbed medications are not possible without more sophisticated pharmacologic and toxicologic testing, certain clues to systemic reactions of the animal can be detected by the skilled observer. An increase in systemic blood pressure may be manifested as piloerection, exophthalmos or epistaxis, while cholinergic stimulation may be shown by an increase in the frequency of urination, salivation, or defecation (fecal pellet count). Changes in urine output or the presence of blood in the urine may indicate toxic actions at the nephron.

TOXICOLOGY OF PROPELLANTS

Of the many types of chemical compounds that have been developed for use as propellants by the aerosol industry, none have enjoyed such wide popularity as the fluorocarbons. Their relatively good chemical stability and low biological lethality, both attributable to the strength of the C–F bond, have been the bases for their widespread usage (9). Fluoroalkanes, such as dichlorodifluoromethane (CCl_2F_2), dichlorofluoromethane ($CHCl_2F$), and chlorodifluoromethane ($CHClF_2$), have been reported to exert toxic effects on the central nervous systems of laboratory animals. In dogs and monkeys tremors and ataxia were produced by approximately 20% concentrations of CCL_2F_2. Mice appear to be more susceptible to the action of the fluoroalkane propellants, showing evidence of toxicity at 4 to 16% concentrations of $CHCl_2F$ and $CHClF_2$, respectively, in the inhaled air (9). Central nervous system toxicity results in guinea pigs and dogs when they are exposed to concentrations of 1,2-dichlorotetrafluoroethane between 14 and 20%. While the lower concentration produced tremors and convulsions in dogs, the upper level of exposure was necessary to evoke similar responses in guinea pigs. The dermal toxicity of the fluoroalkanes has also been shown to be of low magnitude. The symmetrical and asymmetrical isomers of Propellant 114 have been shown to have no dermal toxicity at doses lower than 7500 and 11,000 mg/kg, respectively. It is difficult, at present, to relate the presence of the double bond in fluorocarbons to their toxic potential. In general, fluorinated alkenes are less toxic after inhalation than are chlorinated alkenes (9). For example, $CF_2=CH_2$ has an approximate lethal concentration in rats (ALC) of 128,000 ppm after inhalation, while $CCl_2=CH_2$ has an ALC of 32,000 ppm. The fluoroalkenes appear to exert their lethal effects predominately on the kidneys, whereas the fluoroalkanes are primarily hepatotoxic (9).

Practical application of these data has been made by Calandra and Kay (18). These investigators exposed guinea pigs to repeated discharges (15 sec per exposure) of hair sprays that used a fluoroalkane-ethanol propellant, several times daily for up to 90 days. After exposure to the aerosolized propellant and/or hair spray formulation, the exposure chamber was sealed for 15 min. Observational, hematological, and pathological analyses revealed no significant toxicity that was related to the propellant mixture, which contained a combination of polyvinylpyrrolidone, ethyl alcohol, and fluorinated hydrocarbon.

Several studies have related the use of hair sprays and deodorants to the development of pulmonary granulomatosis and/or thesaurosis in human users (19, 20). Controlled animal experimentation, however, failed to support these clinical observations (21, 22). In one study, Giovacchini et al. (21) exposed beagles to commercial hair sprays in exposure chambers and then periodically sacrificed the dogs and examined them serologically and histopathologically for the development of blood dyscrasias and also thesaurosis, with negative results. In a similar study Draize and his coworkers (22), using rabbits as the test animals, again found no toxic manifestations from any of the hair spray formulations tested. Clayton has recently shown that two commonly used propellants, Propellants 11 and 12, also have a relatively low order of acute inhalation toxicity (23). In a series of experiments at the Haskell Laboratories, E. I. du Pont de Nemours and Company, he showed that Propellant 12 was not lethal at doses of 339,000 ppm, while Propellant 11 produced deaths at concentratios of 183,000 ppm. Mixtures of the two propellants, at concentrations of 130,000 and 115,000 ppm, also produced no deaths in the rat population studied. The deaths from Propellant 11 were attributed to its anesthetic effect.

Although the fluorocarbon propellants exhibit a low degree of toxicity by themselves, they do have detrimental effects in human beings which may affect their ultimate usage. In a series of related experiments, the ability of these chemicals to sensitize the human and/or animal myocardium to catecholamine-induced arrhythmias has brought close scrutinization by the medical profession. Taylor and Harris (24) exposed mice, rats, and dogs to fluorocarbon propellant vapors from aerosolized bronchodilators, used for self-medication by asthmatic patients. They reported that all of the species tested showed an increased sensitivity to asphyxia-induced sinus bradycardia, atrioventricular block, and T-wave depression. It is questionable, however, whether a pledget of cotton in the nebulizer mouthpiece is sufficient to remove all of the isoproterenol, as claimed. The propellants involved were designated as Propellant 12 (dichlorodifluoromethane), Propellant 114 (dichlorotetrafluoroethane) and Propellant 11 (trichloromonofluoromethane). In 1970 Dollery and his coworkers (25) studied the levels of Propellants 11, 12, 113, and 114 in the blood of human volunteers after inhalation from placebo nebulizers. They showed, following analysis by gas chromatography, that the fluorocarbons appeared in the arterial and venous blood of the subjects after inhalation, with a peak blood concentration of 1.7 µg/ml. They also found that, of the four propellants tested, the least volatile, Propellant 11, appeared in the blood in the highest concentrations after inhalation. Even more recently, in a related paper, Reinhardt and his coworkers (26) reviewed the available data concerning the effects of aerosol "sniffing" on cardiac arrhythmias. From experimental studies in dogs, they concluded that the commonly used fluorocarbon aerosol propellants, when inhaled in high concentrations, can sensitize the myocardium to epinephrine, resulting in serious cardiac arrhythmias. In a series of experiments using dogs exposed to many fluorocarbon propellants, including Propellants 11, 12, 22, 114, 115, C-138, 142b and 502, they showed that a 5-min exposure to concentrations of propellants between 0.09 and 50% produced cardiac sensitization to epinephrine in dogs exposed to median and high levels of the propellants. This sensitization by the propellants was shown to be potentiated by hypoxia; fright produced by noise also increased the sensitizing effects of the propellants, a result most evident with Propellant 142b. Presumably the arrhythmias produced in this last study were due to the release of endogenous epinephrine by the high noise levels, because none was administered (26).

TOXICOLOGICAL EFFECTS OF ENVIRONMENTAL AEROSOLS

Although the pharmacologic and toxicologic effects of medicinal and other manufactured aerosols are of primary importance and interest to the reader, the discussion would be incomplete if some mention were not made of the many toxicologic hazards of environmental aerosols, such as those associated with air pollution and with various industrial occupations, such as coal and asbestos mining. Environmental pollution was recognized even by the early Romans, but it was not until recently that the medical hazards of these man-made aerosolized pollutants have been fully appreciated.

Pollution may arise from many sources, but the most frequent causes of difficulty are motor vehicles, heat and power plants (27), and the mining of substances such as asbestos and coal (2). Polluting materials arising from these sources include sulfur dioxide, aldehydes, volatile hydrocarbons, carbon particles, nitrogen dioxide, ozone, and carbon monoxide. Investigations have shown that these substances can interface with many physiological processes, especially of the lungs, resulting in impairment of respiratory function. Bronchospasm and bronchial obstruction from inhalation of sulfur dioxide (28), decreased metabolism of alveolar macrophages and resultant impairment of immunologic capacity from nitrogen dioxide (29), reduction in vital capacity from ozone (29), and wheezing, coughing, and cardiovascular collapse from carbon monoxide are but a few examples of the adverse effects of air pollutants on biological systems (30).

The incidence of respiratory diseases has been shown to increase in areas of excessive air pollution. Surveys indicate that the incidences of chronic otitis media, the common cold, and other respiratory infections are increased in populations frequently exposed to high pollution levels (31). The degree of air pollution and the incidence of chronic bronchitis have also been linked. A more striking association was found between the mortality rate from bronchitis and the fuel burned in a town. The annual mortality rates from chronic bronchitis correlated well with the amount of sulfur dioxide present in the atmosphere (from low-grade, high-residue fuel). For obstructive pulmonary diseases, such as emphysema, the incidence has increased with a rise in the level of air pollutants. The patient with emphysema becomes sicker when he is exposed to irritating air pollutants and improves markedly when put in a room with filtered air. Classical bronchial asthma may even be produced by the common air pollutants, and patients with this condition become progressively worse when the air pollution index rises (32). Many investigators believe that it is the sulfur dioxide pollution that precipitates the asthmatic attack by triggering an allergic reaction in the sensitive individual (31). Although there are many unanswered questions regarding the etiology of lung cancer, at least nine different carcinogens have been isolated from polluted air. Some of the chief sources of one of the more common polluting carcinogens, 3,4-benzpyrene, are fumes from automobile and diesel fuel, soot from wearing automobile tires, incomplete combustion of garbage, and industrial gases (33). It has been estimated that an individual living in a heavily polluted city such as New York is exposed to the same amount of 3,4-benzpyrene from such sources as a person in a rural area who smokes a pack of cigarettes daily (34).

THERAPEUTIC USES OF AEROSOLS

Because of the susceptibility of the respiratory passages to both polluting and medicinal aerosols, it is no surprise that many of the respiratory diseases that have a causal relationship with polluting aerosols are eventually treated with medicinal aerosols.

Although many drugs are currently used in aerosol form for the inhalation therapy of various kinds of respiratory diseases, only a few are marketed as aerosols; the corticosteroids and sympathominetic bronchodilators are the best examples of this type of packaged aerosol medicinal. The remaining drugs, discussed below, are usually available commercially as solutions to be added to the vaporizing fluid in a respirator or intermittent positive-pressure breathing (IPPB) machine. It goes without saying that the convenience of patient, physician, and nurse would be increased if these compounds could be formulated and prepackaged in aerosol form. Varying dosage requirements among patients, however, often make this idea unfeasible.

Water

Of the many substances used in aerosol form by inhalation therapists, water is possibly the mostcommonly employed for inhalation. In upper respiratory tract diseases, such as bronchitis and pneumonia, a cool water mist of relatively large droplets is useful to moisten dried membranes and to decrease congestion. For adequate deposition of moisture in the lower air passages, a fine mist (particle size below 10 μ) is required. These water mists aid in the removal of viscous, obstructive mucus and help to maintain an effective cough reflex. Except for room humidification, steam should not be used for respiratory tract management. Moreover, any very warm aerosol may aggravate vasodilatation and edema in the inflammatory process and could cause death from obstruction in a patient with epiglottitis (35).

Enzymes

Many enzymes, including hyaluronidase, pancreatic dornase, trypsin, chymotrypsin, and streptokinase, have been studied either alone or in combination with antibiotics and bronchodilators to combat various respiratory diseases. Trypsin, administered in aerosol form, has been shown to facilitate expectoration and lower the surface tension of bronchial secretions in asthma and bronchiectasis. Trypsin aerosols have also been used with some success before mechanical aspiration of the bronchi. They have the disadvantages, however, of liberating histamine and causing hypotension; in large doses trypsin may cause bronchial edema and pulmonary hemorrhage. Chymotrypsin, as a dry powder, has been instilled into the lungs with less irritation (10). Pancreatic dornase, a deoxyribounclease, has been shown to be a potent agent when used to increase expectoration, reduce surface tension, and promote liquefaction of the sputum in bronchiectasis, atelectasis, pneumonia, asthma, and emphysema. The best results were obtained with patients who had large amounts of tenacious, mucopurulent secretions (10, 36). In some instances, the addition of a potent bronchodilator to the nebulizer fluid also increased the efficiency of this compound. Because of the protein nature of pancreatic dornase, allergies to this product are a distinct possibility (37).

Detergents

It is claimed that preparations such as Alevaire (0.125% tyloxapol, 2.0% sodium bicarbonate, 5.0% glycerine) lower the surface tension and viscosity of pulmonary secretions, thus facilitating their removal by coughing and ciliary activity (10). Alevaire has a low order of toxicity; in fact, children have slept overnight for years in an Alevaire mist with no reported ill effects (38). Many clinicians, however, believe that products of this type demonstrate no significant advantages over a cool water mist (35). Recent actions by the Food and Drug Administration have resulted in Alevaire being withdrawn from the market (39).

Antibiotics

Because of the predilection of many bacterial infections for the respiratory tract, antibiotics have logically been studied for their local effects on respiratory infections, after administration in aerosol form. Penicillin, streptomycin kanamycin, Chloromycetin, and bacitracin have all been studied after the inhalation of aerosol forms of these drugs, either alone or in combination with simultaneous systemic administration. Both penicillin and procaine penicillin have been used in aerosol form for 'the localized treatment of respiratory infections and also of asthma associated with or complicated by a bacterial infection (10). Streptomycin aerosols, alone or in combination with hyaluronidase and/or p-aminosalicylic acid (PAS), appear to be effective in tuberculosis, pertussis, and other respiratory diseases (40). Combinations of terramycin and a bronchodilator, in aerosol form, also have therapeutic merit in some respiratory diseases (10).

Bilodeau, Roy, and Giroux (41) studied the effect of kanamycin aerosols in over 200 patients with various types of upper respiratory tract infections. When 250 mg of the antibiotic was used 2 to 3 times daily in combination with a detergent and a bronchodilator (isoproterenol) and administered by an IPPB apparatus, these investigators found the drug to be effective against a wide range of microorganisms, including *Mycobacterium tuberculosis, Hemophilus influenza, Staphylococcus, and Streptococcus*. Administration of kanamycin by this method allowed penetration to the deeper parts of the lungs, where a local effect of antibiotic rather than systemic absorption was credited for the therapeutic merits of this mode of administration. Bronchial irritation was not a problem, and damage to the eighth cranial nerve was not observed because only slight absorption took place, as proved by very low urinary levels (not detectable in bile or blood).

Indirect pulmonary infection due to *Pseudomonas* is difficult to treat. If an attempt to improve therapy, polymyxin B sulfate was administered in aerosol form. Potentially serious complications, such as pruritis, respiratory distress, and acute respiratory acidosis, however, have limited the use of polymyxin B sulfate by this method (42).

Although systemic therapy against bacterial infections is superior to aerosol administration of antibiotics in most patients, studies have shown that combinations of systemic and inhaled administration may be more effective (35).

Decongestants

Phenylephrine (Neosynephrine) in 1/8 to 1/2% concentrations, in a fine-drop aerosol, has been shown to reduce mucosal congestion and improve tracheobronchial drainage (35).

Mucolytics

Of the mucolytic agents currently available, acetylcysteine (Respaire, Mucomyst) is the most widely used. It is the N-acetyl derivative of the naturally occurring amino acid $_L$-cysteine. Acetylcysteine, a white powder with a molecular weight of 163.2, lowers the viscosity of pulmonary mucous secretions by reacting with and opening disulfide linkages in the mucoprotein molecule. The optimum pH for activity is between 7 and 9. Acetylcysteine is used therapeutically to reduce the viscosity of bronchial mucus in diseases such as emphysema, tuberculosis, asthma, and cystic fibrosis. Some of the reported side effects include airway blockade (from the copious quantities of mucus produced—a problem only if the cough reflex is absent or diminished), bronchospasm in asthmatics from irritation, a disagreeable odor that soon disappears, stomatitis, nausea, and rhinorrhea (43). The drug is destroyed by high concentrations of oxygen (44) and antagonizes the action of certain antibiotics, such as penicillin (35).

Tergemist is another mucolytic agent used to liquefy bronchial secretions. It is a combination of the detergent ethasulfate with the mycolytic chemical, potassium iodide (45).

Corticosteroids

The hormones most frequently used in aerosol form are the natural and synthetic steroid derivatives of the adrenal cortex. These compounds, administered via oral or nasal inhalation to reduce the inflammation associated with many types of respiratory diseases, such as asthma, hay fever, and allergic rhinitis, have been studied in the hope that local antiinflammatory effects could be produced without the signs of systemic toxicity associated with other forms of steroid therapy. Unfortunately, this hope for decreased side effects with aerosol administration has not been realized; hyperadrenalcorticalism, associated with long-term steroid therapy, is seen in slightly less than 10% of the patients taking these drugs (46). However, dexamethasone has been used successfully in the treatment of asthma, following the IPPB method of administration, with few side effects reported at doses as high as 3.0 mg 1 day (47). After administration via nasal inhalation, dexamethasone (Decadron Turbinaire) has been effective as an antiinflammatory agent in hay fever, nasal polyposis, and allergic rhinitis. Side effects following nasal administration include irritating headache and light-headedness (48).

Bronchodilators

The catecholamines, isoproterenol and epinephrine, are the most commonly used bronchodilators. Both are short acting and stimulate beta adrenergic receptors in the bronchiolar smooth muscle to relieve the bronchospasms and dyspnea often associated with allergic diseases, bronchitis, and asthma. Although systemic effects, such as tachycardia, palpitations, and coronary insufficiency, are rare when either chemical is used in small doses, frequent use or large doses can cause adverse cardiovascular effects (49). A combination product, Duo-Medihaler, is commercially available; it contains both isoproterenol (a bronchodilator) and phenylephrine (another sympathomimetic drug which stimulates alpha adrenergic receptors) to reduce the edema and congestion of the pulmonary vascular beds that often accompany asthma, emphysema, and bronchitis (50). Poor tolerance to these drugs is a therapeutic problem and increases with the dosage. When the patient becomes refractory to the bronchodilatation produced by one catecholamine, he can safely switch to another, with proper dosage adjustments (49). Although bronchodilators in aerosol form theoretically offer a number of advantages over these drugs systemically administered, because of a high local concentration at the tissue involved, this advantage has never been fully realized in clinical usage. The major problem, as previously stated, is one of overdosage; this occurs frequently because of the penetration of these small molecules through the pulmonary parenchyma into the blood stream, and because of the frequency with which patients, especially children, cough up the drug, only to swallow it (46).

Studies have shown that combination of isoproterenol and phenylephrine in aerosol form (Duo-Medihaler) are more effective against histamine-induced asthma and dyspnea on exertion, and have a longer duration of action, than does isoproterenol alone (50).

Anthihistamines

Because of the etiological role assigned to histamine in the genesis of bronchospasm and other manifestations of allergic diseases, it would appear logical to use antihistamine aerosols in the therapy of asthma, bronchitis, hay fever, and other respiratory diseases associated with increased bronchiolar tone. Toward this end, many such compounds (e.g., diphenhydramine, tripelennamine, and chlorpheniramine) have been studied in aerosol form. However, isoproterenol has been shown to be a more powerful bronchodilator than any of the antihistamines; in addition it is unlikely that any of these drugs could adequately penetrate the respiratory tree, unless administered in conjunction with a potent bronchodilator, because of the irritation and subsequent bronchoconstriction that their inhalation causes (10).

Ergot Alkaloids

Ergotamine tartrate has been clinically evaluated as an aerosol in the symptomatic treatment of both migraine headache and Horton's syndrome with varying degrees of success (51, 52). In one study (51) the oral inhalation of ergotamine tartrate produced effective relief of symptoms in 83% of the patients tested; the dosage required and the onset of action compared favorably with the results obtained from intravenous administration of this agent. Side effects were minimal and consisted of nausea, bad taste, and dizziness. In contrast, Graham, Maleva, and Gramm (52) found that aerosols compared favorably with other routes of administration for ergotamine but preferred either suppositories or parenteral administration if possible.

Prostaglandins

Prostaglandins E_1 and E_2, administered in aerosol form to asthmatic patients, produced increases in forced expiratory volume which compared favorably to those obtained with isoproterenol aerosols (53). There was no change in the heart rate, blood pressure, or electrocardiogram in any of the patients, and while the effects of isoproterenol were more rapid in onset, little difference in the overall therapeutic responses to the two preparations was apparent.

Some newer therapeutic uses postulated for medicinals in aerosol form include the administration of radiopaque contrast media and immunization against influenza. Blaug, Fischer, and Miller (54) demonstrated that liquid laryngography and bronchography in dogs, after the inhalation of propyliodone aerosols (median particle size, 2 μ), produced good outlining of the larynx but did not completely outline the bronchial tree. During the A_2 influenza epidemic in the winter of 1967, commercially available inactivated influenza virus vaccine, bivalent, was administered both by aerosol and subcutaneously to human volunteers. When compared to saline-treated controls, the aerosolized vaccine protected almost 3 times more effectively against the influenza virus than did the subcutaneously administered vaccine (55).

REFERENCES

1. N. B. Slonim and J. L. Chapin, *Respiratory Physiology*, C. V. Mosby Company, St. Louis, Mo., 1967, pp. 4–165.

2. T. F. Hatch and P. Gross, *Pulmonary Deposition and Retention of Inhaled Aerosols*, Academic Press, New York, 1964, pp. 11–25.

3. C. J. Lambertsen, Mechanical and Physical Aspects of Respiration, in V. B. Mountcastle (Ed.), *Medical Physiology*, Vol. I, C. V. Mosby Company, St. Louis, Mo., 1968, pp. 613–626.

4. A. W. Ham, and T. S. Leeson, *Histology*, 4th ed., J. B. Lippincott Company, Philadelphia, 1961, pp. 670–688.

5. H. D. Landahl, On the Removal of Air-borne Droplets by the Human Respiratory Tract. I: The Lung, *Bull. Math. Biophys.*, **12** (1950), 161.

6. W. R. Milnor, Pulmonary Circulation, in V. B. Mountcastle (Ed.), *Medical Physiology*, Vol. I, C. V. Mosby Company, St. Louis, Mo., 1968, pp. 209–218.

7. J. J. Sciarra, Aerosol Inhalation Therapy, *Aerosol Age*, **5** 9 (1960), 34.

8. S. M. Blaugh and A. W. Karig, Oral Inhalation Aerosols, *Amer. J. Hospi. Pharm.*, **24** (1967), 603.

9. J. W. Clayton, Fluorocarbon Toxicity and Biological Action, *Fluorine Chem. Rev.*, **1** (1967), 197.

10. L., Dautrebande, *Microaerosols*, Academic Press, New York, 1962.

11. J. H. Draize, Dermal Toxicity, in *Appraisal of the Safety of Chemicals in Foods, Drugs, and Cosmetics*, Association of Food and Drug Officials of the United States, Austin, Tex., 1959, pp. 46–59.

12. A. L. Delaunois, L. Dautrebande, and C. Heymans, Method for Administering Micromicellar Aerosols to Guinea Pig Isolated Lungs, *Arch. Int. Pharmacodyn.*, **108** (1956), 238.

13. L. Dautrebande and C. Heymans, New Studies on Aerosols. VIII: A Method for Recording Changes in Volume of Guinea Pigs' Excised Lungs after Breathing Constricting or Dilating Aerosols, *Arch. Int. Pharmacodyn.*, **122** (1959), 448.

14. J., Stalport, Medicinal Aerosols. VIII, *Arch. Int. Pharmacodyn.*, **71** (1945), 248.

15. Percutaneous Absorption, Eye and Skin Irritation and Sensitization, *Principles and Procedures for Evaluating the Toxicity of Household Substances*, National Academy of Sciences-National Research Council, Washington, D. C., 1964, pp. 8–13.

16. C. O. Ward, Toxicologic Evaluation of Cosmetics, *Drug Intell.*, **4** (1970), 49.

17. Inhalation, *Principles and Procedures for Evaluating the Toxicity of Household Substances*, National Academy of Sciences-National Research Council, Washington, D. C., 1964, pp. 14–19.

18. J. Calandra and J. A. Kay, Inhalation of Aerosol Hair Sprays, *Drug Cosmet. Ind.*, February 1959.

19. M. Bergmann, I. J. Flance, and H. T. Blumenthal, Thesaurosis Following Inhalation of Hair Spray, *New Engl. J. Med.*, **258** (1958), 471.

20. M. A. Nevins, G. H. Stechel, S. I. Fishman, G. Schwartz, and A. C. Allen, Pulmonary Granulomatosis: Two Cases Associated with Inhalation of Cosmetic Aerosols, *J. Amer. Med. Assoc.*, **193** (1965), 266.

21. R. P. Giovacchini, G. H. Becker, M. J. Brunner, and F. E. Dunlap, Pulmonary Disease and Hair Spray Polymers: Effects of Long-Term Exposure in Dogs, *J. Amer. Med. Assoc.*, **193** (1965), 118.

22. J. H. Draize, et al, Inhalation Toxicity Studies of Six Types of Aerosol Hair Sprays, *Proc. Sci. Sect. Toilet Goods Assoc.*, **31** (1959), 28.

23. J. W. Clayton, Toxicity, in P. A. Sanders (Ed.), *Principles of Aerosol Technology*, Van Nostrand Reinhold Company, New York, 1970, pp. 372–390.

24. G. J. Taylor, and W. S. Harris, Cardiac Toxicity of Aerosol Propellants, *J. Amer. Med. Assoc.*, **214** (1970), 81.

25. C. T. Dollery, G. H. Draffan, D. S. Davies, F. M. Williams, and M. E. Conolly, Blood Concentrations in Man of Fluorinated Hydrocarbons after Inhalation of Pressurized Aerosols, *Lancet*, Dec. 5, 1970, pp. 1164–1166.

26. C. F. Reinhardt, A. Azar, M. E. Maxfield, P. E. Smith, and L. S. Mullin, Cardiac Arrhythmias and Aerosol "Sniffing," *Arch. Environ. Health*, **22** (1971), 265.

27. G. B. Ferris and N. R. Frank, Air Pollution and Disease, *Anesthesiology*, **25** (1971), 470.

28. Air Pollution and Its Effects on the Respiratory System, Suppl. to *Allerg. Patient Resp. Dis. News*, **7** (1966), 1.

29. M. Neiburger, Meterological Aspects of Air Pollution in Relation to Biological Responses, A.M.A., Air Pollution Research Conference, Los Angeles, Calif. March 1966.

30. J. R. Goldsmith, Effects of Air Pollution on Man, *Conn. Med.* **27** (1963), 455.

31. *Air Pollution and Respiratory Disease*, Mead Johnson Laboratories, Evansville, Ind., 1969, pp. 17–37.

32. L. D. Zeidberg, R. A. Prindle, and E. Landau, The Nashville Air Pollution Study, *Amer. Rev. Resp. Dis.*, **84** (1961), 489.

33. P. Kotin and H. L. Falk, The Role of Environmental Agents in the Pathogenesis of Lung Cancer, *Cancer*, **12** (1959), 147.

34. V. G. MacKenzie, The Extent of Human Exposure to Air Pollution, *Arch. Environ. Health.* **9** (1964), 596.

35. S. Segal, Inhalation Therapy, *Can. Med. Assoc.*, **92** (1965), 77.

36. R. Spier, E. Witebsky, and J. R. Paine, Aerosolized Pancreatic Dornase and Antibiotics in Pulmonary Infections, *J. Amer. Med. Assoc.*, **178** (1 61), 878.

37. P. Raskin, Bronchospasm after Inhalation of Pancreatic Dornase, *Amer. Rev. Resp. Dis.*, **98** (1968), 697.

38. Package Insert, Alevaire, January 1967.

39. Anonymous, Alevaire: Notice of Withdrawal of Approval of New Drug Application, *Fed. Register*, **36** (1971), 18336.

40. E. Egidi and M. Malaspina, Streptomycin in Therapy of Complicated Bronchial and Pulmonary Disease, *Minerva Pediatr.*, **8** (1965), 45.

41. M. Bilodeau, J. C. Roy, and M. Giroux, *Ann. N.Y. Acad. Sci.*, **132** (1966), 870.

42. G. Marschke and A. Sarauw, *Ann. Interm. Med.,* **74** (1971), 144.

43. Mucomyst, *Physicians' Desk Reference to Pharmaceutical Specialties and Biologicals*, Medical Economics, Oradell, N.J., 1970, p. 897.

44. M. Lande, A Mucolytic Agent: Acetylcysteine in the Prevention and Treatment of Postoperative Pulmonary Complications, *Presse Med.*, **73** (1965), 2491.

45. R. J. Puls, and W. C. Grater, Allergenicity to Mucolytic Agents in Hypersensitive Patients, *West. Med.* **6** (1965), 145.

46. R. L. Harris, and H. D., Riley, "Reactions to Aerosol Medication in Infants and Children, *J. Amer. Med. Assoc.*, **201** (1967), 953.

47. L. Rumble, The Use of Nebulized Steroids, *J. Med. Assoc. Georgia*, **53** (1964), 314.

48. Decadron Tubinaire, *Physicians' Desk Reference to Pharmacetical Specialties and Biologicals*, Medical Economics, Oradell, N.J., 1970, p. 928.

49. E. R. Levine, Inaalation Therapy-Aerosols and Intermittent Positive Pressure Breathing, *Med. Clin. N. Amer.*, **51** (1967), 307.

50. P. Kallos and L. Kallos-Deffner, Comparison of the Protective Effect of Isoproterenol with Isoproterenol-Phenylephrine Aerosols in Asthmatics, *Int. Arch. Allergy*, **24** (1964), 17.

51. L. S. Blumenthal and M. Fuchs, Transpulionary Absorption of Ergotamine Tartrate for Vascular Headaches, *Med. Ann. Distr. Columbia,* **30** (1961), 10.

52. J. R. Graham, B. P. Maleva, And H. F. Gramm, Aerosol Ergotamine Tartrate for Migraine and Horton's Syndrome, *New Engl. J. Med.*, **263** (1960), 802.

53. M. F. Cuthbert, Bronchodilator Activity of Aerosols of Prostaglandins E_1 and E_2 in Asthmatic Subjects, *Proc. Roy. Soc. Med.*, **64** (1971), 15.

54. S. M. Blaug, H. Fischer, and P. J. Miller, Aerosol Bronchography and Laryngography, *Amer. J. Hosp. Pharm.*, **25** (1968), 281.

55. R. H. Waldman, J. Mann, and P. A. Small, Immunization Against Influenza, *J. Amer. Med. Assoc.*, **207** (1969), 520.

LAWS AND
REGULATIONS
GOVERNING THE
DISTRIBUTION OF
AEROSOL PRODUCTS

The distribution of manufactured products of nearly every description, including aerosol products, is controlled in the United States by a variety of federal and state statutes and regulations administered by governmental agencies and departments created for the purpose. In addition, local governmental bodies have also adopted ordinances and regulations for this purpose, and various industry and trade associations publish guidelines for voluntary compliance by those introducing manufactured products into the channels of trade. The purpose of this chapter is to call attention to the laws and regulations (and the agencies administering them) that are of primary concern to manufacturers and distributors of aerosol products. Particular emphasis is placed on the federal laws and regulations, the more important state and local laws being discussed briefly. For those who require more complete coverage of the field, particularly regarding state and local laws and regulations and industry guidelines, we recommend two publications of the Chemical Specialities Manufacturers Association, Inc.: (a) *Laws, Regulations and Agencies of Interest to the Aerosol Industry* (5th ed., 1966), and (b) *Compilation of Labeling Laws and Regulations for Hazardous Substances* (July 1966).*

---

* These publications are available from the Chemical Specialties Manufacturers Association at 50 East 41st Street, New York, N.Y. 10017.

We discuss in this chapter the proper packaging and labeling of aerosol products to comply with the law, particularly the applicable statutes, and set forth certain of the more pertinent provisions of the principal federal statutes, that is, the Federal Food, Drug, and Cosmetic Act; Federal Hazardous Substances Act; Federal Caustic Poison Act; Fair Packaging and Labeling Act; Federal Trade Commission Act; and Federal Insecticide, Fungicide, and Rodenticide Act. Many of the states have enacted statutes paralleling these federal acts for the control of goods distributed only within the state, but not across state lines so as to be subject to the federal statutes.

Most of the applicable laws are related to the packaging and labeling of consumer goods such as aerosol products in order to prevent injury to purchasers or other members of the public who may use such products. Other laws and regulations relate to the transportation and storage of hazardous materials in order to prevent harm to the public.

FEDERAL PACKAGING AND LABELING LAWS AND REGULATIONS

Federal Food, Drug, and Cosmetic Act

The Federal Food, Drug, and Cosmetic Act is administered and enforced by the Food and Drug Administration (FDA), which is a part of the U.S. Department of Health, Education, and Welfare. The FDA promulgates and enforces regulations within the scope of its authority as defined by the act. The act, being a federal statute, applies only to consumer goods moving in interstate commerce in theory, although as a practical matter, since few packaged goods are consumed entirely within the confines of the state in which they are manufactured, it applies to virtually all packaged foods, drugs, and cosmetics sold at the retail level. Inasmuch as a wide variety of food, drug, and cosmetic products are now packaged for sale in aerosol form, the aerosol industry and vendors of aerosol products are finding it essential to familiarize themselves with the FDA regulations governing the labeling and sale of such products.

The requirements of the act and regulations are primarily directed to the product itself (i.e., the food, drug, or cosmetic per se), and to the labeling of the product, rather than to the form in which the product is packaged (e.g., as an aerosol). Therefore we shall not set forth the provisions of the act or implementing regulations extensively since they are very lengthy and have no direct application to aerosol products as such. For those packaging foods, drugs, or cosmetics in aerosol form who wish to consult the act itself, it is published in the United States Code, Title 21, Chapter 9. The regulations under the act may be found in the Code of Federal Regulations, Title 21, Chapter 1. Inquiries as to regulations on specific items may be addressed to the Food and Drug Administration, Department of Health, Education, and Welfare, Washington 25, D.C.

Certain regulations relating to the labeling of aerosol products that are of specific interest to the aerosol industry are, however, set forth below.

Interpretive Statements Re Warnings on Drugs and Devices for Over-The-Counter Sale (Especially on Aerosols)

DRUGS FOR HUMAN USE

DISPENSERS PRESSURIZED BY GASEOUS PROPELLANTS FOR DRUGS FOR EXTERNAL USE

Warning—Keep away from eyes or other mucus membranes. Avoid inhaling. This warning is not necessary for preparations specifically designed for use on mucous membranes.

Where indicated, in order to prevent chilling the tissues, a caution should be included against holding the dispenser too close to the body.

Warning—Contents under pressure. Do not puncture. Do not use or store near heat or open flame. Exposure to temperatures above 130° F. may cause bursting. Never throw container into fire or incinerator.

DRESSING, PROTECTIVE SPRAY-ON TYPE

Warning—In case of deep or puncture wounds or serious burns consult physician. If redness, irritation, swelling, or pain persists or increases or if infection occurs, consult physician. Keep away from eyes or other mucus membranes. Avoid inhaling.

DRUGS FOR VETERINARY USE

DISPENSERS PRESSURIZED BY GASEOUS PROPELLANTS FOR DRUGS FOR EXTERNAL USE

Caution—Keep away from eyes or other mucus membranes. Avoid inhaling. (This warning is not necessary for preparations especially designed for use on mucus membranes.) Do not use or store near heat or open flame. Exposure to temperatures above 130° F. may cause bursting.

Warning—Contents under pressure. Do not puncture. Never throw into fire or incinerator.

DRESSINGS, PROTECTIVE SPRAY-ON TYPE

Warning—In case of deep or puncture wounds or serious burns consult physician. If redness, irritation, swelling, or pain persists or increases or if infection occurs, consult physician. Keep away from eyes or other mucous membranes. Avoid inhaling.

In addition to the above, certain drugs such as Hexadenol and Vibesate normally packaged in aerosol form for spray application to the skin have been made exempt from some of the prescription-dispensing requirements of Sec. 503 (b)(1)(c) of the act as not being necessary for the protection of the public. The regulations providing these exemptions, however, lay down a number of requirements as to purity, strength, and labeling which should be consulted in detail. For example, it is required that the labeling of Hexadenol give adequate directions for use and warn against use on serious burns or skin conditions, prolonged use, and spraying in the vicinity of the eyes, mouth, nose, or ears.

Similar requirements are made for Vibesate, which is defined as a mixture of copolymers of hydroxyvinyl chloride acetate, sebacic acid, and modified maleic rosin ester. Such Vibesate preparations, to be exempt from the prescription drug requirements, must be packaged in a form suitable for self-medication by external application to the skin as a spray and should contain no other prescription drug. The preparation must also meet standards of identity, strength, quality, and purity and must contain no more than 13% Vibesate by weight. Furthermore, the preparation must be labeled with adequate directions for use by external application as a dressing for minor burns, cuts, or other skin irritations. The labeling must also include clear warning statements against use on serious burns or infected, deep, and puncture wounds, against spraying near the eyes or other mucous membranes, and against inhalation of the preparation.

The foregoing regulations for the specific products Hexadenol and Vibesate are illustrative of the type of regulations existing under the act which are generally applicable to all drug products. In general, much less stringent regulations apply to food and cosmetic products, and, as noted above, such regulations are directed primarily to the purity and labeling of the product itself, rather than to its form, and thus are not within the scope of a work on aerosol products.

Federal Hazardous Substances Labeling Act

The Federal Hazardous Substances Labeling Act is administered by the Secretary of Health, Education, and Welfare through the Food and Drug Administration. The purpose of the act is to prevent injury to consumers by requiring that hazardous substances moving in interstate commerce in containers intended for distribution to consumers for household use be properly packaged and labeled to minimize the possibility of injury.

The act defines a "hazardous substance" as any substance or mixture of substances that is toxic, corrosive, an irritant, a strong sensitizer, or flammable or that generates pressure through decomposition, heat, or other means, if such substance or mixture of substances may cause personal injury or substantial illness during or as a proximate result of any customary or reasonable handling or use, including forseeable ingestion by children. The act does not, however, cover economic poisons subject to regulation under the Federal Insecticide, Fungicide, and Rodenticide Act, or foods, drugs, or cosmetics subject to the Federal Food, Drug, and Cosmetic Act. Also excluded are substances intended for use as fuels when stored in containers and utilized in heating, cooking, or a home refrigeration system.

The term "corrosive" as used in the act means any substance that, in contact with living tissue, will cause destruction of the tissue by chemical action, but does not refer to chemical action on inanimate surfaces. As noted previously, the main thrust of the act is to prevent personal injury, not property damage.

The term "strong sensitizer" means a substance that will cause on normal living tissue, through an allergic or photodynamic process, a hypersensitivity that becomes evident on reapplication of the same substance.

The term "extremely flammable" applies to any substance that has a flash point at or below 20°F. The term "flammable" applies to any substance that has a flash point above 20°F up to and including 80°F except that the flammabilities of solids and the contents of self-pressurized containers (such as aerosol products) are established by special regulations applicable to such products, which should be consulted.

The act requires that the label or any other printed or graphic matter on the immediate container for the goods include the name and place of business of the manufacturer, packer, distributor, or seller; the common or usual name or the chemical name of the hazardous substance or each component that contributes substantially to the hazard; the signal word "DANGER" on substances that are extremely flammable, corrosive, or highly toxic; the signal word "WARNING" or "CAUTION" on all other hazardous substances; an affirmative statement of the principal hazard or hazards such as "Flammable," "Vapor harmful," "Causes burns," "Absorbed through skin," or similar wording describing the action to be followed or avoided; instructions when necessary or appropriate for first-aid treatment; the word "Poison" for any highly toxic hazardous substance; instructions for handling and storage of packages that require special handling or storage; and the statement "Keep out of the reach of children" or its practical equivalent. These statements must be in English in conspicuous and legible type contrasting by typography or color with respect to other printed matter on the label.

The act prohibits introduction into interstate commerce of any product not labeled as described above, or the alteration, mutilation, destruction, obliteration, or removal of the whole or any part of the required label which would result in the package becoming "misbranded," that is, not in compliance with the act. The act also prohibits the use of deceptive disclaimers, such as "Harmless" or "Safe around pets," which tend to negate any of the label statements required by the act.

Certain specific provisions of the act of particular application to aerosol products are set forth in full below.

§191.15 Method for determining extremely flammable and flammable contents of self-pressurization containers.

(a) Equipment required. The test equipment consists of a base 8 inches wide, 2 feet long, marked in 6-inch intervals. A rule 2 feet long and marked in inches is supported horizontally on the side of the base and about 6 inches above it. A paraffin candle 1 inch or more in diameter, and of such height that the top third of the flame is at the height of the horizontal rule, is placed at the zero point in the base.

(b) Procedure. The test is conducted in a draft-free area that can be ventilated and cleared after each test. Place the self-pressurized container at a distance of 6 inches from the flame source. Spray for periods of 15 seconds to 20 seconds (one observer noting the extension of the flame and the other operating the container) through the top third of the flame and at a right angle to the flame. The height of the flame should be approximately 2 inches. Take three readings for each test, and average. As a precaution do not spray large quantities in a small, confined space. Free space of previously discharged material.

§191.15 Method for determining flashpoint of extremely flammable contents of self-pressurized containers.

The apparatus used is the Tagliabue Open-Cup Flashpoint Apparatus as described in §191.13. Some means such as Dry Ice in an open container is used to chill the pressurized container. The container, the flash cup, and the bath solution of the apparatus (brine or glycol may be used) are chilled to a temperature of about 25°F. below zero. The chilled container is punctured to exhaust the propellant. The chilled formulation is transferred to the test apparatus and tested in accordance with the method described in §191.13.

§191.62 Exemption from full labeling requirements.

(a) Any person who believes a particular hazardous substance in a container intended or suitable for household use should be exempted from full label compliance otherwise applicable under this act, because of the size of the package or because of the minor hazard presented by the substance, or for other good and sufficient reason, may submit to the Commissioner a request for exemption under section 3(c) of the act, presenting facts in support of the view that full compliance is impracticable or is not necessary for the protection of the public health. The Commissioner shall determine on the basis of the facts submitted and all other available information whether the requested exemption is consistent with adequate protection of the public health and safety. If he so finds, he shall detail the exemption granted and the reasons therefor by appropriate order published in the FEDERAL REGISTER.

(b) The Commissioner may, on his own initiative determine on the basis of facts available to him that a particular hazardous substance in a container intended or suitable for household use should be exempted from full labeling compliance otherwise applicable under this act because of the size of the package, or because of the minor hazard presented by the substance, or for other good and sufficient reason. If he so finds, he shall detail the exemption granted and the reasons therefor by appropriate order in the FEDERAL REGISTER.

§ 191.63   Exemptions for small packages, minor hazards, and special circumstances.

The following exemptions are granted for the labeling of hazardous substances in containers suitable or intended for household use under the provisions of §191.62:

(a) When the sole hazard from a substance in a self-pressurized container is that it generates pressure or when the sole hazard from a substance is that it is flammable or extremely flammable, the name of the component which contributes the hazard need not be stated.

(b) Laboratory chemicals intended only for research or investigational and other laboratory uses (except those in home chemistry sets) shall be exempt from the requirements of placement provided in §191.101 if all information required by this section and the act are placed with the required prominence on the label panel adjacent to the main panel.

§ 191.110   Self-pressurized containers; labeling.

(a) Self-pressurized containers that fail to bear a warning statement adequate for the protection of the public health and safety may be misbranded under the act, except as otherwise provided pursuant to section 3 of the act.

(b) The following warning statement will be considered as meeting the requirements of section 2(p)(1) of the act if the only hazard associated with an article is that the contents are under pressure:

WARNING—CONTENTS UNDER PRESSURE

Do not puncture or incinerate container. Do not expose to heat or store at temperatures above 120° F. Keep out of the reach of children.

The word "CAUTION" may be substituted for the word "WARNING." A practical equivalent may be substituted for the statement "Keep out of the reach of children."

(c) That portion of the warning statement set forth in paragraph (b) in capital letters should be printed on the main (front) panel of the container in capital letters of the type size specified in section 191.101(c). The balance of the cautionary statements may appear together on another panel: provided that the front panel also bears a statement such as "Read carefully other cautions on . . . panel."

(d) If an article has additional hazards, such as skin or eye irritancy, toxicity, or flammability, appropriate additional front and rear panel precautionary labeling is required.

Federal Caustic Poison Act

This is an act to safeguard the distribution and sale of certain dangerous caustic or corrosive acids, alkalies, and other substances in interstate and foreign commerce, and relates primarily to the adequacy of labeling to warn the public against the inherent dangers of such substances.

In general the act requires that retail parcels, packages, or containers of any dangerous caustic or corrosive substance, as defined in the act, bear a conspicuous, easily legible label or sticker containing the following information:

1. The common name of the substance packaged.
2. The name and place of business of the manufacturer, packer, seller, or distributor.
3. The word "Poison" in 24-point type on a plain background of contrasting color and running parallel with the principal text on the label.
4. Directions for treatment of accidental personal injury by the contents if the package is intended for household use. It is permissible to omit such directions if the package is intended solely for use by manufacturers or wholesalers.

Any package of a substance covered by the act that does not carry a label containing the foregoing information is considered to be "misbranded" and is subject to seizure and confiscation after proper process in a U.S. District Court. If such a substance is considered as misbranded by the court, it may be disposed of at the discretion of the court by destruction or sale or may be released to the owner upon payment of legal costs and provision of a bond to the effect that the substance will not be resold until properly labeled.

The act also provides procedures for the exclusion of misbranded imported packages.

The Federal Caustic Poison Act is administered and enforced by the Department of Health, Education, and Welfare, which has the power to make investigations and inspections, collect samples, and analyze them. If it appears upon inspection and analysis or test that any violation has occurred with respect to the labeling of a container of a dangerous or corrosive substance suitable for household use, the agency will notify the violator and give him an opportunity to be heard on the matter. If it then appears that a violation has occurred, the agency is directed to certify the facts to the proper U.S. district attorney for court action as noted above.

The Secretary of Health, Education, and Welfare has the authority to promulgate regulations for the enforcement of the act, and this has been done under 21 Code of Federal Regulation Part 285 (21 CFR 285)

These regulations relate to the format, wording, and placement of labels on consumer packages, guarantees, collection of samples, analyses, investigations, hearings, procedures for relabeling misbranded goods, and the handling of imported goods.

Federal Fair Packaging and Labeling Act

This act is intended to regulate various aspects of the packaging and labeling of consumer products for the purpose of providing consumers with an accurate statement as to the quantity of the contents of a particular consumer package and sufficient information to facilitate value comparisons between competing products and packages of varying size. Although the law does inhibit deceptive practices, it is also intended to regulate practices that are not deceptive but may limit the consumer's ability to select products offering the best value. The act approaches these goals on an industry-wide scale rather than on a case-by-case basis.

The act is intended to do away with confusing, inconspicuous, incomplete, or nonexistent quantity-of-contents statements on labels; to bring uniformity to the designation of units of weight or fluid measure and thus facilitate comparisons of different packages; to abolish the use of meaningless qualifying adjectives such as "giant" or "jumbo"; to obviate the use of relative size terms such as "small," "medium," and "large" and number-of-servings designations in the absence of meaningful standards of reference; to obviate the imprinting of labeling of consumer packages with expressions implying retail bargains, such as "cents off" or "economy size," which promise price advantages that only the retailer and not the manufacturer can control; to supply information as to the percentages of costly ingredients or active or inert ingredients in order to facilitate price comparisons; and to reduce the proliferation of packages of awkward fractional weights or quantities in order to facilitate price comparisons.

The law provides both mandatory regulation applying on an all-product basis and discretionary regulation applying on a product-by-product basis. More specifically the law requires consumer products to bear a label specifying the following:

1. Identity of the product.
2. Name and place of business of the manufacturer or distributor.
3. Net quantity of contents in terms of weight, measure, or numerical count.
4. Net quantity of a serving when the number of servings is represented.

The act and implementing regulations specify the manner in which the required information is to appear on the label. Exemptions from these requirements are permitted on a product-by-product basis.

The discretionary provisions of the act authorize the administering agencies to issue, on a product-by-product basis, if needed, regulations for the following purposes:

1. Defining standards for describing packages as "small," "medium," or "large" and the like.
2. Controlling the use of "cents off" or other claims to price saving.
3. Requiring the disclosure of ingredient information (nonfoods).
4. Preventing nonfunctional slack-fill packaging.

The Federal Fair Packaging and Labeling Act is administered by the Secretary of Health, Education, and Welfare through the Food and Drug Administration as to foods, drugs, medical devices, and cosmetics, and by the Federal Trade Commission with respect to all other consumer products.

In general the act covers consumer products sold at retail, although in doubtful cases the act and any regulations promulgated under it should be consulted to determine whether or not a particular product is covered. The act defines consumer commodities as including any food, drug, device, or cosmetic as defined in the Federal Food, Drug, and Cosmetic Act, and any other article, product, or commodity that is distributed for sale at retail for consumption by individuals or use by individuals for personal care or in the performance of services originally rendered within the household and that is consumed or expended in the course of such consumption or use. The act does not apply to alcoholic beverages, biological animal products, containers, prescription drugs, economic poisons, meat and meat products, poultry and poultry products, seeds, or tobacco and tobacco products. Most of the foregoing excluded products are, of course, covered by other federal laws or regulations.

In addition to the foregoing excluded products, complete or partial exemptions from the act have been made under the food regulations for foods sold from bulk containers; candy of the "penny" type or individually wrapped weighing 1/2 oz or less; random packages; single-serving-size packages for use by passenger carriers, institutions, and restaurants; transparent wrappers or containers; and tray pack display packages. Other specific products may be excluded from the requirements for disclosing the identity of the product, the name and place of business of the manufacturer or distributor, the net quantity of contents, or the net quantity of a serving, if the regulating agency can be persuaded that such requirements are impracticable in the specific instance in question.

The act and regulations require compliance by interstate distributors, retailers and wholesalers, exporters and importers, and common carriers.

The act itself and food regulations under the act became effective on December 31, 1967, for all new food packages, new label designs, and labels being reordered, and on July 1, 1968, for all food packages and labels being introduced into interstate commerce. However, extensions of time were granted on an individual-case basis to prevent hardship. The Federal Trade Commission promulgated regulations to become effective July 1, 1969, but subsequently postponed the effective date indefinitely, pending the outcome of court actions raising issues with respect to these regulations. These or modified regulations will be put into effect in due course on at least 30 days' notice of the new effective date.

Federal Trade Commission Act

This act created the Federal Trade Commission in 1914 for the purpose of regulating trade and preventing unfair competition or deceptive acts and practices in commerce. The Commission has a working relationship with the Federal Food and Drug Administration and the Department of Agriculture to proceed in cases of deceptive advertising of foods, drugs, cosmetics, and pesticides or other economic poisons and to regulate misleading labeling as to the net quantities of contents of containers for such products. As mentioned previously, the commission also now has joint authority with the Food and Drug Administration to enforce the Federal Fair Packaging and Labeling Act. The commission attempts to eliminate unfair methods of competition and to protect the public by means of administrative hearings and cease and desist orders, and by encouraging voluntary compliance with its regulations or adherence to standards worked out by the regulated industries and the commission.

It will be seen, therefore, that the Federal Trade Commission and its regulations have broad influence on entire industries and classes of goods, including but by no means limited to aerosol products. Just to mention a few of the commission's activities, it promotes free and fair competition through the prevention of price-fixing agreements; boycotts; combinations in restraint of trade, that is, mergers or agreements tending to lessen competition; and other unfair or deceptive practices by labeling, advertising, or otherwise.

More specifically, the commission, engages in formal litigation leading to mandatory orders against offenders. Such proceedings are conducted within the administrative tribunals of the commission itself in proceedings similar to other court proceedings. A case is initiated by a formal complaint charging a person or business entity with a violation of a statute administered by the commission. In some cases the charges are not contested. In other cases the charges may be contested; but if they are found to be true after the hearing provided, an order is issued requiring discontinuance of the unlawful practice in question. More often, however, compliance with the law is obtained through voluntary or cooperative action by means of the trade practice conference procedure, through individual agreements, or through informal administrative correction of minor violations.

The trade practice conference procedure involves inviting industry representatives to meet with personnel from the Federal Trade Commission in a conference during which problems arising in the industry are discussed. On the basis of information and proposals presented at the conference, the commission drafts a set of proposed rules for acceptable practices in the industry involved. These rules are then distributed for study and comment, and opportunity is given for objection by all concerned. Thereafter, the commission promulgates final rules for regulation of the industry involved, specifying in detail just what practices are considered to be unfair and in violation of the law. Thereafter, the commission polices the industry in accordance with the rules thus promulgated and provides informal guidance for those who request aid in complying with the rules.

An example of the commission's work of direct interest to the aerosol industry is the promulgation of a trade regulation rule, which became effective May 21, 1969, relating to failure to disclose the hazards of inhaling quick-freeze aerosol spray products designed for the frosting of beverage glasses. Such products, which contain Fluorocarbon 12 (dichlorodifluoromethane), had been found to be lethal when misused by adolescents who intentionally inhaled the fumes. The trade practice rule laid down is as follows:

In connection with the sale in commerce, as "commerce" is defined in the Federal Trade Commission Act, of quick-freeze aerosol spray products containing Fluorocarbon 12 (dichlorodifluoromethane) designed for the frosting of beverage glasses, it constitutes an unfair or deceptive act or practice to fail to include the following warning in bold and conspicuous type on the labels of such products. "WARNING: DO NOT INHALE. USE ONLY AS DIRECTED. DEATH MAY RESULT FROM INHALING THIS PRODUCT."

Federal Insecticide, Rodenticide, and Fungicide Act

This act is broader in scope than some others discussed in this chapter in that it applies not only to products intended for household use but to all economic poisons, including those for manufacturing, agricultural, or other application. The act does not apply, however, to deodorants, bleaches and cleaners, or disinfectants intended for use on the living body of man or other animals, since such products are not considered to be economic poisons. The act is administered by the U.S. Department of Agriculture, which has promulgated regulations for its enforcement and supplements these regulations from time to time by interpretations setting forth additional detailed regulations and directions for compliance.

Manufacturers or marketers of economic poisons should consult the text of the act itself, the regulations, and the existing interpretations, as well as similar state laws on the subject. An excellent source of these materials is *Compilation of Economic Poisons (Pesticides) Laws*, published by the Chemical Specialties Manufacturers Association.

Basically the act requires registration of all labels used on economic poisons, including insecticides, fungicides, and herbicides, these terms being defined broadly in the act itself, and still more broadly in the regulations and interpretations that have been promulgated by the USDA. Forms for the registration of economic poison labels may be obtained from the USDA, as well as directions for their use and regulations as to the size, content, and use of the required labels. Any product sold without an approved label is subject to seizure.

Since many of the interpretations are directed specifically at aerosol or other pressurized products, the rest of this discussion consists of a summary of these interpretations.

Interpretation 15, which relates to Liquid and Pressurized Household Insecticides Acceptable for General Application, has been amended and revised and is now in effect under 21 Code of Federal Regulations Part 362.113 (21 CFR 362.113). This interpretation applies to contact household fly sprays of the mineral oil solidus pyrethrum type. Such products are ordinarily sold as solutions, emulsions, suspensions, or pressurized products designed for use in undiluted form by the consumer although some such products that require dilution are marketed. These products generally have a petroleum distillate base with auxiliary solvents necessary to maintain solution at low temperatures. The usual propellants employed in such products are Propellant 11 (trichloromonofluoromethane) and Propellant 12 (dichlorodifluoromethane). Propellant 12 may be used alone or in mixtures with Propellant 11, methylene chloride, or methyl chloroform. This interpretation covers products that are primarily non deposit-forming, as opposed to those intended to form a substantial deposit.

Part (b) of this interpretation recites in considerable detail acceptable ingredients and the maximum or minimum permissible concentrations of these ingredients in such products, as well as acceptable combinations of the active ingredients. It should be noted, however, that this section has not as yet been revised to delete DDT, TDE, and Lindane from the list of acceptable ingredients, although the use of these substances has been banned.

Part (c) sets forth an example of an acceptable form of ingredient statement for a product of this type, with directions for its use. Interpretation 5 should also be consulted for further information regarding the preparation of acceptable label ingredient statements.

Part (d), Basic Insecticidal Value, is divided into Subpart (1), dealing with petroleum distillate sprays, which sets a minimum standard of insecticidal strength of value for such sprays; Subpart (2), which relates to aerosol-type products; and Subpart (3), relating to pressurized space and contact sprays. Subpart (2) notes that pressurized formulations classified as "aerosols" are usually marketed in dispensers ranging in content from a few ounces up to 5 lb. The majority of such products, however, are packaged in sizes of 12 to 16 oz. These products generally contain 80 to 85% of propellant gas, usually a combination of Propellants 11 and 12, although methylene chloride or methyl chloroform is frequently substituted for Propellant 11. Subpart (2) specifies that such products should have knockdown and insecticidal values comparable to those of the current Official Test Aerosol dispenser of the Chemical Specialties Manufacturers Association. These dispensers may be obtained from the *association* at 50 East 41st Street, New York, N.Y. 10017.

Any testing procedure that accurately compares the knockdown and toxicity of the test aerosol with the reference standard will be considered. The official method of the Chemical Specialties Manufacturers Association, published in the 1959 edition of the *Blue Book and Catalogue*, as previously noted, will be accepted, provided that the results demonstrate that the product is no less effective in 5-, 10-, and 15-min knockdown and 24-hour mortality intervals than the comparison formulation when tested against houseflies at the same dosage or less.

This method of testing may not be considered adequate, however, if claims and directions for killing insects other than flies are proposed or if any new or unusual ingredient or insecticidal usage is involved. In any test, the spray from aerosol dispensers should be in a finely divided form, in which 80% or more of the individual spray particles have a mean diameter of 30 μ or less and none of the spray particles has a diameter exceeding 50 μ. Products that do not have the necessary biological activity when tested by the specified methods or that dispense a coarser type of spray should not be represented as being "aerosols." Full information on the proposed claims and directions should be filed in all such cases. It will be necessary for the applicant to submit data to establish the safety of any new or unusual chemical ingredient or pesticidal treatment that is proposed. The usual practice is to consult with the Public Health Service of the Department of Health, Education, and Welfare on such matters.

The pressurized space and contact sprays of subpart (3) differ from the aerosol-type products primarily in that their biological performance is of a lower order and is slower in effect on the insects sprayed. Standards for such products are specified in this subpart.

Part (e) deals with labeling statements with respect to directions for the use of such products as "fly sprays" and against other named household insects, particular directions being given for each type of insect.

Part (f) relates to cautionary and warning statements required on the product labels to protect the public from injury, including adequate directions for use consistent with this object. Detailed precautions for protection of the user of various pesticidal ingredients may be found in Interpretation 18 and should be consulted in developing an adequate warning statement for any product. It is noted that products of this type should be kept out of the reach of children and pets and should not be used in the presence of open flame or sparks if containing a petroleum distillate. In addition, because of the nature of these products this section requires the label to bear the following statement:

WARNING: Contents under pressure. Do not puncture. Do not use near heat or open flame. Exposure to temperatures above 130° Fahrenheit may cause bursting. Never throw container into fire or incinerator.

It is the responsibility of the registrant, according to this interpretation, to provide precautionary labeling that will be adequate, if complied with, to prevent injury to persons handling or using the product. Where products are of an unusually dangerous nature because of extreme flammability or exposure hazard, they should be considered specially and additional precautionary labeling developed in cooperation with the Director, Plant Pest Control Division, Agricultural Research Service, U.S. Department of Agriculture, Washington 25, D.C.

Parts (g), (h), and (i) of Interpretation 15 deal with deterioration of the product, grade classification of fly sprays, and unwarranted claims, respectively.

Part (j) is concerned with the registration of these products. Submission to the USDA of duplicate labels, circulars, or other literature accompanying the product, complete information concerning the composition of the product, and data regarding its use against any named pests if not conventional is required.

Regulations similar to those discussed above for non-deposit-forming products are given separately in Interpretation 23 for Liquid, Powdered, and Pressurized Household Insecticides Acceptable for Depositing Insecticidal Chemical Residues. Products of this type are generally more dangerous than their non-deposit-forming counterparts simply because they leave a residue liable to expose children, pets, or others to the poison or its fumes. These regulations are intended to minimize such hazards.

Interpretation 24 relates to Claims for Safety and Nontoxicity on Labeling of Economic Poisons. This interpretation is based on the provision of the act that an economic poison or device is misbranded if the labeling bears any statement, design, or graphic representation relative to the economic poison or its ingredients that is false or misleading in any particular, 7 U.S.C. (2)(1). Furthermore [21 CFR 362.14(a)(5)], unwarranted claims as to the safety of an economic poison, such as "safe," "nonpoisonous," or "harmless," without a qualifying phrase like "when used as directed," are prohibited. An exception is made regarding products determined to be nontoxic to human beings or pets by the USDA. Specific regulations, including size of type on the label, are given to implement these rules.

National Commission on Product Safety

Pursuant to Public Law 90-146; 81 Stat. 466, the U.S. Congress, in its concern for the safety of consumers, has established a National Commission on Product Safety. This commission has held and is continuing to hold hearings on the adequacy of industry self-regulation and testing in assuring the safety of household products. More specifically, the hearings will cover the following subjects:

A. Voluntary self-regulation by industry.

    (i) Safety responsibilities, quality control, and standards of individual companies; hazards due to obsolescence.

    (ii) Adequacy of standards-making activities of trade associations and regional and national standards organizations; the "consensus method" of adopting standards; levels of standards relating to specific products; representation of consumers in the standards-making process.

   (iii) Certification programs by industry to indicate compliance with safety standards; acceptance by companies of voluntary standards.

   (iv) Antitrust implications of voluntary standards-making and certification programs.

   (v) Comparison of industry and government standards-making procedures.

B. Laboratory testing.

   (i) Equipment; staff; competency; and sources of revenue.

   (ii) Independence of policy-making and technical functions; reporting of test results in terms of consumer safety; ability to cause modification of submitted products for safety.

   (iii) Use and public understanding and acceptance of seals of approval or certifications.

   (iv) Submission of products by companies; the effect of failure to meet standards.

   (v) Current and future role of government in product testing for safety.

Among the specific questions already discussed at these hearings is the safety of glass bottles and other containers used by consumers in and around the household, including pressurized containers; the nature, cause, frequency, and severity of injuries associated with such containers; and the resulting liability of manufacturers.

FEDERAL TRANSPORTATION AND STORAGE REGULATIONS

The U.S. Department of Transportation controls the shipment of aerosol products under its authority to govern the transportation in interstate and foreign commerce of all explosive and other dangerous articles. The department has published regulations for this purpose under its Tariff 19, the pertinent parts of which are 71 to 79.

Tariff 19 provides regulations for the preparation of explosives and other dangerous articles, including aerosol products, for transportation by common carriers by rail freight, rail express, rail baggage, highway, or water; the construction of containers, including packing, weight, marking, labeling where required, billing, and shipper's certificate of compliance with the regulations; as well as cars, loading, storage, billing, placarding, and movement by carriers by rail (Section 71.1).

Any person who knowingly delivers to any common carrier any explosive or dangerous article, including aerosol products, without properly marking such goods in accordance with the regulations or informing the carrier in writing of the "true character" of the goods (i.e., that they are dangerous articles of a particular type) is liable to fine or imprisonment or both, under the regulations (Section 71.2). The purpose of the regulations, of course, is to prevent accidents, and thus the carrier must be advised of the dangerous nature of the goods so that he can take proper precautions.

A Bureau for Safe Transportation of Explosives and Other Dangerous Articles, usually referred to simply as the Bureau of Explosives, has been established to make inspections and investigations and to consult with manufacturers and shippers for the purpose of obtaining information for use as a basis to establish further regulations to promote the safe transportation of explosives and other dangerous articles (Paragraph 71.3a).

Specifications for shipping containers and methods of packing and shipment, as well as other regulations, are considered and prescribed from time to time by orders of the department, effective as warranted by conditions (Paragraph 71.3b).

Of the Department of Transportation regulations, those employing the word "must" are mandatory; those expressed as "should" are recommendations only.

The regulations apply to carrier transportation by water as well as land (Section 71.11), and to export shipments when transported to the port of embarkation by common carrier (Section 71.12).

The types of commodities covered are listed in Part 72, aerosol products being classed as nonflammable compressed gases (Nonf. G) (Section 72.5). It is possible, of course, that any particular aerosol product may fall into other classifications as well as the basic class for aerosols, for example, flammable liquid (F. L.), oxidizing material (Oxy. M.), corrosive liquid (Cor. L.), flammable compressed gas (F. G.), poison gas or liquid, Class A (Pois. A.), poison liquid or solid, Class B (Pois. B.), or tear gas, Class C (Pois. C). Aerosol products are considered to be compressed gases, not otherwise specified (n.o.s.) or Nonf. G., as noted above, and the exemptions and packing of such products are covered in Sections 73.302, 73.304, 73.305, and 73.306. Aerosol products must be shipped under a green label of the type specified (see p. 659), and the maximum quantity in one outside package by rail express is 300 lb. Products such as paint, enamel, lacquer, stain, shellac, varnish, metallic (aluminum, bronze, gold) paint, wood filler, liquid, and lacquer-base liquid are classed as flammable liquids (F. L.), and exemptions and packing are covered by Section 73.118.

These products require a standard red label (see p. 660), and the maximum quantity permitted in one outside package by rail express is 55 gal. An aerosol product such as a spray paint would therefore be classed as Nonf. G. (aerosol) and F. L. (paint) and would have to comply with both regulations in order to be properly packaged and labeled, the red label indicating the higher degree of danger of the flammable paint in comparison to the nonflammable aerosol.

The proper shipping designation (i.e., the name from the prescribed list denoting the character of the product) must appear on the outside shipping container in Roman type, not italics. In other words, a shipping carton containing not over 300 lb of aerosol product, if to be shipped by rail express, must be marked "Nonf. G." or any higher designation required by the contents (Section 72.1).

A standard label is also required by Section 72.3. For an aerosol product this label, in green, is as follows:

If the product is a paint, the following red label is required:

Inasmuch as the foregoing labels must bear the shipper's name, they are not available from the government and each shipper must arrange for the printing of his own labels.

A flammable liquid is defined as any liquid that gives off flammable vapors at or below a temperature of 80°F as determined by a standard test (Section 73.115). The regulations specify a minimum outage or fill space in containers of flammable liquids (Section 73.116). Certain containers of flammable liquids are exempted from full compliance; that is, unless the liquid is one for which no exemption is permitted under Section 73.5, it can be exempted from the specifications for packaging, marking, and filling if packed in metal containers of not over 1-qt capacity, each packed in a strong outside container, or in containers of not over 1-pint capacity packed in a strong outside container. If the containers are to be shipped by water, however, the name of the contents must be placed on the outside. Flammable liquids not otherwise provided for and having flash points below 20°F are treated under Section 73.119. Detailed specifications are provided for the containers' outer packages under other regulations for specific materials under Subpart C.

Compressed gases are treated under Subpart F, where they are defined as gases having an absolute pressure exceeding 40 psi at 70°F or, regardless of the pressure at 70°F, having an absolute pressure exceeding 104 psi at 130°F, or as any flammable material having a vapor pressure exceeding 40 psi absolute at 100°F as determined by a specified test method. The term "Flammable compressed gas" is also defined, as are other types and conditions of compressed gases. Many gases are listed by name and detailed specifications given for containers, quantities, outside containers, and so forth.

Poisonous articles such as insecticides are dealt with under Subpart G. Poisons are divided into Class A: extremely dangerous poison, Class B: less dangerous poison, Class C: tear gas, and Class D: radioactive materials (Section 73.325). The extremely dangerous poisons are named in Section 73.326. Less dangerous poisons of Class B, which are those most likely to be found in aerosol products, are defined in Section 73.343 as poisons so toxic to man as to afford a hazard to health during transportation. Standards of oral toxicity, toxicity on inhalation, and toxicity by skin absorption are given to aid in defining Class B poisons. The packaging of Class B poisons is covered by Section 73.344, which specifies minimum outage or free space in the containers, as for the compressed gases discussed above. Section 73.345 defines exemptions for poisons of Class B in certain types of containers of specified sizes, and Section 73.346 provides detailed specifications for other poisons not specifically covered elsewhere.

The marking and labeling of explosives and other dangerous articles, including aerosol products, are covered under Subpart H, dangerous articles being defined as packages containing flammable liquids, flammable solids, oxidizing materials, corrosive liquids, compressed gases, and poisons. Such materials must be marked, unless exempted, with the proper shipping name as shown in the commodity list, referred to previously, in Section 72.5 (Section 73.401). Each package of dangerous articles must show the name and address of the consignee except when in carloads and truckloads or less-than-truckloads handled by a motor vehicle and not requiring transfer from one motor carrier to another. Additional shipping information not inconsistent with Parts 71 to 79 may be shown on a container if so desired, but no such label or marking shall be of such design, form, or size as to be confused with the required label. The required label is red, yellow-white, or green, depending on the contents of the container as specified in Section 73.402. For example, a red label is required on flammable liquids; a white label, on corrosive materials. A green label as described in Section 73.408(a)(2) is required on containers of non-flammable compressed gases (aerosols) except when exempted from the regulations by Section 73.302 above. If a nonflammable compressed gas is also a Class A poison or a radioactive material poison (Class D), the "poison gas" label or "radioactive materials" label must also be applied to the package. For Class B poisons, a poison label as prescribed in Section 73.409(a)(2) must be applied in addition to the green label.

Labels authorized for the shipment of explosives and other dangerous materials by air as shown in Sections 73.405-412(b) and 73.414(c) may be used in lieu of labels otherwise prescribed for surface transportation to or from an airport. Such shipments must be tendered with a signed certificate in a specified form [Section 73.402 (14)(b)(1)].

Labels applied to packages offered for transportation by rail express, rail baggage, or other forms of transportation for which a certified shipping order, bill of lading, or other shipping paper is not required must show the shipper's name in printing, stamping, or writing underneath the certificate thereon.

An important exemption for Class B poisons, as compared to Class A, C, or D poisons, is that labels are not required on less-than-carload shipments by motor vehicle by public highway when the articles are readily identifiable by reason of the type of container or when the container is plainly marked to indicate its contents, and (1) the shipment is transported from origin to destination without transfer between vehicles and (2) the shipper or his employees are in direct control of the loading, transporting, and unloading or actually perform these functions.

The required labels are furnished by the shipper and must be applied to the part of the package bearing the consignee's name and address. The labels are diamond shaped and are of specified size and color, as illustrated in Sections 73.405-409.

Part 77 of the regulations, which essentially duplicates the material given above, applies specifically to shipments made by common, contract, or private carriers by public highway as opposed to railway transportation. Part 77 should be consulted directly, therefore, for the specific regulations applicable to highway transportation.

Dangerous articles packed, marked, labeled, and loaded in conformity with the regulations of the Board of Transport Commissioners for

Canada may be accepted by common carriers in the United States (Section 77.805).

For protection of the public against fire, explosion, or other or further hazard, with respect to shipments of explosives or other dangerous articles by motor vehicle by common or contract carrier, the carrier is required to report certain emergencies to the Bureau of Explosives, 60 Vesey Street, New York, N.Y. The enumerated emergencies include (1) instances of packages discovered in transit not properly prepared in accordance with the regulations, and (2) motor carrier accidents involving damage to containers of dangerous materials that requires repacking the materials (Section 77.807).

It is stated that nothing in the regulations in Parts 71 to 79 shall be construed to nullify or supersede any existing regulations of state or municipal bodies regarding transport through urban vehicular tunnels used for mass transportation. As will be seen in the next section, various state and local authorities do prescribe regulations to prevent serious accidents in tunnels from hazardous materials in transit (Section 77.810).

Fires or explosions occurring in connection with the transportation, or storage on a carrier's property, of explosives or other dangerous articles, as well as leaking, broken, or seriously damaged containers, must be reported promptly by the highway carrier to the bureau in order to aid in preventing future accidents of the same nature. A form for such reports is prescribed by Section 77.814.

The Bureau of Explosives will examine aerosol samples for guidance in complying with the regulations at a charge of $40.00 each if the samples are submitted to:

> Bureau of Explosives
> Association of American Railroads
> 63 Vesey Street
> New York, N.Y. 10007
> Mr. T. C. George, Chief Inspector

Labels for use in shipping dangerous materials may be purchased from the Bureau of Explosives at the above address by specifying the exact type of label and the quantity desired.

STATE, MUNICIPAL, AND OTHER LOCAL REGULATIONS

In addition to the laws and regulations so far discussed, which are primarily on the federal level, at least 20 states also have enacted laws and implementing regulations applicable to the labeling of hazardous substances in order to protect the consuming public. To a substantial degree these laws parallel the federal ones, and indeed are often virtually verbatim copies. Such laws are primarily intended to cover products made and distibuted only within the state, since the corresponding federal laws cover products moving in interstate commerce. Although in theory federal laws apply only to goods in interstate commerce, because of the nature of modern commerce virtually all products move across state lines to a degree sufficient to confer jurisdiction on the federal level. Therefore many of the corresponding state laws are more or less redundant.

Inasmuch as the standards of the federal laws and regulations are generally adequate to protect the public, a manufacturer that distributes goods in compliance with the federal laws will in most cases substantially comply with the state laws as well. Keeping track of the welter of laws and regulations of the 50 states, not to mention those of other municipal and local governments, is, of course, a very substantial burden that it is impracticable for any but substantial manufacturers to sustain. For this reason many manufacturers and distributors attempt to comply fully with the federal laws and regulations and proceed on the assumption that, having done so, they will not be in serious violation of any state law. This policy is, of course, less than desirable, and it is suggested that a prudent manufacturer or distributor take steps to ensure compliance with the laws of the state of manufacture at least. This is particularly true when local conditions are such as to warrant consumer protection with regard to particular products that do not present a problem on the national level sufficient to have stimulated regulatory attention.

In this regard, it is worthy of note that even a few municipalities—Buffalo, New York, and Cleveland, Ohio, to mention only two—have adopted regulations pertaining to aerosols. Moreover, the Fire Department of the City of New York has adopted regulations for pressurized products and has established a procedure for the registration of such products. In general, these regulations require, in type of a specified size, prominent label warnings such as "Extremely flammable," "Flammable," or "Combustible," coupled with the phrase "Keep away from heat or flame," on aerosol products employing combustible propellants, the exact warning depending on the degree of danger as determined by established test procedures. A flash point test is also provided.

In addition to state and municipal laws, the U.S. Post Office Department, the U.S. Coast Guard, various tunnel authorities, trade associations, and other instrumentalities of government and industry have regulations applicable to aerosol products.

In view of this situation, it is obvious that a prudent manufacturer should consult with legal counsel familiar with the applicable federal, state, municipal and other laws and regulations before shipping or marketing any aerosol product.

BIBLIOGRAPHY

Agencies and Regulations of Interest to the Aerosol Industry, Compilation of Labeling Laws and Regulations for Hazardous Substances, and Compilation of Economic Poisons (Pesticides) Laws, all available from The Chemical Specialities Manufacturers Association, Inc., 50 East 41st Street, New York, N.Y. 10017

Guide to Packaging Law, available from *Modern Packaging Encyclopedia,* McGraw-Hill, Inc., 1301 Avenue of the Americas, New York, N.Y. 10019.

Digest of Current Labeling Laws and Regulations (relating to foods), available from the National Confectioners Association of the U.S., Inc., 36 South Wabash Avenue, Chicago, Ill. 60603.

Compilation of State Laws Affecting Proprietary Drug and Related Industries, The Proprietary Association, 1700 Pennsylvania Avenue, Washington, D.C. 20006.

1962 National Paint, Varnish and Lacquer Association Recommended Precautionary Tables, The National Paint, Varnish and Lacquer Association, Inc., 1500 Rhode Island Avenue, N.W., Washington, D.C.

George T. Scriba and Norman I. Hearn, Aerosol Packaging-Labeling Laws, *Soap and Chemical Specialties,* McNair-Dorland Company, 101 West 31st Street, New York, N.Y. 10001.

U.S. Postal Manual, Sections 125.22c, 125.23a, and 125.23C.

Regulations, The Federal Aviation Agency.

Regulations, U.S. Coast Guard.

CHAPTER TWENTY-FOUR

THOMAS CIFELLI AND
WILLIAM E. HEDGES

AEROSOL PRODUCTS
LIABILITY

In Chapter 23 we considered statutory laws enacted by the federal and state legislatures and other governmental subdivisions. Chapter 23 also deals with regulations established under the federal and state statutes, all of which should be considered in manufacturing and marketing an aerosol product.

In the present chapter, we consider the nonstatutory law that has developed over the years bearing on the potential liability of manufacturers and distributors of products, particularly aerosol products, to those who purchase or use such products or who may be otherwise injured or sustain losses by them. In general, the nonstatutory law of the United States is based on and has grown out of the common law of England, with the notable exception of the state of Louisiana, the law of which has developed from the Napoleonic Code of France. The common law of England is a body of law that is primarily judge-made, that is, it has developed on a case-by-case basis, each case serving as a precedent for later related cases. Although the first settlers brought the common law of England with them, that law has been substantially modified over the intervening years as has been found necessary on a case-by-case basis to suit the differing conditions in this country. Moreover, inasmuch as each of the original American colonies and, later, each state, district, territory, and possession established its own separate system of courts, the basic common law has developed differently in each of these jurisdictions. Therefore, whereas certain rules of law are much the same throughout the United States, with respect to other matters it is possible for the courts of one state to arrive at a decision diametrically opposite to that reached in another state on the same facts. Although efforts are being made to minimize such anomalous situations by the promulgation of uniform laws for enactment by each state legislature, many important questions of law are still resolved differently in the various jurisdictions.

In view of this situation, it is impossible, within the limitations of a work such as this, to delineate precisely all of the applicable laws of each of the states, territories, districts, and other jurisdictions. Therefore it will be our primary purpose to point out the types of legal questions that arise and to provide illustrations of the ways in which these questions have been resolved in at least some jurisdictions. We cannot emphasize too strongly that the statements presented here are not intended to be relied on as definitive solutions in particular situations. Competent legal counsel should always be consulted to avoid potential questions of product liability or to resolve actual questions that arise. Our purpose here, then, is to aid those responsible for the development of aerosol products in anticipating such problems and in avoiding or minimizing them by adopting suitable safeguards, after consultation with legal counsel.

As must be apparent to anyone who reads the newspapers, trade journals, or other sources of current news, it is becoming increasingly common for the courts to award compensation to parties who have sustained damages, whether in the nature of personal injury or economic loss, on account of defective products. Therefore it has become virtually imperative for manufacturers, intermediate distributors, and retailers to take suitable precautions to minimize their potential liability for the use or misuse of the products they make and sell.

Whereas recovery for personal injury or economic loss caused by defective products was originally based in common law on a breach of contract between the seller and the buyer or on clear negligence of the manufacturer or seller resulting in harm or economic loss to the buyer, an ever-increasing tendency is now developing in our courts to award damages to injured parties regardless of the existence of a contract between the seller and the buyer or of any fault of the manufacturer, distributor, or retailer. In other words, although not yet the law in every state or in every situation, there is a readily recognizable trend toward holding the manufacturer, distributor, or seller strictly liable to the ultimate consumer in any case where an injury is sustained that can be traced to use of the product.

Inasmuch as the law develops slowly and the great growth of the aerosol industry has occurred only relatively recently, relatively few cases involving aerosol products have actually been litigated in the courts. The state of the applicable law may be discerned, however, from cases involving other types of products, whether closely related to aerosols or not, since the legal principles involved are the same. We shall, therefore, examine the origins, development, and present state of the law of products liability as it applies to aerosol products, referring to decided legal cases involving other types of products to illustrate applicable legal principles where necessary in the absence of similar cases actually concerning aerosol products.

For those wishing a more complete treatment of the law of products liability than is possible here, we recommend the *Products Liability Reporter*, published by the Commerce Clearing House, Inc. This *Reporter*, which is a loose-leaf compilation of the law and current developments in this field, has been drawn on heavily by us as a source of basic information and citations to decided cases pertinent to aerosol products.

We treat the question of manufacturers' and distributors' liability for negligence first and then consider the contract–warranty theories of liability.

NEGLIGENCE

The term "negligence" is in common usage in the law to refer to fault, actual or imputed, on the basis of which an injured party can recover damages for personal injury or economic loss. Negligence is conduct that involves an unreasonably great risk of causing damage or, more fully, conduct that falls below the standard established by law for the protection of others against unreasonably great risk of harm. The amount of care demanded by the standard of reasonable conduct must be in proportion to the apparent risk. As the danger becomes greater, the actor is required to exercise a commensurate degree of care.

The obligation to pay damages for negligent acts is not always limited to injuries and damages or consequences that might reasonably have been anticipated. In recent years, some courts have adopted a rule that eliminates the need to demonstrate that the particular consequences of the negligent act could have been foreseen. In such jurisdictions the negligent party is liable for the results of his wrongful act occurring in the ordinary course of events as long as the act is a substantial factor in bringing about the injury or loss.

It is fundamental that, to constitute negligence from which liability arises, there must be proof of resulting damage. When there has been an injury resulting in death from negligence, it has become established practice to employ life expectancy tables in conjunction with the decedent's earning power to establish the amount of damages recoverable.

In order to recover damages from a manufacturer or distributor, however, it is generally necessary, with certain exceptions, to show that actual negligence was a factor, that is, the mere occurrence of the accident and resulting injury is not ordinarily sufficient to establish liability. To establish a case of negligence and to fix liability on a defendant, it is necessary for the injured party or plaintiff to prove some fact that is more consistent with negligence than with the absence of it.

In the following sections, we deal with manufacturers' and distributors' liability in cases where the charges involved specific acts of negligence in the formulation, production, or inspection and testing of products. In still other sections, we consider negligence in the packaging of products, particularly in connection with the containers and the inspection and testing of such containers, and with negligence in labeling the packages by failing to provide adequate instructions for use, cautionary statements for potentially harmful or dangerous products, and directions for first-aid treatment of injuries caused by the products.

Before going into these matters, however, we deal briefly with the requirement, or lack of it, of privity of contract, as it applies to liability for negligence.

Privity of Contract and Its Loss of Importance

At one time the general rule was that manufacturers and other sellers were not liable for injury to persons having no contractual relation with them. By "contractual relation" we mean to include not only formal written contracts or express oral contracts, but also implied contracts such as arise between a seller and purchaser, for example, the implied contract that food served in a restaurant or sold in an aerosol dispenser is fit for human consumption. However, almost as soon as this rule of law was announced in an early case, the requirements of justice in particular cases compelled courts to make exceptions. One of these related to food and beverage products, but another important exception was made for things described as "imminently," "intrinsically," or, more often, "inherently" dangerous, such as explosives, highly flammable substances, and poisonous drugs. Despite these exceptions, for many years there remained a large field in which the rule of nonliability in the absence of privity of contract was applied. Clearly, many types of aerosol products would probably have been considered to fall within the "inherently dangerous" exception, and it would have been questionable, at best, to argue that lack of a contract between buyer and seller precluded recovery from the manufacturer or seller for injury to the buyer of such an aerosol product.

However, in 1916 the landmark case of *MacPherson* v. *Buick Motor Co.* was decided, in which Judge Cardozo, then a member of the New York Court of Appeals (but later a justice of the U.S. Supreme Court), wrote an opinion holding an automobile manufacturer liable, notwithstanding lack of privity of contract, to a remote purchaser for an injury caused by the manufacturer's negligent failure to inspect a wheel having a discoverable defect. The doctrine of the MacPherson case is now almost universally accepted as the law. This case resulted in bringing *all* products that are dangerous if defective into the "inherently dangerous" exception. The exception has now become enlarged to the point where the original general rule applies only to cases in which some ordinarily innocuous product happens to cause injury. It would seem, therefore, that no contract between the buyer and seller would be required to permit the buyer to recover for injury caused by any aerosol product, unless normally innocuous contents were involved.

As noted previously, purveyors of food and beverages have always been held to a special responsibility under the common law, and therefore, when negligence is alleged in regard to such products, the manufacturer cannot take refuge in a defense of lack of privity of contract. The following specific cases are cited as illustrative:

A plaintiff who became ill after consuming a cola drink containing black, slimy matter was held to be entitled to recover from the bottler even though the cola was purchased from an intermediate dealer and no privity of contract existed.

A plaintiff who consumed a candy bar containing a worm was held to have a good cause of action in negligence against the manufacturer, although privity of contract was lacking.

An ultimate consumer was held to have a good cause of action against the manufacturer of bread in which a nail was embedded, although the bread was purchased from a retailer.

Although none of these examples involved an aerosol product, it is clear that similar results would have been reached in cases involving aerosol products with similar defects.

Because of the general acceptance of the rule of the MacPherson case, discussed above, lack of privity of contract between manufacturer and consumer is usually unsuccessful as a defense to an action brought by an injured consumer. Although the MacPherson decision originally stood for the proposition that privity of contract is not required when a product, if negligently manufactured, involves an unreasonable risk of harm to human beings, it has been extended by degrees in some jurisdictions to permit recovery even for property damage. This case has also been extended in some jurisdictions to uphold liability, without privity of contract, for injuries caused by normally very innocuous products. However, in other jurisdictions the rule requiring privity of contract has been invoked to deny recovery for property damage, and in some cases, in certain jurisdictions, because the product was not one involving an unreasonable risk of harm, liability for personal injuries has been denied through application of the rule.

The following cases illustrate the variety of holdings that may be encountered in different states:

Lack of privity of contract between a cattle raiser and a feed manufacturer was held not to be a bar to the former's action to recover for the death of some 45 cattle because feed sold by the defendant was contaminated with anthrax bacilli.

Because there was no proof that a brand of detergent was inherently or imminently dangerous, a purchaser was held to have no cause of action against the manufacturer for contact dermatitis allegedly sustained through use of the product. Lack of privity of contract was a bar to the action.

The purchasers of defective fertilizer were held to have no cause of action for damage to their corn crops against the manufacturer because the manufacturer of products not inherently dangerous when free from defects was held to be under no duty to exercise care in their manufacture for the safety of persons with whom he has no privity of contract.

A purchaser sued the manufacturer of perfume for damages when the application of the defendant's product to her skin resulted in a second-degree burn. The court held that, in a suit for damages caused by the negligence of the manufacturer, privity of contract is not necessary in order for plaintiff to recover.

It can be seen from these cases that, depending on the jurisdiction and the facts in the case, privity of contract may or may not be required to recover for injuries or loss due to negligence.

Manufacture of the Product

FORMULATION AND PRODUCTION

Inasmuch as ordinarily prudent manufacturers of aerosol products are careful to formulate their products in such a way as to preclude the possibility of personal injury through their use, few cases based on a manufacturer's negligence in formulating a product have actually been litigated. It is apparent from the case law in general, however, that in the event of injury or economic loss to a consumer due to the use of a product negligently formulated, the courts would assess liability if the existence of negligence and damage resulting from the negligence could be shown.

Negligence in the formulation of a product can take many forms. For example, a dangerous or harmful material may be incorporated in the product, either by intention or by accident or inadvertence (e.g., by contamination of an ingredient). The possibility of the inclusion of broken glass or decayed or other unwholesome matter in a food product is obvious. Less obvious perhaps is the possiblity of undesirable and unrecognized chemical reactions or changes of the ingredients. The inclusion of highly toxic, caustic, flammable, or other hazardous ingredients, intentionally or unintentionally, would certainly constitute negligence if inappropriate to the product in question and its intended use. Contamination of aerosol products for use in treating plant life with substances inimical to such life could result in economic loss, giving rise to liability for negligence in formulation of the product.

Manufacturers are required to use reasonable care, that is, a degree of care commensurate with the risk of harm involved, to keep their products free from hidden defects so as to avoid unreasonable risk of harm to users and others, if injury to such persons is a foreseeable or likely consequence of the manufacturer's negligence. When negligence is alleged against a manufacturer, the injured party normally must prove (1) that the act of negligence amounted to a breach of a duty owed him by the manufacturer, (2) that the injury was a foreseeable consequence of the manufacturer's negligence, and (3) that the negligence was the proximate cause of the injury. It should be noted, however, that when foreign substances are found in sealed food-product or beverage containers, an inference of negligence in manufacture is permitted by the courts, removing the burden of proof from the injured party.

The following cases illustrate the type of reasoning the courts could be expected to apply in cases involving negligent formulation and production of aerosol products:

A purchaser who became ill from botulism poisoning after eating some corned beef from a can distributed by a defendant meat packer and packaged by a subsidiary corporation in Brazil recovered against the distributor for negligence. There was evidence that the organism causing the poisoning could not have existed in the can unless the can was defective, or unless the packer, in the process of manufacture, had failed to use the proper means to destroy the organism.

Partners in a cattle-raising business recovered a judgment against a feed manufacturer and its distributor for substantial damage to their herd. The circumstantial evidence indicated that the cottonseed food pellets were contaminated with chlorinated naphthalene contained in a lubricant that was applied to the rollers of the machines manufacturing the pellets.

The purchaser of a pound of coffee who swallowed small particles of glass after eating a handful of the ground coffee and was injured thereby was held to have a good cause of action against the manufacturer, which he alleged had negligently permitted the glass particles to remain in the product. The court held that, although eating coffee may not be a use which the manufacturer might expect, the question was one to be resolved by a jury.

A shampoo manufacturer was held not to be liable to a person injured when the shampoo applied to her hair ignited as she attempted to light a cigarette, because the evidence presented was not sufficient to establish that manufacturing had been negligent or that the shampoo was inherently and imminently dangerous.

The purchaser of a bar of soap allegedly containing a needle or small, round piece of steel, which entered the plaintiff's hand while he was using the soap, had no cause of action in negligence against the manufacturer. It was alleged that the soap had been negligently and carelessly manufactured. The court held, however, that because there was no evidence of breach of a duty imposed by law, and because no actionable negligence was shown, no cause of action could arise. The unintentional dropping of a needle into a mixture was held to be a remote possibility and an extraordinary occurrence.

From these cases, it may be seen that varying decisions are reached by the courts of different jurisdictions, and even by courts of the same jurisdiction in different cases, depending on the facts.

INSPECTION AND TESTING

The presence of a known danger or of a danger reasonably to be foreseen makes vigilance a duty. Consequently, manufacturers are required to make such inspections and tests in the course of the manufacturing process and afterward as are reasonably necessary to ensure a safe product. Failure to inspect or test, if an available test would have disclosed a defect likely to cause injury to a user or other person, renders the manufacturer liable, as does negligence in actually making the inspections or tests. The degree of care required of manufacturers in conducting inspections and tests varies in proportion to the risk of harm reasonably to be anticipated. Thus, when the risk of harm is great, searching tests and inspections are required; however, when harm is not a foreseeable consequence of the failure to inspect, inspections and tests are not mandated. In any event, the duty is probably discharged if the manufacturer has used the latest testing and inspection techniques to discover defects.

In one case, the evidence was sufficient to sustain the finding that a cosmetics manufacturer failed properly to test a nail base coat containing toluene sulfonamide formaldehyde resin, a known sensitizer, before marketing the product. Accordingly, the manufacturer was liable for damages to a consumer who developed nail dermatitis.

Packaging

The manufacturer, having used the proper degree of care in the formulation, production, inspection, and testing of a product, must continue to exercise appropriate care in the packaging of the product, the specific amount of care depending on the potential of the product for harm. Such care includes the selection of safe containers, inspection and testing of the containers, and labeling of the containers with adequate instructions for use of the product, cautionary statements regarding the harmful or dangerous propensities of the product, and even instructions for first-aid or other treatment for foreseeable injuries that may be caused by the product.

CONTAINERS

Selection

It is clear, of course, that a manufacturer having the duty to prevent foreseeable harm to the purchaser–consumer must use reasonable care in the selection of containers for aerosol products that are reasonably safe for the intended purpose. In general, injuries or damages attributable to containers for aerosol products can result from latent defects in the containers as obtained from the supplier, defects introduced in handling by the manufacturer of the aerosol product, and defects caused by mishandling of the filled containers in transit, in the hands of the distributor or by the purchaser–consumer. Although excessive pressure in pressurized containers could also lead to possible injury or damage, because of the nature of aerosols, which have a constant relatively low vapor pressure, unless a manufacturer were so negligent as to employ containers entirely unsuitable for aerosol products, the likelihood of injury or damage from excessive pressure is rather remote, except, of course, for the possibility of leakage from defective containers.

As we have already noted, courts do not ordinarily permit an inference of negligence from the mere occurrence of an accident causing injury or damage, such as an explosion or leakage of an aerosol container. Therefore the injured party must ordinarily introduce evidence or the testimony of an expert tending to prove negligence on the part of the manufacturer or distributor in order to recover. In the exceptional case, however, no such proof is needed, as we shall see later when we discuss the exception to the general rule under the legal doctrine of *res ipsa loquitur*. In some jurisdictions, when it can be shown in court that there was no mishandling by the purchaser–consumer which could have caused the accident, an inference of negligence on the part of the manufacturer or distributor is permitted. Such an inference must be overcome by proof that the manufacturer or distributor was not negligent.

The variety of rulings that may result from injury or loss due to defective containers is illustrated by the following cases:

A plaintiff was held to have a good cause of action against the manufacturer of a bleaching liquid for injuries resulting from a bottle bursting from excessive pressure when she removed the cap. Evidence was introduced at the trial tending to prove that the excessive pressure was due to the bottler's negligence.

A cafeteria employee who received an eye injury from an exploding carbonated beverage bottle was denied recovery from the manufacturer of the bottle since the mere fact of the explosion of the bottle was not of itself sufficient to establish negligence in its manufacture.

A plaintiff was held to have a good cause of action against the manufacturer of a liquefied natural gas tank that ruptured and allowed gas to escape, causing fire and explosions and killing a workman. It was shown that the manufacturer had designed, constructed, and recommended the tank in question despite the fact that it was considered more dangerous than other available tanks.

Inspection and Testing

The manufacturer of an aerosol product may also be subject to liability for injuries or loss to users of its product if the manufacturer can be shown to have been negligent in inspecting or testing aerosol containers obtained from others to ensure that they would be safe for their intended use when filled. However, the mere breaking, explosion, or leaking of an aerosol container does not necessarily imply that the manufacturer (i.e., the filler of the container) was negligent in failing to discover the defect. The person suffering the loss or injury is required to show that there was no mishandling of the container after it left the hands of the person charged, that is, the manufacturer or distributor. If this can be done, the manufacturer may be held liable, since sound containers do not rupture under normal conditions of use, giving rise to the inference that the container in question was defective when it left the hands of the manufacturer. In order to establish the manufacturer's negligence and recover damages, the injured party must also show that a reasonable or practicable inspection of the container would have disclosed a flaw.

Some courts have recognized a distinction between the duty of care owed to the consumer by the container manufacturer and the corresponding responsibility by the filler of the aerosol product. This is based on the superior knowledge of the container manufacturer as to the nature and effect of defects in the containers, as compared to the filler, who merely adds the product to the containers. In such a case, if the filler makes reasonable tests customary in the trade, it may escape liability imposed on the manufacturer of the container for latent defects of which the latter should have been aware but which the filler could not reasonably have detected or, if detected, would not have realized had potential for harm. Most courts, however, have treated the filler as though it were also the manufacturer of the container and have found liability for injury due to defective containers regardless of the nature of the defects.

The following are some illustrative cases:

An employee of the purchaser of a drum of acid, who was injured when a plug blew out of the drum and acid spurted on him, was permitted to recover damages against the supplier of the drum, since evidence was introduced to show negligence in employing a used drum with insufficient threads in the bung hole to hold the plug in place under normal handling conditions. A routine inspection would have revealed this defect.

The manufacturer of a bottle of perfume was held not liable to a purchaser who was injured when the bottle broke on opening. The evidence failed to establish that the bottler had negligently failed to inspect the container or that the manufacturer knew it contained a flaw which rendered it unsafe for its intended use.

An injured plaintiff was held to have a good cause of action against a bottling company for harm caused by explosion of a bottle of carbonated beverage due to internal pressure and a defect in the bottle that could have been discovered. The bottler was held to have a duty to exercise maximum care to guard against defects that might cause an explosion.

Where a bottle was shown to have substantial differences in wall thickness from point to point, it was held that the bottle was defective and that inspection should have uncovered the defect. The defendant was held liable for negligence in failing to properly inspect the bottle, resulting in injury to plaintiff.

In another exploding bottle case, the injured party recovered from the bottle manufacturer after having established that the latter had failed to employ test methods that would uncover defects other than those capable of causing only trivial harm. However, despite the finding that the bottle manufacturer was liable for failure to make adequate tests, the bottler of the carbonated beverage was not held to be liable for the injury, since the bottler had purchased the container from a reputable manufacturer and had made tests customary in the trade to discover defective bottles.

In a case where a beer bottle exploded, injuring a tavern employee, the bottler was held liable on evidence showing that it had failed to use sufficient care in inspecting bottles before filling and had used defective bottles.

An injured party failed to recover damages for injuries resulting from the breaking of a milk bottle in a case where there was no evidence to establish what caused the bottle to break. The court held that the bottler's duty was limited to the discovery of defects by reasonable inspection and to the use of reasonable care in the handling of the bottles.

An injured party recovered damages from a bottler in an explosion case upon proof that the bottler had used bottles repeatedly for years without testing them and that cases of bottles were customarily handled without due care by the bottler's employees.

In a case where a defective carboy caused a bottle of acid to break, causing injury, the seller of the acid was held liable (although not the manufacturer) because it had a duty to discover the defective condition of the carboy in the exercise of ordinary care for the safety of others.

In a case involving an action for injuries caused by flying glass from bottles dropped when a carrying carton ruptured, the court held that the bottler was not liable for negligence since the evidence showed that the cartons were inspected before reuse and no evidence that the carton in question had not been damaged after it left the bottler's control was presented.

LABELING

Manufacturers and distributors of aerosol products in making up their packaging must consider not only the safety of the containers as noted above, but also the labeling of the products to warn of their dangerous propensities and to provide adequate instructions for the use of the products to prevent injury or damage through improper application. Furthermore, for products that are toxic or harmful to human beings or animals it is also necessary to give adequate directions for first aid or other treatment in case of injury in order to minimize or avoid liability.

A product that is entirely safe when used properly may be extremely dangerous if used improperly. Therefore a manufacturer or distributor who fails to give adequate notice to the consumer of the potential of the product for harm, or directions as to how to use it so as to avoid injury or loss, may be held to be liable for negligence.

Cautionary Statements and Instructions for Use

Existing case law generally demands that, to avoid liability in negligence, manufacturers and distributors must warn purchasers and users of their products not only of any known hazardous propensities of the products, but also of any limitations in regard to the products or their uses that might not be known to consumers and that could result in injury or loss if not taken into account. This duty to warn is imposed not only for inherently and obviously hazardous substances, but also with respect to substances that are normally innocuous if used properly but may become hazardous if used improperly, or for other than the intended purpose, or may become dangerous if acted upon by a foreseeable extraneous force or agency. A standard legal reference, the authoritative *Restatement, Second, Torts* states the rule as follows:

> Chattel Known to be Dangerous for Intended Use—One who supplies directly or through a third person a chattel for another to use is subject to liability to those whom the supplier should expect to use the chattel with the consent of the other or to be endangered by its probable use, for physical harm caused by the use of the chattel in the manner for which and by a person for whose use it is supplied, if the supplier
>
> (a) knows or has reason to know that the chattel is or is likely to be dangerous for the use for which it is supplied, and
>
> (b) has no reason to believe that those for whose use the chattel is supplied will realize its dangerous condition, and
>
> (c) fails to exercise reasonable care to inform them of its dangerous condition or of the facts which make it likely to be dangerous.

A manufacturer or distributor who places on the label of his product a warning adequate to meet the standards set by the rule quoted above will be considered to have discharged his duty and will not be held liable for negligence except in unusual circumstances. It must be remembered, however, that if the warning on the label is misleading or ambigous, or if the instructions for use of the product are insufficient to prevent injury or loss, the requirements of the rule are not met. The text of the cautionary statement or instructions for use should be such as to enable any person of reasonable intelligence to avoid the hazards of the product or its misuse. It is obvious, of course, that the cautionary statements should be placed on the label in such a way and with such prominence as to be read and noted by a reasonably prudent person, that is, the type should be of appropriate size in proportion to that of the rest of the label, and any directions for safe use should not be obscured by inclusion without emphasis in general instructions.

Although no cases of this type involving aerosol products have been litigated to date, the principle is well illustrated by decided cases on various other types of products, such as those cited below with respect to a drain cleaner and sulfuric acid.

A manufacturer of a drain cleaner containing caustic soda and powdered aluminum metal was found to have been grossly negligent and to be liable for injuries in failing to warn the public on the label that the product was very dangerous when added to water in that it produces substantial heat, a corrosive chemical, and potentially explosive hydrogen gas.

In a case involving injury from the explosion of a drum of sulfuric acid, the producer was held liable for failure to label the drum to warn those who might reasonably be expected to handle it of the hazards involved.

The degree of care that should be exercised by manufacturers and distributors of aerosol products containing inherently hazardous substances such as corrosives, toxic compounds, and flammable materials is commensurate with the hazardous propensities of the material, that is, the more dangerous the material, the stronger is the warning required to meet the manufacturer's or distributor's duty to the public. In general, an adequate label should be designed to have a reasonable likelihood of attracting the attention of a person of normal prudence, and the text of the label should convey a warning understandable to the average person and an indication of the nature and extent of the hazards involved.

For example, a label that merely cautions the user not to breathe the fumes of the product or to use it only in a well-ventilated space has been held to be inadequate if the product contains carbon tetrachloride, since this material is poisonous and may cause serious illness or even death. In other words, merely giving directions that, if followed, would avoid injury may not be sufficient to avoid liability if the product is dangerous, since users may not follow the directions unless they are also aware of the consequences of failing to follow them.

The following illustrative cases may help to clarify the requirements for inherently hazardous substances:

In a case involving a cement paint, the label merely warned that the paint was alkaline and might be injurious to sensitive skin. This warning was held to be inadequate, since as a matter of fact the paint had the potential to cause almost immediate blindness. The court noted that the mild warning statement in effect amounted to an assurance of safety rather than an adequate warning of the real danger.

In another case, a warning on the label of a container of furniture polish to the effect that it might be harmful if swallowed, especially by children, was inadequate because it was included in small type with general instructions for use.

In the case of a newspaper pressman who died as a result of use of a cleaning fluid containing benzol, the court held that his wife had a good cause of action against the manufacturer for failure to label the container with a warning as to its poisonous contents and the potential hazards of the use of the cleaning fluid.

A manufacturer of an antifreeze compound, ethylene glycol, was held liable to a mechanic for $210,000 for permanent injuries resulting from lack of a label warning of the danger of breathing the toxic fumes.

A manufacturer of a pesticide, Parathion dust, was held liable for the deaths of two Puerto Rican farm laborers, semiliterate in English, on the ground that the product should have been labeled with a skull and crossed bones to warn such persons of the dangerous nature of the product.

The distributor of a hair tint containing a very dangerous coal-tar derivative was held liable to a consumer who developed a rare and normally fatal disease, since the product failed to carry a label warning users of the danger of systemic absorption of the compound on continued use.

The duty to warn of the hazards of prescription drugs, unlike the responsibility in regard to other products, is owed not to the consumer–patient but to the prescribing physician. Drug manufacturers must give doctors timely warning of any undesirable side effects that come to their attention in order to permit a physician to safeguard his patients' health.

In the pharmaceutical field, the manufacturer of the drug MER/29 has been held liable for negligence in marketing the drug, which allegedly caused a user to develop cataracts. Since, as noted above, in the drug field, the primary duty is owed to the physician, it would seem that the manufacturer was actionably negligent in failing to give timely warnings to physicians or to withdraw the drug from the market, after it learned from animal experiments and data from the Mayo Clinic that the drug appeared to cause cataracts in human users.

Sales as Negligence

In certain cases, the mere act of introducing a product into the channels of trade for sale to the public may constitute negligence for which recovery may be had for personal injury or loss occasioned by the product.

This is the case only with inherently dangerous products, however, such as those that are highly toxic, flammable, or explosive. When products of this type are involved, the injured party may recover damages without proof of negligence on the part of the manufacturer other than the fact of sale. Examples of such products include highly poisonous hair dyes, fuels for home use containing gasoline, and clothing sized with highly flammable nitrocellulose.

As an example of a case involving a hair dye, a beauty parlor operator was awarded damages from the distributor of a shampoo tint that caused dermatitis to her hands when she used it to color a customer's hair. The evidence tended to show that the distributor knew or should have known that the product contained a sensitizing ingredient and that the product had been represented as safe in every respect.

In a case involving a spray-on deodorant, where to the manufacturer's knowledge less than 1 person in 150,000 developed any allergic reaction on use of the product, the manufacturer was not held liable for negligence in failing to warn of the remote possibility of such an allergic reaction.

Where a 6-year-old child used a lanolin hair spray, purchased as a perfume, by spraying it on her hair and dress and was subsequently injured when the hair spray was ignited by a candle, the manufacturer was not held liable, since the product was safe for its intended purpose and there was no duty to warn of its flammability when improperly applied. This case probably should be restricted to its particular facts, and it is doubtful whether it would be considered a valid precedent in most jurisdictions.

Statutory Violations as Negligence

When particular conduct violates a statute such as the Federal Food, Drug, and Cosmetic Act or the Federal Hazardous Substances Labeling Act, consumers and the general public are usually provided with a higher and surer degree of protection than is afforded by the common law. Where a statute prescribes a standard of conduct in order to protect consumers against a particular type of risk, and a consumer is injured because of the manufacturer's or seller's failure to meet that standard, the courts will consider the standard in determining civil liability. Although the courts are not unanimous in their judgments, the majority rule is that violation of such a statute constitutes negligence as a matter of law.

As examples of statutory violations that have led to liability in negligence the following cases are cited:

When a manufacturer of bromine failed to label the container in accordance with the regulations of the Interstate Commerce Commission regarding shipment of explosives and other hazardous materials, and the container of bromine exploded at the terminal of a common carrier, injuring an employee, the manufacturer was held liable in negligence for breach of its duty to the employee to comply with the statute by properly marking and labeling the container.

Similarly, a manufacturer of a fungicidal liquid was held liable in negligence to a farmer for loss of the latter's orchard by defoliation on the ground that the product had not been labeled to warn against this possiblity, as required by a state statute relating to insecticides, fungicides, and rodenticides.

Submanufacturers' Negligence

Although there is little or no legal precedent with respect to fillers of aerosol containers who do not themselves manufacture the products in question, by analogy to decided cases involving manufacturers who assemble mechanical components furnished by others it would appear that such fillers would probably be held to owe a duty to the consumer to make reasonable inspections and tests to discover latent defects in the products or components obtained from others. Although there is some precedent to the contrary, the fact that an assembler or filler of aerosol products has obtained the components from a reputable manufacturer would not, in the majority of jurisdictions, relieve the filler of the duty to test and inspect the components to the degree reasonably necessary to ensure a safe product. There is also precedent in certain jurisdictions, contrary to the majority rule just expressed, to the effect that the compounder or filler has the same liability as if it had manufactured the components. Clearly, therefore, the safest policy is to take every precaution to determine that the components filled in the aerosol product are safe for their intended use.

One who sells a product as his own, although it was actually made by another, is known as an "ostensible manufacturer." He is held to the same degree of care and resulting liability as if he had actually manufactured the product.

A supplier of a component for use by others in compounding an aerosol product may also be held liable for any injury or loss resulting from the component supplied, if the supplier was negligent in such a way as to involve an unreasonable risk of injury or loss to the party suffering the injury or loss. This usually requires knowledge of the use for which the component is purchased. Such knowledge need not be directly communicated to the supplier of the component by the compounder, however, if the circumstances are such that the supplier could reasonably be held to know what the component was to be used for, or was given sufficient information to impose a duty to inquire as to the use in order to avoid injury to the consumer.

As an example of this type of situation, a distributor who sold gas containing excessive moisture, knowing it was to be used in a furnace, was held liable for negligence when the furnace malfunctioned and exploded. In the same case, the distributor had commingled gas obtained from several manufacturers and sold it as its own, and could not, therefore, avoid liability by arguing that it relied upon reputable sources of gas.

Misrepresentation

Fraudulent misrepresentation with respect to the nature of goods has long been recognized as legally actionable, and there is no question but that the established principles of the law in this area would be applicable to fraudulent misrepresentation of the utility or effectiveness of an aerosol product. In recent years a trend has been developing in the case law to recognize negligent misrepresentation as well. By this is meant a misrepresentation that the goods are safe or free from hidden defects, when the manufacturer or distributor lacks reasonable grounds to believe that this is so and also knows that, if it is not, injury or loss may result to the purchaser or consumer. In other words, there is such disregard for the safety or property of the purchaser as to amount of actionable negligence, although actual knowledge of lack of safety or utility of the product sufficient to constitute fraud is absent. This is particularly the case when the seller's negligent misrepresentation is such as to lull the purchaser into a false sense of security unwarranted by the seller's actual knowledge. It may be appreciated, therefore, that, when there is lack of knowledge of crucial facts or when important facts are ignored, a negligent misrepresentation, if flagrant, may support an inference of fraud, although actual, knowing misrepresentation of fact is absent.

The general rule is that misrepresentation of a material fact by the seller, by means of advertising, labeling, or otherwise, imposes strict liability on the seller for damages to the innocent purchaser, consumer, or user. This principle is illustrated by the following cases:

In the MER/29 drug case cited previously, the manufacturer allegedly untruthfully represented, both orally and in advertising in medical journals and promotional literature, that the drug was practically nontoxic, remarkably free of side effects and perfectly safe. In fact the MER/29 allegedly caused cataracts. At the time that the manufacturer was making the above representations, it had knowledge of animal experiments indicating that blood and eye changes were caused by the drug in most animals tested. Although the manufacturer had no actual knowledge that the drug would cause cataracts in human users, it would appear that the animal tests had afforded sufficient notice so that the assertion that the drug was safe was made under such circumstances as to raise the question of negligent misrepresentation.

In another drug case, a woman who became blind after taking the drug dinitrophenol on her doctor's prescription was held to be entitled to damages from the manufacturer, which had represented in newspapers, medical journals, and circulars that the drug would aid in relieving obesity and was harmless, when in fact the manufacturer knew that it had potential to cause blindness. This appears to have been a clear case of fraudulent misrepresentation because of the manufacturer's direct knowledge of the potential harm, as opposed to a reckless disregard for harm in the absence of knowledge. The mere fact that the drug had been prescribed by a physician was held not to absolve the manufacturer from liability under these circumstances.

Disclaimers

It is common practice for manufacturers and distributors to attempt to limit their liability by the use of written disclaimers on product labels and invoices, and such disclaimers are generally given effect on the ground that two parties may enter into a contract on any terms to which they mutually agree. Inasmuch as such agreements are drawn by the manufacturer, and the buyer has no opportunity, as a rule, to negotiate their terms, they are normally construed rigidly against the manufacturer, as is the case with other legal documents drawn by only one party. It should be noted, moreover, that disclaimers for the purpose of releasing the manufacturer from liability for damages resulting from breach of warranty, express or implied, generally are not considered to relieve the manufacturer from the consequences of its own negligence. There are cases also where an exculpatory agreement attempting to waive claims has been held ineffective because the sale of the product was prohibited by a statute. Clearly, a manufacturer cannot disclaim liability for violation of a statute, by means of a statement on the label of the product.

In one case, involving an insecticide, the disclaimer "Seller makes no warranty of any kind, express or implied, concerning the use of this product. Buyer assumes all risks of use or handling, whether in accordance with directions or not" did not relieve the manufacturer of liability due to negligence for damage to a crop of cabbage.

Res Ipsa Loquitur

Res ipsa loquitur (the Latin phrase means "the thing speaks for itself") pertains to a narrow class of cases constituting an exception to the general rule that the injured party must prove negligence on the part of the manufacturer or distributor in order to be entitled to an award of damages. The term is applied to a legal doctrine or rule of law to the effect that, when the thing which caused the injury is under the exclusive control of the manufacturer (or distributor) and the accident is one that would not have happened in the ordinary course of events if those in control had used proper care, the accident itself provides reasonable evidence of lack of such care, making it unnecessary to prove the act of negligence. In practice, "exclusive control" amounts to the article being under the control of the person charged at the time the negligent act took place. Proof is required that the article was not altered after it left the manufacturer before the rule of *res ipsa loquitur* can be applied.

The classic *res ipsa loquitur* situation is a foreign body in a container, such as a mouse in a bottle of carbonated beverage. It is clear that the doctrine would apply similarly to aerosol products, although few if any such cases have been litigated.

The doctrine of *res ipsa loquitur* clearly applies when a foreign body is actually canned or embedded in an article purchased by the consumer in such a way that it could not have gotten there after the product left the control of the manufacturer. In this situation the responsibility of the manufacturer cannot be questioned, and the doctrine applies. Continuity of control is obvious in such cases, eliminating the need to prove that the product was not contaminated after it left the manufacturer.

The doctrine is illustrated by the following cases:

When a can of spaghetti that had been improperly filled too full exploded when heated before opening according to the instructions on the can, the injured consumer was entitled to recover against the food processor on the doctrine of *res ipsa loquitur*.

When a consumer broke a tooth on a metal object embedded in a piece of candy, it was clear that the candy had not been contaminated after it left the manufacturer; therefore the injured party was allowed to recover under the doctrine of *res ipsa loquitur* without proof of negligence by the manufacturer.

Most jurisdictions also recognize that the doctrine of *res ipsa loquitur* applies to exploding container cases, which have been discussed previously, if it can be shown that the containers were not damaged or mishandled after they left the manufacturer's plant, that is, the manufacturer of the final product, not of the container itself. A minority view is held in a few jurisdictions that the injured party must also prove some act of negligence on the part of the manufacturer in order to recover, for example, failure to inspect the containers for fitness for use. This rule makes recovery somewhat more difficult, since the injured party must show some negligence, and *res ipsa loquitur* is generally resorted to only when no evidence of negligence is available.

WARRANTY

As we have already seen, the principal basis, other than negligence, for recovery of damages for injury or loss caused by defective products rests on the theory of a contract or warranty between the seller, whether a manufacturer, distributor, or retailer, and the purchaser–consumer, user, or other person sustaining injury or loss.

Although in some cases there is a formal written contract between the seller and the buyer of goods, more frequently there is nothing more than a casual bargain and sale. In either case, the warranty given the buyer by the seller may be express or implied, written, oral, or merely imputed from circumstances.

A warranty may be defined as a representation, express or implied, with respect to the character or quality of the product sold. A warranty normally arises as an incident of the contract of sale of the goods. In any legal action based on a warranty, the party seeking to recover must prove (1) that the warranty existed, (2) that the warranty was broken, and (3) that the breach of warranty was the proximate cause of the injury or loss sustained by the purchaser or user.

Privity of Contract

In its beginnings in early English common law, the concept of the warranty obligation grew out of the tort of deceit, a tort being a civil wrong as opposed to a crime (e. g., trespass). An action based on a warranty did not, however, require *scienter* or guilty knowledge, on the part of the seller, of the unfitness of the goods, which would be a required element of recovery in an action for deceit.

An alternative legal remedy for what we would now call a breach of warranty developed out of the common law action of *indebitatus assumpsit*, which was predicated on a contractual relationship existing between the buyer and the seller of goods that implied a promise by the seller to pay the buyer fair value for any failure on the part of the former to fulfill the contract. Inasmuch as this legal procedure was the simpler of the two, it came into more common usage, with the result that recovery in warranty suits came to be predicated largely on contract rather than tort or personal injury principles. It was soon recognized that one who was not a party to a contract could not maintain an action for its breach, and consequently the courts reasoned that one who was not a party to the sale giving rise to the contract could not properly claim a breach of warranty when the goods proved defective. In other words, the consumer buying an article from a retailer could not recover from the manufacturer for his injuries on a breach of warranty theory, since he had no "privity of contract" with the manufacturer.

Injuries caused by defective products are often sustained by persons who did not actually purchase the product in question and who do not, therefore, have a contractual relation with the seller, whether a retailer, distributor, or manufacturer. Purchase of an aerosol product at the retail level creates a contractual relationship between the actual purchaser, perhaps a housewife, and the retailer, but not between the seller and other members of the purchaser's family, for example, or house guests or others to whom the goods may be lent or given. For this reason, under the common law only the actual buyer could recover damages sustained for injury or loss caused by a defective product, and only the retailer was liable. Others injured by the product could not recover damages.

Despite the absence of a contractual relationship, however, injured persons have been increasingly successful in recent years in persuading courts to hold manufacturers and distributors liable by convincing the courts (1) that their cases come within a judicially recognized exception to the privity rule, (2) that an exception ought to be created for the particular situation, or (3) that the privity rule ought to be abandoned as outmoded in the particular jurisdiction in question.

EXCEPTIONS TO THE PRIVITY RULE

The traditional rule that privity of contract is indispensable to recovery of damages for breach of warranty, express or implied, was first challenged in cases involving sales of unfit foods and beverages. The "food exception" thus established has now been extended by some courts to include such products as animal foods, hair waving preparations, detergents, and even wearing apparel. As a matter of fact, the requirement of privity has been relaxed or abandoned altogether in a number of important jurisdictions. Such decisions have usually, but not always, involved products that are hazardous if defectively manufactured. The courts have also increasingly "found" privity between manufacturers and remote consumers because of advertising and product labeling. In many instances, consumers and users other than the actual buyer have been afforded the protection of manufacturers' and sellers' warranties.

Some courts and many important states have now gone so far as to admit frankly that they have completely abandoned the old requirement for privity and are now imposing strict liability on manufacturers and distributors of defective products.

Comestibles

Sellers of food and beverages were held to a special responsibility even under the old common law, and even the earliest American decisions usually imposed strict liability on the seller of unfit food in favor of the purchaser. Decisions permitting recovery against manufacturers, processors, and sellers of food products by persons not in privity with them began to appear 50 to 60 years ago. Although the courts have not always agreed on the basis for recovery in such cases, two common theories are (1) that there is an "implied representation" that food is safe for human consumption, and (2) that a "warranty runs with the product" and thus inures to the benefit of the consumer despite lack of privity. Other courts have imposed strict liability as a matter of public policy. Pure food and drug statutes, both state and federal, have also provided remedies against unfit foods and beverages.

Other Products

A few courts have extended the food products exception to include a number of other products. Although they have not always concurred in their reasons for extending the food privity exception, deceptive labeling and advertising have been significant factors; other courts have taken the view that there is no sound reason for permitting recovery in food cases and denying it in others, particularly when the product involved is as dangerous as contaminated food.

In the movement to abandon the privity requirement for other products, the decision of the New Jersey Supreme Court in *Henningsen* v. *Bloomfield Motors, Inc.*, 32 N.J. 358, 161 A. 2d 69 (1960), constitutes a landmark as one of the first decisions rejecting outright the requirement of privity of contract in regard to all classes of products. It is appropriate to note the reasoning of the court in reaching its decision:

In ... recent times a noticeable disposition has appeared in a number of jurisdictions to break through the narrow barrier of privity when dealing with sales of goods in order to give realistic recognition to a universally accepted fact. The fact is that the dealer and the ordinary buyer do not, and are not expected to, buy goods, whether they be foodstuffs or automobiles, exclusively for their own consumption or use. Makers and manufacturers know this and advertise and market their products on that assumption .... The limitations of privity in contracts for the sale of goods developed their place in the law when marketing conditions were simple, when maker and buyer frequently met face to face .... With the advent of mass marketing, the manufacturer became remote from the purchaser, sales were accomplished through intermediaries, and demand for the product was created by advertising media. In such an economy it became obvious that the consumer was the person being cultivated. Manifestly, the connotation of "consumer" was broader than that of "buyer." He signified such a person who, in the reasonable contemplation of the parties to the sale, might be expected to use the product. Thus, where the commodities sold are such that if defectively manufactured they will be dangerous to life and limb, then society's interest can only be protected by eliminating the requirement of privity between the maker and his dealers and the reasonably expected ultimate consumer. In that way the burden of losses consequent upon the use of defective articles is borne by those who are in a position to either control the danger or make an equitable distribution of the losses when they do occur.

In recent years the high courts of Arkansas, California, Connecticut, Florida, Illinois, Iowa, Kansas, Kentucky, Michigan, Minnesota, Mississippi, Missouri, New Jersey, New York, North Carolina, Ohio, Oklahoma, Oregon, Pennsylvania, South Carolina, Tennessee, Texas, and Washington have rejected the rule requiring privity of contract in actions against manufacturers and remote sellers to recover for breach of implied warranty. Rejection of the privity requirement by the courts in these important states has led a number of such courts to admit candidly that they are imposing strict liability on sellers of defective products. The legislatures of two states, Arkansas and Virginia, have enacted laws depriving manufacturers of defective products of the defense of privity. Usually, but not always, the decisions rejecting or limiting the privity requirement have involved products, such as airplanes, automobiles, and power tools, that are dangerous if defectively manufactured. Although very few cases have as yet arisen involving aerosol products, it seems clear that the privity requirement would also be abandoned with regard to such products if inherently dangerous because of improper manufacture. Examples of cases of this type involving various other products are given below.

Arkansas has enacted a statute that reads as follows:

The lack of privity between plaintiff and defendant shall be no defense in any action brought against manufacturer or seller of goods to recover damages for breach of warranty, express or implied, or for negligence, although the plaintiff did not purchase the goods from the defendant, if the plaintiff was a person whom the manufacturer or seller might reasonably have expected to use, consume, or be affected by the goods.

Privity of contract cannot be invoked as a defense of an action for breach of warranty by the servant or the employee of the original purchaser. Therefore a welder who sustained injury in an explosion because of a defect in an oxygen cylinder was permitted to recover damages from the cylinder supplier.

In another case, the manufacturer of a hair waving and setting lotion packaged and sold in aerosol cans was held to have a case for breach of warranty of fitness against the manufacturer of the cans, despite the absence of privity of contract, to recover for loss of profits and for damage to its name and good will. The cans had corroded or rusted, forcing plaintiff to refund substantial sums of money to its customers. Privity of contract was not necessary because the can manufacturer was aware of the purpose for which the product was intended, and knew of the user's reliance on the cans being fit for this purpose.

A distributor and a manufacturer of a hair waving solution were each found liable for breach of an implied warranty to a beauty parlor patron who developed sores on her head because of irritating chemicals in the solution.

In another case, absence of privity of contract did not preclude a customer of a beauty salon from recovering damages from the distributor of a hair rinse that caused injury to her scalp.

In still another case, the purchaser of a home permanent who was injured by it was found to have a proper cause of action against the manufacturer of the product for breach of implied and express warranties despite the absence of privity of contract.

Express Warranty

An express warranty is an assertion of fact or promise by the seller relating to the quality of goods which may tend to induce the buyer to purchase the product. This type of warranty may arise from advertising, sales literature, product labeling, or even oral statements. A description of the product and its properties or submission of a sample constitutes a warranty that the goods sold will conform to the description and sample.

Under the common law, it was necessary for the plaintiff to show that he *relied* upon the warranty in order to prevail. Under the Uniform Commercial Code, which has now been adopted by most jurisdictions, however, this is no longer a required element of the plaintiff's case.

It is not necessary to the creation of an express warranty that formal words such as "warrant" or "guarantee" be employed, or that there be any specific *intention* to make a warranty.

Personal injury actions, based on the performance of goods inconsistent with the advertising, literature, and labeling of the goods, for products such as detergents, hair preparations, and insecticides for household use, among other things, have resulted in findings of liability or potential liability on the part of manufacturers. Although here again there are few decided cases involving aerosol products, the same legal principles would apply.

The following cases are illustrative of such personal injury cases:

An injured purchaser of household detergent was held to have a good cause of action against the manufacturer for breach of express and implied warranties, since the product was purchased in reliance on extensive advertising claims that the detergent could be safely used for household cleaning tasks.

A consumer was permitted recovery of damages against the distributor of a hair dye for an allergic reaction due to the use of the dye. The accompanying instructions did not require a patch test to determine hypersensitivity to the dye; therefore, there was a breach of warranty because the defendant warranted the dye to be safe if used in accordance with the instructions.

Advertisements by a seller and product labeling by a manufacturer that a particular hand cream was " . . . hospital tested and proven. It is new. It is safe. It works . . . " constituted an express warranty which was breached when a purchaser of the cream developed pain, swelling, and blistering of the hands a few hours after using the product.

A consumer prevailed against a manufacturer for injuries suffered as a result of the poisonous effect of an insecticide that had settled on various portions of her person and caused inflammation and boils of a serious and recurrent nature. The container was labeled with assurances that the product, a spray, was harmless to human beings but deadly to insects. The labeling was considered to constitute an express warranty to the user that the insecticide, when used in the manner directed, was harmless to human beings, and a person suffering injury because of the breach of the warranty was entitled to recover damages. The warranty was effective even though the product was purchased from a distributor.

The purchaser of a permanent wave product recovered damages under an express warranty in a case where the manufacturer had induced the sale by advertising claims of gentleness and quality.

The user of a concentrated proprietary solution was held to have a good cause of action for breach of express warranty against a pharmacy for severe burns. The facts were that, after surgery, the injured party was given a prescription or order for "Aqueous (named) Solution 1: 750" for external use. Plaintiff's mother went to defendant's pharmacy to fill the prescription, and an employee there sold her a packaged drug from a shelf, remarking, "This is the same thing." The drug sold was a concentrated solution that severely burned plaintiff when she applied the chemical to her body.

Property damage claims, based on the performance of goods inconsistent with express warranties, have also been successful, as the following cases illustrate:

A complaint alleging that there were traces of herbicide in an insecticide having a label which purported to list all active ingredients but included no herbicides stated a good cause of action for damage to a crop.

A paper products manufacturer was held liable to the operators of a nursery for damage to young citrus trees when its product, represented in a catalog as waterproof, long lasting, and suitable as a wrapping for the boles of young trees, proved unfit after only short exposure to the weather.

An animal vaccine manufacturer was held liable, on the basis of a circular, to a rancher for the loss of several thousand sheep because of failure of its product to provide effective immunization against disease.

Cases of this type involving aerosol products are quite rare, since such products, being primarily intended for home use, do not have the inherent potential for extensive property damage. However, in the event that an aerosol product did cause property loss, whether through fire or damage to valuable antique furniture, for example, the legal principles would be the same as those illustrated in the following cases:

When the quality of the first lot of soap purchased was the basis of the buyer's acceptance of the seller's offer to sell a second lot, there was an express warranty of quality, and the seller was liable for any damages resulting from deficiencies in quality of the second lot.

A vaccine manufacturer breached an express warranty when it represented in a circular that the bacterin would give effective immunization against a lamb disease if the animals were vaccinated 10 to 12 days before being put on full feed, and the lambs subsequently died from the disease. There was also evidence supporting a finding that the defendant was liable for breach of an implied warranty of fitness where the defendant's agent induced the plaintiff to use the vaccine by representing it to be "as good as any obtainable."

A chemical manufacturer was found liable for breach of an express warranty when its agent sold 2,4,5-T, a hormone herbicide, to a strawberry raiser on the representation that it was "perfectly safe for strawberries." The buyer applied the herbicide by a conventional method when the strawberry plants were dormant, but the chemical destroyed not only weeds but also the strawberry plants.

A hog producer was able to recover damages against the manufacturer of an insecticide for breach of express warranty in a case where the manufacturer's agent had represented the insecticide as safe for use around animals, including pigs, and the plaintiff had relied on this statement. In actual fact, the product caused the deaths of more than 200 pigs.

An oral statement to a buyer that the seller stood behind the paint sold and that the buyer might look to him rather than the manufacturer if the product was defective was held to amount to an express warranty. When the paint peeled off the walls after application, the buyer was found to be entitled to recover against the seller for breach of the express warranty.

PRIVITY OF CONTRACT DEFENSE

We have seen that at common law there could be no recovery from a manufacturer by an ultimate consumer who purchased goods from a retailer, second manufacturer, or other intermediary, since the intermediary broke the chain of contract and there was no "privity of contract" between the manufacturer and the ultimate consumer. As has been explained in detail, this is no longer the law in many jurisdictions. Instead, although the courts have been reluctant to classify statements in advertisements and on product labels as warranties, it is now established in many jurisdictions that the seller's statements may be relied on both by an immediate buyer and by a subpurchaser, for instance, a retailer and his customer, to establish a warranty in actions brought for the breach of such representations. Therefore the trend is to brush aside the manufacturer's traditional defense of lack of privity of contract, and to hold that warranties through advertising or product labels operate in favor of the ultimate consumer.

RESTATEMENT OF THE LAW OF TORTS

Although per se it is not the law, the well-known restatement of the law of torts is highly regarded as a guide to what the law is, should be, or may become in the future. The restatement on this subject takes the following position:

Misrepresentation by Seller of Chattels to Consumer—One engaged in the business of selling chattels who, by advertising, labels or otherwise, makes to the public a misrepresentation of a material fact concerning the character or quality of a chattel sold by him is subject to liability for physical harm to a consumer of the chattel caused by justifiable reliance upon the misrepresentation even though.

(a) it is not made fraudulently or negligently, and,

(b) the consumer has not bought the chattel from or entered into any contractual relation with the seller.

Caveat:

The Institute [promulgating the statement] expresses no opinion as to whether the rule stated in this Section may apply

(1) where the representation is not made to the public, but to an individual, or

(2) where physical harm is caused to one who is not a consumer of the chattel.

Implied Warranty

Basically there are two types of implied warranty: the implied warranty of fitness and the implied warranty of merchantability. These warranties are created by operation of law, as opposed to contract, and are governed largely by statute. The former is created when the buyer makes known to the seller the purpose for which the goods are required, thereby giving rise to an implied warranty that the goods shall be reasonably fit for that purpose, at least in jurisdictions having a statute to this effect. The latter type, the implied warranty of merchantability that the article sold is reasonably fit for the general purpose for which it was manufactured, is implied if the seller is a merchant handling goods of the kind in question, at least in jurisdictions having a statue to this effect.

The implied warranties of fitness and merchantability are in many cases difficult to separate. In actual practice it is seldom necessary to do so, since the courts often treat them interchangeably, although in rare circumstances a case may succeed on one theory and not on the other.

The implied warranties of description and sample of the goods have now been made express warranties in most jurisdictions, that is, those that have adopted the Uniform Commercial Code, referred to above.

FITNESS

In most jurisdictions, statutes have been enacted which provide that, when an article is sold, an implied warranty arises automatically that the article is fit for a particular purpose if the buyer can establish that he has relied on the judgment of the seller, who knows the purpose for which the product is purchased. This rule has been incorporated in the Uniform Commercial Code, for example, which, as we have seen, has been adopted in most jurisdictions. Common law jurisdictions that have no statute covering the subject concur in holding that a manufacturer warrants the fitness of his goods for a particular purpose if he has been expressly or impliedly informed of that purpose by the buyer. However, in the common law jurisdictions there is no unanimity as to whether an implied warranty of fitness exists when a seller other than the manufacturer is responsible for the sale.

An injured consumer, in order to recover for breach of an implied warranty of fitness, must prove that he expressly or impliedly made known to the seller the purpose for which the article was purchased. This is seldom an issue in such a case, however, since this knowledge is generally imputed to the seller by the fact that goods, such as food, for example, have only one legitimate use. However, in cases where knowledge of the specific use cannot be imputed to the seller, this may be a good defense to such an action.

The general rule is subject to the exception that there is no warranty when an examination by the buyer would have disclosed the injury-causing defect in the goods.

Proprietary Sales

At common law, which still governs these matters in a few jurisdictions, there is no implied warranty of fitness when an article has been sold by its patent or trade name, on the theory that in such a case the buyer has asked about and knows the nature of the goods purchased. Under the Uniform Commercial Code, which as we have seen is now the law in most jurisdictions, there is no such provision exempting proprietary goods from implied warranties of fitness. However, this does not materially change the law, since the buyer must still show that he relied on the seller's judgment to furnish suitable goods, in order to recover on the basis of an implied warranty.

In one illustrative case, the fact that an article had a trade name did not conclusively preclude the existence of an implied warranty of fitness, since the product was sold for the specific purpose of inhibiting the growth of mold in cheese and failed to do so.

MERCHANTABILITY

As we have seen above, although the warranty of merchantability is recognized in jurisdictions where the common law still prevails, in such cases it is generally limited to sales made by growers and manufacturers. Under the Uniform Commercial Code now in effect in most jurisdictions, the law is as follows:

(1) Unless excluded or modified (Section 2-316), a warranty that the goods shall be merchantable is implied in a contract for their sale if the seller is a merchant with respect to goods of that kind. Under this section the serving for value of food or drink to be consumed either on the premises or elsewhere is a sale.

(2) Goods to be merchantable must be at least such as

(a) pass without objection in the trade under the contract description; and

(b) in the case of fungible goods, are of fair average quality within the description; and

(c) are fit for the ordinary purpose for which goods are used; and

(d) run, within the variations permitted by the agreement, of even kind, quality and quantity within each unit and among all units involved; and

(e) are adequately contained, packaged, and labeled as the agreement may require; and

(f) conform to the promises or affirmations of fact made on the container or label if any.

(3) Unless excluded or modified (Section 2-316) other implied warranties may arise from course of dealing or usage of trade.

Exceptions to warranties of merchantability are similar to those discussed above in regard to fitness statutes.

Proprietary Sales

The "patent or other trade name" rule is generally not applied to the warranty of merchantability.

Most jurisdictions also recognize, as another exception to the rule, occasions when the warranties of fitness and of merchantable quality coexist. For example, a warranty of fitness for a particular purpose may be the equivalent of a warranty of merchantability, in which event a recovery may be founded upon either. In a leading case illustrating this principle, Judge Cardozo held that the Uniform Sales Act "trade name" exception did not bar a purchaser's action against a grocer for injuries sustained because of a pin in a loaf of bread even though she had asked for the bread by brand name, made her own choice, and used her own judgment. Since the buyer had specified what she wanted, there was no warranty of fitness. However, there was a warranty that the bread, bought by description from a seller who dealt in such products, was of merchantable quality. Since the bread did not comply with this warranty, the warranty was breached, and the purchaser was entitled to judgment.

Liability to Third Parties

Where the privity rule has been waived or modified by the creation of an exception, a question arises as to how far warranty protection is to be extended. In other words, what persons or class of persons, other than the actual buyer, may take advantage of, or sue for breach of, a manufacturer's or seller's warranty? For example, are members of the buyer's family, his employees, his friends, or his guests, if injured, also entitled to the warranty protection that has been afforded the purchaser himself?

Jurisdictions that have recognized the food products exception to the privity rule and those that have extended the exception to other products or have abandoned the privity requirement altogether, thereby permitting warranty actions against manufacturers, ordinarily extend the privity exception to include members of the purchaser's family. A buyer's employees also have been afforded warranty protection, as have guests, and those who borrow or receive the goods as a gift. It is apparent, therefore, that the trend of recent decisions has been to extend warranty protection to all persons within the distributive chain whose injury by a defective product is reasonably foreseeable. For the most part, however, courts have been reluctant to extend this protection to persons outside of the distributive chain, that is, to those described as bystanders, although such persons might otherwise have an actionable claim based on negligence.

In jurisdictions where the requirement of privity is strictly enforced, an injured purchaser cannot recover in warranty against a manufacturer. And, formerly, strict enforcement of the privity rule usually deprived members of the purchaser's family of a warranty action against even a retailer, because they were not in privity of contract with the seller. However, some courts have ameliorated the harshness of the doctrine by employing theories of agency and third-party beneficiary to bring into privity injured persons whose relationship to the party actually in privity is such that the warranties are viewed as running in the injured person's favor. Creating artificial privity through the use of the agency theory, for example, enlarges the seller's responsibility but does not affect the manufacturer's or distributor's liability. Accordingly, the requirement of privity was satisfied in a breach of warranty action against the retail seller of pork sausage containing trichinae, the sausage having been purchased by the plaintiff's child while acting as the plaintiff's agent.

Uniform Commercial Code

In states where the Uniform Commercial Code is effective, the requirement of privity has been eased somewhat, affecting significantly the former decisional law in a number of such states. Although not affecting the privity requirement as it applies to manufacturers, the Uniform

Commercial Code specifically extends the seller's warranties to members of the buyer's family, household, and guests. Section 2-318 provides:

A seller's warranty whether express or implied extends to any natural person who is in the family or household of the buyer or who is a guest at his home if it is reasonable to expect that such person may use, consume or be affected by the goods and who is injured in person by breach of the warranty. A seller may not exclude or limit operation of this section.

Seven states, through variations from the official text of Section 2-318 of the Uniform Commercial Code, extend the seller's express and implied warranties to any injured person who may reasonably be expected to use, consume, or be affected by the goods. These states are Alabama, Colorado, Delaware, South Carolina, South Dakota, Vermont, and Wyoming. In addition, statutes in Arkansas and Virginia have gone one step further and have abolished the defense of privity in warranty actions against manufacturers as well as sellers in cases where the injured person was one whom the defendant might reasonably have expected to use, consume, or be affected by the goods.

INDEX